Foundations of Statistical Inference

Proceedings of the Symposium on the Foundations of Statistical Inference prepared under the auspices of the René Descartes Foundation and held at the Department of Statistics, University of Waterloo, Ontario, Canada, from March 31 to April 9, 1970.

Edited by

V.P. Godambe
Professor in Charge of the Symposium

D.A. Sprott
Chairman of the Statistics Department
Dean of the Mathematics Faculty

Holt, Rinehart and Winston of Canada, Limited
Toronto, Montreal

Edited by

V.P. Godambe
Professor in Charge of the Symposium

D.A. Sprott
Chairman of the Statistics Department
Dean of the Mathematics Faculty

II

PREFACE

A Symposium on Foundations of Statistical Inference was organized at the Department of Statistics, University of Waterloo, during the Spring of 1970. The ten-day Symposium, March 31 to April 9, was attended by about forty experts in the field from various parts of the world. This volume contains the papers read at the Symposium along with their discussion.

The subject "Statistical Inference" touches every aspect of human thinking, so that contributions have been made from the entire spectrum from abstract philosophy, through pure mathematics to applied statistics. Indeed, along with the professional statisticians participating in this Symposium were, for example, philosophers, physicists and biometricians. Even with this variety, the gathering at the Symposium was not fully representative of all aspects of statistical thinking. A glance at this volume shows, however, the divergent views among the different participants.

The Symposium arose essentially out of the symposium on "Foundations of Survey-Sampling" organized at Chapel Hill by North Carolina University at the suggestion of Professor V.P. Godambe. Its success confirmed the idea, already contemplated by us, that it was now time for one on a broader subject like "Foundations of Statistical Inference". Indeed, as far as we can tell, until this Symposium at the University of Waterloo, never in the history of Statistics have so many experts come together for such a period of time to discuss its Foundations.

It is a great pleasure for us to record our indebtedness to persons and institutions whose generous help greatly assisted this Symposium. The necessary finances came from the Mathematics Faculty of the University of Waterloo, from the University of Waterloo Research Grant Programme, from the National Research Council of Canada, and from the Canada Council. The advice of Professors G. A. Barnard and S. Geisser regarding the programme was valuable. Professors C. F. A. Beaumont, J. G. Kalbfleisch, W. F. Forbes and C. Springer were particularly helpful in administration matters. We should also like to thank Professors K. D. Fryer, J. Gentleman, K. Shah, C. Springer, M. Thompson, J. Whitney, and Messrs. Abbas, Hanley and Sekkappan for their help; the secretarial assistance was provided by Miss Bennewies, Mrs. Comeau and Mrs. Long. We would also like to thank the publishers Holt, Rinehart and Winston of Canada, Ltd. for their co-operation and for the many ways in which they expedited the publishing of these Proceedings.

Editors

V. P. Godambe
Professor in Charge of the Symposium

D. A. Sprott
Dean of the Mathematics Faculty

CONTENTS

The papers follow the order in which they were read at the Symposium.

FOUNDATIONS OF BEHAVIORISTIC STATISTICS*

Jerzy Neyman
University of California, Berkeley

1
Introduction

The description of the theory of statistics involving a reference to behavior, for example, *behavioristic statistics,* has been introduced to contrast with what has been termed *inductive reasoning.* Rather than speak of inductive reasoning I prefer to speak of *inductive behavior.* As far as I can remember, the term *behavior* referring to a statistical procedure was first used in 1933 in a joint paper with Egon S. Pearson [1]. I believe that the term *inductive behavior* was first used [2] in 1938. Actually, the French version of this term *comportement inductif* was used. Later, *statistical decision making,* another term meant to designate the same concept, was introduced by Abraham Wald [3].

The motivation behind this terminology stems from the continuing connection between probability and statistics, on the one hand, and research in science on the other. With an overwhelming frequency, modern science is concerned with categories of objects which, while satisfying a common definition, vary considerably in those characteristics that are relevant to the study performed. As a result, any attempt at a mathematical treatment must involve assumptions that some given characteristic X of the objects studied is a random variable with some particular, known or unknown, distribution.

In astronomy we may be concerned with galaxies, some classified as ellipticals and others as spirals. The subject of study may be the mass of galaxies and contemporary literature contains assertions that elliptical galaxies are

*This paper was prepared with the partial support of U. S. Army Research Office (Durham). Contract DA-ARO-D-31-124-G1135.

more massive than spirals. In this form, the statement sounds deterministic and could be interpreted to mean that all the elliptical galaxies, and also all the spirals, have some fixed masses $M(E)$ and $M(S)$, respectively, and that $M(S) < M(E)$. In actual fact, of course, the statement *elliptical galaxies are more massive than the spiral galaxies* is no more than a convenient abbreviation of something much more complicated, referring to the distributions of mass among galaxies belonging to the two categories mentioned. When speaking of elliptical galaxies we speak of a population of objects that cannot be studied in its totality but only through samples. In addition, the accessible samples are all biased, with a preference for galaxies that are easier to observe, perhaps because of their relative size, or because they are nearer to us than the others, etc. Obviously, a detailed mathematical study of the realm of galaxies is possible only in probabilistic-statistical terms. Among other things, such a study must involve the analysis of the biases in samples accessible for direct observation, that is of supposedly random mechanisms, through which, for example, more massive galaxies of either type are more frequently observed than the smaller ones.

While the above example refers to astronomy, the situation with other sciences is very similar. As a consequence, modern sciences may be contemplated in terms of a long sequence of situations, say $\{S_n\}$, each involving a separate random variable X_n, usually a vector, representing a characteristic of an individual in a population Π_n studied in the situation S_n. Frequently, there is not just one population but several, considered simultaneously, say $\Pi_{n_1}, \Pi_{n_2}, \ldots, \Pi_{ns_n}$. In the astronomical example above, Π_{n_1} may be elliptical and Π_{n_2} spiral galaxies. X_{n_1} and X_{n_2} may designate masses of these two types of galaxies. The assertion that the ellipticals are more massive than the spirals may then mean $E(X_{n_1}) > E(X_{n_2})$, or some other formula relating to the true distribution of the two random variables considered. Thus, in addition to involving the populations Π_{n_i} and the distributions of some random variables X_{n_i}, the situation S_n involves also the *true state of nature* which is an excellent term introduced by Wald [3]. This true state of nature, say σ_n, is unknown and, at best, we may postulate the knowledge of the set, say Σ_n, of states of nature which in the situation S_n are possible. (Naturally, this postulate may be unrealistic just as any other general assertion made about the outside universe). The problem in situation S_n is then to use the sample of values of the random variables considered which is made available by observations in order to select out of the set Σ_n of possible states of nature the one on which to base our future actions. (For example, in our future studies of galaxies should we act on the assumption that the mean mass $E(X_{n_1})$ of ellipticals is greater than the mean mass $E(X_{n_2})$ of spirals?)

As nicely described by Wald, the above circumstances bring under consideration a novel element of study, the set D_n of possible decisions or actions. Also, any such decision $d \in D$, being based on the sample of observable random variables, may in some sense be *right, wrong* or *very wrong*. Thus, in an effort at a *mathematization* of the discussion, it is unavoidable to consider a loss function (another term due to Wald), say $L(d, \sigma)$, representing the loss that we would suffer by adopting the decision $d \in D_n$ while the true state of nature is the specified $\sigma \in \Sigma_n$.

Incidentally, while the above behavioristic point of view on research in science, including the loss function, was brought into modern statistical theory in the 1930's [1], [2], [4] and excellently formulated by Wald a decade or so later, the actual origin of the underlying ideas goes back to the beginning of the nineteenth century. The loss function is already present in Laplace. Later on, with a remarkable clarity it is discussed by Gauss [5] as a basis for his treatment of the theory of least squares. Regretfully, while the method of least squares has been used for generations, the conceptual basis of this method was forgotten and had to be rediscovered in the 1930's. Actually, I owe the information about the ideas of Gauss to Professor Erich L. Lehmann.

The next essential conceptual element of the modern theory of statistics is the statistical decision function, say d_n (X), defined on the space of sample values of the observables X, with values in D_n. This concept is, of course, familiar: if I observe $X = x$, then the decision to be taken is d (x) ϵ D. Subject to the various assumptions of regularity, the properties of a decision function d (X), and in particular, its relative desirability, are characterized by the expectation of the loss that one might suffer by adopting d (.) as a guide for one's inductive behavior, E $\{$ L $[$ d (X), $\sigma]$ $\}$. This function, defined for all $\sigma \epsilon \Sigma$, might be called the performance characteristics of the particular decision function $d(.)$.

On occasion, very rarely in my personal experience, the true state of nature σ may be appropriately considered as a random variable with values in Σ. In this case, the performance characteristic of d (.) is random variable itself and the relative goodness of d (.) may well be characterized by its *risk*, the expectation with regard to the distribution of σ. In all that follows, the true state of nature σ will be considered as an unknown point in the space Σ not as a random variable.

Now, with the above collection of basic concepts, we may proceed to the formulation of what, in my opinion, is the general problem of behavioristic theory of statistics: *with reference to any given situation S_n, and to all the data pertinent thereto, to define an optimal statistical decision function $d_n(.)$ and to determine it, if it exists; if the originally defined optimal decision function does not exist (for example, uniformly most powerful tests exist but very rarely), define a compromise best statistical decision function, etc.*

In the main text of the paper three recent instances of the above procedure are described as illustrations. The first of them, due to Herbert Robbins and to his several collaborators, represents a remarkable achievement likely to influence theoretical-statistical and also substantive research in many domains of science. The other two examples are much more modest but, I hope, will prove interesting.

2

*Test of a Composite Hypothesis Against a Composite
Alternative Having Power Equal to Unity*

Consider an experimental situation S in which the true state of nature is characterized by a single parameter ξ capable of assuming values in some interval (a,b) with $a < b$. For each $\xi \in$ (a,b) the observable random variable has a known distribution. The hypothesis H to test asserts $\xi < \xi_0$ (or $\xi \leq \xi_0$), with $a < \xi_0 < b$. The alternative is $\xi \geq \xi_0$ or $\xi > \xi_0$. The situation is such that, if H happens to be true, no particular action is required. On the other hand, if H is false, it is most desirable to take a certain action which may be termed corrective action.

Practical situations of this sort may be illustrated by the production in a factory. As long as the process of manufacture is characterized by the mean value $\xi \leq \xi_0$ of a certain characteristic X of the product, the situation is satisfactory and no changes are necessary. On the other hand, if the mean value of X becomes $\xi > \xi_0$, it is imperative to stop the process and to readjust the machines.

As another example, I wish to mention the evaluation of two competing medical treatments of a disease. Both treatments A and B are supposed to be effective, but there is the question of undesirable aftereffects. In particular, if A is a more or less established treatment, a randomized pair experiment may be considered to determine whether the frequency of bad aftereffects of A is greater than that of B. Should this be the case, one would want to replace A by B.

Obviously, in the two cases, but particularly in the second example, the desirable property of the test of H is as high a power as practicable, perhaps with some neglect of the probability of rejecting H when true.

The honor of having first noticed that the power of the test of H can be made equal to unity for all values of $\xi > \xi_0$ seems to belong to George Barnard [6]. In the publication *British Standard 3704*, dated 1964, he proposed a continuing sampling scheme of manufactured products ensuring that, if the probability ξ of faulty product equals or exceeds the preassigned critical value $p = .01$, the continuing process of inspection will detect the fact with the probability one. Also, the test may be so adjusted that, for any preassigned $\xi_1 < p$, the probability of unjustly rejecting H, which asserts $\xi < p$, while in actual fact $\xi = \xi < p$, is just as small as desired.

To me, the discovery of this possibility represents a remarkable achievement. However, I am even more impressed by the independent findings published since 1967 in a series of papers by Darling and Robbins [7] and by Robbins and Siegmund [8]. With reference to the same hypothesis H the sequential test obtained ensures the power of unity, uniformly for all values of ξ exceeding the limit specified by the hypothesis tested, just as in the case of Barnard. In addition, however, the Darling-Robbins test ensures that the probability of rejecting H when true does not exceed the preassigned limit α, whatever may be ξ consistent with H. Figure 1 is intended to illustrate the relationship between the three interconnected sets of findings: the now familiar Barnard-Wald systems of acceptance sampling, found some time during World War II, perhaps about 1943, the Barnard test of H, ensuring the power of unity with a preassigned level of significance at some chosen $\xi_1 < \xi_0$, and the Darling-Robbins test. In all three panels the quantity measured on the horizontal axis is the parameter under test, ξ. The quantity on the vertical axis is $P(\xi)$ the probability of rejecting H that $\xi < \xi_0$ (or $\xi \leq \xi_0$) plotted against the true value of ξ.

4

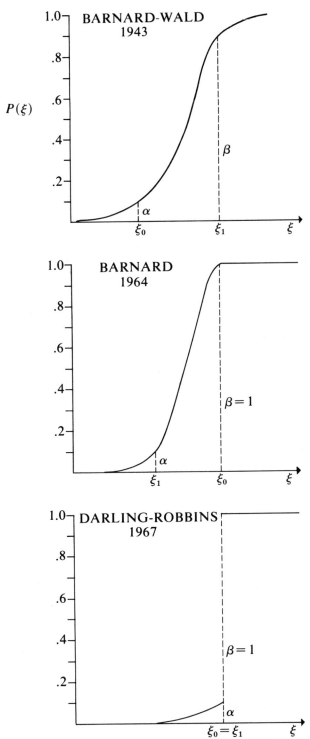

Figure 1. Three stages in the development of sequential tests of the hypotheses asserting $\xi < \xi_0$ (or $\xi \leq \xi_0$).

In the situation described in the first panel, it is desired to arrange that if the true $\xi \stackrel{<}{=} \xi_0$, then the hypothesis asserting this relation be rejected with probability $P(\xi)$ not exceeding the preassigned level α; also it is arranged that, if $\xi \geq \xi_1 > \xi_0$, then $P(\xi)$ is at least equal to the preassigned $\beta > \alpha$. However, $\beta < 1$. The second panel presents the situation of the Barnard test just described. Here, the preassigned power is $\beta = 1$ and it is guaranteed that it is attained at $\xi = \xi_0$ and maintained for $\xi > \xi_0$. However, the significance level α can be guaranteed for a preassigned value ξ_1, necessarily less than ξ_0 and the power function $P(\xi)$ is continuous between ξ_1 and ξ_0. The Darling-Robbins procedure, depicted in the third panel of Figure 1, brings together the two values ξ_1 and ξ_0, so to speak, so that the resulting power function is discontinuous of $\xi = \xi_0$: to the left of this point it is constantly less than the preassigned α; to the right of this point it is exactly unity.

According to Wald [9], the problem of sequential analysis, considered as a means of decreasing the amount of acceptance sampling, was suggested to him in 1943 by Milton Friedman and W. Allen Wallis. His first classified report appeared in the same year. In his book, Wald quotes George A. Barnard [10]. There is no date, but apparently, the two reports were prepared at about the same time. It will be noticed that, in both Barnard's and Wald's treatment of sequential experimentation only the economy of repeating the observations, compared with the fixed size samples, is emphasized. The extra possibilities in the infinitely dimensional sample space, not available in finite dimensions appear to have been first noticed by Charles M. Stein [11]. Of course, power identically equal to unity is not attainable with fixed size samples. I find it interesting to speculate about the reasons why, after the invasion of infinitely dimensional space (with sequential analysis) in about 1943, it took the statistical community more than two decades to travel from the first to the third panel of Figure 1. The mathematics involved is not very difficult and I suppose that the decisive moment in the progress is the audacity of the research workers in asking questions. Compared to Barnard and, to an even greater degree, compared to Robbins and his colleagues, the rest of us who worked on tests after World War II, must be a timid lot!

As far as I know, after his coup in the *British Standard 3704*, Barnard let the matter drop. On the other hand, the work of Robbins and his colleagues continues to develop covering cases of different distributions, on the one hand, and a variety of statistical procedures on the other. The item I described is intended for illustration only and, compared to the totality of results already obtained, is just a small incident.

3

Concept of β-optimal, Level α Tests

The second item I wish to bring to the attention of this conference is a recent achievement of a very young man Dr. Robert Davies, a New Zealander, whom we had the pleasure of having with us in Berkeley up to last summer.

A new idea, particularly a new idea proposed by an unknown young man, must take some time to reach the printing press. Thus, the paper of Davies, submitted by him to the *Journal of the Royal Statistical Society* had to wait a considerable time for the decision of the Editors, but is now expected to appear before very long. Its origin is in the study we conducted of the rain stimulation experiments (by *cloud seeding*) conducted in a number of countries. The variability of natural rainfall is proverbial and it happens that, in order to have a reasonable chance of detecting even a rather strong effect of cloud seeding, an experiment must last some 5 or 10 years. As a result, and because most meteorologists have little affinity with the concept of the power of a test, the completed proper experiments are frequently inconclusive. Yes, there is a difference between the *seeded* and *not seeded* precipitation, but overwhelmingly the difference is not significant, at least not significant by the usual standards.

In these circumstances, it is natural to consider not just each completed experiment separately from others, but to study all of them jointly. We are familiar with the comprehensive test of several independent hypotheses invented by R. A. Fisher. For each particular hypothesis (or for each experiment) compute the appropriate significance P. Then compute $-2 \log P$ and sum these quantities over all the experiments considered. In the absence of any effects in any of the experiments, the sum $-2 \Sigma \log P$ must be distributed as a χ^2. Egon S. Pearson became interested in this test and, with reference to a particular set of alternatives proved that it has a property of optimality.

Under pressure of empirical studies we used Fisher's test more or less routinely, but then we conceived some doubts. The point is that among the experiments analysed there were some that lasted many and some that lasted only a few years. Also, the details of the experiments varied so that the calculated estimates of the power appeared very different. Yet, Fisher's comprehensive test treats all these experiments, so to speak, on par. Considerations of this kind suggested that, through an appropriate formulation of the problem, one might deduce a more satisfactory test. Originally an effort was made to use the multivariate version of optimal $C(\alpha)$ tests [12]. Later, I attempted a new formulation of the problem and the results are described in [13]. The general set up is as follows. We consider a number s of independent experiments. The possible effect of a treatment in the nth experiment is designated by ξ_n. If $\xi_n = 0$, there is no effect. Otherwise, it is admitted that ξ_n may well be positive or negative. The essential point in the set up is that the numbers ξ_n are considered as particular values of a random variable Ξ for which the following structure is postulated

$$\Xi = \xi_o + W \sqrt{\eta} \qquad (1)$$

where ξ_0 and $\eta \geq 0$ are unspecified constants and W is a random variable with expectation zero and variance unity. It follows that Ξ has expectation ξ_0 and variance η. With this set up, and with no further restriction imposed on W, the hypothesis of no effect of treatment in any of the s experiments considered is equivalent to the combination of two subhypotheses. One of these asserts that $\eta = 0$, meaning that in all the experiments the effect of treatment is the same, namely ξ_0; the other subhypothesis is that $\xi_0 = 0$.

Each subhypothesis presents an independent interest and each offers the possibility of deducing the optimal $C(\alpha)$ test. Tests of this category have the property of being asymptotically locally optimal. This means that as ξ_0 (or η) diverges from the value zero specified by the hypothesis tested, the asymptotic power increases at least as fast as that of any other test of the same category.

I must admit that I attached high hopes for the tests deduced, particularly because they take account of the number of observations in each of the experiments studied, and also of certain other characteristics of these experiments. In the meantime, Davies became interested in the problem and embarked on an independent investigation. In particular, he used the digital computer in order to investigate the power of the new tests as compared to that of the comprehensive test of Fisher. The results were disappointing to me. Unavoidably, for small ξ_0 and η, the new tests were more powerful than Fisher's test. But then, if the intended level of significance is, say, $\alpha = .05$, and for some alternative hypothesis one of the tests yields the power $\beta = .06$ while the other yields $\beta = .07$, the difference between the two is of no practical importance. On the other hand, when the hypothesis tested is seriously false, with either ξ_0 and/or η substantial, it appears that Fisher's test had definite advantages over the optimal $C(\alpha)$ test. This circumstance suggested to Davies a novel definition of *compromise optimality*. This is as follows.

Fix an arbitrary level of significance α, between zero and unity and an arbitrary value β of power $\alpha < \beta < 1$, perhaps $\beta = .80$. Next consider the error in the hypothesis tested which, with a given test T, would ensure the preassigned level α and the preassigned power β. Let $\xi(\alpha, \beta | T)$ designate this error in the hypothesis tested. Then, the definition of "β-*optimal, level α*" test of a given class C applies to that test $T^* \epsilon C$, for which $\xi(\alpha, \beta | T^*)$ is a minimum.

In my opinion, the above concept is a very fruitful one, particularly because Davies managed to devise a method of construction that, under certain conditions, yields the desired solution. In applying his idea to the problem of the comprehensive test for a system of s independent experiments, Davies started with a set-up similar to that in the above equation (1). However, in order to be able to ensure not only the preassigned level α, but also the preassigned power β, it was necessary for him to strengthen the restrictions imposed on the variable W.

As I see it, the achievement of Davies represents a good example of research in behavioristic theory of statistics. In an experimental situation S, we visualise the properties of the desired statistical decision function that, for one reason or another, appear optimal to us. This being done, we seek the corresponding decision function. In the case studied by Davies, it appeared desirable to reach a reasonable preassigned value of the power, so to speak, as soon as possible, perhaps with the unavoidable neglect of power for *small errors* in the hypothesis tested. If it is agreed that this particular property of the test is something desirable in some specified situation, then Davies' test is the solution.

At this point it is appropriate to mention that, while Davies' summary test appears frequently more desirable than the one I deduced, the advantages over Fisher's test are substantially less pronounced. Also Fisher's test has the advantage of the remarkable simplicity of calculations involved.

My last example refers to the continuing studies of tests that, in a sense, are symmetric. The original idea and the early developments are described in [14]. The results of calculations reported below are due to Mrs. Rose Ray.

Consider the situation where a substantive scientist, perhaps an experimental meteorologist, asserts that a treatment T, devised by himself, has a beneficial effect $\xi > 0$ on some characteristics of the experimental units. For example, the experimental units may be cloudy days, the treatment T may be cloud seeding and ξ may stand for the percent of difference between average seeded and average not-seeded precipitation,

$$\xi = 100\,(S\text{-}NS)\,/\,NS \qquad (2)$$

with the obvious meaning of the symbols involved. The claims of the experimenter are subjected to a randomized test and, taking into account what is known about the distribution of precipitation per day, it is natural to seek a statistical test of the hypothesis $\xi = 0$ that is the most powerful. The difficulty is that, while the hoped for values of ξ are positive, it is admitted that they may also be negative. Also, for both the practical and for purely scientific purposes, it is quite essential to detect whether, in some specified conditions, ξ is positive or negative.

Considerations of this kind led to the idea of symmetric tests (labeled strongly symmetric) defined by the property of their power function (either exact, or asymptotic) $\beta\,(\text{-}\,\xi) = \beta\,(\,\xi)$. Of course, in many *bookish* examples this property is easily achieved. For example when we deal with samples from two populations, known to be normal with the same variance but with uncertain means, then the two-sided t-test has the desired property of strong symmetry and is optimal. However, in the overwhelming variety of applied problems, the situation is much more complicated and the text-book tests prove either of doubtful power or of doubtful symmetry or both. Also, one can easily visualize the indignation of an experimentalist if he learns that the statistical test suggested to him has a power function with $\beta(-\xi) > \beta(\xi)$ when $\xi > 0$.

With reference to this category of situations optimal $C(\alpha)$ tests were deduced under appropriate conditions and it was found that the *strong symmetry* $\beta(-\xi) = \beta(\xi)$ can be achieved by two different criteria with asymptotic power functions $\beta_1(.)$ and $\beta_2(.)$. One part of Mrs. Ray's study is concerned with the set of conditions under which $\beta_1(\xi) \geq \beta_2(\xi)$ or vice versa. Another part of the study compares the two strongly symmetric tests with the third test labeled *weakly symmetric*. Really, as will be seen below, the weakly symmetric test may be entirely asymmetric, so that the label chosen is inappropriate. The definition of the weakly symmetric test requires that for $\xi > 0$ the derivative at $\xi = 0$ of the sum, say $\beta_3(\xi) + \beta_3(-\xi)$, be a maximum, compared to that of any other test of the same category.

The motivation for considering the tests possessing this property is the theoretical (and also the practical) possibility that particular cases may exist where, at least for a set of values of $\xi > 0$ judged important, the following inequalities may hold:

$$\beta_1(-\xi) < \beta_2(-\xi) < \beta_3(-\xi) > \beta_3(\xi) > \beta_2(\xi) > \beta_1(\xi). \qquad (3)$$

In other words, it was anticipated that, in some identifiable cases, the weakly symmetric test may be both distinctly asymmetric and yet uniformly more powerful than either of the two strongly symmetric tests. Rose Ray found that cases of this kind actually do exist. Figure 2, referring to a particular problem of experiments with *memory boost* [14], illustrates some of her findings.

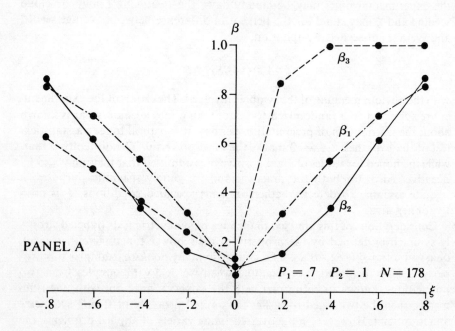

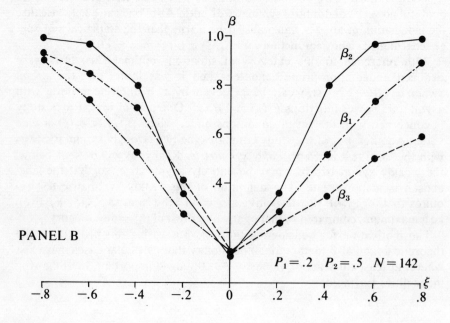

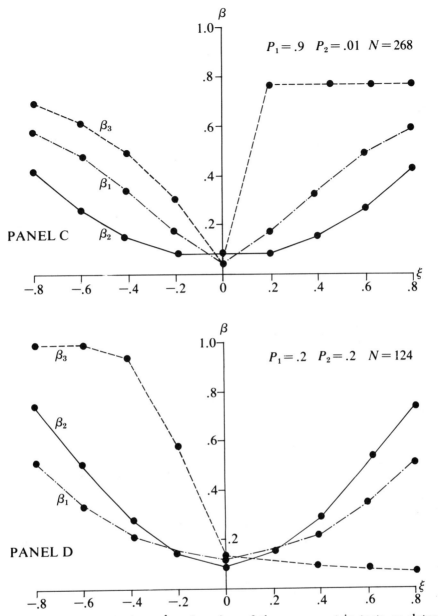

Figure 2. Monte Carlo power function of three symmetric tests as determined by the values of the nuisance parameters P_1 and P_2.

Each of the four panels of Figure 2 is the result of 1000 independent Monte Carlo simulations of an experiment on memory boost. The filled circles in the figure represent empirical frequencies with which the hypothesis tested $\xi = 0$, was rejected when the true effect of the treatment was represented by the indicated values of ξ, from -.8 to +.8. The empirical power functions of the three tests, two of them strongly and the third weakly symmetric, are marked by β_1, β_2, and β_3, respectively. One of the purposes of the Monte Carlo simulation was to see whether, with moderately large numbers N of hypothetical experimental rats, the actual power of the three tests behaves approximately as indicated by the asymptotic formula. In each case, the tests were adjusted so as to conform with the same chosen level of significance $\alpha = .10$. The number N of experimental units was determined to ensure the asymptotic power of one of the three tests to be $\beta = .80$ in some (easily identifiable) cases at $\xi = .4$ and in other cases at $\xi = .8$.

The symbols P_1 and P_2 indicated in Figure 2 represent the nuisance parameters, and it will be seen that the relative advantages of the three tests depend very much on their values. Panel A illustrates a case where the first of the strongly symmetric appears preferable to the other tests, with β_1 almost uniformly greater than β_2. It is true that for positive ξ the function β_3 is very much larger than β_1, but for negative ξ the inequality is reversed. In panel B the second of the two strongly symmetric tests is the better. Panel C illustrates a case where β_3 is uniformly greater than β_1 and β_2, but is markedly asymmetric. Finally in panel D the power of the weakly symmetric test is much larger than those of the two competing strongly symmetric tests, but this advantage is limited to negative values of ξ. For positive ξ, the values of β_3 are even less than the preassigned $\alpha = .10$!

Figure 2 illustrates several important questions that confront an experimenter and the cooperating statistician.

(a) In a given experimental situation there may be several alternative tests of the same composite hypothesis, all maintaining the preassigned level of significance. Which test to use? Obviously, the proper answer must depend on the power functions of the competing tests.

(b) However, as illustrated in the four panels of Figure 2, the power functions of the three tests considered depend sharply on the nuisance parameters P_1 and P_2, the values of which are not known. What should one do?

Of course, there is no sure answer to this question. However, if one is informed of the various possibilities, including the optimal situation, then in some cases at least, this optimal can be approached through the adjustment of the experimental design and of the method of analysis. For example, one of Rose Ray's findings is that in order to attain large power of the tests considered, one should aim at a relatively small value of $P_1 > 0$ and at a large value of $P_2 < 1 - P_1$. This general qualitative specification can be approached in the experiments considered.

5
Concluding Remarks

The above three examples illustrate the concept of behavioristic theory of statistics. In many instances human behavior has to be determined on the basis of experiments or of observations, the results of which are subject to variation that is impossible to ignore. An effort is made to substitute for real experiments and real observation certain hypothetical stochastic mechanisms with probabilities idealizing the relative frequencies. Naturally, some of these stochastic models are grossly inadequate, but some others do reproduce the observable frequencies to a satisfactory degree.

The available models are then used to deduce rules of behavior (or statistical decision functions) that, in a sense, minimize the frequency of wrong decisions. It would be nice if something could be done to guard against errors in each particular case. However, as long as the postulate is maintained that the observations are subject to variation affected by chance (in the sense of frequentist theory of probability), all that appears possible to do is to control the frequencies of errors in a sequence of situations $\{S_n\}$, whether similar, or all very different.

Notes

1. Neyman, J. and Pearson, E. S., "On the Problem of the Most Efficient Tests of Statistical Hypotheses," *Philos. Trans. Roy. Soc. London, A*, 231, 289-337, 1933.
2. Neyman, J., "L'estimation Statistique Traitée Comme un Problème Classique de Probabilité," *Actualitées Scientifiques et Industrielles*, 739, 25-27, 1938.
3. Wald, Abraham, *Statistical Decision Functions*, New York, John Wiley and Sons, Inc., 1950.
4. Neyman, J. and Pearson, E. S., "The Testing of Statistical Hypotheses in Relation to Probabilities *a priori*," *Proceedings of the Cambridge Philos. Soc.*, 29, 492-510, 1933.
5. Gauss, C. F., *Abhandlungen zur Methode der Kleinsten Quadrate von Carl Friedrich Gauss*, German translation by A. Barsch and P. Simon, Berlin, 1887.
6. Barnard, George, "Practical Applications of Tests with Power One," *Bulletin, International Statistical Institute*, in press.
7. Darling, D. A. and Robbins, Herbert, series of notes published in the *Proceedings of the National Academy of Sciences*, beginning with volume 57, 1188-1192, May 1967.
8. Robbins, Herbert and Siegmund, David, "Confidence Intervals and Interminable Tests," *Bull. Int. Stat. Inst.*, in press.
9. Wald, Abraham, *Sequential Analysis*, New York, John Wiley and Sons, Inc., 1947.
10. Barnard, G. A., "Economy in Sampling with Reference to Engineering Experimentation," *Technical Report, Series R*, Number QC/R/7, British Ministry of Supply, Advisory Service on Statistical Method and Quality Control.
11. Stein, Charles M., "A Two-Sample Test for a Linear Hypothesis Whose Power is Independent of the Variance," *Annals of Mathematical Statistics*, 16, 243-258, 1945.
12. Davies, Robert B. and Puri, Prem S., "Some Techniques of Summary Evaluations of Several Independent Experiments," *Proc. Fifth Berkeley Symposium Math. Stat. and Prob.*, V, 385-388, 1965.
13. Neyman, J., "Experimentation with Weather Control," *Journal of the Royal Statistical Society, A*, 130, 285-326, 1967.
14. Neyman, J., "Statistical Problems in Science. The Symmetric Test of a Composite Hypothesis," *Journal of the American Statistical Association*, 64, 1154-1171, 1969.

COMMENTS

G. A. Barnard:

First, a point of detail: I think it is reported in Kendall and Stuart's treatise that Bernard Welch suggested examining tests which were not locally but, instead, remotely powerful and this seems to have bearing on Dr. Davies' useful idea.

Second, Dr. Neyman asks why I did not proceed with further studies on tests of power one. For one thing, I was preoccupied with the other practical applications, for example, to the supervision of weights and measures. Here I failed to persuade those concerned to take up the idea, though we are to try again later this year and perhaps we may make some progress. For another thing, I certainly failed to see the far-reaching implications which Robbins and Darling did see, and this was partly due to the relative crudeness of my mathematical apparatus compared with the very delicate treatment Robbins and Darling use. Finally—and perhaps this is instructive for our conference— I did hope to publish a longer paper embodying the idea; but I had the ambition to devise an argument which would lead to a rational choice of the quality level to be insisted upon. The articles in question were contraceptives. Insisting on a very high quality would put up the price and to this extent might well increase, rather than decrease, the number of unwanted babies. You can imagine many other factors that would have to be considered. I am afraid I completely failed to resolve the problem and the final figure was selected on the basis of vague judgment. We may reflect on whether it is worthwhile to devise sophisticated statistical tools in situations where the parameters of the problem are so ill determined. We should bear this in mind in connection with our controversies. By all means let us argue vigorously, but let us bear in mind that the differences between the practical procedures we finally arrive at are quite likely to be within the tolerance range of the data in our problems.

D. J. Bartholomew:

It may be of interest to report a remark which Robbins made in the course of his Wald lectures in 1969 about the test which has power one. The problem is to test the hypothesis $H_0 : \theta \leq 0$ against the alternative $H_1 : \theta > 0$, where θ is the mean of a normal distribution with unit variance. The remark was that the test has the same form as a Bayesian test based on (a) the rejection of H_0 if $\Pr[\theta < 0 \text{ observations}] < \epsilon$ for some small ϵ, (b) a prior distribution for θ with a concentration of probability at $\theta = 0$.

This certainly provides food for thought for those who feel that the Bayesian and frequentist interpretations of a procedure should be compatible.

I. J. Good:

On page 7 of Dr. Neyman's paper he refers to the combination of independent significance tests or, as I call them, *significance tests in series,* and he points out that some tests deserve more weight than others. I should like to draw attention to a 1955 paper concerning the weighted combination of significance tests (*J. Roy. Statist. Soc., B*, 17, 1955, 264-265). In that paper I proposed that instead of multiplying together the tail-area probabilities one could multiply together powers of the tail-area probabilities and that the powers selected could depend on the weights that we wish to give to the various tests. There is a difficulty here in deciding what powers are appropriate, but whatever weights are selected it is at any rate practicable to find the distribution of the statistic explicitly. Some comments on how to select these weights were made in a later paper on the subject of combining parallel significance tests (*J. Am. Stat. Ass.*, 53, 1958, 799-813). The proposal of using a product of powers is admittedly somewhat *ad hoc*, but it is about as natural as any *ad hoc* method can be. After all, even the multiplication of the tail-area probabilities is somewhat *ad hoc*, as Fisher knew. His attitude was that people were already doing it and misinterpreting the result, so he provided the correct distribution for them. One might also select a variety of different sets of weights and use the harmonic mean of the tail-area probabilities obtained by the various different sets.

As a historical matter, sequential analysis using the probability ratio test was introduced independent of Wald by Turing in 1940 for an important classified application. It was in this connection that he invented the expressions *factor in favor of a hypothesis* for the ratio of the final to initial odds and *deciban* as a unit of measurement for the log-factor.

The idea of Robert Davies is somewhat analogous to a Bayesian technique I sometimes call *marginalism*; that is, "we often more or less consciously think as if we had formulated the non-null hypothesis so as to give it a reasonable chance of winning on the size of sample we are apt to have available." (*J. Am. Stat. Ass.*, 64, 1970, page 61). Although this distorts the prior distribution it has the merit that if the evidence is overwhelming the distortion does not matter and if the evidence is inadequate to prove anything, then the distortion again doesn't matter. It is only in marginal cases that the choice of the prior makes much difference. Hence the name marginalism. It is a trick that does not give accurate final probabilities, but it protects you from missing what the data is trying to say owing to a careless choice of prior distribution.

O. Kempthorne:

I am rather inclined to the view that Dr. Neyman's work on seeding and rainfall is not directed to the relevant scientific and technological questions. Are averages relevant? To take the case of penicillin, the fact that penicillin works in some very high proportion of a class of problems (and hence has a *good* average effect) should not lead us to forget that penicillin therapy has caused, I am told, the death of some individuals.

With regard to the combination of tests of hypotheses, I believe the work of M. Zelen of some years ago is highly relevant. In Zelen's case, the approach was apt because he was concerned with interblock and intrablock information on treatment effects. I may also add that many (including my colleague, Dr. Zyskind) have worked on the combination of information with regard to estimation and to separate the problem of estimation and tests can be regarded as spurious in logic.

I suggest that in most of the applications of statistics, we do not have random samples. I give views on this in my essay.

I am glad to see that Dr. Neyman is becoming aware of the morass of the theory of unbiassed tests. I am of the opinion that this morass was recognized by R. A. Fisher some forty years ago. His words did not, however, convey meaning to Dr. Neyman, in contrast to some others.

Dr. Neyman has made great contributions to statistics. I believe, however, that he has not weighed adequately (if at all) several of the criticisms of the doctrine that one should pick an α and then maximize power. Perhaps my remarks in the discussion of Cornfield's paper in *Biometrics* recently and several remarks in my essay for this symposium will communicate a meaning; but the prognosis is not good.

REPLY

Before commenting on the discussion of my paper, I wish to offer three items of appreciation.

Professor V. P. Godambe deserves hearty compliments on the organization of this Symposium. It provided a most valuable opportunity for personal contacts among representatives of a great variety of schools of thought and, in due course, beneficial effects of mutual influences are bound to be reflected in the further development of our discipline.

I am very grateful to Professor Henry E. Kyburg, Jr., the Chairman of the session at which I spoke, for his introductory remarks and for the friendly feelings reflected in his much too flattering description of my status among statisticians.

Hearty thanks are due to several speakers who pointed out omissions of relevant references to previous work. I particularly regret the oversight of the early paper on sequential procedures published in 1943 by Walter Bartky (*Ann. Math. Stat.,* Vol. 14, 363-377). Quite likely, this is the first paper on the subject ever published in an open journal. Professor L. M. LeCam tells me that he has seen notes of Bartky's lectures, delivered at the Bell Telephone Laboratories some time before World War II, possibly in the late 1920's, in which the elements of sequential sampling were described.

Several speakers at the Symposium expressed the desirability of an effort at a unification or, at least, of a harmonization of the various schools of thought. I am not sure that I fully agree with this idea. An effort at some compromise among several points of view is clearly indicated at a meeting of individuals who are bound together and face the necessity of a joint action: perhaps a political group or, possibly, senate of a university, etc. The purpose of a scholarly symposium is different. Here a gain in clarity of thinking appears of paramount importance. To this end, the emphasis on and the delineation of items of divergence among the several points of view appears just as important as the description of points of agreement or convergence. The following remarks are formulated in this spirit. They refer primarily to the four written contributions I have before me, by Professors Barnard, Bartholomew, Good and Kempthorne.

(a) It is a pleasure to join Professor Barnard in his appreciation of Darling's and Robbins' mastery of the many mathematical tools needed in the modern research in probability and statistics. I wish my own tool box were anywhere near theirs.

As to the influences that inspired Robert Davies in his work on β-optimal, level α test, there must have been many and I expect that the relevant references are given in his paper.

I am in full sympathy with Barnard's regrets about the multitude of problems and preoccupations. Like Barnard I also hope to write something comprehensive and fail to do so for reasons similar to his. However, these regrets are moderated by the awareness that, if there were only a few problems floating around, our lives would have been duller than they actually are.

(b) The historical remark of Professor Bartholomew is very interesting. Regretfully, I could not attend all the Robbins' Wald lectures and missed the point on the possibility of the Bayesian approach to the main problem treated.

(c) The discussion of my paper from the Bayesian point of view included a remark by Professor Rubin to the general effect that I might qualify for the classification as a Bayesian. At the time I was surprised, but who knows? David Blackwell is emphatic on being a Bayesian, and I never heard or read anything by him to which I would not willingly subscribe. Also a brief conversation I had at this Symposium with Professor Dempster, whom I believe to be a declared Bayesian (if I am mistaken, I apologize humbly!), showed that our views on certain matters are quite close. If I understood Professor Dempster correctly, he agreed with me that the ultimate aim of all of us is to decrease the frequency of erroneous decisions or judgments. Finally, among the "Twenty-Seven Principles of Rationality" distributed by Professor Good among the participants in the Symposium, there is one, Principle 4, which reads: "the principle of rationality is the recommendation to maximize expected utility". With a degree of interpretation, this principle might be considered equivalent to my "ultimate aims". There are, then, some points of convergence between the Bayesian or the Neo-Bayesian school of thought and my own stand. However, there are also some points of divergence. In particular, I cannot subscribe to quite a few points in Professor Good's writings.

The principal of these points of difference is, I think, a degree of dogmatism which I sense in Professor Good's papers. Occasionally, this dogmatism is, so to speak, explicit as, for example, in his title "Twenty-Seven Principles of Rationality". This title implies that, if anyone fails to conform with the principles formulated by Professor Good, he will not deserve the description of being a rational being. The same element of dogmatism is present in other writings of Professor Good where it appears in a somewhat disguised form, in his description of something being "reasonable", etc. Somehow, dogmatism is foreign to my thinking.

Of the three references (which I shall label I, II, and III) in Professor Good's contribution to the discussion of my paper, it is paper II that is particularly relevant to the present comments. [Paper I is essentially limited to an important mathematical result which, as acknowledged in Paper II, has been obtained earlier by several authors, including the ubiquitous Robbins and, later, Darling]. Paper II includes the consideration of a case, labeled B, in which a statistician is faced with the problem of testing a hypothesis H_0 against an alternative H_1. Several alternative tests of H_0 are available, say T_1, T_2, ..., T_n, which it is admitted are *inefficient*, with the qualification that "the word 'efficient' need not be interpreted in any precise sense here." These tests have already been applied to the same data E and yielded significance probabilities P_1, P_2, ..., P_n. The problem is how to arrive at the final conclusion regarding H_0. Professor Good offers a piece of advice which is permissive in character: "If the statistician can think of nothing better to do... it is *reasonable* for him" (italics mine) to base his judgment on the weighted harmonic mean $\lambda(\omega)$ of the calculated P_1, P_2, ..., P_n.

I bring out these details for the reason that they illustrate a point of sharp divergence between the stand of Professor Good and my own thinking. In spite of its permissive character, Professor Good's advice is dogmatic and has nothing to do with the ultimate aim of behavioristic statistics. *If one really desires* to act with low probabilities of errors (or so as to maximize expected utility), some action in the desired direction is clearly indicated. This action would involve the consideration of existing possibilities, more or less on the lines of my Sections 3 and 4, and the selection of the test which fits best the statistician's subjective feelings about the relative importance of the various errors that he wishes to guard against. In particular, as in the case of Fisher's comprehensive test, one of the original tests T_1, T_2, ..., T_n might prove to have acceptable properties.

(d) Professor Kempthorne's discussion of my paper contains emphatic and unqualified disapprovals of two sections of my research activity, a reference to the work of Zelen, for which I am grateful, and some remarks about the available samples being nonrandom. This latter point being entirely irrelevant, the following comments are restricted to disapprovals. The remarkable circumstance is that these disapprovals exhibit Professor Kempthorne's lack of information about the subjects discussed.

With reference to weather modification experiments, the tell-tale item is Professor Kempthorne's comparison with the hypothetical (I am told!) problem of penicillin. It indicates that, at least at the time of writing, Professor Kempthorne was entirely unaware of the state of affairs in weather modification studies and of my stand in the matter. The literature on the subject is quite extensive and those not interested have no obligation to read it. [Several review articles have been published recently in the *Review of the International Statistical Institute*, Vol. 38, No. 1, 1970]. But then why make pronouncements, whether favorable or condemning, on subjects in which one is not interested and is not informed?

With reference to testing hypotheses, Professor Kempthorne wrote quite a tirade on the theory of unbiased tests. The trouble is that unbiased tests are not even mentioned in my paper and the tests that are discussed can be biased or unbiased, depending on the circumstances of the particular problem. Obviously, Professor Kempthorne must be not-quite-informed what unbiased tests really are and got mixed up. My comment is, as above, that, unless one is interested, one should not feel compelled to read the theory of testing hypotheses. However, why make pronouncements on subjects on which one is not informed?

Naturally, I feel downtrodden by Professor Kempthorne's pessimistic prognosis on my preceptive ability. The actual situation must be even worse than Professor Kempthorne describes because, contrary to his hopeful presumption, I continue to be unaware of any "morass" in the theory of unbiased tests. Now, isn't this just too bad!

During the Waterloo Symposium Professor Kempthorne's vivacity provided the audience with quite a few moments of merriment which many of us enjoyed. It would have been nice if this were combined with a degree of sense of responsibility restricting Professor Kempthorne's pronouncements to those subjects on which he is informed.

Note added in proofs. The recent advances in sequential analysis briefly mentioned in my paper are outstanding and here references to earlier work are particularly important. In this connection I wish to quote from the handout of Robbins' Wald Lecture I, delivered at the 1969 Annual Meeting of the IMS, as follows: "The main ancestors known at this time to me are listed below in [1], [6], and [7]." The three references are:

[1] Blum, J. R. and J. Rosenblatt, "On Some Statistical Problems Requiring Purely Sequential Sampling Schemes," *Ann. Inst. Stat. Math.,* 18, 351-354, 1966.
[6] Fabian, V., "A Decision Function," *Czech. Math. Jour.,* 6, 31-41, 1956.
[7] Farrell, R. H., "Asymptotic Behavior of Expected Sample Size in Certain One Sided Tests," *Ann. Math. Stat.,* 35, 36-72, 1964.

WHEN IS INFERENCE
STATISTICAL INFERENCE?
M.S. Bartlett
University Of Oxford

Summary

The thesis is argued that a statistician *qua* statistician should either confine his inferences to statistical inferences in the sense indicated in this paper, or at least make it clear when his inferences are not so classifiable.

The arguments for and against prior probabilities, likelihood ratios, and alternative methods of estimation are discussed and illustrated in this context.

1
Introductory Remarks

I have become more pessimistic over the years about the possibility of complete agreement over the principles of *statistical* inference. The difficulties are partly terminological and semantic, as exemplified by differences of opinion of the meaning and range of the words *statistics* and *statistical;* but over and above this there is the almost inevitable controversy over the nature and methods of all inductive inference. Statistical inference is certainly at least part of all inference; hence the wider controversy muddies the waters of the narrower stream. Statisticians and others interested in the statistical inference problem often seem to me to have refused to recognize the open-endedness of the problem by adopting one of two attitudes: either (a) they may ignore or belittle the inductive inference aspect of the statistical problem, when they discuss statistical *hypotheses* as if one of them at least were true (we can more realistically treat all those so far enunciated as false), or (b) (like H. Jeffreys and L.J. Savage) they are so attracted by the apparent unity and comprehensiveness of the Bayesian approach to all induction that they do not always recognize or admit that statisticians have a field of discourse that is not identical with the universe of all discourse (except that, curiously enough, their examples are often so dominated by statistical examples that I have claimed that they use the more robust statistical component of these problems to bolster up the weaker non-statistical part). One could classify these two approaches as (a) partial precision by unjustifiable segregation, (b) spurious over-all precision created by analogy. My pessimism over the existence of a final complete solution explains my reluctance to add further to the spate of words that have been already written on this subject, a spate that should have warned us of the possible absence of any unique acceptable solution. I have suggested above that we have all been in danger at times of evasion and omission, and if this current round of discussion is to be more than a merry-go-round it could be because the participants are prepared to face difficulties frankly and impartially rather than indulge in what has been at best a debate between protagonists and at worst a slanging match.

Some of the omission stems from the historical accidents of our education and obligations, so that mathematicians discussing Bayesian theory often concentrate on the mathematics without bothering to ask themselves what they are talking about, statisticians undertaking statistical analyses for a scientific paper or for a client cannot confuse the immediate task by a digression on philosophical fundamentals. This limitation of their horizon, whether deliberate or not, is often quite right and proper. It may, as I have remarked before, be compared with the routine work of the physicist or biologist, who would merely bore his audience if he dragged the whole question of scientific method into his work at every conceivable opportunity. It is not a valid criticism of his work to say he has not discussed its foundations; it would only be a valid criticism if it were established that he had never considered these. Of course to compare the statistician with the physicist or biologist is to claim that there are recognizable phenomena that may be classified as *statistical,* just as there are phenomena classified as physical and phenomena classified as biological. Such classification is inevitably bedevilled both by woolly and overlapping frontiers, and by the philosophical conundrum of separating fact and theory; in the case of statistics it has not apparently been accepted by everybody. However, like Hacking (1965), Edwards (1969) and others, I propose to accept the existence of sex-ratios, expectations of life, genetic re-combination fractions and the like, as much as velocities, electrons, living organisms and cells. This necessitates the use of statistical *probabilities* if we are to have any kind of mathematical theory, and a theory of statistical inference which is not purely deductive I would define as concerned with inference about such probabilities.

This attitude raises various questions and points which need enunciation and discussion. First of all, it is of course often necessary in practice to draw rather vague inferences about statistical quantities, just as about other quantities, because of the vague nature of the data – an example would be a rough prediction in epidemiology or meteorology, because of the heterogeneous or impure character of the data. The next point to raise is whether in appropriate, more homogeneous or *pure* statistical situations it is possible to employ specific techniques which would not be available elsewhere and, if so, whether they are intrinsically statistical in their interpretation or can be applied to non-statistical questions, such as: What will happen at a unique trial? There is an obvious and major difficulty here which affects my own general attitude to these questions. This is how we recognize the appropriate pure statistical situation. To someone who is not prepared to consider such a situation the question does not arise, but if phenomena are not to be classified at all, life is going to be hard indeed. However, I must remind you of my *caveat* that in agreeing to isolate the statistical problem we must not ignore its general background. This is why I have never claimed that statistical methods or inferences are more than a partial answer to some of our real questions, and we must never forget this (cf. Skellam, 1969). It seems to me that my attitude is nearer to R.A. Fisher's earlier outlook, when he emphasized the reduction of data, the sampling properties of estimates, etc. than to his later attitude, which was closer to that of writers like G. Barnard, Edwards and Hacking, when they discuss artificial examples which seem to me at times to be in danger of being over-academic and narrow. My own reaction is still to formulate questions statistically in statistical situations while recognizing the limitations of some of the answers, and to regard an answer to a non-statistical question in a statistical situation as no more than a convention (which may be interesting, but no more necessarily logically acceptable than any other inductive inference). Questions and answers in non-statistical situations I would regard as non-statistical by definition, and hence outside my terms of reference.

This, I think, gives me a certain consistency of approach associated with a refusal to mislead by spurious over-all precision, and a realization that we do not all seem to see eye to eye on these matters. As this statement of my own attitude is possibly unclear without illustration, I propose to add remarks on some of the more specific controversial techniques with a view to reiterating my own position. Any apologies for repetition are qualified by the comment that others demonstrably do not always bother to read, or at any rate to take in, what one has previously said, so that some degree of redundancy in one's own writings seems not only more justifiable, but almost essential for adequate communication.

2
Prior Probabilities

In a review of the interesting little book by Maynard Smith on *The Mathematics of Biology*, I objected to one example and its solution, both seeming to me incompatible with the interpretation of probability as relative frequency given in the text of the book. I quote below the relevant portion of my review (1969a):

> The last example on page 70 has nothing to do with probability as frequency, and requires detailed comment. The author says it nearly wrecked a conference on theoretical biology, but also says that it yields at once to common sense or to Bayes' theorem. If it is so simple, this does not seem to say much for the conference participants; moreover, as a question on Bayes' theorem, the answer given may legitimately be questioned. The example is as follows:
>
> > Of three prisoners, Matthew, Mark and Luke, two are to be executed, but Matthew does not know which. He therefore asks the jailer 'Since either Mark or Luke are certainly going to be executed, you will give me no information about my own

chances if you give me the name of one man, either Mark or Luke, who is going to be executed.' Accepting this argument, the jailer truthfully replied 'Mark will be executed'. Thereupon, Matthew felt happier, because before the jailer replied his own chances of execution were $2/3$, but afterwards there are only two people, himself and Luke, who could be the one not to be executed, and so his chance of execution is only $1/2$.

Is Matthew right to feel happier?

The individuals to be executed are already determined and known to the jailer, who is therefore justified in passing on as irrelevant to Matthew, the name of either Mark or Luke as one of these. However, it is said that Matthew is prepared in the absence of this certain knowledge to allocate numerical probabilities to possible eventualities as indicated. Is he being consistent in his assessments? Suppose the prior probabilities are denoted in general by p, q and r respectively for Matthew's, Mark's or Luke's survival. Then the posterior probability of his own survival after the jailer's information is from Bayes' theorem changed from p to $p'=pP/(pP+r)$, where P is the probability of the jailer naming Mark if both Mark and Luke are to be executed. Note that *to the jailer* either p or r is 1, and in either case (whatever P) there is for him no change in p. Now *if* Matthew assigned values $p=q=r=1/3$, $P=1/2$, then also for him $p'=P$, and he was behaving inconsistently in claiming $p'=1/2$. But he is quite justified (consistently with the wording of the example) in first assigning $p=1/3$, $p'=1/2$, provided he then chooses $r/P=1/3$. His assessments are now all consistent, and, as $p'>p$, he has a perfect right to feel happier!

If the example proves anything, it is the indefiniteness of subjective prior probabilities.

Professor Lindley, one of the most whole-hearted Bayesians in England, objected to my comments, and made the following reply (1969):

In the course of a book review (*Nature*, 221, 291; 1969) Professor M.S. Bartlett discusses the problem of three prisoners, two of whom are to be executed. His comments merit further discussion. He first remarks that if, as has been reported, the problem nearly wrecked a conference on theoretical biology and yet yields at once to Bayes's theorem, it does not say much for the conference participants. This seems a little hard on the theoretical biologists who will typically have learnt their probability from a member of the frequentist school who, if he mentioned Bayes's theorem at all, will have played it down as of minor interest. The fault surely lies with the statistician, not the biologist.

The second point is more material. Bartlett draws the conclusion that subjective probabilities are indefinite. The basis for this assertion seems to be that the prisoner, Matthew, is entitled to feel happier if P (the probability of the jailer naming Mark if both Mark and Luke are to be executed) equals $3r$. If $r>1/3$, the Bayesian analysis shows that this is impossible so that Matthew is inconsistent. If $r=1/3$, this requires $P=1$. Taking $r=1/3$ as a reasonable value, this shows that Matthew's elation is only justifiable (to him!) if he believes that when there is a choice, the jailer will always name Mark. So I would say to Matthew, "If you feel elated it is equivalent to your assuming this preference on the jailer's part". Matthew would typically reply that he has no reason for thinking the jailer has such a preference: therefore, I say, he has no reason for feeling happier. The subjective probabilities must cohere and their value lies in doing just this – in this example in establishing coherence between Matthew's happiness and his opinion of the jailer. This seems to me to lead to a definite conclusion of some value, contrary to what has been said.

One or two other writers joined in, and I also had a personal communication from Harold Jeffreys, but none of these appear to me to have thrown any new light on the problem, and I do not propose to quote them all here. I merely add my own published reply (1969b) to Lindley:

Professor Lindley (*Nature,* 221, 594; 1969) is, like Matthew, entitled to his opinion, but I do not see that his comments in any way dispose of my criticisms. First of all perhaps I could protest at the irrelevant remarks in his first paragraph, which includes an inaccurate reporting of what I quoted Professor Maynard Smith as saying, namely, that the problem 'yields at once to common sense or to Bayes's theorem'. By omitting the phrase 'to common sense or' Professor Lindley distorts, quite unfairly, the implication in my own remarks.

To come now to the problem, this is unsatisfactory partly because of the third-person reporting in it of Matthew's opinions as data to be used in any numerical evaluations. These data included Matthew's assessments of his own probability of execution as changing from 2/3 to 1/2, so that he had a right to feel happier in particular if it could be shown that he was being consistent. Professors Maynard Smith and Lindley apparently think he had no such right, but my claim that this view 'may legitimately be questioned' was illustrated by my demonstrating how Matthew could be consistent in his assessments. Lindley introduces the further assumption that $r=1/3$ 'as a reasonable value', and concludes that P, the probability of the jailer naming Mark if both Mark and Luke are to be executed, would then have to be 1. Dismissing this possibility, Lindley rejects the consistency of Matthew's assessments. However, even if I accept Lindley's value $r=1/3$ (which Matthew may not have done), Matthew has still a 'right to feel happier' on the weaker basis that his second assessment should be at any rate $<2/3$, which leads to the condition $P>1/2$. I see nothing extraordinary in such a belief by Matthew about P, especially if he recalls the order of precedence of Mark and Luke in the New Testament (the jailer may not be a Christian, but Matthew may be unaware of this).

Of course I am not arguing against the explicit use of prior probabilities when they are well-determined in the statistical sense, such as with the prior probability of like-sex twins being monozygotic, or in other empirical uses of Bayes's theorem as advocated by H. Robbins and other writers.

3
Likelihood Ratios

If, like Hacking and others, we concentrate on the likelihood ratio as the relevant sample quantity to discriminate between two hypotheses (note that its relevance and sufficiency may be deduced from Bayes's theorem or directly, as was first done, I think, by B.L. Welch), we have yet to say how it is to be used. I agree with Hacking that without Bayes's theorem we have to decide what direction gives the greater plausibility to one as against the other alternative, but the direction (which would follow in any event from cases where Bayes's theorem has a frequency interpretation) is not one that I have ever heard queried, so it is not an unreasonable axiom. Hacking suggested in his book that the axiom does not follow from long-run frequency—his counter-example* was not, however, too convincing, for he appealed to a minimax solution which made the long-run solution obviously absurd. I discussed quantitative aspects of the long-run solution further in my review (1966), but agreed that this further discussion still did not itself justify the universally accepted solution to his problem. The more relevant comment appears to me to be supplied by Hacking himself, that someone "may say that it is habit that makes you suppose that if there were to be no long run, there would still be a reasonable guess". I do not object to the axiom that the hypothesis with the greater likelihood has the greater *support;* indeed, this is merely a definition of support, but different people have different reasons for their acceptance of such an axiom, and I do not see that either a Bayesian background of uniform prior probabilities, or a statistical background of long-run assertions based on likelihood ratios, can be dismissed as irrelevant if an individual chooses to make use of one of them. Indeed, the acceptance of a maximum likelihood estimate is for me at least qualified by the associated statistical problem, which is simplest in the standard large-sample case. Thus a statement on the support for any statistical hypothesis against another is a statistical inference; but how we make use of this result, for example, in a particular case, is not.

Edwards has proposed extending the range of support (so far defined in terms of likelihoods) to cover prior knowledge. He says "The prior support for one hypothesis or set is S if, prior to any experiment, I support the one against the other to the same degree that I would if I had conducted an experiment which led to experimental support S in a situation in which, prior to this conceptual experiment, I was unable to express any preference for the one over the other". This proposal is an ingenious one which allows forms of inference intermediate between ones consistent with a more complete statistical attitude and with the Bayesian approach. However, while this new concept of prior support has, by definition, the numerical properties of experimental support, this seems to me still reminiscent of the Bayesian's introduction of prior probabilities with the numerical properties of statistical probabilities; in both cases, we cannot avoid the subjectivity introduced, as Edwards himself points out in the case of prior support. He gives an application of his ideas to a genetical example, where an inference about an unknown probability p of a mouse being homozygous for a certain gene locus depends on missing information about the type of parental mating. If two alternatives about this are allocated equal support, it proves possible to make two alternative statistical inferences A and B with a known numerical support in favour of one of them. While this inference is weaker (and more complicated) than the orthodox Bayes solution for the entire problem, it of course still hinges on the subjective allocation of equal prior support; indeed, Edwards's proposal is not limited to the use of *equal* prior supports, and the allocation of unequal prior supports would seem to me to have nearly as much arbitrariness as unequal prior probabilities.

My own predilection for statistical interpretation to inferences made in sta-

---

* A green ball is drawn from an urn which has three possible constitutions: (a) 99 green balls, 1 red; (b) 2 green, 98 red; (c) 1 green, 99 red. What is the best supported hypothesis: (a), (b) or (c)? (You are told that the frequency with which any colour is drawn, with replacement, from an urn is proportional to the number of balls of that colour in the urn.)

tistical situations is of course close to the behaviouristic outlook associated with Neyman. This standpoint has been the occasion of much ridicule from Bayesians and even Fisherians, but I will emphasize again (speaking for myself at least) that the recording of, say, a confidence statement does not imply that we confuse the inference for a unique occasion with a statistical inference anymore than we confuse the statistical probability of an event with the realisation at the next trial.

I do not propose to discuss here in detail the distinction between confidence and fiducial intervals, except to reiterate my personal view (1965) that the original fiducial interval concept introduced by Fisher was identical in interpretation with a class of confidence interval, and that it was only later than the statistical interpretation of a fiducial interval was specifically rejected by Fisher.

Returning to the recent emphasis on the use of the likelihood function as such, I would agree that this must (of course only if the statistical specification of the model is correct!) contain the relevant information on unknown parameters θ_i ($i=1\ldots k$), but I would also wish to know the sampling situation. For example, when it is proposed to accept values of the parameters such that the likelihood ratio of the function at a parametric point θ_i ($i=1\ldots k$) to the maximum value is not less than a fixed constant, how do we assign the constant? With the knowledge that $-2(L-L_{\max})$ is in appropriate problems asymptotically a chi-square with degrees of freedom equal to k (L being log likelihood), we can fix the critical value to correspond to a confidence region of known probability. This is rather crude, and we can often improve on this considerably by a study of the derivatives $\partial L/\partial \theta_i$, but if this asymptotic theory is invalid it seems to me that the situation requires further specific appraisal.

Consider the following problem. It is admittedly of the kind I have labelled over-academic but it does illustrate in an idealized form a natural epidemiological problem discussed by Almond (1954). A simple multiplicative birth-and-death process starting with n_0 individuals becomes extinct after N transitions. What can be inferred about the unknown parameter $p = \mu/(\lambda + \mu)$, where μ is the death-rate, λ the birth-rate? The likelihood function is fairly readily

$$Cp^{\frac{1}{2}(N+n_0)} q^{\frac{1}{2}(N-n_0)}, \tag{1}$$

which appears very well-behaved, but we must be very wary of the sampling situation. If $p > \frac{1}{2}$, all is well, and while N is random and not at our choice, the asymptotic sampling situation may be investigated in detail for n_0 not too small. However, if $p \leq \frac{1}{2}$, we have the remarkable result first established by O'N. Waugh (1958) that *conditional on extinction occurring* the distribution of N is identical with the case $p > \frac{1}{2}$ *with p and q interchanged*. The chance of extinction is $(p/q)^{n_0}$, and the likelihood function similarly factorizes to

$$(p/q)^{n_0}(Cq^{\frac{1}{2}(N+n_0)} p^{\frac{1}{2}(N-n_0)}). \tag{2}$$

Thus any sampling theory would now yield a comparable inference about q/p, *given extinction*, coupled with the initial factor for the chance of extinction. This factor would of course be small for n_0 not too small, but it seems instructive to keep it separate, especially if the possibility of *selection* for a case of extinction could not be ruled out. For small n_0, the asymptotic theory would not be applicable, but confidence statements would still be available, even if rather complex, in view of such results as (to take a trivial case with $n_0 = 1$ and $N = 1$)

$$P(N > 1) < 0.1 \text{ if } p > 0.9;$$

and

$$P(1 < N < \infty \mid N < \infty) < 0.1,$$

with

$$P(N < \infty) < 0.11, \text{ if } p < 0.1.$$

Incidentally, the distribution of $P(N)$ for $n_0 = 1$ and $p > \frac{1}{2}$ is determined by its probability-generating function

$$\Pi(z) = [1 - (1 - 4pqz^2)^{\frac{1}{2}}]/(2pq) \tag{3}$$

with mean $2q/(p - q)$ and variance $4pq/(p - q)^3$.

4

Statistical Properties of Estimates

The view that it is important to know the statistical properties of our estimates may be emphasized in relation to another type of situation examined in recent years, for example, in connection with the estimation of a multivariate mean by Stein (1962), but, as I pointed out in the discussion on Stein's paper, effectively ventilated some time ago in the context of factor analysis.

In vector and matrix notation, let $z = M_0 f_0 + M_1 f_1$, where z is a set of test scores, f_0 the group factors and f_1 the specifics. We estimate f_0 for a particular person with scores z by the equation (1937)

$$\check{f}_0 = J^{-1} M'_0 M_1^{-2} z,$$

where

$$\begin{cases} J = M'_0 M_1^{-2} M_0, \quad E_1\{\check{f}_0\} = f_0, \\ E_1\{(\check{f}_0 - f_0)(\check{f}_0 - f_0)'\} = J^{-1}, \end{cases}$$

E_1 denoting averaging over all possible sets of tests comparable with the given set in regard to the information on the group factors f_0.

By contrast, Godfrey Thomson's regression estimates (1939 p. 305) were specified by

$$\hat{f}_0 = K\check{f}_0, \quad K^{-1} = I + J^{-1},$$

where

$$\begin{cases} E\{\hat{f}_0\} = E\{f_0\} = 0, \\ E\{(\hat{f}_0 - f_0)(\hat{f}_0 - f_0)'\} = I - K, \end{cases}$$

E denoting averaging over all persons in the (standardized) population.

In the particular case of only one common factor, g, say, J reduces to the quantity

$$S = \Sigma_i \, r_{ig}^2 / (1 - r_{ig}^2)$$

and we have

$$\sigma^2(\breve{g}) = 1/S, \; v(\hat{g}) = 1/(1+S) < \sigma^2(\breve{g}),$$

where $v(\hat{g})$ refers to the *mean square error of* $\hat{g}$, the appropriate measure for comparison because the $\hat{g}$ estimate is biased towards zero. Note that while $v(\hat{g})$ is less than $\sigma^2(\breve{g})$, it is an average over persons of the quantity

$$\frac{g^2}{(S+1)^2} + \frac{S}{(S+1)^2},$$

and this will be greater than $\sigma^2(\breve{g})$ for some persons. When the different statistical properties of $\hat{f}$ and $\breve{f}$ are elucidated, it is possible to choose one in preference to the other in a given situation and for a given purpose. A concentration on one or more particular persons would favour $\breve{f}$; a general investigation on the entire population would favour $\hat{f}$. However, as I pointed out at the time, one set is obtainable from the other, so that from the standpoint of the reduction of data they are still equivalent.

5
Conclusion

If I could sum up very briefly my own outlook, it is that statistics as a subject cannot from its very nature be concerned with single individuals or events as such, and consequently I do not see that *statistical* inference can either. This is not to say that in the statistical domain the most relevant choice of statistical framework for the problem has not to be sought, nor that the inevitable transition from the statistical situation to a specific individual or event is to be overlooked. The first of these two tasks is still in the statistical domain. The second is not; statisticians are mere human beings when struggling with it, and should not confuse others by claiming otherwise.

Acknowledgment

I am indebted to the editor of *Nature* for permission to include the quotations from that Journal, and also, of course, to Professor Lindley in the case of his own letter.

References

1. Almond, J., "A Note on the X^2 Test Applied to Epidemic Chains," *Biometrics,* 10, 459, 1954.

2. Bartlett, M.S., "Methods of Estimating Mental Factors," *Nature,* 141, 609-610, 1938.
3. Bartlett, M.S., "R.A. Fisher and the Last Fifty Years of Statistical Methodology," *Journal of the American Statistical Association,* 60, 395-409, 1965.
4. Bartlett, M.S., "Review of 'Logic of Statistical Inference', by I. Hacking," *Biometrika,* 53, 631-633, 1966.
5. Bartlett, M.S., "Review of 'Mathematical Ideas in Biology', by J. Maynard Smith," *Nature,* 221, 291, 1969a.
6. Bartlett, M.S., "Probability and Prejudice: Correspondence," *Nature,* 221, 786, 1969b.
7. Edwards, A.W.F., "Statistical Methods in Scientific Inference," *Nature,* 222, 1233-1237, 1969.
8. Hacking, I., *Logic of Statistical Inference,* Cambridge University Press, 1965.
9. Lindley, D.V., "Probability and Prejudice Correspondence," *Nature,* 221, 594, 1969.
10. Maynard Smith, J., *Mathematical Ideas in Biology,* Cambridge University Press, 1968.
11. Skellam, J.G., "Models, Inference, and Strategy," *Biometrics,* 25, 457, 1969.
12. Stein, C.M., "Confidence Sets for the Mean of a Multivariate Normal Distribution," *Journal of the Royal Statistical Society, B,* 24, 265, 1962.
13. Thomson, Godfrey H., *The Factorial Analysis of Human Ability,* University of London Press, 1939.
14. Waugh, W.A. O'N., "Conditional Markov Processes," *Biometrika,* 45, 241, 1958.

COMMENTS

G.A. Barnard:

I would like to support Professor Bartlett's point, that statistical inference is concerned with inference about what he calls statistical probabilities. At the same time, because the statistician is by experience accustomed to handling probabilistic concepts and inferences, he surely has rather more than the ordinary experience in connection with more general problems of induction. Of course, there are many other issues in scientific method than probability considerations, and with these the statistician has no special claim.

I agree that a dangerous feature of the Bayesian approach is that the message in the data may become confused with prior assumptions which are inadequately supported.

The problem of the jailer surely involves limited ambiguity of the type familiar to solvers of the *Times* crossword puzzle. The question seems to be, whether we are entitled to deduce, from the jailer's acceptance of Matthew's argument, that the jailer knows that Matthew knows that he is equally likely to name Mark and Luke, given that both are to be executed. I refrain from commenting further, for fear of incurring further criticism for discussing examples which are over-academic.

I.J. Good:

Professor Bartlett's thesis is that a statistician should either use large samples or should otherwise mention that he is merely a man and not a 100% statistician. I would rather say I am 100% a man and not merely a statistician.

The advice can be generalized: Statisticians should be *honest* and in particular they should make it clear when they use appreciable personal judgment, namely in *most* practical applications of statistics! There is always some personal judgment, even in physics. As J. Bronowski said: "There is no God's eye view of nature..., only a man's eye view" (*The Identity of Man,* Natural History Press, N.Y., 1965, page 37). Of course sometimes the subjective element is small.

So while I agree with Professor Bartlett up to a point, and perhaps more than he realizes, for he is a reasonable man, I feel that his thesis is liable to be misleading. This is because it appears to draw a clear line between statistical inference and reasoning in general. It might encourage the compartmentation of knowledge and of universities. Each research worker should "do his thing" and if this lies in "no-man's land" then *so much the better*. I am surprised that Professor Bartlett, whose own interests are so wide ranging, should be in danger of encouraging compartmentation.

Consider the consequences of giving an exact and narrow definition of statistical inference. We might find that the only people who could become professors of statistics would be those who concentrate all their efforts in this narrow field. This would discourage statisticians from thinking in a broad frame of reference, and statistical inference would tend to be stultified. It is as if geometry had been defined as the Euclidean geometry of three-dimensional space, or even of only two.

Human reasoning is more important than statistics, and Professor Bartlett clearly believes this since he says he had "never claimed that statistical methods or inferences are more than a partial answer to some of our real questions". So let's beware of defining ourselves out of existence.

O. Kempthorne:

I applaud the great bulk of what Dr. Bartlett says. However, I attach more importance to the individual case than he does. I give a few remarks on this in my own essay. I have to insist that prediction for the *individual case* is critical. It is what the "ball game" is about. To treat this case I have to embed the individual case in a hypothetical population of repetitions. This is where a subjective element intrudes, and I see no escape from this problem.

I do wonder seriously, and not facetiously, what Dr. Bartlett means by saying that sex-ratios and so on *exist*. I would hate statistics to become involved in the existentialist mire, but I believe the underlying obscurity *cannot* be avoided.

D.V. Lindley:

Since Bartlett has considered the prisoners' example that he and I discussed publicly in *Nature*, it is perhaps appropriate that I should say something. I did not, as Bartlett asserts, "reject the consistency of Matthew's assessments". All I am saying is that, if Matthew feels happier as a result of the jailer's statement, then it implies that he thinks that, if Mark and Luke are both to be executed, then the jailer is more likely to report Mark than Luke. The implication in this last sentence follows from the assumption that Matthew wishes to be consistent. (If he does not, then he has no case for using probabilities at all.) As Neyman has argued earlier in this conference, this is a free country, and consequently Matthew is entitled to his opinions provided they are consistent. Furthermore consistency is a valuable tool because it provides a check on one's opinion. Thus, in this example, most people would see no reason for the jailer's preferential reporting, so most people would not feel happier. "This seems to me to lead to a definite conclusion of some value."

The real issue between us is, I suppose, the use of subjective probabilities (as opposed to frequency probabilities) in inference. I have many difficulties with the frequency approach but the principle one is that I find it often useless. In most practical situations that come my way it is necessary to make a statement appropriate to that occasion, as when an assertion about the value of a treatment is required. It seems unnatural and unnecessary to imbed this assertion in a sequence of statements. For 250 years opponents of the frequency school have asked for a description of the sequence into which a given statement should be imbedded without getting a clear answer. As an industrialist once said to me: "Lindley, I don't want to know I'll be wrong only 1 in 20 times, I want to know if I'm wrong *now*."

On a technical point in the paper: The discussion of the birth-and-death process is terse and I have possibly misunderstood Bartlett, but when extinction is not certain and yet has been observed he appears to be conditioning the probabilities on extinction. If so, I would only point out that this can be dangerous, as the examples in the literature show.

REPLY

In answer to Dr. Good, the situation is rather easier in research institutes; in universities compartmentation is to some extent a fact to be accepted, and in particular it seems to me important to distinguish statistical inference from all inference when teaching in a department of statistics. Otherwise one ought to be in a department of philosophy. In my opinion one need not worry over any danger of having too narrow a field; the range of statistics even when restricted in the sense indicated in my paper is very wide indeed.

I accept Prof. Lindley's request for coherency where possible in subjective assessments as my own discussion of the prisoners' example shows, but might remind him that coherency can be achieved by assessing the prior probabilities after the posterior probabilities. With regard to the birth-and-death process, my main intention was to cite an example where bifurcation does not permit the usual large-sample ergodic properties to be used.

Dr. Kempthorne asks whether sex-ratios really exist. I think the problem of existence is too deep to settle here!

EVENTS, INFORMATION PROCESSING, AND THE STRUCTURED MODEL

D. A. S. Fraser

Universities of Hawaii and Toronto

Summary

Consider a probability space (E, A, P) and a statistic t from E to X. A realized value t_0 from the statistic gives information concerning the antecedent value on the probability space and, in effect, replaces the distribution P by the conditional distribution given $t = t_0$. If, however, the value t_o comes from some unknown statistic ϕ in a class $\Phi = \{\phi\}$, then, typically, there is still information concerning the realized value on the probability space. This paper examines the nature of such information and characterizes the classes Φ that provide a valid basis for conditional probability. Some multivariate and nonparametric models are examined in detail.

1
Introduction

The *classical model* of statistics is a class C of probability distributions on a space X. The *structured model* of statistics is a probability distribution D on a space E and a class Φ of transformations from the space E to a space X. A classical model C and a structured model (D, Φ) are fundamentally different. As mathematical structures they involve *different objects*, and as models in an application, they function differently: a user of C has *partial* knowledge concerning the source distribution, and *full* information concerning the realized value; a user of (D, Φ) has *full* knowledge concerning the source distribution, but only *partial* information concerning the realized value. In the model (D, Φ) the class Φ acts as a *processor* of information concerning a realized value from D. This paper considers such processors Φ and characterizes those that are effectively equivalent to a single statistic.

Consider a measurable space (E, A) and a class $\Phi = \{\phi\}$ of measurable transformations to a measurable space (X, B). Let x (in X) be a value produced by some unknown transformation ϕ_0 (in Φ) from an unknown e_0 (in E). What information do x and Φ provide concerning e_0? Or more generally, what kind of information is produced by an *information processor* Φ? Consider some examples.

Let E be R^2 with typical element (e_1, e_2); let Φ have a single element ϕ with $\phi(e_1, e_2) = e_1 + e_2$; and consider x_0 from Φ. The value x_0 from Φ provides information concerning the antecedent (e_1, e_2): *the antecedent (e_1, e_2) is in a set $S_{x_0} = \phi^{-1}x_0$, a set generated by the map ϕ^{-1} into the σ-field A on E.* The image of ϕ^{-1} is a partition of E: im $\phi^{-1} = \{S_x : x \in R^1\}$ where $S_x = \{(e_1, e_2): e_1 + e_2 = x\}$. *Any (e_1, e_2) in the set S_{x_0} when processed by Φ would produce the same information, and any (e_1, e_2) not in S_{x_0} when processed by Φ would provide different information.* The processor *partitions E*, and treat *identically* all points in any set of the partition.

Let $E = E_1 \times E_2^3$ where $E_1 = \{AA, Aa\}$, $E_2 = \{Aa, aa\}$; let Φ have a single element ϕ, the phenotype map operating coordinate by coordinate: $\phi(AA) = A$, $\phi(Aa) = A$, $\phi(aa) = a$; and consider $x_0 = (A, A, A, A)$ from Φ. The value x_0 from Φ provides information concerning the antecedent e: *the antecedent e is in the set $\{(AA, Aa, Aa, Aa), (Aa, Aa, Aa, Aa)\}$, a set generated by the map ϕ^{-1}.* The image of ϕ^{-1} is the partition $\{\{(AA, e_1, e_2, e_3), (Aa, e_1, e_2, e_3)\}: (e_1, e_2, e_3) \in E_2^3\}$. Again the processor *partitions E* and treats *identically* all points in any set of the partition.

Let E be R^2 with typical element (e_1, e_2); let $\Phi = \{[\Phi] : \phi \in R^1\}$ where $[\phi](e_1, e_2) = (e_1 + \phi, e_2 = \phi)$; and consider (x_{10}, x_{20}) from Φ as produced by some unknown ϕ_0 in Φ. The value (x_{10}, x_{20}) from Φ provides information concerning the antecedent (e_1, e_2): *the antecedent (e_1, e_2) is in a set of the class*

$$\{[\phi]^{-1}(x_{10}, x_{20}) : [\phi]^{-1} \in \Phi^{-1}\},$$

a class indexed by maps Φ^{-1} into A. The maps Φ are 1–1, and hence the sets are single points; the class of sets has union set

$$S_{x_{10}x_{20}} = \{(x_{10} - k, x_{20} - k): k \in R^1\}.$$

Any (e_1, e_2) in the set $S_{x_{10}x_{20}}$ when processed by Φ would produce the same information. For consider: a typical (e_1, e_2) in $S_{x_{10}x_{20}}$ is $h^{-1}(x_{10}, x_{20})$ with h^{-1} in Φ^{-1}; the output from the processor would be $\phi_0 h^{-1}(x_{10}, x_{20})$; this value from the processor would provide the information: *the antecedent (e_1, e_2) is in a set of the class*

$$\{[\phi]^{-1} \phi_0 h^{-1}(x_{10}, x_{20}): [\phi]^{-1} \in \Phi^{-1}\} = \{[\phi]^{-1}(x_{10}, x_{20}): [\phi]^{-1} \in \Phi^{-1}\};$$

the reexpression follows from the group-like properties of Φ. And any (e_1, e_2) not in the $S_{x_{10}x_{20}}$ when processed by Φ would clearly provide different information. The processor *partitions* E and treats *identically* all points in any set of the partition.

Let E be $R \times (R - \{0\})$ with typical element (e_1, e_2); let $\Phi = \{[\phi] : \phi \neq 0\}$ where $[\phi] (e_1, e_2) = (e_1 + \phi, e_2 / \phi)$; and consider (x_{10}, x_{20}) from Φ. The value (x_{10}, x_{20}) from Φ provides information concerning the antecedent (e_1, e_2): *the antecedent (e_1, e_2) is in a set of the class*

$$\{\{(x_{10} - \phi, x_{20}\phi)\} : [\phi]^{-1} \in \Phi^{-1}\},$$

a class indexed by maps Φ^{-1} into A. The maps Φ are $1-1$ and hence the sets are single points; the class of sets has union set

$$S_{x_{10}x_{20}} = \{E \cap \{(e_1, e_2) : e_1 x_{20} + e_2 = x_{10} x_{20}\}.$$

Different points in the set $S_{x_{10}x_{20}}$ when processed by Φ would have different second coordinates and would produce different sets $S_{x_1 x_2}$. Thus the processor Φ leads to information presentations; *but* any two different points for which that information is true lead by the processor to *different* presentations. The information obtained from the processor needs qualification against properties of the processor: the apparent information is not the real information.

In the preceding example the processor does not partition E. Examples can also be constructed where the processor partitions E but does not treat identically all points in each set of the partition.

3

Separable Processors

The preceding section has indicated how apparent information from a processor may in fact contain distortions or biases as assessed against properties of the processor.

In the fourth example the processor did *not* induce a partition on the initial space. With a distribution on the initial space various conditional distributions could be formally constructed by choosing lines $e_1 x_2 + e_2 = x_1 x_2$ to form various partitions adjacent to the line $e_1 x_{20} + e_2 = x_{10} x_{20}$. And even in cases where the processor does induce a partition and identifies a set in the partition, there remains the possibility that the processor could produce *different* information from different points in the identified set. With a distribution on the initial space the nominal conditional distribution on the identified set of the partition would then be invalid.

For the application of conditional probability, information must identify a set within a prescribed partition, and must provide no differential information within such a set. This section characterizes such processors. They are called *separable processors*: the known can be separated set-theoretically from the unknown.

Let Φ be a processor, a class of measurable transformations from a measurable space (E, A) into a measurable space (X, B). Suppose an unknown ϕ_0 in Φ produces an outcome in X from some unknown in E. And suppose the processor Φ is separable: *each of the possible E-values that appear in an information presentation would produce, if processed by Φ, the same information presentation.*

Consider an outcome x from the processor Φ. The properties of the processor identify the antecedent E-value as an element in a set in a class

$$\{\phi^{-1}x: \phi^{-1} \in \Phi^{-1}\},$$

a class which is indexed by maps into the algebra A on E. The union set $S = \cup \phi^{-1} x$ with ϕ^{-1} in Φ^{-1} is the set of possible antecedent E-values. A value in S and a map in ϕ would produce an outcome in X; let T be the set of such possible outcomes in X. By the separable property any two X-values in T must induce the same set S on E. Hence the sets S form a partition $P_1 = \{S\}$ of E, the sets T form a partition $P_2 = \{T\}$ of X and there is a $1-1$ map $t: P_1 \rightarrow P_2$ such that e in S implies ϕe in tS for all ϕ in Φ (that is: $\phi e \in tS(e)$ for all ϕ, e).

The separable property applied to the union sets S has identified the $1-1$ correspondence between P_1 and P_2. It suffices then to consider a particular S and the corresponding T. Consider an outcome x in T. The properties of the processor identify the antecedent E-value as an element in a set in the class

$$A^0 = \{\phi^{-1}x; \phi^{-1} \in \Phi^{-1}\}$$

which is indexed by Φ^{-1}. These sets may not constitute a partition of S; but they induce a partition. Suppose for the present that S is reduced modulo this partition. The class A^0 must be independent of x in T.

Let ϕ_*^{-1} be any element of Φ^{-1}; im ϕ_*^{-1} is a partition of S; for any such partition of S let i_* map each set of the partition into one of its elements; let $\hat{\phi}_*^{-1} = i_* \phi_*^{-1}$. The maps $\hat{\phi}_*^{-1}$ can be used to produce a possible antecedent S-value from any outcome. The resulting information presentation is

$$\{\phi^{-1}\phi_0\hat{\phi}_*^{-1}x : \phi^{-1} \in \Phi^{-1}\} = \{hx : h \in \Phi^{-1}\phi_0\hat{\phi}_*^{-1}\}$$

which is the same class A^0 but now indexed by the maps $\Phi^{-1}\phi_0\hat{\phi}_*^{-1}$. By the separable property it follows that

$$\Phi^{-1}\phi_0\hat{\phi}_*^{-1} = \Phi^{-1} \ ;$$

hence

$$\hat{\Phi}^{-1}\phi_0\hat{\phi}_*^{-1} = \hat{\Phi}^{-1} \ ;$$

hence

$$\phi_0\hat{\Phi}^{-1}\phi_0\hat{\phi}_*^{-1} = \phi_0\hat{\Phi}^{-1}$$

or

$$\Psi\psi = \Psi$$

where $\psi = \phi_0\hat{\phi}^{-1}$ is a typical element of the class $\Psi = \phi_0\hat{\Phi}^{-1}$ of maps from T into T. The class Ψ contains the identity $\phi_0\hat{\phi}_0^{-1}$ and is closed under right composition: it is a transitive group of transformations of T onto T. And it must be independent of ϕ_0. Thus it follows from the separable property that the class Φ^{-1} of transformations from T to A^0 is invariant under the group Ψ on T; or that the class Φ from S to T is invariant under the group Ψ on T, $\psi\Phi=\Phi$ for all ψ in Ψ.

Thus a separable processor Φ from (E, A) to (X, B) induces a $1-1$ map t from the partition P_1 of E to the partition P_2 of X; induces a group of transformations Ψ on X whose orbits are the sets of the partition P_2; and is invariant under the group Ψ on the outcome space: $\psi\Phi=\Phi$ for all ψ in Ψ. The processor Φ is then equivalent to the statistic t.

Now suppose there is a probability distribution D on the space (E, A). An observed value x_0 from a separable processor is equivalent to an observed value $t_0=T(x_0)$ from the statistic t. The initial distribution D describing a realized E-value is then replaced by the conditional distribution of D given t_0 as the description of the E-value antecedent to x_0.

4
Homogeneous Separable Processors

Consider a separable processor Φ with partition $P_1=\{S\}$ on E, with partition $P_2=\{T\}$ on X, with induced map t from P_1 to P_2, and with group Ψ on X. Now suppose that the processor Φ is *homogeneous*: Φ does not partition more finely than any component ϕ in Φ. The processor Φ leads to a class A^0 on any S. The sets of A^0 are sets $\phi^{-1}x$. If the induced partition is not finer than that of any ϕ, then A^0 is a partition of S and there is a corresponding partition P_0 of E: $P_0=\mathrm{im}\ \phi^{-1}\phi$ for each ϕ in Φ. A homogeneous processor Φ consists of elements that produce the same functional reduction; that is, $\mathrm{im}\ \phi^{-1}\phi$ is independent of ϕ in Φ.

Let u be the canonical mapping from E to the partition P_0: $u=\phi^{-1}\phi$. And let $\bar{\phi}$ be ϕ reexpressed on P_0: $\phi=\bar{\phi}u$. Then the maps $\bar{\phi}$ in $\bar{\Phi}$ are $1-1$ from P_0 to X.

The group $\Psi=\{\psi\}$ on X consists of elements $\phi_0\hat{\phi}^{-1}=\bar{\phi}_0\bar{\phi}^{-1}$ and hence consists of all elements $\bar{\phi}_1\bar{\phi}_2^{-1}$ with ϕ_1, ϕ_2 in Φ. The group Ψ induces a corresponding group $H=\{\theta\}$ on P_0 which consists of all elements $\bar{\phi}_1^{-1}\bar{\phi}_2$ with ϕ_1, ϕ_2 in Φ. Let ϕ_* be some reference element in Φ. Then a general element in Φ can be written $\bar{\phi}_*\theta u$ where θ is in H. And the corresponding element in Ψ is $\bar{\phi}_*\theta\phi_*^{-1}$. Thus $\Psi=\bar{\phi}_*H\bar{\phi}_*^{-1}$.

In summary: a homogeneous separable processor Φ can be factored as $\bar{\phi}_*\ H\ u$, where u is the canonical map to the partition P_0 of Φ, H is a group on that partition P_0, and $\bar{\phi}_*$ is some reference element of Φ reduced modulo the partition P_0.

5
The Structural Model

The structural model (Fraser, 1968) is a structured model (D, Φ) in which (E, A) is a copy of (X, B) and Φ is an exact group of transformations on X. The conditions were verified (Fraser, 1968) for the validity of the conditional distribution describing the inaccessible E-value. This conditional distribution together with the processor output combine in fact to give a distribution, the *structural distribution*, for the unknown element in the processor Φ.

James Bondar (1967) has examined a generalization in which the requirement of exactness on the group is omitted. The structural distribution becomes a distribution on a right coset space in Φ.

Andrew Kalotay (1968) has investigated conditions on nonsingular linear transformations Φ such that Φ generates a partition regardless of sample size: that $\Phi^{-1}\Phi$ is a group of transformations on E.

R. J. Beran (1970) considers a model (D, Φ) in a nonparametric context: the distribution D is *unknown*; the transformations in Φ form a parametric class. He takes a formal conditional distribution given information available and inverts it to obtain a distribution on a partition of Φ.

Sections 3 and 4 have characterized processors Φ that provide valid conditional distributions for an inaccessible E-value in a structured model. For corresponding models, let:

A *separable structured model* (D, Φ) is a structured model in which $\Psi = \Phi\hat{\Phi}^{-1}$ is a group of transformations on X. Let t be the $1-1$ map from the induced partition P_1 on E to the partition P_2 on X. Then an inaccessible E-realization from D is described by the distribution D conditional on t being equal to the observed processor output.

A *homogeneous structured model* (D, Φ) is a structured model in which Φ is a homogeneous separable processor. Let P_1, P_2 and t be as defined for the separable structural model. Let P_0 be the partition of E induced by Φ (that is, the partition of any element of Φ); and let u be the canonical map from E to P_0. Let $\Psi = \overline{\Phi}\overline{\Phi}^{-1}$ be the induced group on X, and let H be the induced group on P_0. Then an inaccessible E-realization from D is described by the distribution D conditional on t being equal to the observed processor output.

A general element in Φ can be represented as $\overline{\phi}_*\theta u$ where θ is in H. This representation gives a $1-1$ correspondence between H and Φ. The conditional distribution describing the E-realization (say e) remains valid *in conjunction with* the equation $x = \overline{\phi}_*\theta u e$ containing the observed processor output. Let H_v be the isotropy subgroup of v in P_0 and for a specified isotropy subgroup H_v let $\theta = \theta_1\theta_2$ be a factorization having θ_1 in H_v. The equation $x = \overline{\phi}_*\theta u e$ can be rearranged

$$ue = \theta_2^{-1}\theta_1^{-1}\overline{\phi}_*^{-1}x = \theta_2^{-1}\overline{\phi}_*^{-1}x,$$

where H_v is the isotropy group of $v = \theta_*^{-1}x$. There is a $1-1$ correspondence between the values of ue and the right cosets of H_v as indexed by θ_2. The conditional distribution describing e induces a distribution for ue, which in turn reflects onto the right coset space of H_v by means of the relabelling given by the equation $\theta_2 ue = \overline{\phi}_*^{-1}x$.

In the language of the structural model, the separable structured model admits a valid error distribution, and the homogeneous structured model admits both a valid error distribution and a valid structural distribution.

It seems appropriate then to refer to the separable structured model as a *semi-structural model*.

And, in the direction taken by Bondar but not to the extent taken by Beran, it seems appropriate to refer to the homogeneous structured model as a *structural model*.

6
A Normal Structural Model

Consider a simple normal example used by Bondar as a paradigm for his structural generalization. Let $E=R^1$ and D be the standard normal distribution; let $\Phi=\{[\mu,\sigma]: \mu \in R, \sigma \in R^+\}$ where $[\mu,\sigma]\ e=\mu+\sigma e$. The isotropy subgroup of a point x is $\Phi_x=\{[x-cx,c]: c \in R^+\}$; a transformation $[\mu,\sigma]$ can be factored: $[\mu,\sigma]=[x-\sigma x,\sigma][\delta,1]$ where $\mu=x+\sigma(\delta-x)$, $\delta=x+(\mu-x)/\sigma$.

The equation $x=[\mu,\sigma]e=\mu+\sigma e$ can be rewritten as

$$[\delta,1]^{-1}[x-\sigma x,\sigma]^{-1}x=e,$$
$$[\delta,1]^{-1}x=e.$$

Thus δ, a label for the right coset $\Phi_x[\delta,1]$ in the group Φ, has a normal structural distribution with mean x and the standard deviation 1.

Now consider a multivariate analogue of the preceding example. Let $E=R^{pn}=\{E\}$ where

$$E=\begin{bmatrix} e_{11} \cdots e_{1n} \\ \cdot \quad\quad \cdot \\ \cdot \quad\quad \cdot \\ \cdot \quad\quad \cdot \\ e_{p1} \cdots e_{pn} \end{bmatrix}=\begin{bmatrix} \underline{e}_1{}' \\ \cdot \\ \cdot \\ \cdot \\ \underline{e}_p{}' \end{bmatrix},$$

and D be the standard normal distribution in R^{pn}. Let $\Phi=\{[\mu,\Gamma]: \mu \in R^p, |\Gamma_{p\times p}|>0\}$ be the identity component of $GL(n,R)$ with $[\mu,\Gamma]\ E\ \mu\underline{1}'+\Gamma E$.
Consider the case $n=1$. The isotropy subgroup of a point X is

$$\Phi_X=[X,I]\{[0,C]: |C_{p\times p}|>0\}[-X,I]$$
$$=\{[X-CX,C]: |C|>0\}.$$

A transformation $[\mu,\Gamma]$ can be factored: $[\mu,\Gamma]=[X-\Gamma X,\Gamma]\ [\delta,I]$ where $\mu=X+\Gamma(\delta-X)$, $\delta=X+\Gamma^{-1}(\mu-X)$.

The equation $X=[\mu,\Gamma]E$ can be rewritten as

$$[\delta,I]^{-1}[X-\Gamma X,\Gamma]^{-1}X=E,$$
$$[\delta,I]^{-1}X=E.$$

Thus δ, as a label for the right coset $\Phi_X[\delta,I]$ in the group Φ, has a p-variate normal structural distribution with mean at X and with identity covariance matrix.

Now consider the case $n=p$. The isotropy subgroup of a point X (say with $|X|\neq 0$) is

$$\Phi_X = [0, X]\{[\underline{a}, I - \underline{a}\underline{1}'] : |I - \underline{a}\underline{1}'| > 0] [0, X]^{-1}$$
$$= \{[X\underline{a}, (X - X\underline{a}\underline{1}')X^{-1}] : |\overline{I} - \underline{a}\underline{1}'| > 0\}$$

where the first row, for convenience, may involve elements from $GL(n, R) - \Phi$. A transformation $[\mu, \Gamma]$ can be factored

$$[\mu, \Gamma] = [X\underline{a}, (X - X\underline{a}\underline{1}')X^{-1}]I_i(X)[0, B]$$

where B is $p \times p$ with $|B| > 0$ and $I_i(X) = [A(X), C(X)]$ has two values $I_1(X)$, $I_2(X)$ chosen so that

$$I_1(X)X = \begin{bmatrix} 1 & & & & 0 \\ & 1 & & & \\ & & \cdot & & \\ & & & \cdot & \\ & & & & \cdot \\ 0 & & & & 1 \end{bmatrix}, \quad I_2(X)X = \begin{bmatrix} -1 & & & & 0 \\ & 1 & & & \\ & & \cdot & & \\ & & & \cdot & \\ & & & & \cdot \\ 0 & & & & 1 \end{bmatrix}.$$

The equation $X = [\mu, \Gamma]E$ can be rewritten as

$$[0, B]^{-1}I^{-1}(X)[X\alpha, (X - X\underline{a}\underline{1}')X^{-1}]^{-1}X = E$$
$$[0, B]I^{-1}(X)X = E.$$

Thus (B, i) as a label for right cosets of Φ_X in Φ has probability $1/2$ at each value of i and the standard normal distribution in R^{p^2} truncated to $|B| > 0$ for B.

7

A Nonparametric Structural Model

Beran (1970) has suggested a range of nonparametric structural-type models: a sample space from an error variable of unknown form; a parametric class of transformations to response variables. None of his examples seems amenable to the ordinary structural analysis. In fact, nonparametric models rather generally seem to evade reasonable formulation as structural or semi-structural models.

A notable exception however seems to be the simple case of a sample from an unknown distribution on the real line. Let Φ be the class of positive homeomorphisms: $(0, 1) \to R$; let $E = (0, 1)^n$; let $X = R^n$; and suppose $\phi(e_1, \ldots, e_n) = (\phi(e_1), \ldots, \phi(e_n))$ for ϕ in Φ. Let D be the uniform distribution on E with respect to Lebesgue measure.

The class $\Phi^{-1}\Phi = \Psi$ is the group of positive homeomorphisms $\Psi: R \to R$. And the class $\Phi\Phi^{-1} = H$ is the group of positive homeomorphisms $\theta: (0, 1) \to (0, 1)$ or equivalently on R^n and on $(0, 1)^n$.

For simplicity delete the diagonal hyperplanes (equal coordinates) from E and from X. This bypasses the question of ties. And the deletion conveniently partitions E into P_1 and partitions X into P_2. The probability content of any set in P_1 is $1/n!$

39

The transformations ϕ are all $1-1$; hence the partition P_0 is the finest partition of E; that is, into points themselves.

The isotropy subgroup $\Psi_{\underset{\sim}{x}}$ of a point $\underset{\sim}{x}$ in X is the set of homeomorphisms τ in Ψ that transform $\underset{\sim}{x}$ into $\underset{\sim}{x}$. Let $N=\{\eta\}$ be a minimal class of homeomorphisms in Φ such that the union of the $\Psi_{\underset{\sim}{x}}\eta$ is Φ. A transformation ϕ can then be factored $\phi=\tau\eta$ where $\tau \in \Psi_{\underset{\sim}{x}}$ and $\eta \in N$.

The equation $\underset{\sim}{x} = \phi\underset{\sim}{e}$ can be rewritten as

$$\eta^{-1}\tau^{-1}\underset{\sim}{x} = \underset{\sim}{e}$$
$$\eta^{-1}\underset{\sim}{x} = \underset{\sim}{e}.$$

It follows that the structural distribution of η has $\eta^{-1}x$ uniformly distributed over the region of E having the same order statistic value as $\underset{\sim}{x}$. But $\phi^{-1}x$ is the distribution function of $x = \phi e$ where e is uniform $(0, 1)$. Thus the values of the distribution function at $x_1, \ldots, x_n$ have a structural distribution that is uniform over the permissible values for a distribution function (same ordering as the corresponding x's).

Now let μ be the p-fractile of the distribution of x: $\phi p = \mu$. The class of transformations ϕ having a specified p-fractile μ has *left* coset form — when mapped into H, or into Ψ; the structural distribution, on the other hand, is on right cosets. For this special case, however, any left coset determines a *set* of right cosets; and the possible sets form a partition of right cosets. Specifically the structural distribution gives

$$\Pr\{x_{(i)} < \mu < x_{(i+1)}\} = \Pr\{\eta^{-1}x_{(i)} < p < \eta^{-1}x_{(i+1)}\}$$
$$= \Pr\{e_{(i)} < p < e_{(i+1)}\}$$
$$= \binom{n}{i}p^i(1-p)^{n-i}$$

where $(e_{(1)}, \ldots, e_{(n)})$ is the order statistic of $(e_1, \ldots, e_n)$. Thus *the p-fractile falls in* $(-\infty, x_{(1)}), (x_{(1)}, x_{(2)}), \ldots, (x_{(n)}, \infty)$ *with binomial probabilities* q^n, $npq^{n-1}, \ldots, p^n$ *respectively*.

Suppose now that we have a second structural model duplicating the preceding model but with no connection between the unknown ϕ_1 used in the first model application and the unknown ϕ_2 used in the second model.

Let η_1 be the coset index from the first model and η_2 be the coset index *in* the second model. Then $(\eta_1^{-1}x_{(1)}, \ldots, \eta_1^{-1}x_{(n)})$ has the uniform distribution over the observed rank statistic region on $(0, 1)^n$, and $(\eta_2^{-1}y_{(1)}, \ldots, \eta_2^{-1}y_{(m)})$ has an independent uniform distribution over the observed order statistic region on $(0, 1)^m$. These distributions are in large measure incompatible: the difference in medians, for example, devolves to nothing because of probability in the end regions: $(-\infty, x_{(1)}), (x_{(n)}, \infty), (-\infty, y_{(1)}), (y_{(N)}, \infty)$. One marginal distribution does seem of interest and is accessible: the values of $G = F_2^{-1}F_1$ at $x_{(1)}, \ldots, x_{(n)}$ where $F_1 = \phi_1^{-1}$, $F_2 = \phi_2^{-1}$ are the distribution functions of $x = \phi_1 e$, $y = \phi_2 e$ respectively. The probability of any particular ordering of $\eta_1^{-1}x_{(1)}, \ldots, \eta_1^{-1}x_{(n)}, \eta_2^{-1}y_{(1)}, \ldots, \eta_2^{-1}y_{(m)}$ is $1/\binom{m+n}{n}$. Hence: $1/\binom{m+n}{n}$ *is the probability attached to each inter-ordering of* $G(x_{(1)})$, $\ldots, G(x_{(n)})$ *with* $y_{(1)}, \ldots, y_{(n)}$.

References

1. Beran, R.J., *The Third Hawaii Conference on System Sciences*, Honolulu, Abstract, 1970.
2. Bondar, James, *Invariance and Structural Duality*, University of Toronto, unpublished thesis, 1967.
3. Fraser, D.A.S., *The Structure of Inference*, New York, Wiley, 1968.
4. Kalotay, Andrew, *Events and Groups in Structural Inference*, University of Toronto, unpublished thesis, 1968.

ADDENDUM

[After the oral outline of my paper, Professor Kempthorne remarked that he had read parts of *The Structure of Inference* and had read the *Biometrika* review by Professor Lindley; he asked concerning my views on that review.]

The *Biometrika* review raises questions concerning the structured model and questions concerning contemporary inference. I will comment on what seem to be the principal points of disagreement.

1

Consider a piece of laboratory equipment that produces values in accordance with a t-distribution on 7 degrees-of-freedom; the piece of equipment and a sealed container are housed in a laboratory together with a technician to operate the equipment. Suppose the technician obtains n values from the equipment and keeps the values concealed from analysts outside the laboratory. By almost all current texts on probability and statistics, the analysts would describe the concealed values in terms of probability: independent t-distributions on 7 degrees-of-freedom for the n values.

Now suppose the technician transmits to the analysts outside the laboratory, information concerning the n values. Again by current texts on probability and statistics, the analysts would describe the concealed values in terms of probability: the joint t-distribution, but now conditioned by the information received. Conditional probability, in fact, exists and is useful precisely for such cases—concealed realized values from a random source, the realization being necessary to provide the information, and the concealment being necessary so the information is partial.

Professor Lindley allows such probability descriptions for concealed realized values. The following problem is given in the first chapter of his book on probability (Lindley, 1965): "A male rat is either doubly dominant (AA) or heterozygous (Aa) and, owing to Mendelian properties, the probability of either being true is ½. The male rat is mated with a doubly recessive (aa) female....Suppose of three offspring all exhibit the dominant characteristic. What is the probability that the male rat is doubly dominant?" The *equipment* for this example has two parts: an initial mating producing a male rat—*either AA or Aa*; and a subsequent mating of the male rat producing three offspring —*each either Aa or aa*. The genotype AA or Aa of the male rat cannot be identified by physical characteristics—it is *concealed*. The genotype *Aa* or *aa* of an offspring can be identified—it is *transmitted* to the observer by physical characteristics of the animal. The information concerning the three offspring conditions a joint distribution and gives a conditional distribution describing the concealed genotype of the male rat.

Bayesian analysis quite generally allows such descriptions: a joint distribution for a parameter and a response; an observed value of the response; conditioning of the joint distribution by the information available; the resulting conditional distribution as the description of the concealed parameter value.

2

Consider again the laboratory and the *n* concealed values, but with no information transmitted to the analysts. Suppose the technician goes to the sealed container and removes its contents, a value θ of interest to the analysts and about which they have no information; the technician adds θ to each of the concealed values obtaining $x_1,\ldots,x_n$; he transmits the *n* values $x_1,\ldots,x_n$ to the analysts together with the information that they are θ-translates of the concealed values.

The Bayesian approach to statistics would present θ as a realized value from some prior distribution, and it would present the *x*'s as realized values from a *t*-distribution located at θ. Then—with an introspective, or an *ad hoc*, or a strategic choice of prior—the Bayesian method produces a posterior distribution for θ conditional on the *x*'s. Or—without a specific choice of prior—the Bayesian method produces a class of distributions for θ conditional on the *x*'s.

What has happened to the piece of equipment inside the laboratory? What has happened to the *n* concealed values from the equipment? Where has a generator for θ come from?

The Bayesian approach has *removed* a distribution that describes something present in the actual context. It thus ignores the piece of laboratory equipment and ignores the *n* concealed values. Indeed, it does not need these for Bayesian analysis. The Bayesian approach has *added* a distribution for something that is not present in the actual context, a source for the unknown θ. The introduction of a prior by hypothesis, thus, has two functions in the Bayesian approach: it produces an answer—a distribution for θ, and it produces a justification—for the neglect of information that is given.

Statistics is concerned with inference in science. The Bayesian method of inference ignores what is and substitutes what is not—in the name of knowing what is.

The example involves a piece of laboratory equipment and an unknown θ. It could equally be a measuring instrument and a physical quantity, or a bioassay with a drug of unknown strength (Fraser and Prentice, 1970), or a chemical process with internal interaction and outwardly expressed response, or any of a broad range of problems involving *error* and a *quantity*. In fact, in almost all texts on statistics the first model of substance after the binomial, Poisson, and normal is the regression model: $y = \alpha + \beta_1 x_1 + \ldots \beta_r x_r + e$. The accompanying description refers to *error* in the application, and the model is expressed in terms of *error*.

Professor Lindley proffers the Bayesian method of inference: "The Bayesian argument, with (prior distribution) instead of (sample space) still seems to me to be the best currently available."

In *The Structure of Inference,* however, the analysis uses the probability description of the concealed values and examines the information provided by the *x*'s.

3

The Structure of Inference has five chapters concerned with a structural model and with generalizations of the structural model. The structural model has two components: an *error distribution* that can describe a source of error (such as the laboratory equipment, or a measuring instrument, or the response of an animal to a randomly chosen dosage, or the internal interactions in a physical or social process); and a *quantity* θ that describes how such error manifests itself in the outward response expression. A detailed analysis of the structural model is included.

A later chapter considers the classical model, a model that describes only response values. Simple analysis stimulated by that for the structural model produces the likelihood function *together with* the distribution for the likelihood function. This leads then to those current methods that are in harmony with such a preliminary reduction. The analysis does not endorse the likelihood principle.

The structural model is distinct from the classical model, as distinct as a description of two things (error and response) is distinct from a description of one thing (response).

Error was recognized explicitly by the probabilists some two hundred years ago. But they did not notice that they could and should condition on available information concerning error values.

Recent statistics *does not recognize error explicitly*—not to the extent of putting it effectively into the statistical model; as a consequence it struggles with many problems that need not be.

Experimental design does acknowledge error. Indeed experimental design acknowledges that the generation of pure error by randomization is the key to the legitimacy of the experimental conclusions. It should be noted that the Bayesian approach has little concern for this randomization or for the related tests of significance.

Structural method of inference? The book has a structural *model.* The analysis, however, is simple and direct. It uses only the definition and meaning of conditional probability; there is no need for the many principles and reduction concepts of the classical model. Structural *analysis*, then, is but the straightforward use of conditional probability—as in Professor Lindley's genetic example. True, closer attention needs to be focussed on information concerning realized values and on how such information is transmitted. But this attention is long overdue.

4

Professor Lindley seems not to admit error as an explicit component of the structural model. He does discuss a *range* of structuring: *from* the Bayesian minimum of x alone, θ in Ω, and $p\,(x\!:\!\theta)$; *through* the classical model x in X, θ in Ω, and $p\,(x\!:\!\theta)$; *to* the interpretation of the structural model as the classical model with an additional "special structure to the problem through an intimate relationship between the two spaces (sample and parameter)." The structural model is treated as a classical model with certain special transformation properties.

This is apparent in Professor Lindley's "version of the argument" where, for example, a function of error is changed to a function of the response having a fixed distribution, an ancillary statistic. The changes in the argument lead to difficulties. He is concerned that the concept, *ancillary*, is not mentioned in *The Structure of Inference* and that other standard phrases (sufficiency, exhaustiveness, conditionality, pivotal quantity, perhaps) are not mentioned. They are not mentioned for a simple reason: they are neither needed for the argument nor relevant to the argument.

The structural model and the genetic example have: a *known* distribution, a realized value, and *partial* information concerning that value. In Professor Lindley's version of the argument, this is changed to: a *partially known* distribution, a realized value, and *full* information concerning that value.

The structural model contains more and needs less.

5

The absence of standard concepts leads Professor Lindley to enquire why *probability* itself should not be omitted and replaced by reference to a "positive function integrating to one".

The Bayesian prior is introduced as an instrument of analysis — to obtain a solution to a statistical problem. Professor Lindley treats the presence of error as the use of an instrument of analysis. For the structural model, however, *error* is part of the *given*; it is *part of the model*, it describes something objective, such as the piece of laboratory equipment, or a measuring instrument, or the response of an animal to a randomly chosen dosage.

In consequence of this objective nature of error, *probability* is the central concept; it is treated with the strictness that should be accorded classical frequency-based probability.

If current analyses of some applied problems have neglected to model objective error, then in Professor Lindley's words the current analyses do "not use a satisfactory probability specification".

The absence of the standard statistical phrases and the absence of reference to where they occur in the literature leads Professor Lindley to raise a suspicion — that "there are difficulties which the author wants to hide". No. The methods of analysis are the simple direct methods that are appropriate to the genetic example.

When the structural model is appropriate in an application, *it should be used*. When only a classical model can be justified, then the use of a structural model would be exploratory or conditional on — *suppose there is a source of error such as in this model*.

6

Professor Lindley differentiates classical and Bayesian analyses. Each has the outcome x and the density $p(x\!:\theta)$ with θ in Ω. Classical analysis adds a sample space X; Bayesian analysis adds a prior distribution $p(\theta)$. Either X or $p(\theta)$; "to any arbitrariness in the prior distribution lies a corresponding quality in the choice of sample space".

The various sample spaces X would correspond to various ways of identifying or partitioning the possible outcomes in the physical context being described. On the other hand, the various $p(\theta)$ would correspond to various subjective assessments concerning the unknown value. More typically, with $p(\theta)$ viewed as an instrument of analysis, $p(\cdot)$ would appear as a variable in the Bayesian conclusions.

The identification of X is the identification of how the data are generated in the physical context. The identification of $p(\theta)$ is the identification of the investigator's feelings concerning the quantity being investigated or is the choice of a compatible function for integration purposes.

The selection of X and the selection of $p(\theta)$ do not enter inference on comparable footings. The analytic fact that the introduction of the latter makes the former unnecessary should make the latter doubly suspect.

7

The Bayesian view that $p(x: \theta)$ can be available for the observed x without regard to a sample space X needs consideration. If $p(x: \theta)$ or $p(x:\theta)dx$ can be calculated for the model, then the normalization for each θ must come from somewhere: from components in the physical context; from a developing stochastic sequence; from somewhere where normalization is present.

For example, with (x, n) in a typical binomial context, there would be a sequence of successes or failures, or of 1's and 0's. The Bayesian does not care whether x arose from predetermined n or n from predetermined x. But he does need the developing sequence with probabilities for 0 and 1 in order to obtain the probability for the observed (x, n).

For the general case, some developing sequence has led to the observed x. Two questions arise. First, would certain developing sequences bypassing the value x have been overlooked so that confrontation with the investigator would not have occurred? Let C denote the event corresponding to such sequences and let $P(C: \theta)$ be the corresponding probability. The effect of this *external selection* is then to change the likelihood for x from $p(x: \theta)$ to $p(x: \theta)/[1 - P(C: \theta)]$. Second, would certain developing sequences intersecting the value x have been overlooked so that confrontation with the investigator would not have occurred? Let D denote the event corresponding to such sequences and let $P(D: \theta)$ be the corresponding probability. The effect of this *internal selection* is then to change the likelihood for x from $p(x: \theta)$ to $[p(x: \theta) - P(D:\theta)]/[1 - P(D:\theta)]$.

For the example, with observed $(x, n) = (1, 2)$, the sequence $(0, 1, \ldots)$ might have produced the actual outcome and yet the sequence $(1, 0, \ldots)$ might have been one that would have missed investigation due to some selection factor. In the simple Bernoulli case the effect would show essentially in the denominator. But if the probability of success varies along the sequence then the effect could also show in the numerator, with typically greater effect.

The effect of both external and internal selection is to change an apparent likelihood $p(x: \theta)$ to an actual likelihood.

$$\frac{p(x-\theta) - P(D:\theta)}{1 - P(C \cup D:\theta)}.$$

In general the interpretation of an "observed" x can depend very heavily on the selection factors influencing how that x was generated. Experimental design has found it fundamental to avoid such selection influences — by randomization of treatments to experimental units. It is a measure of Bayesian methodology that such selection influences are ignored, or are neglected.

8

Probability theory, in general, shows little concern for how a realized value is generated or how information concerning a realized value is processed and transmitted. The effects of selection — of how a realized value is generated — were mentioned in the preceding section. The effects of processing and transmission will be examined briefly here.

In probability theory an *event* is elementary or is a set of sample points. Books on probability theory seldom examine in detail the meaning or significance of *event* in applications. Consider the initial example with the piece of laboratory equipment and a technician in the laboratory; let $n = 2$.

Suppose the technician reports to the analysts "There is at least one positive value" — after having noted that the first t value is positive. To complete the picture suppose he would have reported "There is at least one negative value" if the first value had been negative.

The analysts, on the basis of the information received, can identify the concealed signs as being one of $\{(+, +), (+, -), (-, +)\}$. Treating this as though it were an event, they would assign total probability $3/4$ and would assign conditional probabilities $1/3, 1/3, 1/3$ for the three inaccessible sign configurations.

On the other hand, if they obtain knowledge as to how the received information was generated and processed, then they would identify the concealed signs as being one of $\{(+, +), (+, -)\}$. Treating this as an event, correctly, they would assign total probability $1/2$ and conditional probabilities $1/2, 1/2$ for the two possible sign configurations.

It is not enough for the analysts to know the set of outcomes for which the information is true. Rather they must know the set of outcomes that would have produced the actual information, and more.

Thus, transmitted information alone is not enough for probability assessment of concealed realizations. One must also know *how* the information was generated and processed; in more formal terms one must know the *function* that produced the information received.

If the Bayesian feels he can assign probabilities for an unknown on the basis of information available, then the content of the example suggests he must assess not only the information but also the antecedent production and processing of the information. It may well be that the Bayesian assumption of personal probabilities for everything is self contradictory.

Probabilities for an unknown: a topic treated quickly in standard texts and treated largely in subjective terms in the Bayesian development. An early section in *The Structure of Inference* is concerned with probabilities for concealed realized values — *from a source with known distribution*. Professor Lindley finds the section "incomprehensible: leastwise, that is not strictly true because it seems to be a simple misunderstanding of the meaning of conditional probability, but I am sure Fraser is not capable of that so a mystery it must remain". The example, a combinatorial example, was unfortunately chosen. It does seem to be the case, however, that the Bayesian approach goes directly from information concerning an unknown to probabilities concerning an unknown. In the case of a random source, this means treating information *prima facie* as an event. And it is reinforced by Professor Lindley's reaction to the section, a section whose goal is the distinction: *information* as opposed to *the event that produced the information*. It is a distinction that is long overdue in probability, and one of critical importance for inference.

9

Professor Lindley notes that the distribution for θ from a structural model can be obtained more easily from the classical model by "multiplying the likelihood by a function of θ that can be interpreted as a prior density" — the strategic choice of the left invariant prior. He remarks that in the structural framework it has a justification that is lacking in the Bayesian.

For marginal likelihood the situation is more complicated. Professor Lindley indicates that "It is the integration of the likelihood with respect to this prior that provides the marginal likelihood (for λ) available for inferences when the full structural model [for (θ, λ)] is not present". This holds for simpler cases but not for more complex cases. In cases with further complexities there are open issues concerning the definition of likelihood.

For consider the structural model,

Error distribution: $g(y{:}\lambda)f(e)\,dy\,de$

Response production: $y = y, \, x = T_{y\lambda}\,\theta e$

with the additional parameter λ. θ is an element of a group G, and λ is a parameter that affects the distribution of y and affects the generation of x at each y value. For simplicity suppose the range of e is G and de is left invariant.

The structural distribution for θ is

$$f(\theta^{-1}T_{y\lambda}^{-1}x)\,\Delta\,(T_{y\lambda}^{-1}x)\,\Delta\,(\theta^{-1})\,d\theta$$

where $\Delta(\theta)$ is the ratio of left to right invariant differentials. The marginal likelihood function for λ is

$$g(y{:}\lambda).$$

The likelihood function for (θ, λ) is

$$g(y{:}\lambda)f(\theta^{-1}T_{y\lambda}^{-1}x)\left|\frac{\partial T_{y\lambda}^{-1}x}{\partial x}\right|.$$

Integration with respect to $\Delta(\theta^{-1})d\theta$ produces

$$g(y:\lambda)\Delta^{-1}(T_{y\lambda}^{-1}x)\left|\frac{\partial T_{y\lambda}^{-1}x}{\partial x}\right|.$$

In general, this is *not* the likelihood function $g(y:\lambda)$ from the informative variable y; rather it is that likelihood function with additional factors that depend on x, y, λ in general. To avoid these factors, the prior itself would need to depend on x,y,λ.

The use of a left invariant prior for θ can reproduce marginal likelihood in simpler cases. But, in more complex cases as illustrated, the strategic choice of prior would need to depend *on the parameter* and on the observed x, y. To have a prior for one parameter depend on the other is complicated but not improper. But to have the prior depend on the observed outcome is hazardous and in fact improper, for it is no longer a prior and it is no longer supported by Bayesian analysis — the conditional of x given θ multiplied by the conditional of θ given x is generally not the joint of (x,θ).

Allowing priors to depend on observed values opens many possibilities. The ordinary Bayes solutions include the structural solution as a particular case (and reassures the Bayesian concerning structural analysis or concerning Bayesian analysis). The generalized Bayes solutions could embrace all distributions and would thereby include the solutions from *any* proposed method of inference.

It is of interest that Box and Cox (1964) have used a prior that depends on an observed outcome. Their success, however, reflects much more their skills than it does qualities of the generalized Bayes.

It is also of interest that likelihood methods alone — likelihood, weighted likelihood, integrated likelihood — seem able to do all that the Bayes methods can do. The simpler framework has the advantage that results have less chance of being mistaken for probabilities in the ordinary frequency-based sense.

10

Professor Lindley examines two structural models for a bivariate problem. He notes that the two models give different distributions for a variance and asks "Which is right?" Starting with different things it is natural to obtain different things. So the question becomes, "Which model is right?" The answer would be — the one for which there is adequate supporting evidence. Or, if both models are presented tentatively then the question devolves to questions of testing — whether one of the models describes the application — or — which model describes the application.

If the models are indistinguishable in terms of the variables examined, then there are characteristics that are not *identifiable* in terms of the variables examined. It could be appropriate then to seek outside discriminating evidence or to use the models in an exploratory manner quoting with each the appropriate assumption.

11

The major portion of the development in *The Structure of Inference* is concerned with tests of hypotheses, with distributions of test statistics, with distributions for concealed error values, and with marginal distributions for characteristics of concealed error. Only *a small portion of the development* is concerned with the terminal expression of a structural distribution for a parameter.

The statistical profession is highly polarized — those who cannot tolerate a distribution for a parameter — and those who obtain such distribution in rich profusion. The polarization does not happily dispose attention to the intermediate ground — particularly to whether a stronger model can support classical frequency-based probabilities for parameters.

Consider again the laboratory with $n = 1$ observation obtained by the technician from the piece of laboratory equipment. For simplicity suppose a positive t value is taken as $e = +1$ and a negative t value as $e = -1$. The technician removes the unknown θ from the sealed container, and adds it to the e value obtaining $x = 127.3$; he then transmits the value x to the analysts together with its manner of composition from the concealed e and the unknown θ :

$$\theta + e = 127.3 \, .$$

To give another context, suppose θ is a location on a survey route and that a measuring instrument produces error as just described. Suppose the analysts and some scientific observers go to location 127.3.

Before the production $x = 127.3$ the analysts would present

$$\Pr(e = -1) = \tfrac{1}{2} \, , \qquad \Pr(e = +1) = \tfrac{1}{2}$$

as the description of the concealed error value e.

After the production of $x = 127.3$ the analysts would acknowledge that information concerning the concealed e has not been changed; they would still present

$$\Pr(e = -1) = \tfrac{1}{2} \, , \qquad \Pr(e = +1) = \tfrac{1}{2}$$

as the description of the concealed error value e.

If the error value e is -1, then the true value of θ is 128.3; if the error value e is $+1$, then the true value of θ is 126.3 The *meaning* to the scientific observers at location 127.3 is clear:

$$\Pr(\theta = 128.3) = \tfrac{1}{2} \, , \qquad \Pr(\theta = 126.3) = \tfrac{1}{2} \, .$$

12

Meaning may mean money. Consider the following gambling operation. A well balanced symmetrical die is shaken and tossed in a concealed place. Its value is observed electronically and stored for a five minute betting period. At the end of the period one of six lights is turned on; the lights are in a vertical bank on a display panel and the bottom is turned on if the die showed one, the second bottom light if two, ..., the top light if six. To add challenge during the betting period, the gambling house chooses a challenge number when the die is thrown.

The challenge number is immediately displayed on the panel beside a particular light, beside the light that will indicate the result of the toss; but simultaneously the house displays numbers beside *each* light, the numbers forming a sequence from top to bottom that increases by +1 per position. During the betting period, the gamblers can place bets as to which of the six displayed numbers is the challenge number.

If the gamblers have no information concerning the challenge number on a performance, then they are betting in effect on the toss of the die and their betting should be based on probabilities $1/6, \ldots, 1/6$ for the six possible challenge numbers.

If the gamblers feel they have information concerning the challenge number on a performance then they might modify their betting practice. In the long run they could then be substantially behind depending on the reliability of the information they feel they had on the component performances. In the *reductio-ad-absurdum* extreme in which the house used the same challenge number continually, the smart gambler would close in on it very quickly.

By the same token if the house feels they have information concerning systematic or organized betting practice of a gambler, then by judicious choice of challenge numbers they could be substantially ahead or behind depending on the reliability of the information they felt they had on the component performances.

The betting situation, in case of no information concerning the challenge number, can be described by a simple structural model. Note that on any performance the six positions on the display panel will have the same total: possible value for die plus possible challenge number. The error distribution ($1/6$ for each possibility) describes the value on the concealed die and also describes the *concealed* challenge number. The conclusions from the structural model concerning the challenge number are isomorphic to those that would be obtained from a random choice of challenge number from the six possibilities displayed.

For a scientific application of the structural model it is assumed that there is no information concerning the value of the quantity being investigated. If there is informal information available, then there can be good scientific grounds for deliberating omitting it — to see what the process or experiment itself says concerning the value of the quantity. Much scientific effort in some places goes into excluding possible bias from informal information; in the present context the possible bias is easily avoided by the simple operation of excluding the informal information.

The viewpoint associated with the structural model is: description should center on 'where the action is', on the primary variation, and should not withdraw from that description to a model that describes only outward or terminal responses. With the structural model the response is, in fact, not even referred to as a conventional random variable.

Professor Lindley expresses preference for the use of measure theoretic language for clarifying distinctions in the use of probability: "A view that regards probability as a measure over a σ-algebra of sets and a random variable as a transformation from that to the reals may be highbrow and pedantic to some but does at least distinguish between $x(\omega)$ as a random variable and θ as a parameter, and make one realize that something drastic has to be done to make θ into $\theta(\omega)$."

The terminal probability measure in a Bayesian analysis does not arise from a transformation from the *given* probability space.

Bayesian analysis has a probability space $(\Omega \times X, A \times B, Q)$. Let Q^1 be the probability measure on (Ω, A) induced by the canonical projection $\pi_1 : \Omega \times X \to \Omega$; and let P_θ^2 (P_x^1) be the conditional probability measure on $(\Omega \times X, A \times B)$ given θ in Ω (x in X) where it is assumed there is a determination at each value of θ (x). The probability space (Ω, A, Q^1) is the prior model representing the Bayesian's personal assessment of the unknown; the statistical model $\{(X, B, Q_\theta^2 : \theta \in \Omega\}$ is the description of the process under investigation. Let θ be the identity map: $\Omega \to \Omega$; let x be the identity map: $X \to X$; then $\omega = (\theta, x)$ is the identity map: $(\Omega \times X) \to (\Omega \times X)$.

In an application, the Bayesian has information concerning the realized value ω_0 from $(\Omega \times X, A \times B, Q)$; he has the value $x_0 = \pi_2(\omega_0)$ of the canonical projection $\pi_2 : \Omega \times X \to X$. The observed x_0 identifies the set $\pi_2^{-1} x_0$ in the partition $\pi_2^{-1} X$ on $\Omega \times X$. This identification produces a conditional probability measure Q_{x_0} on $A \times B$: it produces a *new* probability space $[\Omega \times X, A \times B, Q_{x_0}]$.

For given x_0 the points in the identified set $\pi_2^{-1} x_0$ are in 1 - 1 correspondence with the points in Ω, with the actual point ω_0 in correspondence with the true parameter value $\theta_0 = \pi_1(\omega_0)$. This 1 - 1 correspondence produces a probability measure on A, the Bayes posterior measure Q_B; it produces a probability space (Ω, A, Q_B). Given the datum x_0 and the model $(\Omega \times X, A \times B, Q)$, the interpretation for this probability space is: by isomorphism, it is equivalent to view the true parameter value θ_0 as having arisen randomly from (Ω, A, Q_B).

Just as in the gambling example, the Bayesian model places possible θ values in 1 - 1 correspondence with points in the probability space. With reliable probabilistic information concerning the choice of θ, if one is prepared to bet on the sample space value, then one is in fact betting on the θ value.

Structural analysis has a probability space (X, B, P); and it has a space Ω (with σ-algebra A) of 1 - 1 measurable maps $\theta : (X, B) \to (X, B)$ with necessary continuity properties. The probability space (X, B, P) describes the internal error of the process being investigated; the class Ω describes the production of the response from the internal error. Let e be the identity map: $X \to X$ and let x designate the composite map $\theta e: X \to X$.

In more informal notation this would be presented as

Error generation: $f(e) de$

Response production: $x = \theta e \ \theta \in \Omega$

where e is the identity map on $X = R^n$ and Ω is an analytic group of transformations: $R^n \to R^n$. The σ-algebra B is the Borel σ-algebra, the differential de presents Lebesgue measure on B, and $P: B \to R$ is given by

$$P(B) = \int_B f(e) de.$$

In an application the scientist has information concerning the realized e_0 from (X, B, P): he has the value $x_0 = \theta_0 e_0$ of the map $\theta_0: X \to X$ in which θ_0

is the unknown true value of the parameter θ. The observed x_0 identifies the set $G^{-1}x_0$ in the partition $G^{-1}X$ on X. This identification produces a conditional probability measure P_{x_0} on B: it produces a *new* probability space (X, B, P_{x_0}).

For given x_0, the points in the identified set $G^{-1}x_0$ are in 1 - 1 correspondence with the points in Ω, with the actual point e_0 in correspondence with the true parameter value θ_0 (assume G is exact on X). This 1 - 1 correspondence produces a probability measure on A, the structural posterior measure P_s: it produces a probability space (Ω, A, P_s). Given the datum x_0 and the model $((X, A, P), \Omega)$, the interpretation for this probability space is: by isomorphism, it is equivalent to view the true parameter value θ_0 as having arisen randomly from (Ω, A, P_s).

Just as in the gambling example the structural model places possible θ values in 1 - 1 correspondence with possible error values. With no information concerning θ, if one is prepared to bet on the error values, one is in fact betting on the θ value. The structural distribution is a posterior distribution. It arises with a specific kind of model in the absence of prior information.

References

1. Lindley, D.V., "Review," *Biometrika*, 56, 453-456, 1969.
2. Fraser, D.A.S., *The Structure of Inference*, New York, John Wiley and Sons, 1968.
3. Fraser, D.A.S. and Prentice, R., "Randomized models and the Dilution and Bioassay Problems," submitted to *Annals Mathematical Statistics*, 1970.
4. Box, G.E.P. and Cox, D.R., "An Analysis of Transformations," *J. Roy. Statist. Soc. B*, 26, 211-243, 1964.

COMMENTS

I.J. Good:

In order to exemplify his theory in a simple manner Dr. Fraser tossed a coin with which he had associated +1 with heads and –1 with tails, and denoted the outcome by a random variable e, $e = \pm 1$. He then thought of an integer n and informed us that $n + e = 3132$. We were not supposed to have any information about n. He asked for an estimate of the probability of $e = 1$. Good: "One half".

[Dr. Fraser then inferred from this that we should now believe $P(n = 3133) = \frac{1}{2}$, $P(n=3131)=\frac{1}{2}$.]

Good: "In that case our prior probabilities for $n = 3133$ and $n = 3131$ must have been equal, and, by the same argument, the prior probabilities of all odd numbers must be equal, and likewise of all even numbers. In other words a Bayes postulate has been inserted unobtrusively".

O. Kempthorne:

I was extremely mystified by a reference Dr. Fraser made in his oral presentation (but *not* in his written presentation) to some work I have done on randomization and finite model theory. Perhaps my work is relevant. I would certainly be interested in Dr. Fraser expanding by the *written word* what he has in mind. I am totally skeptical, but am willing, of course, to be convinced.

D.V. Lindley:

The difficulty I find myself in when discussing Fraser's work is surely one of communication. We are simply not communicating with one another. I feel that it is important that we try to surmount this difficulty because much of what he says makes very good sense to me and his results are, to me, intuitively satisfying.

My main hurdle lies in the basic formulation of the problem. Not only has he dispensed with terms like ancillary and fiducial, he has even avoided using probability. This is not quite true, for he uses a density function, but he could have spoken of a non-negative function integrating to one and thereby avoided any stochastic term. The rules of probability are not employed. In my view, by doing this he has obscured the issue and perhaps even made a blunder.

To discuss this it is necessary to use what in England is referred to as *Annals (of Mathematical Statistics) language*. I have a higher opinion of the Annals than most of my colleagues at home and the language often seems essential to the clarification of a problem. In this language probability is understood in the context of an abstract space Ω of elements ω with an associated σ-algebra A. Probability is σ-positive set function over (Ω, A) with $p(\Omega) = 1$. A random variable $x(\cdot)$ is a map from Ω to the reals which is measurable using A and Borel sets. Now in Fraser's notation (which I believe to be inadequate) a random variable is referred to as x or X rather than $x(\cdot)$, and Ω is entirely omitted. Now we are told that e has a (known) distribution, so is a random variable, properly written $e(\cdot)$ The observation x is presumably also random so we may write $x(\cdot)$. Finally there is an operator ϕ belonging to a class Φ, with the property that $\phi e(\cdot) = x(\cdot)$. In this formulation I fail to see how ϕ can become a random variable, that is a map from Ω to Φ. In the orthodox model there is a class p_θ of measures indexed by a parameter θ. Equally θ cannot become a random variable without some further assumption. The Bayesian version is to make this assumption explicit. Would Fraser oblige by translating into Annals language?

REPLY

To Professor Lindley:

Bayesian analysis has a probability space $(\Omega \times X, A \times B, Q)$. An observed x_0 identifies a *set* in a *partition* on $\Omega \times X$, a set and partition generated by the projection map. This identification produces a conditional probability measure Q_{x_0} on $A \times B$: *it produces a new probability space* $(\Omega \times X, A \times B, Q_{x_0})$.

For given x_0, the points in the identified set are in 1 - 1 correspondence with the points in Ω, with the actual point in correspondence with the true parameter value. This 1 - 1 correspondence produces a probability measure on A, the Bayes posterior measure Q_B: it produces a probability space (Ω, A, Q_B). Given data and model, the interpretation for this probability space is: by isomorphism it is equivalent to view the true parameter value as having arisen randomly from (Ω, A, Q_B).

Structural analysis has a probability space (X, B, P) and it has a space Ω (with σ-algebra A) of 1 - 1 measurable maps: $(X, B) \rightarrow (X, B)$. An observed x_0 identifies a *set* in a *partition* on X, a set and partition generated by the orbital map. This identification produces a conditional measure P_{x_0} on B: *it produces a new probability space* (X, B, P_{x_0}). This space provides a basis for many of the standard classical inference methods, if desired.

For given x_0, the points in the identified set are in 1 - 1 correspondence with the points in Ω, with the actual value in correspondence with the true parameter value. This 1 - 1 correspondence produces a probability measure on A, the structural posterior measure P_S: it produces a probability space (Ω, A, P_S). Given data and model, the interpretation for this probability space is: by isomorphism it is equivalent to view the true parameter value as having arisen randomly from (Ω, A, P_S).

The extra structure for Bayesian analysis describes the personal feelings of the Bayesian. The extra structure for structural analysis describes the source of variation (as opposed to its manifestation in the response).

To Professor Good:

The committed Bayesian believes that he can describe any unknown by a probability distribution, his prior distribution for that unknown. I will call this the Universal Bayesian Assumption. I do not accept this assumption and there are enough statisticians who do not accept it that it is not an article of faith in the statistical profession. This reply is not the place to argue the relevance of the assumption but I will note that the principles of experimentation were found necessary primarily because of the unreliability of such personal assessments.

Bayesian analysis of a statistical model uses acceptable probability theory to establish: if there is a prior distribution, then there is a posterior distribution.

The operating principle in Professor Good's contribution is: if there is a posterior distribution, then there is a prior distribution.

Without the Universal Bayesian Assumption, I do not see any grounds for this operating principle.

With the Universal Bayesian Assumption I do not see any grounds for this operating principle. For, different Bayesians have different priors, hence different posteriors and an objective posterior — when priors are excluded on grounds of their personal and mutually-contradictory nature — is not covered by consequences of the Universal Bayesian Assumption.

To Professor Kempthorne:

My oral and written presentation was concerned with the identification of events for concealed outcomes from a random process, a problem with relevance in probability theory and statistical inference. After the presentation,

Professor Kempthorne mentioned my book, *The Structure of Inference* (Wiley, 1968), and enquired concerning my thoughts on Professor Lindley's review of that book (*Biometrika*, 1968, 453-6). As part of my reply I noted that explicit error is present in many areas of application and that such error should then be specifically represented in the model, as for example in the structural model. I remarked that randomization was a case where error was explicitly injected into the experimental situation and that I was sure Professor Kempthorne viewed such randomization as fundamental. I do not know whether Professor Kempthorne is skeptical concerning randomization or concerning theoretical assessment of randomization.

MODEL SEARCHING AND ESTIMATION IN THE LOGIC OF INFERENCE

A. P. Dempster
Harvard University

Summary

It is argued in Section 1 that the postdictive mode of probabilistic reasoning from data is associated mainly with searching for a model while the predictive or Bayesian mode is associated with uncertain estimation of the values of unknowns. The extent to which this pattern can be broken, whether by frequentist estimation or by Bayesian significance testing, is also examined. The interaction between the two modes is considered in Section 2 where it is argued that a type of confusion afflicting our ability to estimate parameters from given data effectively limits the complexity which can be built into a model. A rudimentary theory is proposed under which a balance would be obtained by trading away goodness of fit for improved clarity of estimation. Confusion arises especially when it is necessary to estimate many parameters. In Section 3, two devices are described which seek to reduce confusion through reducing the number of parameters, which means working through models for the behavior of the parameter values. The simple random effects model is considered in detail, and it is suggested that careful attention must be given to the point in the logic where model searching leaves off and estimation begins. Finally, the hope is expressed that the devices of Section 3 will be capable of widespread development and application to highly parameterized multivariate models.

1
Patterns of Applied Inference

I believe that the theory of inference makes contact with real data sets primarily through two separable types of logical activity which I have called *postdictive* reasoning and *predictive* reasoning (Dempster, 1964). Accordingly, I believe that an understanding of applied inference requires an appreciation of the contrasting roles of the two types of reasoning. In postdictive reasoning the moving force is surprise at the occurrence of a highly improbable event, where the surprise engenders doubts about the theory which implied the improbability of the event. In predictive reasoning, one attempts to portion out relative degrees of probability to various hypotheses. Postdictive reasoning is essentially negative: it denies hypotheses without, in itself, confirming alternatives. Predictive reasoning, in contrast, supplies relative probability weights to various hypotheses. Subject to acceptance of an overall model, high probability indicates confirmation or positive support for a hypothesis.

The distinction made here is closely related to the difference between a significance level and a posterior probability. The advent of frequency interpretations for significance levels has made the explicit postdictive interpretation unfashionable, but the postdictive mode is easily found in writing from the 19th and early 20th centuries, and I believe it to be inescapable if the application of significance tests to specific data sets is to be understood. Posterior probabilities on the other hand are clearly meant for normal predictive interpretation.

Pushing the parallel lines of thought further, it can be argued that the two functions of searching for models and of estimating the values of unknowns are separated logically by their association respectively with postdictive and predictive modes of thought. I do not wish to overstate the role of the logic of inference in these functions. Important elements of knowledge and understanding of data sets and the wider contexts into which they fit are required, to the point that the probabilistic manipulations of inference are like the tip of an iceberg visible especially in the writings of mathematical statisticians. Nor is it easy to separate the functions. Since the very definition of most quantities which are estimated requires a framework going beyond immediate observations, the model building function must have a measure of precedence, but the two processes generally go hand in hand, or cyclically with complex interrelations.

Sometimes formal descriptions make one process appear like the other. For example, if one accepts a null hypothesis stated as $\mu_1 = \mu_2$, then is one not estimating the value of $\mu_1 - \mu_2$ to be zero? A little reflection shows that any acceptance of a parametric hypothesis implies an infinity of precise quantitative judgments, as when one assumes normal populations one assumes precise knowledge of the relative sizes of all differences between pairs of population quantiles. It is a distortion of terminology to describe as estimates such byproducts of the model building process. The term estimation is more appropriate when different possible values of specific unknowns are consciously contemplated, a circumstance predicated on the existence of a model which renders such unknowns meaningful.

To argue that the postdictive and predictive modes of reasoning are in one-one correspondence with the functions of model searching and estimation is automatically to run counter to the natural bias of theorists in favor of a unified theory of inference. Thus, according to the *bon mot* ascribed to Doog, "To a Bayesian, all things are Bayesian." Similarly, the widely favored Neyman-Pearson approach of evaluating "procedures" by operating characteristics does not differentiate between hypothesis testing and estimation in a fundamental way. And even if one rejects the formal decision-theoretic structure of Neyman and Pearson, there remains a widespread feeling that informal reasoning based on sampling distributions is appropriate both for hypothesis testing and for estimation.

I do not believe that either the Bayesian approach or the sampling distribution approach to unity is a total error, but I do find that subtle issues are involved which compromise parts of both schools, so that a mixed viewpoint becomes desirable. Specifically, one must reckon with the weaknesses of sampling distribution methods for estimation and of Bayesian methods for significance testing.

Sampling distributions have been widely used for estimation since Bernoulli (1713) introduced the binomial distribution for estimating a frequency and Simpson (1755) introduced the distribution of a sample mean for estimating a quantity measured with error. The sampling distribution of an estimate, or equivalently of the difference between an estimate and a true value, is generally dependent on a set of unknown parameter values including the particular value being estimated. While this dependence is awkward, the greatest weakness in the use of sampling distributions for estimation is that the sampling distribution of an estimate has a prospective probability interpretation only *before* the value of an estimate is observed. If, according to a sampling distribution, one accepts that

$$\Pr(\, |\, \hat{\theta} - \theta \, | < c) = .99, \tag{1.1}$$

the directly interpretable meaning is that the assertion $|\, \theta - \hat{\theta}\, | < c$ has degree of certainty .99 before θ is observed. Thus, when θ becomes known, one can say that either $|\, \theta - \hat{\theta}\, | < c$ or an improbable event with probability .01 has occurred. This postdictive assertion is very different from a predictive assertion that $|\, \theta - \hat{\theta}\, | < c$ with degree of certainty .99 where θ is a specific observed number.

The distinction just noted was blurred by Laplace (1813), but was remarked upon by a number of his readers including DeMorgan (1837). In more recent times, the issue came sharply into focus when Fisher and Neyman took different paths in attempting to legitimize the use of intervals deduced from the inversion of sampling distributions. Neyman stressed frequency interpretations showing that a long series of repetitions of an interval like $|\, \theta - \hat{\theta}\, | < c$ as in (1.1) will describe a true assertion with frequency very close to .99. The statement $|\, \theta - \hat{\theta}\, | < c$ with a fixed observed θ is called a confidence statement with confidence coefficient .99, but its defining property asserts only that the statement is one of a long sequence of repeated trials of which .99 are true. The statistician with a specific interval of θ values $\hat{\theta} - c < \theta < \hat{\theta} + c$ has no more than the postdictive interpretation described above, namely that $\hat{\theta} - c < \theta < \hat{\theta} + c$ or an improbable event has occurred. At first, Fisher's writing about fiducial intervals seemed closely to parallel that of Neyman on confidence intervals, but gradually Fisher made it clear that he did mean fiducial probability to be ordinary predictive probability given the data. Fisher rightly rejected Neyman's argument as inadequate for predictive purposes and set out to discover the conditions and limitations which could lead to predictive validity for intervals based on the inversion of sampling distributions. I think that most scholars would agree that Fisher failed to establish broadly acceptable conditions. To cite but one example of the difficulty of Fisher's task, consider the plausible requirement that he who assigns predictive odds of 99 to 1 in favor of an interval $| \theta - \theta | < c$ should be willing to offer the choice of sides in a bet at these odds. For intervals based on Student's t, however, which Fisher took to be central examples of fiducial intervals, procedures for the choice of sides exist (Buehler, 1959) which guarantee an unfair game in the long run whatever sequence of θ values is presented. Confidence intervals likewise, especially those which fail to condition on prominent but irrelevant features of the observation, are prone to fail this test in a damaging way, thus reinforcing the warning not to misassign a posterior probability interpretation to a confidence coefficient.

Nevertheless, it remains an open question whether credible post-observation predictive validity can be achieved via sampling distributions. Fisher's intuition, as opposed to his specific theories, may have an important kernel of rightness. As a matter of practice, there are many situations where widely acceptable Bayesian probabilities seem unattainable by the approved channel of large sample theory. As suggested later in this paper, it can happen that available sample sizes, even quite large, may be quite unable to cope with large parameter spaces in the sense of reducing the effect of a prior distribution. Still, in these same situations there may be widely trusted inversions of certain sampling distributions. I suspect that such trust may often be warranted, for several possible reasons. One reason is that the absence of appealing competing techniques may leave one in a *de facto* state of insufficient reason, with no reason to think that another statistician with the same data could operate a noticeably unfair selection procedure if offered a choice of bets, and hence no reason to reject the sampling distribution inversion. A second reason is that new models and new Bayesianly acceptable techniques may appear which back up the sampling distribution intuitions.

Consider next the possibilities for using Bayesian ideas in tests of significance. Many statisticians feel a strong intuitive pull towards conditional tests, or tests based on sampling distributions computed conditionally, given certain aspects of the data. Bayesian testing goes a step further by basing tests on posterior distributions computed conditionally given all of the data. The logical key is still reaction to a small probability, but, instead of reacting postdictively to a probability held predictively meaningful only before the data are observed, the Bayesian tester reacts to a small posterior probability, which is predictively meaningful after the data are observed. The fundamental appeal of Bayesian testing, as of Bayesian methods generally, comes from continuing to work with predictively interpretable probabilities after the observations are in hand.

Under the sampling distribution approach, the theory of what constitutes a good test depends on the formulation of precise alternative hypotheses, but actually carrying out a test requires only sampling distributions specified by a null hypothesis. In contrast, a Bayesian test rests on a posterior distribution over a collection of null and alternative hypotheses, so that a precondition for carrying out the test is a precise specification of alternative hypotheses. This requirement is both a strength and weakness of Bayesian testing. On the positive side, Bayesian testing can lay claim to impeccable logical foundation *when alternative hypotheses are not in dispute*. On the negative side, Bayesian testing cannot be a discursive exploratory tool aimed in part at discovering plausible alternative hypotheses, since it operates within a closed system. Consequently,

Bayesian testing necessarily has a restricted role within the larger field of testing.

There appear to be two main ways to specify a posterior probability whose smallness could be grounds for suspicion of a null hypothesis, and which thereby would define a Bayesian test. The first of these is simply to compute the posterior probability of the null hypothesis, while the second is to compute a tail area. Bayesian theorists beginning with Laplace, and quite explicitly Poisson (1837, sections 41, 42), have concentrated on the first way, and have ignored the second, probably because the latter seems tied to the sampling distribution formulation of testing and has no obvious purely Bayesian motivation. I shall argue below for the proposition that the second approach is much more useful than the first.

There is no Bayes posterior probability of a null hypothesis without a prior probability of that null hypothesis. In principle, however, the concept of a prior probability of a null hypothesis clashes with the very concept of a null hypothesis as a tentative stopping place on the way to devising a model in accord with the known facts. One knows that no null hypothesis could ever be true, but hopes to find one that fits and illumines the facts. Even to conceive of a numerical degree of certainty attached to such an entity is awkard, letting alone the practical question of specifying the number. In practice, when the data are such that judgments of significance are borderline, the dependence of the posterior probability of the null hypothesis on its prior probability will be crucial, and the user will be led to agonize over a largely fictitious aspect of his prior knowledge at the expense of constructively examining his available data.

When parameter spaces are continuous, a null hypothesis of the form $\theta = \theta_0$ can have positive posterior probability only if a discrete spike of prior probability is placed on $\theta = \theta_0$. For example, a prior probability of $\frac{1}{2}$ that $\theta = \theta_0$ is a common choice, with the remaining probability continuously distributed over alternative θ values. A technical peculiarity exists regarding diffuse priors over alternative θ values. A diffuse prior is often conceived as a uniform distribution over $(-T, T)$ where T is a large but unspecified constant. In ordinary Bayesian inference, with no discrete spike at θ_0, the choice of T is often unimportant, but it does not remain unimportant when there is a spike at θ_0. If the prior distribution is defined by the continuous density $p/2T$ on $-T < \theta < T$ and the discrete density q at θ_0, with $p + q = 1$ and if this prior distribution is combined with a continuous likelihood function $L(\theta)$, then the posterior distribution has the continuous density $K p L(\theta) / 2T$ on $-T < \theta < T$ and the discrete density $K q L(\theta_0)$ at θ_0, where

$$\frac{1}{K} = \int_{-T}^{T} \frac{p L(\theta)}{2T} d\theta + q L(\theta_0). \tag{1.2}$$

Assuming that $\int_{-\infty}^{\infty} L(\theta) d\theta = L < \infty$, it follows that

$$K q L(\theta_0) \doteq \cfrac{1}{1 + \cfrac{1}{2T} \cdot \cfrac{p}{q} \cdot \cfrac{L}{L(\theta_0)}} \tag{1.3}$$

for large T, showing that the posterior probability of the null hypothesis has

60

a nontrivial dependence on T and moreover tends to unity as $T \to \infty$. A conventional choice of prior density element $d\theta$ effectively means choosing $T = \frac{1}{2}$. This choice could very well be absurd, and in any case is arbitrary if the choice of scale is arbitrary, for if $\phi = c\theta$ then prior densities $d\theta$ and $d\phi$ are inconsistent. To escape the difficulty one needs a prior density for θ representing genuine prior knowledge of θ, which I believe to be nonexistent in a genuine significance testing situation where the question is whether or not θ is an empirically justifiable conception.

Alternatively, one can give up the prior spike at θ_0 and with it the possibility of a posterior probability of the null hypothesis. This leads to the second approach to testing under which diffuse priors retain operational meaning and judgments of significance depend on posterior probabilities of the events $\theta > \theta_0$ and $\theta < \theta_0$. If either of these posterior tail areas is small, the implication is that the null hypothesis is placed under suspicion. The second type of Bayesian test relies on a choice of meaningful parameters such that a predisposition exists in the mind of the data analyst to regard an extreme value of the added parameter θ as unusual and thence suspicious. The posterior distribution of θ provides the calibrating device for judging when an outlying value rate is being judged extreme. The comparison with ordinary significance tests is interesting, since, as I view them, most practically used tests are chosen not because of optimality properties, which are based on somewhat unconditional averages, but because the meaning of the selected test statistic suggests deviation from the null hypothesis when an extreme value appears. Again, calibration is provided, in the case of ordinary tests by the sampling distribution of the test statistic. The advantage of the Bayesian version is that the calibrating probabilities are completely conditioned on the observed data. The disadvantage, as remarked above, is that the tester must lead with a system of alternative hypotheses, which effectively begs the question of estimation. In principle, therefore, I believe that Bayesian testing, even of the tail area variety should be reserved for a second pass at testing, after a first pass of ordinary testing has helped suggest a family of alternative hypotheses.

A limitation on the second type of Bayesian test is that the Bayesian analysis should be reasonably objective, meaning reasonably robust against variations in the diffuse prior distribution, and meaning in turn that the sample of data must be large enough to fix the free parameters reasonably precisely. I suspect, however, that in the current state of the art the requirement of ample data applies to all estimation procedures, non-Bayesian as well as Bayesian, that are really trustworthy. Another limitation or difficulty arises when the null value θ_0 lies at the extreme of the range of parameter values, as happens for example when the null hypothesis of Poissonness is located within the family of negative binomial distributions. In this case, the Bayesian process of integrating out can still be applied to the nuisance parameters, but another device based on examining the marginal likelihood $L(\theta)$ must be applied to the final parameter of interest. I would argue that single parameter inferences are not in general difficult and that the reduction to $L(\theta)$ is the more sensitive step in the analysis.

It is worth noting that in large samples the Bayesian test based on $\theta > \theta_0$ or $\theta < \theta_0$ will tend to agree with the sampling distribution test based on $\hat{\theta} - \theta_0$ where $\hat{\theta}$ is a BAN estimator. I find this to be an important reason for accepting the standard asymptotic tests, though I rarely have any feeling for the degree of built-in protection against variation in diffuse priors.

With careful handling, Bayesian tests of the second type appear to be necessary and desirable tools for the model building process where a sizable family of null and alternative hypotheses can be accepted as initial limits on the process. Of course, it often happens that only a null hypothesis is fixed, and one hopes that the data will suggest alternative hypotheses. Ordinary tests may help in the process of identifying believable peculiarities which could suggest alternative hypotheses. Then Bayesian tests aimed at the suggested alternatives could be carried out, as refinements of the original ordinary tests.

The foregoing review of more or less familiar ideas was designed to create a background for the viewpoints expressed in Sections 2 and 3 below.

2

Towards a Theory of Complementarity

If a dual perspective on inference is accepted, with legitimate roles for post-dictive and predictive reasoning, it becomes natural and desirable to seek out principles governing when to leave off the model searching process and to pick up the estimation process. Such principles will lead to a better understanding of the complementary nature of the two modes, a complementarity which is obscured by the superficial unity of an exclusively Bayesian or exclusively frequentist position.

Two views of model searching need to be contrasted at the outset. One view represents the goal as the isolation of truth in the form of a theoretical construct which governs or explains the state or course of some observed phenomena. The second view does not necessarily quarrel with the goal just described, but sees it so distant as to be virtually meaningless, and certainly of little use as an operating principle. Some may hold that the beauty of science lies in a conjunction of truth and simplicity, and certainly there are way stations along the path of science which give that impression, but I would hold with the second view that the real world, even the real world of some quite restricted scientific phenomenon is endlessly complex. In particular, the models commonly used by statistical data analysts and modellers sit on very precarious ledges, with uncharted paths to the next ledge, and are very far from any triumphant peak.

The real life process of constructing a model may be roughly conceived as operating between a floor and a ceiling where increasing altitude means increasing *complexity* of the model. The floor pushing complexity upward consists in the need for the model adequately to *fit* the known facts of the phenomenon under study. The ceiling, however, is a much less well understood logical construct which I should like to label *confusion*. Confusion arises because too many dimensions of an overly complex model are insufficiently determined by the available facts, so that predictions and insights from the model are muffled. I believe, in other words, that there are inherent limitations on the process of estimation which imply that restricted fixed data simply cannot determine a broadly acceptable posterior distribution on the many parameters of a highly complex model. Thus, the processes of model searching and estimation interact through the determination of a comfortable altitude of complexity at which the available supply of factual data can be reasonably absorbed and interpreted.

The concept of confusion as a ceiling is not widely recognized in statistics. Nevertheless, I believe that the current state of the theory of estimation does bump constantly, if unknowingly, against the ceiling. Frequency theory, for instance, offers as its ultimate distillation the advice that admissible procedures are best, which leads mainly to Bayes. Bayesians are weakened by the rapidly shifting effects which slight variations of diffuse priors can have when complex models are applied to restricted data. It is a commonly heard Bayesian dictum that a model should be *as big as an elephant*, presumably to be sure of staying above the floor, but what of the ceiling? Occam's razor, on the other hand, counsels staying as close to the floor as possible: *no complexity without necessity*.

It would be of interest to construct a mathematical theory of inference which would reflect the principle of floor and ceiling just described. The theory would present the model builder with two kinds of loss, one resulting from adopting an oversimplified model which fits badly, and the other resulting from adopting a model so complex that its parameters start to lose credible identification in the data. I shall not attempt to describe such a theory in detail, but only to discuss some of the considerations and difficulties involved.

First, as regards fit, it is clear that data may fit a model well on many criteria, and will usually fit badly on other criteria. Some subjective weighting of different aspects of fit is necessary to reach a single criterion of fit which would have an inverse monotone relation to the loss due to bad fit. How this loss should be calibrated relative to the observed significance level of the criterion is an open question. It is also worth noting that fit need not simply mean fit to data in hand, but may require fit to prior knowledge as well. For example, prior experience may assert strongly that several different populations have different parameters, so that an adopted model might not be considered adequate unless separate parameter structures were provided for the different populations. In a first pass at theory construction, I would be inclined to introduce formally only the element of fit to data.

Although the concepts of complexity and confusion are far from mathematically precise, there are some obvious monotonicity relations. Thus, complexity is a function of a model which increases as more free parameters are appended to the model, or, equivalently, as the model entertains a wider range of hypotheses. Similarly, confusion is a function of both model and data which increases either as the model becomes more complex with fixed data, or as data decreases relative to a fixed model. The latter monotonicity principle is a general version of what I called an uncertainty principle in connection with my theory of generalized Bayesian inference (Dempster, 1968). But I now feel that confusion is a more appropriate term than uncertainty, for confusion is an uncertainty in an uncertainty, where the latter uncertainty is the uncertainty defined by probability itself. In generalized Bayesian inference, the objective is to assess an upper bound P^* and a lower bound P_* on the posterior probability of each particular event. The difference P^*-P_* becomes a measure of confusion relative to a particular event. It would be of interest to make checks of this measure of confusion against the monotonicity principle of confusion mentioned above, but very little of this has been attempted. Also, since some events may have ill-determined probabilities while other probabilities are unambiguous, there is a need for overall measures of confusion and for theory which develops mathematical properties of these measures. For present purposes, however, it will suffice if at least the possibility of a mathematically representable concept is made plausible.

Many models are cherished in science, even though they clearly do not fit the facts, because the insight which they provide is believed to outweigh the attendant discomfort from bad fit. Such models are often quite simple, perhaps even qualitative in nature, and not much troubled by confusion. These models illustrate scientific insight functioning to provide a negative loss due to confusion. I believe, however, that cases of positive loss due to confusion are increasing in frequency and severity as our technical capacity to gather and process statistical information steadily expands. In Section 2, I have proposed that we learn better how to quantify such losses. In Section 3, I shall discuss some devices for reducing these losses.

3
The Problems of Many Parameters

3.1 Manifestations of the Problem

Consider a hypothetical sample of 200 animals each measured on a set of 10 continuously varying characteristics. Suppose that the 10 marginal distributions show a variety of symptoms of nonnormality, such as skewness and long-tailedness, sometimes to the left and sometimes to the right, so that no systematic pattern of simple transformations to normality seems possible. Suppose also that the scatterplots of pairs of variables display lumpiness and suggestions of curvilinear relations. Standard statistical methods applied to such data could do no more than suggest fitting a 10 dimensional normal distribution to the data, probably after some transformations on margins, and hoping for the best. A more adventuresome spirit might propose fitting a more highly parameterized model to achieve better fit, let us say a population whose log density is a fourth degree polynomial rather than the second degree polynomial of the normal model. A quick count shows why the proposal is adventuresome; the normal model already has the substantial number 55 of parameters, but the fourth degree model has 990 parameters.

Although I know of no formal theoretical analysis, my judgment suggests that estimating a 10 dimensional normal from a sample of 200 is on the fringes where confusion starts to appear, perhaps most severely in the estimation of the elements of the covariance matrix and its inverse. It can hardly be doubted, however, that attempting to fit a 990 parameter fourth degree model would result in something virtually meaningless even assuming that a fitting procedure could be numerically carried out. In other words, the confusion ceiling is very quickly encountered in multivariate analysis if the attempt is made to surmount the restriction to normality. Accordingly, I believe that something fundamental has limited the theory of continuous multivariate analysis so narrowly to assumed normal populations. The difficulty is not merely that the derivation of sampling distributions becomes intractable for more general models, as Kendall (1968) seems to imply, for how would one escape the haze of nuisance parameters even if any desired sampling distribution could be instantly computed? The trouble is also highly visible under a conventional Bayesian formulation, for who would place any faith in a diffuse prior distribution over 990 dimensions?

A more accessible model for illustrative purposes consists of observables $X_1, X_2, \ldots, X_k$ and parameters $\theta_1, \theta_2, \ldots, \theta_k$ where $X_1-\theta_1, X_2-\theta_2, \ldots, X_k-\theta_k$ can reasonably be conceived before sampling as independently and identically normally distributed with known mean zero and known variance σ^2. For large k, this model of many means defines a simple and almost canonical form of the many parameter problem. Associated inference procedures from every standard viewpoint are well known, but to my knowledge the difficulties of estimation with this model have not been specifically interpreted as manifesting the phenomenon of confusion. I believe, however, that the striking paradox of Stein (1959) carries the important message that estimation of $\theta_1, \theta_2, \ldots, \theta_k$ from $X_1, X_2, \ldots, X_k$ via a predictively acceptable posterior distribution incorporating *a priori* ignorance will often be very far from confusion-free. In brief, Stein showed that the flat prior Bayesian approach implies that $\Sigma_1^k \theta_i^2$ has a posterior distribution centered about the mean $\sigma^2(k + \Sigma_1^k X_i^2/\sigma^2)$ with standard deviation $\sigma^2(2k + \Sigma_1^k X_i^2/\sigma^2)^{\frac{1}{2}}$, while inversion of the sampling distribution of the natural statistic $\Sigma_1^k X_i^2$ leads to a confidence distribution for $\Sigma_1^k \theta_i^2$ roughly centered about $\sigma^2(-k + \Sigma_1^k X_i^2/\sigma^2)$ with spread similar to that of the posterior distribution. If k is reasonably large and $\Sigma_1^k X_i^2/\sigma^2$ not too large, the two approaches to estimating $\Sigma_1^k \theta_i^2$ can differ sharply, to the point of nonoverlapping intervals having posterior probability or confidence coefficient, respectively, close to unity.

The paradox is striking because it undermines many standard procedures connected with the analysis of variance. If one accepts standard confidence procedures for multiple comparisons, as determined by a region R of sampling probability $1-\alpha$ in the space of $X_1-\theta_1, X_2-\theta_2, \ldots, X_k-\theta_k$, then one is being consistent with flat prior Bayesian inferences, since the latter assign posterior probability $1-\alpha$ to R. On the other hand, standard procedures for estimating variance components are analogous to estimating $\Sigma_1^k \theta_i^2$ by inverting the sampling distribution of $\Sigma_1^k X_i^2$. If the Bayesian yardstick of consistency is adopted, meaning that inferences should be consistent with some posterior distribution, then the two types of analysis of variance procedures cannot both be accepted in principle, and in practice will often produce noticeably discrepant results.

For a frequentist, Bayesian inconsistency is not in itself damning, but it is interesting to note that Charles Stein has been a leader in proposing simultaneous estimation procedures for $\theta_1, \theta_2, \ldots, \theta_k$ which are more nearly in line with the standard variance component procedures. Stein's motivation derives from the observation in Stein (1956) that the almost universally trusted point estimators $X_1, X_2, \ldots, X_k$ for $\theta_1, \theta_2, \ldots, \theta_k$ are inadmissible relative to total squared error as a loss function. The remedy proposed by James and Stein (1961) is to use

$$\theta_i = \left(1 - \frac{k-2}{\Sigma_i^k X_i^2}\right) X_i \tag{3.1}$$

for $i = 1, 2, \ldots, k$, which amounts to pulling each raw estimator X_i back towards the origin, using a scale factor based on all of the observations. Stein (1962) further proposed finding confidence regions for $\theta_1, \theta_2, \ldots, \theta_k$ centered about the estimated vector $\theta_1, \theta_2, \ldots, \theta_k$ defined in (3.1). Alternatively, one could shrink towards an arbitrary point instead of the origin as in (3.1). The natural frequentist advice would be to shrink towards that point which seems *a priori* most plausible. Shrinking each X_i towards the mean of all the X_i may also be plausible in some circumstances, as suggested by equations (3.2) and (3.4) below.

Since admissible procedures are generally Bayes's procedures, Stein's approach is not unrelated to a Bayesian approach, which in this case implies specifying the model more completely by assigning a joint distribution to $\theta_1, \theta_2, \ldots, \theta_k$. In the discussion following Stein (1962), several speakers made explicit the point which Stein left implicit, namely that pull-back procedures like (3.1) follow naturally from certain randomness assumptions on $\theta_1, \theta_2, \ldots, \theta_k$. Whether or not the procedures are viewed as Bayesian depends on the partly artificial questions of whether or not the distribution of $\theta_1, \theta_2, \ldots, \theta_k$ is explicitly introduced, and, if it is introduced, on whether or not it is called a prior distribution.

In Section 3.2 I shall pursue the model with random $\theta_1, \theta_2, \ldots, \theta_k$ as an illustration of one method of parameter reduction: instead of k parameters $\theta_1, \theta_2, \ldots, \theta_k$ one fits a small number of free parameters in the distribution of $\theta_1, \theta_2, \ldots, \theta_k$. The immediate point is the interpretation that without some device in the model to soften the implications of complete ignorance of the k parameters, it is unlikely that any precise inference procedure about many means $\theta_1, \theta_2, \ldots, \theta_k$ can be widely agreed upon.

Since many covariances are generated more rapidly than many means, it should be expected that confusion will seriously affect the estimation of covariances among k variables unless sample size is of order k^2. A paradox was demonstrated in Dempster (1963) regarding covariance estimation with sample sizes of order k. See Dempster (1969, pp. 370-1) for another view of this paradox which in some ways parallels the Stein (1959) paradox on many means.

Problems which arise in the context of multivariate normal sampling will inevitably appear in broader fields of multivariate analysis, where variables are categorical as well as continuous, and especially where several variables are observed at successive points of time. Model building in these areas requires special attention to the feasibility of meaningfully estimating proposed parameter sets.

3.2 Parameter Reduction I: Random Parameters

One approach to many parameters with limited data is to declare the situation impossible. A more optimistic attitude holds that models can be simplified so that available data may speak through them with acceptable fit and clarity. The general principle is drastically to reduce the number of free parameters when confusion threatens. Two specific parameter reduction devices are discussed in Sections 3.2 and 3.3, respectively.

Prior knowledge and understanding of a situation may dictate a model with a large number of functionally independent parameters, meaning undetermined quantities which cannot plausibly be reduced to functions of a small number of more basic parameters. But it remains possible to direct the probability modelling process at the parameters themselves. The objective is to assign the original parameters a distribution depending on fewer and more basic parameters which can be estimated from the data with reasonable assurance.

The canonical example here is an extension of the many means example of Section 3.1. In addition to specifying that $X_1-\theta_1$, $X_2-\theta_2$, ..., $X_k-\theta_k$ are independently $N(0,\sigma^2)$ distributed, one specifies that θ_1, θ_2, ..., θ_k are independently $N(\alpha, \beta^2)$ distributed and independent of $X_1-\theta_1$, $X_2-\theta_2$, ..., $X_k-\theta_k$. If α, β and σ are assumed known, then estimation of θ_1, θ_2, ..., θ_k is characterized by the conditional distribution of θ_1, θ_2, ..., θ_k given X_1, X_2, ..., X_k. Accordingly θ_1, θ_2, ..., θ_k are viewed *a posteriori* as independent $N(\theta_i, \tau^2)$ where

$$\theta_i = \left(\frac{1}{\beta^2}\alpha + \frac{1}{\sigma^2}X_i \right) \bigg/ \left(\frac{1}{\beta^2} + \frac{1}{\sigma^2} \right) \tag{3.2}$$

for $i = 1, 2, \ldots, k$ and

$$\frac{1}{\tau^2} = \frac{1}{\beta^2} + \frac{1}{\sigma^2}. \tag{3.3}$$

Writing (3.2) in the form

$$\theta_i = \alpha + [X_i - \alpha] \frac{\frac{1}{\sigma^2}}{\frac{1}{\sigma^2} + \frac{1}{\beta^2}} \tag{3.4}$$

shows that the estimators θ_i for $i = 1, 2, \ldots, k$ are shrunk back towards the prior mean α by a factor which is the same for all i.

A more realistic version of the above model would assume that α, β and σ were unknown *a priori* and therefore to be estimated from the data. In order that σ^2 can be estimated, the data must be augmented, for example by having n measurements X_{i1}, $X_{i2}, \ldots, X_{in}$ on each θ_i where $X_{ij} - \theta_i$ are assumed independently $N(0, \sigma^2)$ distributed for $i = 1, 2, \ldots, k$ and $j = 1, 2, \ldots, n$. Then the means $\bar{X}_i = \Sigma_1^n X_{ij}/n$ play the roles of the X_i in the original model while the role of σ^2 is taken over by σ^2/n. The resulting model is now known as the random effects model for the analysis of variance, and has a long history. The model is implicit in Fisher's (1921) discussion of the intraclass correlation coefficient, and Novick and Thayer (1969) recently remarked that a version of equation (3.2) appears in Kelley's (1923) textbook. Frequentists are comfortable with random effects models when it is natural to regard θ_1, θ_2, ..., θ_k as randomly drawn from some population. The term prior distribution can then be avoided, but if that term is admitted for the initial distribution assigned to $\theta_1, \theta_2, \ldots, \theta_k$, then the use of equation (3.2) with estimated α, β and σ^2 becomes an example of an empirical Bayes technique in the terminology of Robbins (1956). The technique itself is clearly much older than the label empirical Bayes.

The obvious strategy of replacing α, β and σ in equation (3.2) with point estimates $\hat{\alpha}$, $\hat{\beta}$ and $\hat{\sigma}$ cannot be faulted when the estimators are sufficiently precise that their errors add negligible amounts to the mean squared error τ^2 of the $\hat{\theta}_i$ as defined in equation (3.3). Otherwise, the choice of particular estimators $\hat{\alpha}$, $\hat{\beta}$ and $\hat{\sigma}$ becomes a vexing question, already symptomatic of confusion. A logically more satisfying strategy would be to average the estimator in equation (3.2) over the posterior distributions of α, β and σ corresponding to a range of prior distributions which could plausibly be regarded as diffuse. One could then investigate dependence on the choice of diffuse priors, first of the estimates themselves, and second of the frequentist operating characteristics of the estimators. Various Bayesian alternatives have been studied by Box and Tiao (1968a), Hill (1965), Novick and Thayer (1969), Portnoy (1969), Stone and Springer (1965), Tiao and Tan (1965), and no doubt by others. I feel that comparisons among Bayesian alternatives are capable of greatly illuminating confusion over the estimation of α, β and σ, but they are unlikely to produce estimates which are robust against confusion when samples are small enough that confusion is real. Accordingly, I shall sketch below another method of inference which brings out an explicit representation of confusion.

I believe that too much emphasis is being placed on the estimation phase of inference in the approaches just described. Estimation via conditional probability is a powerful tool, but it needs to be used sparingly because the conditions for its use are difficult to satisfy. On the other hand, the less potent tool of postdictive reasoning is much more broadly applicable, and applied, although it can be used with confidence only in the process of model searching. I suggest therefore that we shift the burden somewhat from the ambitious confirmatory techniques of estimation, back to being content with laying out classes of hypotheses which are plausible and do not clash postdictively with observed data. There can be little point in anguishing about prior distributions representing relative uncertainties about hypotheses which are themselves speculative hypothetical entities and very much dependent on the limitations of data in hand.

Very simply, then, I would propose to lay out intervals of α and β^2/σ^2 which are acceptably consistent with the data in the sense of passing simple significance tests, or equivalently of belonging in certain confidence regions. The limits on α and β^2/σ^2 then translate into intervals on the $\hat{\theta}_i$ determined by equation (3.2). These I would interpret roughly as upper and lower bounds on posterior means or medians, accompanied by a remark that those willing to introduce a range of plausible prior distributions may achieve tighter bounds. There is nothing deeply original about this technique, but I do wish to stress the conscious choice of a mix of postdictive and predictive ideas, that is, postdictive determination of an acceptable range of α and β^2/σ^2 with predictive estimation via equation (3.2).

For concreteness, consider the hypothetical data of Table 1 giving pairs of observations on 20 means. Suppose that the random effects model is

Table 1. Pairs of measurements on 20 means ($k=20$, $n=2$).

100	104	121	114
103	110	112	120
135	128	96	91
105	111	99	102
100	93	103	108
75	82	107	109
111	119	115	116
103	102	90	93
91	90	112	102
120	132	117	116

accepted, say after checking that the observed distributions of 20 sums and 20 differences look reasonably normal and independent. The grand mean estimating α is 106.425 while the variance of the 20 means yields a mean square of 162.19 on 19 degrees of freedom estimating $\beta^2 + \sigma^2/2$. These yield an equi-tailed 95% confidence interval (100.465, 112.385) for α from Student's t. Similarly, the sum of squares of 20 differences yields a mean square of 18.625 on 20 degrees of freedom estimating σ^2. The ratio of the two mean squares on 19 and 20 degrees of freedom, respectively, yields an equi-tailed 95% confidence interval (.856, .977) for the shrinkage factor $(2/\sigma^2)/(2/\sigma^2 + 1/\beta^2)$ which appears in the modified equation (3.2) required for $n = 2$ instead of $n = 1$. Table 2 shows the actual θ_i from which the data were simulated, together with the unshrunk sample mean estimates, and the posterior mean interval estimates proposed in the preceding paragraph. It may be noted that the actual total squared error from the use of X_i is 249.25, which reduces to 163.14 if the midpoint of $[\theta_{i*}, \theta_i\star]$ is used.

Table 2. Simulated θ_i together with sample mean estimates and posterior mean interval estimates based on the data in Table 1.

θ_i	$\bar{X}_i$	$[\theta_{i*}, \theta_i\star]$	θ_i	X_i	$[\theta_{i*}, \theta_i\star]$
107	102	[101.8, 103.5]	108	117.5	[115.1, 117.4]
109	106.5	[105.6, 107.4]	113	116	[113.8, 115.9]
129	131.5	[127.0, 131.1]	95	93.5	[93.7, 96.2]
108	108	[106.9, 108.6]	104	100.5	[100.5, 102.2]
101	96.5	[96.6, 98.8]	106	105.5	[104.8, 106.5]
79	78.5	[79.0, 83.4]	111	108	[106.9, 108.6]
112	115	[112.9, 114.9]	117	115.5	[113.3, 115.4]
106	102.5	[102.2, 103.9]	94	91.5	[91.7, 94.5]
91	90.5	[90.7, 93.7]	110	107	[106.1, 107.8]
123	126	[122.3, 124.0]	112	116.5	[114.9, 116.4]

My purpose in Section 3.2 is not to resolve the detailed analysis of the random effects model, but rather to make the general suggestion that the construction of random models for many parameters be much more widely attempted by theoretical statisticians. The chief obstacle is the entrenched mental habit of many statisticians to shy away from random models which lack a plausible real world sampling interpretation. For example, the random effects model above would be regarded as acceptable only if the k means referred to entities (for example, animals or families or blocks) which could be conceived as random samples of larger collections of entities of the same type. I believe, however, that careful weakening of the entrenched mental habit is justifiable, and that substantial benefits will accrue from such weakening by permitting statisticians to make more meaningful use of complex highly parameterized models.

The conventional wisdom which limits random models to situations having sampling interpretations is based on the belief that a model must be at least conceptually testable under repeated trials before it can be entertained. I find it necessary to separate the wheat from the chaff in this widespread frequentist belief. Although frequentists often welcome the label objective, every frequentist model rests on a necessarily subjective judgment that a particular set of elements were drawn from a larger collection of elements in a way which favored no elements over any other, that is, according to an equiprobabilistic scheme. It is from this judgment alone that the possibility of a frequency interpretation follows. To the extent that actual repeated trials are manifested in the data, they are the cornerstone of statistics, on which the very possibility of inference depends. But to the extent that frequencies are hypothetical they are chaff which can be thrown away, *provided the necessary kernel of a symmetry judgment is maintained.* It is my belief that such symmetry judgments can be made, to be sure with some speculative content, much more widely than frequentist ideology would permit. These judgments do not immediately solve problems, but they do open the door to more widespread attempts to capitalize on smooth behavior of parameter sets by means of probability models which capture that smoothness.

To illustrate, return once more to the random effects model. If it can be agreed, at least as a rough working principle, that the state of initial ignorance of the k entities and their associated $\theta_1, \theta_2,..., \theta_k$ places all of the entities on an equal footing, then it is but a short step to assuming a probability model for $\theta_1, \theta_2,..., \theta_k$ which is formally equivalent to a random sampling model. The famous theorem of De Finetti (1937) to the effect that symmetry among entities in every subset of an infinite class of entities implies a mixture of random sampling models is often used to support a Bayesian position. But it also lends support to the search for adequately fitting random sampling models through which the Bayesian dream can be realized. The question then becomes what class of distributions to assign to $\theta_1, \theta_2,..., \theta_k$? The model asserting that $\theta_1, \theta_2,..., \theta_k$ are independently $N(\alpha, \beta^2)$ distributed is a conventional choice for purposes of illustration, and may surprisingly often be adequate in practice, but in real life one would assess the evidence in the data for different forms, and for more or less complexity, depending on the nature and extent of the data.

Another canonical example exists for multinomial sampling, where k unknown theoretical frequencies $p_1, p_2, \ldots, p_k$ can be viewed *a priori* as having complete symmetry regarding knowledge or ignorance. The constraint $\Sigma_1^k p_i = 1$ suggests that the p_i be represented as $p_i = \lambda_i / \Sigma_1^k \lambda_i$ where the λ_i are randomly drawn from some positive distribution. The particular distribution is up for grabs in the model building process, but there is a canonical choice, namely a gamma distribution with unknown shape parameter. This choice implies that the observed frequencies are drawn from a negative binomial distribution, whence one can test the plausibility of the gamma assumption, or, after accepting the gamma hypothesis, estimate its unknown shape parameter. As with the random effects model, I would suggest estimating the shape parameter postdictively and then switching over to Bayesian estimation with a range of posterior distributions for the p_i.

The field of model construction for random parameters has barely been scratched. If means can be viewed symmetrically, so can variances. Models for large correlation matrices are urgently needed, where symmetry under permutations of the variables could often be tried as a foundation. The whole broad field of highly parameterized models for discrete and continuous multivariate data stretches out ahead.

3.3 Parameter Reduction II: Fractional Expenditure of Parameters

A second strategy for reducing a large class of parameters involves setting most of them at null values and attempting to estimate only a salient few. Such null values often arise naturally when models increase in complexity by stages, with a set of parameters being given specified null values at one stage, then being undetermined at the next stage. The key difficulty, especially when the number of parameters involved is large, is to recognize which are the salient few and which merely appear salient due to outlying sampling fluctuations. The question can of course be answered only with uncertainty. In Section 3.3, I shall describe a device aimed at coping with the difficulty. I am currently investigating techniques in this area in collaboration with Dr. Martin Schatzoff of IBM Cambridge Scientific Center and Mr. Bernard Rosner of the Harvard Department of Statistics.

As in Section 3.2, a canonical illustration is provided by the estimation of many means. Suppose that X_1, X_2,...,X_k measure θ_1, θ_2,...,θ_k with independent and identically $N(0, \sigma^2)$ errors. On this foundation, many variants of the estimation problem can be constructed. The simplest *a priori* hypothesis would assert that with a few unknown exceptions all of the θ_i are zero, or at least known and translatable to zero. Such zero hypotheses could be plausible if the θ_i were, say, second order interactions. With actual means θ_i, a more frequently plausible assumption would have all but a few equal to some common value $\bar{\theta}$. Still another variant could have all of θ_1, θ_2,..., θ_k, excepting an unknown few, representable as drawn from a $N(\alpha, \beta^2)$ population, thus linking the device of Section 3.3 with that of Section 3.2.

The proposed techniques mix model building and estimation in a specific way. Data-based modelling is used primarily to determine the number of deviant parameter values which need special attention. Call this number r. In a first pass at the data, some idea of r may be obtainable from rough estimates or from graphical displays such as normal plots. These initial looks will of course point a finger not only at r but at particular suspected parameters. As remarked above, however, it is necessary to try to separate the guilty suspects from the innocent. The device proposed for separation is essentially Bayesian: it is supposed that before the data were examined an *a priori* symmetry assumption held, so that all choices of a set of *r* parameters as candidates for the deviant set were *a priori* equally probable, assuming for the moment a given *r*. One can then compute an *a posteriori* distribution over the possible subsets of *r* parameters, defining probabilities for which subset is the deviant subset on the basis of the data. Thus one obtains evidence on whether or not an apparently deviant subset actually stands up to scrutiny as a candidate for *the* deviant subset. Under each candidate subset, estimates can be computed for the whole set of parameters, and final estimates may be obtained by averaging over the posterior distribution associated with the candidate subsets. These estimates refer to a prespecified *r*. By computing them for an increasing sequence of *r* values, one can have a second look at the choice of *r*, with the hope that fit will rapidly improve as *r* increases and that a plausible level of fit will be achieved while *r* is still small.

Two artificial examples are summarized in Tables 3, 4 and 5, as illustrations of the method. In both examples, a preselected vector θ of 10 means was disturbed by a vector of 10 standard normal deviates to produce the observable vector X. For purposes of display, the elements of X shown in Tables 3 and 4 were ranked according to absolute value.

In the Table 3 example, the general hypothesis is that all but a few of the θ_i are zero, and the $N(0,1)$ error distribution is assumed known. For a prespecified r, each of the $\binom{10}{r}$ subsets I of size r are contemplated as carriers of nonzero θ_i. Further restricting to a specified I, the estimate for θ is given by

$$\theta_{ri}(I) = X_i \quad \text{for } i \in I$$
$$= 0 \quad \text{for } i \notin I. \tag{3.5}$$

Table 3. Estimates θ_r for $r = 0, 1, 2, 3, 4$ referring to artificial X and θ where X-θ is simulated from a $N(0,1)$ distribution. Hypothesis r allows r nonzero means and 10-r zero means.

θ	X	θ_0	θ_1	θ_2	θ_3	θ_4
0	-.32	0	-.0001	-.0005	-.0061	-.0255
0	.53	0	.0002	.0008	.0111	.0460
0	-.70	0	-.0003	-.0012	-.0163	-.0673
0	-.93	0	-.0005	-.0020	-.0261	-.1073
0	1.12	0	.0007	.0029	.0382	.1561
0	1.44	0	.0015	.0056	.0739	.2969
0	1.89	0	.0040	.0155	.2042	.7753
2.5	2.74	0	.0418	.1589	2.0113	2.3970
2.5	3.78	0	1.7102	3.6183	3.7415	3.7629
2.5	3.82	0	2.0120	3.6794	3.7865	3.8052
$\Sigma(X_i - \theta_{ri})^2$		45.027	23.49	14.77	7.94	4.95
$\Sigma(\theta_{ri} - \theta_i)^2$		18.75	6.90	8.27	3.48	4.04

Adopting a uniform prior on nonzero θ_i, the marginal likelihood of I is

$$L(I) \propto \exp(-\tfrac{1}{2}\sum_i [X_i - \theta_{ri}(I)]^2) = \exp(-\tfrac{1}{2}\sum_{i \notin I} X_i^2), \qquad (3.6)$$

which is also proportional to the posterior probability for I in view of the *a priori* symmetry assumption about the choice of I. The final estimate of θ_r is therefore

$$\theta_{ri} = \sum_I \theta_{ri}(I) L(I) / \sum_I L(I), \qquad (3.7)$$

as exhibited in Table 3 for $r=0, 1, 2, 3, 4$. The quantities $\Sigma(X_i - \theta_{ri})^2$ are roughly X^2 measures of fit on $10-r$ degrees of freedom. These show roughly significant drops up to $r=3$, so that a final choice of $r=3$ would be accepted by me as a plausible model, and thence θ_3 as plausible final estimates.

With the Table 4 data, the fitting procedure assumes a common mean $\bar{\theta}$, which is initially unknown, except for a few deviant θ_i values. Accordingly equation (3.5) is altered to

$$\theta_{ri}(I) = X_i \quad \text{for } i \in I$$
$$= (\sum_{j \notin I} X_j)/(10-r) \quad \text{for } i \notin I. \qquad (3.8)$$

The first expression in equation (3.6) remains valid, as does formula (3.7) for the estimates shown in Table 4. Again, the drop in the X^2- like quantities would indicate a plausible final choice of $r=2$.

Table 4. Estimates θ_r for $r=0, 1, 2$, referring to artificial X and θ where $X-\theta$ is simulated from a $N(0, 1)$ distribution. Hypothesis r means, that r means are equal while the remaining $10-r$ are undetermined.

θ	X	θ_0	θ_1	θ_2
0	-.061	.319	-.045	-.295
0	-.134	.319	-.045	-.297
0	-.286	.319	-.046	-.301
0	-.333	.319	-.046	-.302
0	.446	.319	-.044	-.282
0	-.586	.319	-.046	-.310
0	-.664	.319	-.046	-.313
0	-1.420	.319	-.059	-.387
2.828	2.496	.319	.006	1.973
4.000	3.728	.319	3.581	3.701
$\Sigma(X_i - \theta_{ri})^2$		22.33	9.14	2.15
$\Sigma(\theta_{ri} - \theta_i)^2$		20.66	8.18	1.60

The procedures just illustrated provide for fractional expenditure of parameters. If a specified number r of deviant parameter values is allowed, then this number r is spent fractionally over all subsets of size r. A weight associated with each θ_i can be computed by summing the posterior probabilities over subsets I containing i, as exhibited in Table 5 for both examples, thus displaying the amount which each θ_i contributes to the total expenditure of r parameters. For an analysis of this type to maintain credibility, it is desirable that the weights corresponding to the finally selected r should concentrate mainly on r individual θ_i. It is the essence of the technique, however, that the weights should not be arbitrarily reset to unity on the r largest values, because it is the deviations from unity which provide a margin of protection against picking up false deviants.

Table 5.   Weights assigned to individual θ_i for the analyses reported in Tables 3 and 4.

	Table 3 Data				Table 4 Data	
r	1	2	3	4	1	2
Wts.	.000	.002	.021	.080	.002	.021
	.000	.002	.021	.087	.002	.021
	.000	.002	.021	.096	.002	.021
	.001	.002	.027	.115	.002	.022
	.001	.003	.032	.140	.002	.024
	.001	.004	.054	.206	.002	.024
	.002	.008	.108	.410	.002	.026
	.015	.058	.732	.875	.008	.059
	.452	.957	.991	.995	.021	.789
	.527	.963	.991	.995	.957	.992
Totals	1.00	2.00	3.00	4.00	1.00	2.00

Methods of estimation based on formula (3.7) share with those based on equations (3.1) or (3.2) the feature of pulling back the raw estimates X_i towards some common value. Unlike the methods of Section 3.2, the methods of Section 3.3 operate selectively on different values of i so that X_i with low weight are pulled back drastically while those with large weight are altered only slightly. I believe that a correct choice among procedures, for example between the formulations of Sections 3.2 and 3.3, requires sensitivity to the data and clearly cannot be made on the basis of *a priori* considerations alone. It appears therefore that a real data analyst can be neither an extreme frequentist who insists that procedures defined before peeking at the data be adhered to firmly, nor an extreme Bayesian who likewise formulates his complete representation of prior knowledge independently of the data. Both of these fictional types need liberal doses of a model searching attitude before becoming operational.

A near cousin of the above techniques of fractional expenditure is the approach to compensating for outliers proposed by Box and Tiao (1968b). The difference is that Box and Tiao tighten the Bayesian noose a notch at the outset by assuming that the deviant few θ_i were selected *a priori* by binomial sampling with small p, so that r is viewed *a priori* as binomially distributed. The user must then either assess plausible values of p from the data, which is a step deeper and more difficult than assessing r directly, or he must introduce prior knowledge of p. Where prior knowledge of p can be demonstrated, say from a series of similar experiments, its introduction might be considered innocuous, but I remain skeptical on grounds unrelated to knowledge of p. Hidden in my analysis, are implicit diffuse prior densities on the deviant θ_i which are simply dropped from my formula (3.6) when the marginal likelihoods $L(I)$ are computed. I believe this to be acceptable for likelihoods associated with a single r, in line with treating all subsets I symmetrically. The binomial assumption, however, requires weighting together likelihoods associated with different r, and this involves implicit comparison of diffuse prior probability elements in spaces of different dimensions, which is somewhat like comparing a ranch house and a skyscraper by computing their respective ground areas. The difficulty is another version of the difficulty described in Section 1 concerning discrete spikes of prior probability on a null hypothesis. Box and Tiao introduce a parameter k into their prior on θ_i and assess its effect on their estimation procedure. To me, however, direct assessement of r on the basis of fit to data appears less burdened by extraneous difficulties. The issue again is one of where to draw the line between postdictive model building and predictive estimation.

The kind of technique just described in the context of many means is more original and more experimental than the kind of technique described in Section 3.2. Among the various applications contemplated are the traditional problem of detecting outliers, and the analogous problem of assessing contrasts computed from observations on a 2^k factorial design. Another area where fractional expenditure of parameters may find use is in the selection of predictor variables in multiple regression. It is a well attested observation that with limited data improved prediction may result from holding the number of predictors down. But for any permitted number r of regression coefficient parameters even better prediction formulas may result from expending these r parameters fractionally over the available class of predictors. Procedures for selecting variables are in fact selective pull-back procedures, where a subset of the parameters is left unshrunk while the rest are pulled all the way back to zero. I am suggesting here that more flexible selective pull-back procedures should be able to allow for the biasing effect of selecting predictor variables which appear good in the data, and should therefore offer hope for improved prediction schemes.

In a broader scheme of things, there are many model building situations where large sets of parameters come on the scene simultaneously and in a reasonably symmetric way. Examples abound as soon as one steps outside the limits of familiar simple models for structured multivariate data. In these examples, the data may often suggest fitting only a few parameters. In other cases, a simple random model plus a few deviant parameters may give adequate fit. Almost all of these potentially useful techniques remain to be developed.

Acknowledgments

This paper was prepared with partial support of the National Science Foundation under grant GP-8774. The work benefitted from discussion with other participants in the summer seminar held at the Department of Statistics, University of Colorado, Fort Collins, Colorado in August 1969, also with the support of the National Science Foundation.

Bibliography

1. Bernoulli, James, *Ars Conjectandi*, Basel, 1713.
2. Box, G.E.P. and Tiao, G.C., "Bayesian Estimation of Means for the Random Effects Model," *Journal of the American Statistical Association*, 63, 174-181, 1968a.
3. Box, G.E.P. and Tiao, G.C., "A Bayesian Approach to Some Outlier Problems," *Biometrika*, 55, 119-130, 1968b.
4. Buehler, Robert J., "Some Validity Criteria for Statistical Inferences," *Annals of Mathematical Statistics*, 30, 845-863, 1959.
5. De Finetti, B., "La Prévision: Ses Lois Logiques, Ses Sources Subjectives," *Ann. Inst. H. Poincaré*, 7, 1-68, 1837.
6. De Morgan, Augustus, "On a Question in the Theory of Probabilities," *Trans. Camb. Philos. Soc.*, 6, 423-430, 1937.
7. Dempster, A.P., "On a Paradox Concerning Inference About a Covariance Matrix," *Ann. Math. Statist.*, 34, 1414-1418, 1963.
8. Dempster, A. P., "On the Difficulties Inherent in Fisher's Fiducial Argument," *J. Am. Stat. Ass.*, 59, 56-66, 1964.
9. Dempster, A.P., "A Generalization of Bayesian Inference," *Journal of the Royal Statistical Society, B*, 30, 205-247, 1968.
10. Dempster, A.P., *Elements of Continuous Multivariate Analysis*, Reading, Massachusetts, Addison-Wesley, 1969.
11. Fisher, R.A., "On the 'Probable Error' of a Coefficient of Correlation deduced from a Small Sample," *Metron*, 1, 3-32, 1921.
12. Hill, B.M., "Inference About Variance Components in a One-Way Model," *J. Am. Stat. Ass.*, 60, 806-825, 1965.
13. James, W. and Stein, C.M., "Estimation with Quadratic Loss Function," *Proceedings of the Fourth Berkeley Symposium*, 1, 361-379, 1961.
14. Kelley, T.L., *Fundamentals of Statistics*, Harvard, Cambridge, Massachusetts, 1923.
15. Kendall, M.G., "On the future of Statistics — a Second Look (with Discussion)," *Journal of the Royal Statistical Society, A*, 131, 182-204, 1968.
16. Laplace, Pierre Simon, *Théorie Analytique des Probabilités*, Paris, 1813.
17. Novick, Melvin R. and Thayer, Dorothy T., "A Comparison of Bayesian Estimates of True Score," *Research Bulletin* 69-74, Educational Testing Service, Princeton, N.J., 1969.
18. Poisson, S. D., *Recherches sur la Probabilité des Jugements*, Paris, 1837.
19. Portnoy, S. L., "Formal Bayes Estimation with Application to a Random Effects Analysis of Variance Model," *Technical Report* Number 2, N.S.F. Grant GP-8985, Department of Statistics, Stanford University, 1969.
20. Robbins, Herbert E., "An Empirical Bayes Approach to Statistics," *Proceedings of the Third Berkeley Symposium*, 1, 157-163, 1956.

21. Simpson, Thomas, "On the Advantage of Taking the Mean of a Number of Observations in Practical Astronomy," *Philos. Trans. Roy. Soc. London*, 41, Part 1, 82-93, 1755.

22. Stein, C.M., "An Example of Wide Discrepancy Between Fiducial and Confidence Intervals," *Ann. Math. Statist.*, 30, 877-880, 1959.

23. Stein, C. M., "Inadmissibility for the Usual Estimator of the Mean of a Multivariate Normal Distribution," *Proc. 3rd Berkeley Symposium*, 1, 197-206, 1956.

24. Stein, C. M., "Confidence Sets for the Mean of a Multivariate Normal Distribution (with discussion)," *J. Roy. Statist. Soc.*, B, 24, 265-296, 1962.

25. Stone, M. and Springer, B.G.F., "A Paradox Involving Quasi Prior Distributions," *Biometrika*, 52, 623-627, 1965.

26. Tiao, G.C. and Tan, W.Y., "Bayesian Analysis of Random Effects Models in Analysis of Variance I. Posterior Distributions of Variance Components," *Biometrika*, 52, 37-53, 1965.

COMMENTS

D.R. Cox:

It is hard to comment on the very interesting procedures in Section 3 of the paper without knowing what is to be done further with the estimates $\hat{\theta}_i$. Thus, if contrasts among the θ_i's are of main interest, the shrinking towards a general mean affects all estimated contrasts equally and hence for some purposes leaves the qualitative conclusions unchanged. On the other hand, the methods based on an assumed clustering of the θ_i's may accentuate spurious gaps between adjacent means; there may be some connection with the gap tests for analysis of sets of means proposed by J. W. Tukey and others.

A natural, if rather academic, question that is likely to arise is whether there is evidence from the data that a normal distribution for the θ_i's fits better or worse than a grouping into say two sets, at unknown points, θ', θ''. This is a simplified version of a problem arising also in cluster analysis. Now the difference, d, of maximized log likelihoods can easily be computed. Here the difference between different approaches emerges. One might be content with the observed value of d, and in extreme cases this will be adequate, but it seems to me important also to have some idea of the distributions of d when one or other model is true, especially when the two different models have very different numbers of adjustable parameters; the distribution problem is difficult in this case. A Bayesian analysis with flat priors requires a value for the relative heights of the prior densities for the two models.

I.J. Good:

Dr. Dempster raised the question, in his spoken version, of where I stand in relation to physical probabilities. By a theorem of de Finetti, coherent subjective probabilities can be expressed *as if* physical probabilities exist, and I tend to believe they actually *do* exist (see my contribution to the Salzburg conference in *Synthese*, 1969). Dr. Dempster's position seems to be very close to mine.

On page 57 Dr. Dempster refers to the *bon mot* of K. Caj Doog: "To a Bayesian, all things are Bayesian". I am afraid that Dr. Dempster has been misled by the modesty of Doog who should have said "To a Doogian all things are Doogian". As Dr. Bartlett once mentioned, the word *Bayesian* is misleading because there are a variety of different kinds of Bayesians such as savage ones and good ones. I have read everything published by Doog and have now listed twenty-seven of his laws in an appendix to my paper.

I agree with many things in Dr. Dempster's paper, and I admire his ancient historical references, but I thought these should be supplemented by more references to Doog. [In the spoken version there were several references to Good.]

I have a number of further comments. When a confidence interval is given and one says that either the parameter lies in an assigned interval or an improbable event has occurred, there is of course a philosophical difficulty. It is that in a sense every event that occurs in the real world is improbable, as Heraclitus implied two thousand years ago. Similarly Harold Jeffreys emphasized that the trouble with the tail-area probability approach is that it lumps with the event that did occur a lot of other events that did not. Fortunately the tail-area probability approach can be given an approximate informal Bayesian justification as in my 1950 book page 94.

On page 59 Dr. Dempster states that the inversion of certain sampling distributions may often be warranted because of the appearance of new Bayesianly acceptable techniques. He presumably took for granted that a number of examples of this are already in the literature, and especially Jeffrey's inverse probability interpretation of some of Fisher's fiducial arguments, and some further arguments by Lindley. I think it is better to be explicit about this.

On page 60 Dr. Dempster says that Bayesian testing necessarily has a restricted role within the larger field of testing. I think this is misleading because I believe that every method of significance testing, if it is sensible at all, can be given an informal Bayesian justification by making use of the Bayes/non-Bayes compromise. This is an example of what Doog intended by his *bon mot*. He certainly did not mean that tail-area probabilities are complete nonsense.

On page 60 Dr. Dempster criticizes the notion of the initial probability of a null hypothesis. His criticism is somewhat weakened if first one remembers that a null hypothesis is usually intended only to be an approximation in a sense that can in principle be specified. Secondly, if one makes use of Bayes-Jeffreys-Turing factors then the initial probability of a null hypothesis does not come into the argument, although a judgment of it can be incorporated, just as when a significance level is selected.

A formal solution to a part of the problem of *complementarity* is my Sharpened Razor.

The difference between upper and lower probabilities as a measure of confusion was suggested by Good, "How rational should a manager be?" *Management Science*, 8, 1962, 383-393: except that it was called the *vagueness* of the judgment. There was also a suggestion of another possible definition for vagueness, namely, the difference of the upper and lower log-odds. This second definition had the advantage that when one is considering the case of only two mutually exclusive simple statistical hypotheses, then the vagueness is unchanged by the evidence. To quote, "this second definition of vagueness would justify the statement that if the judgment of the final probability of a hypothesis is more precise than that of its initial probability, then the use of Bayes' theorem in the forwards direction would not be very helpful" (see also *Probability and the Weighing of Evidence*, 1950, page 82). I think *vagueness* is a better term than *confusion* in this context.

On page 64 Dr. Dempster points out the difficulty of extending multivariate analysis beyond the usual multivariate normal model because, if the index of the exponential is changed from a quadratic to a quartic, then the number of parameters might increase very greatly. The proposal has been made to generalize multivariate analysis by means of the real stable characteristic functions (see B. Mandelbrot, *Econometrica*, 29, 1961, 517-543; and I. J. Good, *Proc. Roy. Soc. A*, 307, 1968, 317-334). If this becomes practicable then it has the advantage of introducing only one additional parameter.

On page 66 Dr. Dempster mentions that when there are many parameters it is helpful to introduce a distribution that depends on fewer and more basic parameters which can be estimated from the data with reasonable assurance. An example of this approach was the selection of a linear combination of symmetric Dirichlet initial distributions for the problem of finding significance tests for multinomial distributions. Having found that a single symmetric Dirichlet distribution was inadequate as a *prior*, I took the next simplest assumption and the results seemed to be satisfactory. If they had not been then I would have introduced a symmetrical combination of unsymmetrical Dirichlet distributions, but this would have involved the use of more parameters. Of course I am here referring to parameters in a prior or type II distribution rather than in an ordinary or type I distribution.

On page 68 Dr. Dempster uses the expression *Empirical Bayes* as if it were identical with the use of a sampling method for determining an initial distribution, as proposed by von Mises. I think this is historically misleading. As I understand it, the Empirical Bayes method is a method for making estimates of type I parameters while largely eliminating a specific form for the type II distributions, or the priors; but assuming that the prior exists as a *physical* distribution. The type II distribution must not be completely eliminated as in Yate's formula for the probability of the interior of a contingency table conditional on the marginals. For if the type II distribution is completely eliminated no previous *experience* would be relevant, and the existence of previous experiences is an essential feature of the Empirical Bayes technique. The idea of the Empirical Bayes method was introduced by A. M. Turing in 1940, with some assistance from myself, for a particular problem, namely, the estimation of the physical probabilities of species. I wrote this up with elaborations in 1953 (see "On the Population Frequencies of Species and the Estimation of Population Parameters," *Biometrika*, 40, 237-264; and I. J. Good and G. H. Toulmin, "The Number of New Species, and the Increase of Population Coverage, when a Sample is Increased," *Biometrika*, 43, 1956, pp. 45-63). In my opinion it was a clear anticipation of the method of Empirical Bayes (especially for Poisson and binomial distributions) which was expressed in explicit generality by Herbert Robbins. The species paper showed that smoothing techniques are essential in the practical application of Turing's idea. Smoothing is equally essential in the general empirical Bayes technique although it was introduced only in 1966 by Maritz (*Biometrika*, 53, 417-429) who mentioned the species paper.

J. Neyman:

Two particular points related to Professor Dempster's paper come to my mind: (a) the possibility of a misunderstanding and (b) a question.

(a) On page 58 Professor Dempster writes that: "The statistician with a specific interval of θ values $\hat{\theta} - c < \theta < \hat{\theta} + c$ has no more than the postdictive interpretation described above, namely, that $\hat{\theta} - c < \theta < \hat{\theta} + c$ or an improbable event occurred".

Similar statements are occasionally being made by other authors who treat them, more or less, as symbols or conventional summaries of my own way of thinking. I wish to make it clear that such summaries miss the point. This point is clearly stated by Professor Dempster on page 58 when he describes the defining property of confidence intervals.

The observations leading to the estimate $\hat{\theta}$ and the subsequent trouble in calculating $\hat{\theta}$ and c are made for some purpose. The only purpose that one can think of is to determine how to behave in the future with respect to the unknown value of θ. The specific interval $\hat{\theta} - c < \theta < \hat{\theta} + c$, calculated for a confidence coefficient α, selected by the statistician to suit his purposes, gives an unambiguous answer: *act on the assumption that the unknown θ lies between the limits indicated.* If the consumer asks why he should do so, the answer is: (i) if you behave that way, you will be right (approximately) in 100 α per cent of cases. Furthermore, if the theoretical problem of confidence intervals in the case of interest is properly solved, the advice on the frequency of right decision would be supplemented by (ii), a description of optimality of the procedure.

As is well known, the answer (i) can be verified empirically by Monte Carlo simulation of the experiments concerned. Also, the answer (ii) can be empirically illustrated by simulating not just the *optimal* but also some alternative confidence interval corresponding to the same α.

As to the postdictive interpretation: "either an improbable event occurred, or else...", it is obviously true, but is very far from my motivation to deduce confidence intervals. In fact, it is something that I did my best to put on the second plane of statistical thinking. At the spring meeting of the Royal Statistical Society of 1934, when I first tried to bring out the theory of confidence intervals, Arthur Bowley, then one of the leaders of the Society, formulated the statement "either an improbable event occurred or...". The reader may be amused by glancing at the discussion that followed.

(b) The question I would like Professor Dempster to answer is related to the above remarks. If I understood Dr. Dempster correctly, he is not quite appreciative of the properties of confidence intervals, whether in some sense optimal or otherwise, and prefers other methods of dealing with unknown value of θ under consideration. There is no doubt that, in cases where a frequentist prior is known, it is possible to deduce confidence intervals for θ that are, in a sense, narrower than those available without the prior. My question refers to cases where no frequentist prior distribution of θ is available.

In these cases, what are Professor Dempster's preferred methods of estimating θ? Also, are there any experiments possible that could illustrate the advantages of these methods somewhat in the manner of answers (i) and (ii) above?

REPLY

I am grateful to Dr. Good for his detailed comments. In the area of statistical inference, there must be little that anyone has thought about that Dr. Good has not written about, to the point that a computerized information retrieval system would be very helpful to scholars in the area. In a general way, he and I do look similarly at probability, but with considerable differences in emphasis. For example, I find that Dr. Good's classification of probabilities into Types I, II and III, while reflecting important conceptions, is not so clear cut as to bear the weight of elaborate theorizing, such as attempts to distinguish one kind of empirical Bayes from another. Likewise, I do not see the Bayes non-Bayes compromise as a reflection of virtue against human weakness, but rather I see postdictive reasoning as a prime mover towards new hypotheses free and clear of whatever Bayesian arguments may come with the train of new hypotheses. Again, I am astonished that Dr. Good would think that the problem of analyzing a real data-rich multivariate world would be perceptibly eased, except in very rare instances, by the single extra parameter of the stable laws. This, too, says a good deal about our differences in outlook. Finally, on a matter of terminology, I believe that honest confusion can often be recognized as legitimate while vagueness is rarely a virtue.

In reply to Dr. Cox, I can only say that I am interested in the lines of enquiry which he suggests. The intent in section 3 of my paper was mainly to place some new viewpoints on the table, and not to suggest that they solve all problems or that they need not be subjected to analytical scrutiny.

Dr. Neyman's preference for the inductive behavior interpretation of confidence statements over the postdictive interpretation is consistent with his writings on inference over many decades. My preference is for a philosophy of inference which incorporates belief ("believe as though θ were drawn from a certain distribution") as well as behavior ("act as though θ were in a certain interval"), and which allows belief to pertain to particular circumstances rather than to long run frequencies only. Even accepting a thoroughly behaviorist position, I find it a weakness of Neyman's view that he does not feel it necessary to come into line with the standard Bayesian principles of *rational* behavior. A critical weakness of the frequency theory, as applied to any specific action, is that the specific action can generally be embedded in different sequences with different frequency properties. Thus the probabilistic foundation for any specific action floats disturbingly. Feeling as I do, I find myself pushed back to the postdictive interpretation of confidence statements, which everyone agrees is unsatisfactory, but which seems to be the best available outside the predictive or Bayesian scheme.

The answer to Dr. Neyman's question is that I will sail as close to the Bayesian wind as possible. Often it will be necessary to admit a degree of confusion, but often there will turn out to be unanticipated Bayesian avenues open. Such was the major theme of the latter part of my paper.

PROBABILITY AND
INFORMATIVE INFERENCE
Henry E. Kyburg, Jr.
University of Rochester

1
Introduction

There are a lot of ways of skinning the statistical cat and there are a lot of ways of classifying those techniques. One way is to distinguish between those approaches which stem from an epistemological stance and those approaches which are pragmatic in character. Alan Birnbaum [1] has characterized the problem with which the epistemological approach has been concerned as the problem of *informative inference*. The kind of problem with which the pragmatic approach is concerned might be characterized as a decision problem; but if we so characterize it, we must beware of supposing that decision theory is concerned merely with the theory of decision. The words that statisticians use in their professional activities must be treated with circumspection.

To talk about accepting or rejecting hypotheses, for example, is *prima facie* to talk epistemologically; and yet in statistical literature to accept the hypothesis that the parameter θ is less than θ^* is often merely a fancy and roundabout way of saying that Mr. Doe should offer no more than \$36.52 for a certain bag of bolts. Neyman [2], for example, talks, or used to talk, about rejecting hypotheses; however, he made it very clear that this sort of talk was not to be interpreted epistemologically, but only as a kind of shorthand way of talking about inductive behavior — i.e., of making choices among courses of action on the basis of statistical evidence. On such a purely pragmatic approach, there seems to be no point in speaking of *accepting hypotheses* at all; it would be better and clearer simply to say what one means — to talk of buying the bag of bolts or not buying it.

As a general approach to statistics, however, this seems too narrow. It is, one might say, a function of engineering and not of science, to determine whether or not the evidence warrants buying a certain bag of bolts at a certain price. When it comes to general scientific hypotheses (e.g., that *f(x)* represents the distribution of weights in a certain species of fish, or that a binomial distribution with a parameter of ½ represents the behavior of a certain coin in a certain tossing machine) then the purely pragmatic, decision theoretic, approach has nothing to offer us. The classical statistical testing procedures, derived from this viewpoint, are nevertheless employed in psychological, social, medical, etc. sciences. They result in such statements as: *the null hypothesis is rejected at the 2% level of significance.* The import of such a statement is that such and such an empirical hypothesis is (for the time being, pending new evidence) acceptable. It would be a rare and foolish scientist who would claim that a certain hypothesis was acceptable, given certain evidence now, and that that hypothesis would continue to be acceptable regardless of what new evidence came to light. Acceptance is of course tentative, and of course subject to change. Nevertheless, the commonsense import of a scientific paper is generally that *A* causes *B,* or that the incidence of *X* is higher among *C*'s than among *D*'s, or that the distribution of the quantity *Y* among the members of *P* is approximately normal with mean *m* and standard deviation *s.* We use such hypotheses to guide our decisions in a wide variety of circumstances; we use them also as background information in further inquiries. Unless there are very good reasons to the contrary, it seems to me that we ought to try to accommodate this common sense view of the results of scientific inquiry in our theory of statistics. Therefore in what follows, I shall suppose that we do in fact accept some scientific hypotheses (if only tentatively) some of the time.

The main alternative to the hypothesis-testing approach is the Bayesian approach. Here again it is by no means clear from the way people talk and the words that they use whether they are concerned with informative inference or practical decision. The prior probabilities required for the Bayesian framework (and thus the framework itself) may be taken in three ways. First, the prior probabilities it requires may be taken as mere opinions — as reflecting psychological states of the decision maker or inferer. This is the way in which it is taken by Jeffrey [3], de Finetti [4], and Savage [5]. It might be thought that this would lead to a purely pragmatic statistics in which both the input and the output reflect the propensities of individuals to make certain sorts of decisions in certain sorts of circumstances. Although this approach in general eschews the acceptance of hypotheses, it may nevertheless be regarded as a species of informative inference, if we construe the outcome of the inference as, not the acceptance of a hypothesis, but the adoption of a posterior conditional probability function $P(H/E)$ characterizing epistemologically the relevant hypotheses. Second, the prior probabilities the Bayesian approach requires may be taken as reflecting purely logical relations: this is the hopeful approach of Harold Jeffreys [6] and Rudolf Carnap [7], and it leads to a similar epistemological approach to statistical inference. What one obtains, for example, is a posterior distribution which is represented by a function $F(x)$ which, relative to a total body of evidence legislatively characterizes one's epistemological attitude toward a certain parameter in a binomial distribution; i.e., one's degree of belief, under those circumstances, that the parameter lies between *a* and *b* should be $\int_{a}^{b} dF(x)$. The third way in which the Bayesian framework may be taken is that in which the prior probabilities are taken to represent known (or reasonably believed!) empirical frequencies. Here again we have an epistemological sort of statistics. But here we are definitely and explicitly committed to accepting at least tentatively at least some statistical hypotheses.

A brief tabular summary of this very sketchy characterization may come in handy later. We may organize it by conceptions of probability; techniques of inference; and results of inference. Under frequency or empirical conceptions of probability, we find non-Bayesian techniques of inference, except when there are empirical prior probabilities to be fed into Bayes's theorem. The results, so far as informative inference are concerned, are generally *reject H;* under the circumstances that there exist prior probabilities, there may emerge as a result a posterior distribution, *P(H/E).* Subjective conceptions of probability employ only Bayesian techniques of inference; the outcome is always a posterior distribution *P(H/E).* Logical conceptions of probability also employ Bayesian techniques of inference; one outcome is always a posterior distribution *P(H/E),* but on some views (such as those of Hintikka and his students [8]) it is sometimes possible to detach a hypothesis *H* from its evidence and to accept *H.* This depends not only on *P(H/E)* but also on the internal character of *H* and of *E.* A fourth conception of probability, and its implications for statistical inference, will be discussed later.

Table.

Concepts of Probability	Techniques of Inference	Form of Result

empirical	Non-Bayesian
	Bayesian based on
	  accepted statistical
	  knowledge
		reject *H*
		posterior
		  distribution

subjectivist	Bayesian	posterior distribution for *H*; an action
logical	Bayesian	posterior distribution for *H*

2

Probability

The crucial concept in statistical inference is that of probability. The interpretation I am about to offer is not new — I have offered it before [9] — but I hope to make clear how it leads to a unified approach to inference.

Probabilities concern statements. The probability of a certain statement is an interval (the endpoints of which constitute the upper and lower probabilities that some people refer to). The statement *S* has the probability *(a,b),* roughly speaking, if and only if *S* is known to be equivalent to a certain statement of the form,"$x \in z$," x is known to belong to an appropriate reference class y, x is a random member of y with respect to belonging to z, relative to all that we know about it, it is known that the measure of z in y is in the interval *(a,b),* and there is no proper subinterval of *(a,b)* into which this measure is known to fall.

84

This rough characterization refers constantly to what is known. The easiest way to formalize the reference to what is known, is to refer to the statements that embody that knowledge: thus instead of saying that it is known that x belongs to y, to say that the statement, "$x \in y$," belongs to a certain set of statements. It may be objected that one cannot expect universal agreement about what is known — i.e., about what statements may be taken to belong to our body of knowledge and what cannot. This difficulty may arise in two ways. First, different people have had different experiences, different educations, etc., so that what is known to one person may not be known to another. Second, given the same data and background knowledge, two people may still disagree on whether or not some new hypothesis is so well supported that it should be included in their collective body of knowledge. The first source of difficulty is not serious: one of the first requisites of the scientific enterprise is the willingness to share data; and one of the first consequences of the openness with which scientific investigation is conducted is that scientific· data are highly accurate and dependable. Thus so far as data is concerned—that is, the down to earth particulars that constitute our evidence in scientific investigation—there is no reason to suppose that two people may not share a common body of knowledge.

The second source of difficulty — that people may not agree on what hypotheses to accept — has in turn two sources. (Recall that I am speaking of intellectual acceptance; acceptance on the basis of what Birnbaum calls *informative inference*. I am not talking about acceptance in the sense of performing some particular action, in which, of course, ordinary utilities, which might differ from person to person, are involved). One source is that one person may have a higher standard of rigor than another: what is well enough demonstrated for one person, another may regard as insufficiently demonstrated, simply because the second person has higher standards in general. One person may be willing to accept a psychological hypothesis that is *not rejected at the .05 level,* while a fellow psychologist would not want to regard it as a part of his body of knowledge unless it were *not rejected at the .01 level.* Rejection levels provide a rather crude way of indicating the concept of rigor I have in mind, but they suffice to show how that there is no real difficulty here: although two people may not agree on what is an appropriate degree of rigor in a given context (i.e., what is an appropriate rejection level) they may be expected to agree on what is acceptable as knowledge at a *given* degree of rigor.

The other source of difficulty concerning the acceptance of hypotheses as parts of ingredients of our bodies of knowledge is that there simply is no general body of agreement on the logical question of what constitutes cogent evidence, even for a given degree of rigor. By calling it a logical question, I am of course presupposing that standards compelling essentially universal agreement can be given, and that these are standards of the same general character as those standards of deductive cogency that have so profoundly changed the character of mathematics in the past seventy-five years. Perhaps I am wrong about this. I hope not, for it is one of the more attractive myths about science that it sooner or later compels universal agreement. This has not happened yet, particularly when science touches on personal predilections, preferences, or prejudices. Some scientists, in the view of others, have a tendency to leap to conclusions, accepting hypotheses long before there is an adequate amount of evidence supporting them. Some scientists, in the view of others, tend stubbornly to resist new hypotheses, despite the overwhelming evidence in their favor. If such notions as *adequate evidence, overwhelming evidence* and the like are as objective as they pretend to be and as they should be, it should be possible to formalize them. At any rate, that is my goal.

Thus we shall construe a body of knowledge as a set of statements — much easier things to represent than propositions or facts, and of course we know a lot about sets. It should be observed that we do not know very much about bodies of knowledge, however. For example, it is not at all clear whether bodies of knowledge should be construed as being closed under conjunction; if s_1, and s_2, and ... and s_n are all statements that belong to the set of statements comprising our body of knowledge, will the statement consisting of the conjunction of these n statements also belong to it? I would answer *no* on the grounds that I feel practically certain that at least one of the things that I am practically certain is true, is in fact false. Others feel that bodies of knowledge should be circumscribed in such a way that this kind of thing does not happen on the grounds that they would not want the set of statements they accept to entail a contradiction. Such issues will not concern us here. We shall take the set of statements representing a body of knowledge as given, and make explicit any assumptions we make about it.

The first step is to introduce a distinction between statistical hypotheses and probability statements. It is the distinction, familiar to logicians, between object language and metalanguage. Statistical hypotheses are statements in the language of science — the object language — and concern the world out there: the stochastic behavior of dice, the distribution of telephone calls at an exchange, the behavior of radioactive substances, the distribution of shoesizes in an army, and the like. Probability statements, on the other hand, *reflect* our knowledge about that world out there: and that knowledge, as I have suggested, is best represented by the statements in the object language of science. To talk *about* that knowledge therefore requires that we talk about statements in the object language (and indeed about sets of such statements) and that requires what the logicians call a metalanguage. Fortunately we can for many purposes (and at the cost of a little sloppiness) get by with the first and most common technique for mentioning and talking about statements, namely, quotation.

Statistical hypotheses are among the most important ingredients of a body of knowledge. They represent our general scientific knowledge about the world — that knowledge which guides our future actions and choices. Not all statistical statements are equally significant and interesting, however. For example, let S be the set of throws of a certain die that land with the six up, and let T be the unit set consisting of the next throw. It is perfectly true that a plausible measure of the set of sixes among the union of S and T is one or close to one; but this statement cannot be of scientific interest to us. To avoid having to consider such irrelevant statements, let us restrict the set of statistical hypotheses in the following way.

We begin with a set of potentially acceptable reference classes R, subject to the following conditions. These conditions are designed to eliminate from our considerations not only such oddly inappropriate reference classes as the one just mentioned, but also perhaps what Fisher has referred to as "artificially constructed pivotal elements" (p. 120) which he took to lead to inconsistency through false applications of the fiducial method. The construction of R on the basis of a formal scientific language we leave for an appendix.

Condition 1. $x, y, \in R \supset x \cap y \in R$
Condition 2. $x, y \in R \supset x \times y \in R$
Condition 3. $x \in R \supset x^n \in R$ (where x^n is taken as the set of functions from the first n ordinal numbers to R).

We now consider a set of sets D_F. The subscript F suggests that these sets arise from the result of applying functions to objects belonging to reference classes. The construction is again relegated to an appendix. We suppose,

Condition 4. $R \subset D_F$
Condition 5. $x, y \in D_F \supset x \cap y \in D_F$
Condition 6. $x, y \in D_F \supset x \times y \in D_F$
Condition 7. $x \in D_F \supset x^n \in D_F$

In addition, we suppose what must *not* hold in general for R,

Condition 8. $x, y \in D_F \supset x \cup y \in D_F$
Condition 9. $x, y \in D_F \supset x - y \in D_F$

A statistical statement is a statement of the form "$H(x, y, p, q)$" where $x \in R$, $y \in D_F$, and $0 \leq p \leq q \leq 1$. The statement "$H(x, y, p, q)$" is true if and only if the proportion of the members of x that are members of y lies between the limits p and q, or when there is an agreed upon standard measure and x is infinite, when the measure of y in x lies between these limits.

When we talk about a statistical hypothesis belonging to K we shall mean a strongest statistical hypotheses; i.e., if "$H(x, y, p, q)$" is a statistical hypothesis belonging to K, then if "$H(x, y, r, s)$" is also a statement belonging to K, "$H(x, y, p, q)$" entails "$H(x, y, r, s)$".

Although we do not suppose that K is closed under logical deduction — i.e., contains all of the consequences of all of its elements — we shall suppose that its unit subsets are closed under logical deduction. Thus if $S \in K$ and S entails T, then $T \in K$.

It should be observed that statistical statements do not by any means embody all of our statistical knowledge. Even a distribution function or a density function would not, according to the definition, constitute a statistical statement. But of course any such function would constitute, so to speak, shorthand for an infinitely large set of statistical statements of the form specified earlier. Thus if $y(z)$ is a set of objects such that their weight is less than z pounds, and weight in x is distributed normally with unit variance and mean m then "$(z) (H(x, y(z), \Phi (z-m), \Phi (z-m)))$" is a statement, and a true one, but not a statistical statement as construed above, though it becomes one when we replace the variable z by a constant, e.g., "$H(x, y(5), \Phi(5-m), \Phi(5-m))$".

Now how about probability? I shall take probability to be an attribute of statements, as statistical hypotheses represent attributes of sets. Probability will be relative to a body of knowledge: that is, the probability of a given statement may have one value or another, depending on the body of knowledge relative to which that probability is taken. Furthermore, probabilities of statements will be directly related to statistical hypotheses in the following way: the probability of a statement S, relative to a body of knowledge K, will be the interval (p,q), under the following circumstances:

(a) S is known to be equivalent to a statement of the form, $a \in C$; formally, this is to say that the biconditional, "$S \equiv a \in C$" is to be a member of the set of statements K, where a and C are such that

(b) there exists a B, such that it is known that the measure of C in B is the interval (p,q)—i.e., such that "$H(B,C,p,q)$" is a statistical hypothesis in K about B and C—and such that

(c) a is (relative to K) a *random member* of B with respect to C. This condition is not to be construed (as is traditionally the case) as a new statistical hypothesis: it is *not* to be interpreted as asserting that a is as likely to be a as any other member of B, or that a is selected by a method which will produce each member of B equally often in the long run, or any such thing.

I shall refer to these three conditions as, respectively, the equivalence condition, the statistical condition, and the randomness condition. Each of them, and particularly the third, requires elucidation. The first condition stipulates that S is known to be equivalent to a statement of the form "$a \in C$," i.e., that the biconditional "$S \equiv a \in C$" must be a member of the set of statements constituting our body of knowledge. Of course if the biconditional is logically true — if S is *logically equivalent* to "$a \in C$"—then the biconditional will belong to our body of knowledge. (This follows from the unit set closure condition on K, for every statement entails every logical truth.) Such biconditionals may also belong to our body of knowledge on empirical grounds; they may even be quite accidental. Thus we may be interested in the probability that a green ball was drawn from urn number 1 and we may be informed (reliably) that a green ball was drawn from urn number 1 if and only if a black ball was drawn from urn number 2. If we happen to know nothing about the proportion of green balls in urn 1, but we do happen to know that half the balls in urn two are black, this will be a help. If we also happen to know that only ¼ of the balls in urn 1 are green, it will be less than no help. In general, of course, there will be a number of statements of the form, $a \in C$, that S is known to be equivalent to, and we require some means of picking out the (or an) appropriate one.

The second requirement spells out the necessity for the probability interval to be based solidly on *knowledge*, i.e., on a statistical hypothesis in the set of statements we take to represent our body of knowledge.

The third condition is the hardest to be explicit about. Indeed, a full treatment of this condition is beyond the scope of this paper, and possibly beyond that of its author. Nevertheless, there is one clear-cut circumstance under which we wish to deny that, relative to what we know, a is a random member of B with respect to C: namely, when we also know that a is a member of B^*, when we know that B^* is a subset of B, and when we have non-trivial statistical knowledge of B^* and C (i.e., knowledge not of the tautologous form "$H(B^*,C,0,1)$") that conflicts with the knowledge we have of B. This condition for the application of statistical knowledge has been pointed out explicitly by Reichenbach [11]; it is also mentioned explicitly by Fisher [12, p. 32]; and is indeed either implicit or explicit in most discussions of the *application* of statistical knowledge. This single condition for rejecting the randomness of an individual as a member of one class with respect to belonging to another will take us a long way, whether or not it can be made formally sufficient. At any rate, it will suffice for the developments to follow.

We must also define conditional probability. It cannot have the conventional definition, for there is, on the scheme being outlined here, no calculus of probability, other than that yielded as a pale reflection of the measure theoretic calculus of statistical hypotheses. The probability of a statement S, given a statement T, must be clearly be understood as the probability of S, relative to a body of knowledge just like the one we are employing now, except that T is added to it, and all of those individual statements inconsistent with T are deleted from it. The problem then is one of specifying how our body of knowledge is to be modified in the process of adding T to it.

Let "$Cn(X)$" denote the set of consequences of a statement X. We must first free the body of knowledge K of any statements which contradict T; that is, we must delete from K all those statements P, such that P entails the denial of T, or what is the same thing, such that T entails the denial of P. Or, finally, the set of all those P such that negation $- P \in Cn(t)$. Then we add the consequences of T (since these will not contradict any statements in K), and we have our new, conditional, body of knowledge K^*. We might, conceivably want to add more elements: thus we might want to add the consequents of all those conditionals that have consequences of T as antecedents to our new K — but we have enough to keep track of at the moment without considering such refinements, and in order to prepare the way for this we would have to first delete all the negations of such statements from K — and so on . . .

The body of knowledge that is like the body of knowledge K, but for containing T, we denote by $K + T$. According to the discussion just completed,
$$K + T = (K - \{P: \text{negation } P \in Cn(T)\}) \cup Cn(T).$$
The conditional probability of S, given T, relative to the body of knowledge K, is therefore simply the probability of S, relative to the body of knowledge $K + T$.

3

An Example

To fix our ideas, let us take a very careful look at a very simple problem in informative inference. Let us suppose that in a certain population R, either $\frac{1}{4}$ or $\frac{2}{3}$ of the individuals have a certain property P. Then we know that either $\frac{1}{4}$ or $\frac{2}{3}$ of R is P, i.e., we contain in our body of knowledge the statement "$H(R, P, \frac{1}{4}, \frac{1}{4})$ v $H(R, P, \frac{2}{3}, \frac{2}{3})$". This statement in turn entails the statistical hypothesis, "$H(R, P, \frac{1}{4}, \frac{2}{3})$"—in this over-simplified example we suppose we know no more about the measure of P in R than is embodied in the disjunction of the two statistical statements.

We wish to achieve greater knowledge of R and P; one way of expressing our desire is to say we wish to choose between the two statements; another way of expressing it is as the desire to test one hypothesis against the other; yet another way of expressing it is as the desire to use evidence to modify our present opinions through the application of Bayes's theorem. Let H_1 = "$H(R,P,\ ¼,¼)$" and let H_2 be the alternative. One way to get an enlightening comparison among all these ways of looking at the problem of increasing our knowledge, is to look at it as a fixed sample size problem, and to compute the long run errors involved in various rejection rules. A rejection rule will have the form: reject H_1 if there are more than n P's in the sample, where n may vary from zero to the sample size. The rule will be denoted by S_n. Each rule will have two errors associated with it: that of erroneously rejecting H_1, and that of erroneoulsy failing to reject H_1. Given the truth of H_1, we may calculate the long run frequency of error of the first kind, and similarly, we may calculate the long run frequency of error of the second kind. These errors are plotted against each other, for samples of size six and twelve, for all the relevant rules, on the accompanying graphs.

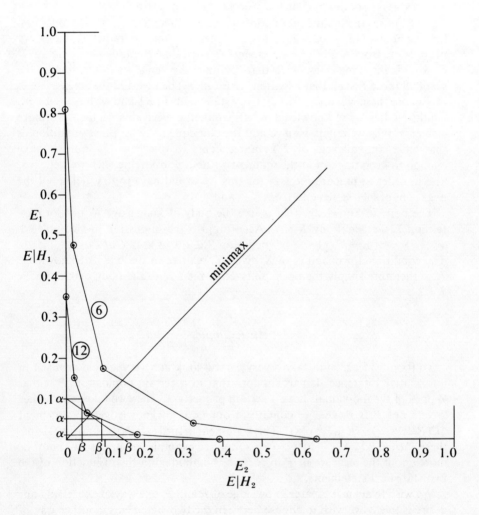

Figure 1.

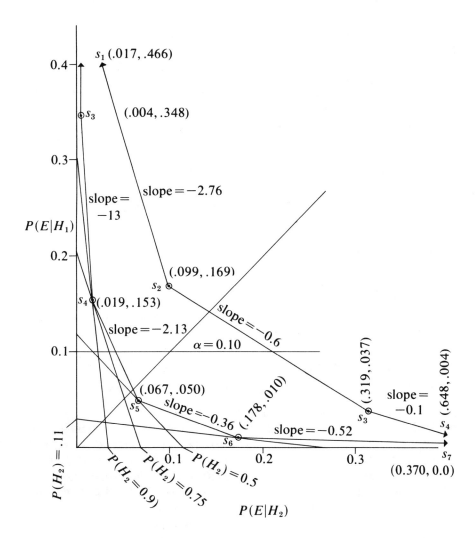

Figure 2.

It should be noticed that before sampling (at least under most circumstances) these long run conditional frequencies of error are genuine conditional probabilities of error. There is nothing in *K,* our body of knowledge, inconistant with H_1; if we add H_1 to our body of knowledge, we add its consequence, that the frequency with which sets of twelve *R*'s contain more than five *P*'s is (say) *f.* Assuming that the sample to be drawn is a random member of the set of all sets of 12 *P*'s with respect to containing more than five *P*'s, the conditional probability that it will contain more than five *P*'s is exactly *f.* If we are following a rule, S_5, which says that we should reject H_1 if and only if there are more than five *P*'s in the sample, we will be led to error (assuming the truth of H_1)if and only if there are more than five *P*'s in the sample; and thus the conditional probability of our being led into error is also exactly *f.*

On these figures we can represent various statistical philosophies. A Bayesian choice of strategy, for example, can be indicated graphically by the point at which a line of slope $-\dfrac{w}{1-w}$ first touches the convex set representing all possible mixed and pure strategies, where w is the prior probability of H_2 and $1-w$ the prior probability of H_1. A minimax choice of strategy attributing equal importance to both sorts of errors would be indicated by the point at which the boundary of the convex set is intersected by a 45° line through the origin. A test of level α and maximum power, is indicated by the intersection of the horizontal line representing an error of the first kind equal to α, and the boundary of the convex set.

Such a diagram can be used (and perhaps most often is used) to exhibit the consistency of these philosophies. There is no conflict among them, so the argument runs, because (for example) to choose S_i as a result of attributing a prior probability of w to H_2 is pragmatically indistinguishable from choosing S_i as a result of deciding to use the most powerful test of level α_w. More generally, let S_M be the strategy selected by the minimax method, and characterized by long run frequencies of errors of either kind equal to (say) $\epsilon : f(E_1) = f(E_2) = \epsilon$. Either S_M will be a pure strategy, or a mixed one. If it is a pure strategy, then there will exist a bundle of lines of negative slope touching the convex set at that point: say the bundle of lines whose slopes lie between $-m_1$ and $-m_2$. These lines correspond to prior probabilities of H_2 in the range between $\dfrac{m_1}{1+m_1}$ and $\dfrac{m_2}{1+m_2}$. In this case, to choose S_M on minimax grounds is just to take the prior probability of H_2 to be between these limits. If S_M is a mixed strategy, it is a mixture of two pure strategies $S_{M'}$ and $S_{M''}$, and there is just one straight line of negative slope (say, m) touching the convex set at S_M. This corresponds to assigning a prior probability of $m/(1+m)$ to H_2; under this assignment the best Bayesian strategy is $S_{M'}$ or $S_{M''}$ or any mixture of the two, including S_M.

In a similar fashion, to adopt the strategy S_M on the basis of a minimax philosophy amounts to the same thing in classical terms as deciding that a significance level of ϵ is satisfactory, and choosing the most powerful test of that size: it turns out to be S_M, and its power, of course, is $1-\epsilon$.

Or let us begin from the classical testing point of view, and choose the most powerful test of size α. This will lead to a strategy S_C. As before, we can find Bayesian arguments which will also justify the choice of S_C. Furthermore, S_C will be the minimax strategy for the case in which we attribute unequal importance to errors of the two sorts: namely in which we regard an error of the first kind as k times as important as an error of the second kind, where $k = \alpha/f(E_2) = f(E_1)/f(E_2)$.

Finally, if we begin with prior probabilities for H_1 and H_2, and are led to the strategy S_B, with error frequencies $f_B(E_1)$ and $f_B(E_2)$, this is the same as using a minimax philosophy, having decided that type I errors are $[f_B(E_1)]/[f_B(E_2)]$ times as important as type II errors, or as selecting the most powerful test of H_2 against H_1 of size $f_B(E_1)$.

So all the philosophies are compatible! So what is all the fuss about?

But the compatibility we have just seen is obviously an illusion. I do not refer to the fact that the arguments become vastly more far fetched when we deal with more complicated problems than that of choosing between two simple alternative hypotheses by means of a test based on a sample of fixed size. I refer to the fact that a person comes to a statistical problem with certain opinions (his degree of belief, for example, that H_2 is true), *and* a certain set of values (the relative importance he in fact attributes to the two sorts of error), *and* perhaps with a certain determination not to commit an error of the first kind more than alpha of the time. To be sure, one can attempt persuasive reconciliation in the case of conflict. Thus one might argue that the distinction between prior probabilities and values is an obscure one, and that a proper behavioral assessment of either will preclude conflict; or that they should be combined into seriosities (Braithwaite's expression, I believe [13]). But this just adds a new source of conflict: initially the scientist may now have one more source of conflict — who is to say that his seriosities will not conflict with both his prior probabilities and his values? One cannot define the conflict away; one can redefine the terms of the conflict so that they are compatible, but in doing so one may only have monkeyed with language, and not with the source of the conflict. It is as if one were to say, "Since you can only pursue one strategy, it must be the case that $S_C = S_M = S_B$; therefore there is no conflict in the philosophies, since they must all indicate the same strategy."

My solution to this potential conflict of philosophies is both trivial and obvious. (In keeping with the great tradition of Western Philosophy, it consists of offering yet another point of view.) We consider first the case where both errors are regarded as equally serious. For the Bayes prior probability w, of H_2, I put a pair of probability limits w_u and w_l which in extreme cases may be the same or may be 1 and 0. On the convex set of possible strategies this pair of numbers either determines a unique pure strategy, $S_{u,l}$ or it determines a pair of pure strategies S_u and S_l such that every strategy which is optimal for a prior probability of H_2 lying between w_u and w_l lies between S_u and S_l, and S_u is the optimal strategy for w_u and S_l is the optimal strategy for w_l. The choice of a strategy is constrained to be $S_{u,l}$, or to lie between S_l and S_u.

In the example previously considered, if the upper probability of H_2 is less than .93, and the lower probability of H_2 is greater than .348, that is, if $.93 > w_u > w_l > .348$, there is only one strategy open to us, and that is S_4; if $.348 > w_u > w_l > .265$, the only strategy open to us is S_5; if $.265 > w_u > w_l > .049$ the only strategy open to us is S_6. On the other hand, if we can do no better than pin down the probability of H_2 to the interval $(0.1, 0.9)$, then all three strategies and their mixtures satisfy the Bayesian constraints. If we have no knowledge at all on the basis of which to construct a probability for H_2 — a fairly rare situation, I believe — then the upper probability of H_2 is 1, the lower probability is 0, and the Bayesian constraints are empty.

I refer to S_u and S_l as constraints because on the view of probability I have offered, to attribute to H_2 the probability (w_u, w_l) is to say that we know (have adequate grounds to accept) that H_2 is equivalent to $a \in B$; that relative to what we know, a is a random member of C with respect to belonging to B; and that we *know* (have adequate grounds to accept) that the proportion of C that are B (or the measure of B in C) lies in the interval $[w_u, w_l]$, i.e., we have the statistical hypotheses $H(C, B, w_u, w_l)$ in our body of knowledge. Given any frequency of B in C, we can calculate the corresponding expectation of error in applying any strategy S. Given any frequency w of B in C, in the interval $[w_u, w_l]$ we can be sure that the expectation of error under any strategy S between S_u and S_l is less than the expectation of error under any strategy outside those limits. Should we choose a strategy S (say, a minimax strategy) that falls outside those limits, we would be following a strategy that has a demonstrably greater expectation of error than some alternative strategy.

The size of the test we want to use imposes an additional — and independent — constraint. If α is larger than the frequency of error of the first kind under strategy S_u, the constraint is empty; if α is smaller than the frequency of error of the first kind under strategy S_l it rules out any strategy as satisfactory; and if α lies between these limits, it corresponds to a strategy S_α, say, and the acceptable strategies are those between S_α and S_l. For example, in the problem we have been considering, if we know only that the probability of H_2 is the interval $(0.1, 0.9)$, the Bayesian constraints allow us to use S_4, S_6, or any strategy between. If in addition, we wish to make an error of the first kind no more than 10% of the time, the strategies open to us are S_5 and S_6 and any mixture of them, and any mixture of S_4 and S_5 that contains no more than roughly half instances of S_4.

These constraints may still leave a choice of strategies (as in the last mentioned case), and the minimax procedure may serve to pick out one of these remaining strategies — e.g., the minimax mixture of S_4 and S_5.

The case in which we do not regard all errors as equally serious introduces nothing new; the axes of the diagram become expected losses rather than frequencies of error.

4
Post Experiment Considerations

There are instances in which consideration of various strategies in terms of their long term frequencies of errors (or expected losses from errors) is appropriate and relevant, in the design of an experiment, for example, or in the design of a testing procedure to be applied over and over again in industrial quality control. In general, however, in the case of informative inference, these considerations cease to be relevant after we have performed the experiment and obtained the data.

Let us suppose that a manufacturer wants to test, repeatedly, whether certain large groups of items contain $\frac{2}{3}$ or $\frac{1}{4}$ oversize. (We need not suppose that *oversize* is a defect; we leave the utilities to one side, in view of the fact that it is clear enough, in principle, how to work them in.) A fast talking statistician convinces him that he should use a minimax test, which in the example we have been at such pains to construct, consists of rejecting the hypothesis $p = \frac{1}{4}$ if a chance event of probability $\frac{8}{73}$ occurs and more than 4 of 12 sampled gadgets are oversize, or if a chance event of probability $\frac{65}{73}$ occurs and more than 5 out of 12 of our sampled gadgets are oversize. Put more simply: reject $p = \frac{1}{4}$ if more than five are oversize; reject $p = \frac{2}{3}$ if less than five are oversized, and if exactly 5 are oversized, reject $p = \frac{1}{4}$ if and only if a chance event of probability $\frac{8}{73}$ occurs.

The first large lot comes out of the machine. We select a sample of 12 items (you may select them by random numbers, if you wish; we will get to randomization in a minute). Seven of them are oversized. The chance event of probability $^{65}/_{73}$ occurs. The minimax rule thus tells us to reject $p = \frac{1}{4}$. It also tells us that by following this rule, in the long run we will be in error about 6 times out of a hundred.

What Alan Birnbaum calls the *Principle of Conditionality,* however, tells us that his error figure is utterly irrelevant: the possible outcomes of experiments which might have been performed, but were not, cannot be relevant to the evidential import of those experiments that were performed.

In view of the fact that in the auxiliary experiment the most probable event occurred, the rule we were in fact following in sample was that we were going to reject $p = \frac{1}{4}$ provided more than 5 of the gadgets in our sample were oversized. *This* experiment is characterized by error frequencies of 0.050 if $p = \frac{1}{4}$ and .067 if $p = \frac{2}{3}$. We can thus say that the interval (.050, .067) characterizes the experiment better than the number .061, once we have observed the outcome of our randomizing device.

We can explain and illustrate this by reference to the conception of probability characterized earlier: before performing the complex experiment, the instance that confronts us is an application of a rule with an error frequency of .061; it is, relative to what we know before we perform the experiment, a random member of the set of applications of that rule. The probability that it will lead us to error is therefore .061, exactly. Once we have performed the auxiliary experiment, however, it is unreasonable to regard the instance that confronts us as a random application of the rule.

Furthermore, in this experiment that we in fact performed, exactly seven of the objects were oversized. Given that we have no information about the prior probabilities of the two alternative hypotheses, this leads to no further specification of the probability of error: that is, the conditional probabiltity of H_1 $(p = \frac{1}{4})$, given 7 oversized objects in the sample, the conditional probability of error is the entire interval $(0,1)$. This is no help at all. Therefore, under these circumstances, the appropriate *probability* of error is just that given by the *test* we have used, namely $(0.050, 0.067)$. We may, of course, recognize that this application of this test belongs to that subclass of applications of the test in which the result is to direct us to reject H_1; it belongs also to that even finer subclass of applications of the test in which seven of the tested objects were oversized. In neither of these subclasses, however, do we have *any* (non-trivial) knowledge concerning the frequency with which the test will lead us astray.

The situation changes abruptly and dramatically as soon as we have any limitations at all on the prior probability of the two hypotheses. As soon as the prior probability of H_1 is no longer the whole interval $(0,1)$ it is possible to compute non-trivial posterior probabilities for the two hypotheses. Given, for example, that we have found 7 oversized objects in our sample of 12, the probability of error, in the application of any rule which directs us to reject H_1 under these circumstances is just the probability of H_1, given that evidence. The following table gives some examples:

Table.

prior probability		strategy followed	P (Error)	P (Error, given observation of n oversize objects)				
of H_1	of H_2			$n = 3$	$n = 5$	$n = 6$	$n = 7$	$n = 9$
(.75,.90)	(.25,.10)	S_6 *	(.027,.052)	$P(H_2)=$ (.126,.303)	$P(H_2)=$ (.049,.133)	$P(H_2)=$ (.232,.475)	$P(H_1)=$ (.158,.278)	$P(H_1)=$ (.0043,.0127)
(.4,.7)	(.6,.3)	S_5 *	(.055,.062)	$P(H_2)=$ (.052,.162)	$P(H_2)=$ (.164,.407)	$P(H_1)=$ (.137,.355)	$P(H_1)=$ (.026,.086)	$P(H_1)=$ (.0006,.0022)
(.1,.3)	(.9,.7)	S_4 *	(.032,.059)	$P(H_2)=$ (.028,.114)	$P(H_1)=$ (.209,.481)	$P(H_1)=$ (.038,.134)	$P(H_1)=$ (.007,.026)	$P(H_1)=$ (.0002,.0006)
(.1,.9)	(.1,.9)	S_5 *	(.052,.065)	$P(H_2)=$ (0013,.1025)	$P(H_2)=$ (.049,.805)	$P(H_1)=$ (.040,.764)	$P(H_1)=$ (.007,.358)	(.0002,.0128)
(0,1)	(0,1)	S_6	(.010,.178)	(0,1)				
(0,1)	(0,1)	S_5	(.050,.067)	(0,1)				
(0,1)	(0,1)	S_4	(.153,.019)	(0,1)				

* When prior probability of hypotheses is given, probability of a given hypothesis, given observation, is independent of strategy.

In order to see just what has been going on in terms of randomness and our body of knowledge, let us formalize our references slightly. Let a, B, and C, as before, be the three objects such that we know that H_1 is equivalent to $a \in B$, and relative to what we know, a is a random member of C with respect to B.
Let S be the set of applications of the mixed strategy.
Let S_i now denote the set of applications of strategy S_i, regarded as a subset of S, as well as denoting the strategy itself.
Let R^{12} be the set of twelve-membered subsets of R, and let iP be the subset of R^{12}, each element of which contains exactly i members of P.
Let $E(S_i)$ be the subset of S_i in which we are led to error.
Let jS_i be the application of the strategy S_i to a sample containing j members of P.
Let s be the particular application we are making in this instance; let r be the sample we have drawn.
Let E1 be the statement that an error of the first kind is committed, and E2 the statement than an error of the second kind is committed.

Stage 1. We know only $(a, s, r) \in C \times S \times R^{12}$

$s \in E(S) \equiv (a, s, r) \in C \times E(s) \times R^{12}$

$H(S, E(S), .061, .061)$

$H(C \times S \times R^{12}, C \times E(S) \times R^{12}, .061, .061)$

(a, s, r) is a *random member* of $C \times S \times R^{12}$ with respect to $C \times E(S) \times R^{12}$

The probability of error before testing is

$P(H_1) = (0, 1) \quad P(H_1) = (.4, .7) \quad P(H_1) = (.1, .9)$
$P(\text{Error}) = (.061, .061) \quad (.061, .061) \quad (.061, .061)$

Stage 2. We know, in addition, that $s \in S_5$.

$$s \in E(S) \equiv s \in E(S_5)$$
$$\equiv (a, s, r) \in C \times E(S_5) \times R^{12}$$

$H(S_5, E(S_5), .050, .067)$
$H(C \times S_5 \times R^{12}, C \times E(S_5) \times R^{12}, .50, .067)$
$H(C, B, .4, .7) \Rightarrow H(C \times S_5 \times R^{12}, C \times E(S_5) \times R^{12}, .055, .060)$
$H(C, B, .1, .9) \Rightarrow H(C \times S_5 \times R^{12}, C \times E(S_5) \times R^{12}, .052, .065)$
(a, s, r) is a random member of $C \times S_5 \times R^{12}$ with respect to
$\quad C \times E(S_5) \times R^{12}$.

The probability of error at this stage of testing is

$P(H_1)$	$= (0, 1)$	$(.4, .7)$	$(.1, .9)$
$P(\text{error})$	$= (.050, .067)$	$(.055, .060)$	$(.052, .065)$

Stage 3. We know, in addition, that $r \in 7P$

$$s \in E(S) \equiv H_1$$
$$\equiv s \in E(S_5)$$
$$\equiv a \in B$$
$$\equiv (a, s, r) \in B \times S \times R^{12}$$
$$\equiv (a, s, r) \in C \times E(S_5) \times R^{12}$$

$H(C, B, .4, .7) \Rightarrow H(C \times 7P, B \times 7P, .086, .026)$
$H(C, B, .1, .9) \Rightarrow H(C \times 7P, B \times 7P, .007, .358)$
$H(C, B, 0, 1) \Rightarrow H(C \times 7P, B \times 7P, 0, 1)$
$H(C, B, .4, .7) \Rightarrow H(C \times S_5 \times 7P, B \times S_5 \times 7P, .086, .026)$
$H(C, B, .1, .9) \Rightarrow H(C \times S_5 \times 7P, B \times S_5 \times 7P, .007, .358)$
$H(C, B, 0, 1) \Rightarrow H(C \times S_5 \times 7P, B \times S_5 \times 7P, .0, 1)$
$H(C \times S_5 \times R^{12}, C \times E(S_5) \times R^{12}, .050, .067)$

At this third stage, two possibilities emerge:
(a) if $P(H_1) = (0, 1)$, (a, r, s) is a random member of $C \times S_5 \times R^{12}$
    with respect to $C \times E(S_5) \times R^{12}$
(b) otherwise, (a, r, s) is a random member of $C \times S_5 \times 7P$ with
    respect to $B \times S_5 \times 7P$.

The probability of error at this last stage of testing is therefore

$P(H_1)$	$= (0, 1)$	$(.4, .7)$	$(.1, .9)$
$P(\text{error})$	$= (.050, .067)$	$(.026, .086)$	$(.007, .358)$

This state of affairs may seem to have an air of paradox: where our information is least, (i.e., where all we can say of the probability of the hypothesis before testing is that it is the whole interval $(0,1)$) our probability assessment after testing is the most precise. When we do know something about the prior probability of the hypothesis, then our proterior probability is less precise.

But it is the appearance only of paradox, as may be illustrated by a hypothetical example. Suppose that we have a classical population of bags of balls, in each of which there are black and white balls. We might know that in the whole population exactly half of the balls are white, and at the same time have some what more limited information about the proportion of white and black balls in various bags. Thus of number seven we might know that between .5 and .7 of the balls are white. Given the knowledge merely that a ball has been chosen from the whole population, it is surely correct to take the probability that it is white to be ½. Given the further knowledge that it was drawn from bag number seven, the appropriate probability is equally clearly the whole interval (.5,.7). Our increase in knowledge has here led to a decrease in the precision of our probability. But that is no paradox and no excuse for not using what knowledge we have. The principle of total evidence which prohibits us from throwing away information to achieve more precise probabilities admits of no non-circular justification; it is simply a principle of rationality.

In general, of course, things go the other way: the more information we have about a population (e.g., as obtained by drawing samples from it) the more precise will our knowledge of its composition be, and correspondingly, the more precise will be our probability knowledge concerning future samples drawn from it.

5

Random Observations

Ian Hacking, in a number of publications, has called attention to the difference between the relevant characteristics of testing strategies before we make a test, and those that are relevant after we have observed the result of a test. He offers a number of examples to support this point; one of them is the following (from *British Journal for the Philosophy of Science,* 15, 1964-65):

	$P(E_1)$	$P(E_2)$	$P(E_3)$	$P(E_4)$
H	0	.01	.01	.98
J	.01	.01	.97	.01

R: reject H if and only if E_3 occurs
S: reject H if and only if E_1 or E_2 occurs.

We are testing H against the alternative J. $P(E_1)$ represents a long-run frequency, or chance, or empirical probability. Rule R and rule S are two alternative rules we are considering; they both have size .01. Rule R has power .97, however, while rule S has a power of only .02. It is clear that before a test, Rule R is to be preferred.

But after we have taken our sample, and observed it to be E_1, we *know* that rule R has led us astray. After the test, it is obvious that rule S is preferable.

In terms of the concept of probability set forth earlier, the explanation is straightforward. Before taking the sample, the probability that rule R will lead us astray is (.01,.03), provided that our application of the rule is a random member of its applications with respect to yielding the truth. The probability that rule S will lead us astray is (0.01,0.98). The former seems preferable to the latter, since the upper bound on error is so much smaller. But after we have observed E_1, the application of either rule is no longer a random member of the set of all its applications with respect to yielding the truth. It is a member of a special subset of those applications: namely, applications in which the experimental result is E_1. And in this special subset of applications, the probability that rule R will yield the truth is (0,0) and the probability that rule S will yield the truth is (1,1).

The likelihood principle, which according to Savage *(The Foundations of*

Statistical Inference, [14] p. 17) "flows directly from Bayes's Theorem and the concept of subjective probability" asserts that "the evidential meaning of any experimental outcome is fully characterized by the likelihood function determined by the observations, without further reference to the structure of the experiment". [15] Savage offers the following example: if λ is the frequency of red-eyed flies in a population, and you have observed six out of a hundred to have red eyes, the evidential import of this observation, according to the likelihood principle, is embodied in the function $(\lambda)^6(1-\lambda)^{94}$. It is embodied in this function whether you have chosen a sample of 100 flies and counted the number that have red eyes, or counted flies until you have accumulated six that have red eyes, or counted flies until you have found 94 that do not have red eyes, or however you have chosen it, so long as the probability of the result is proportional to the function mentioned. This is in flat contradiction to the classical testing approach, according to which these are three quite different experiments.

If one adopts the likelihood principle, optional stopping is never relevant. If one adopts a classical point of view it is always relevant. According to the view adopted here, optional stopping is sometimes evidentially relevant, and sometimes not, according to what we know, and according to whether or not what we know is such as to prevent the sample from being a random one in the appropriate sense.

Thus in testing a sharp null hypothesis, it is in principle possible, given an arbitrary level of significance, to continue testing until that hypothesis is rejected at that level of significance. Such a rejection is obviously irrelevant. Why? Because if that is what a person sets out to do, his sample is clearly not a random one with respect to yielding truth under his rule of rejection. It belongs rather to a very special subclass of samples, in which the proportion of the time that the rule yields truth is known to fall into exactly the same interval as that representing the prior probability that the null hypothesis is false. On the other hand, if an investigator stops testing because he is tired, or because he has run out of money, or because he is sufficiently convinced of the falsity of the null hypothesis, or because it is five o'clock, it is perfectly possible that his arbitrarily limited sample is random with respect to yielding the truth under his rule of rejection, and therefore that his significance level does represent a genuine epistemological probability.

The attitude toward randomization that follows from the view of probability adopted here is, naturally, very similar to that of the subjectivist. A sample which is obtained with the aid of a randomizing device may or may not be random in the sense required for inference. If one is doing an agricultural experiment, and finds as a result of one's randomization that the treated plots run in orderly diagonals across the field, one will do the randomization over again. Observe that we do not *know* that such an experiment will be misleading, or even have any reason to believe it will be misleading. Thus it is still a little difficult, on purely subjectivistic grounds, to find a reason for rejecting the cogency for such an experiment. But on the view presented here, such an experiment is always imperiled by the possibility that one *might* find, for example, that fertility gradients ran diagonally across the field, and *then,* relative to this new and expanded body of knowledge, the sample would not be random in the required sense. It is to guard as best we can against such future eventualities that we randomize; and it is to guard against present defects in our experimental design that we appropriately take such randomization to be subject to check against the body of knowledge we have now.

In short, the point of view adopted here argues that neither long-run frequencies, nor the slavish adherence to mechanical rules, nor unsubstantiated opinion, should be allowed to interfere with the reasoned application of factual knowledge to the particular instances of inference with which we are concerned.

Notes

1. Alan Birnbaum, "On the Foundations of Statistical Inference," *Annals of Mathematical Statistics,* 32, pp. 414-435, 1961.
2. Jerzy Neyman, "The Problem of Inductive Inference," *Communications on Pure and Applied Mathematics,* 8, pp. 13-45, 1955.
3. Richard Jeffrey, *The Logic of Decision,* New York, McGraw Hill, 1965.
4. Bruno de Finetti, "Foresight: Its Logical Laws, Its Subjective Sources," in Kyburg and Smokler (eds.), *Studies in Subjective Probability,* New York, John Wiley & Sons, 1964.
5. L.J. Savage, *The Foundations of Statistics,* New Jersey, Prentice Hall, 1954.
6. Harold Jeffreys, *Scientific Inference,* Cambridge, Cambridge University Press, 1957.
7. Rudolf Carnap, *The Logical Foundations of Probability,* Chicago, University of Chicago Press, 1962.
8. Jaakko Hintikka and Risto Hilpinen, "Knowledge, Acceptance, and Inductive Logic," in Hintikka and Suppes, *Aspects of Inductive Logic*, Amsterdam, North Holland Publishing Company, pp. 1-20, 1966.
9. *Probability and the Logic of Rational Belief,* Middletown, Wesleyan University Press, 1961.
10. This question, often formulated in terms of a lottery, was the subject of a conference last year at the University of Pennsylvania.
11. Hans Riechenbach, *The Theory of Probability,* California, University of California at Berkeley, 1949.
12. R.A. Fisher, *Statistical Methods and Scientific Inference,* New York, Hafner Publishing Company, 1956.
13. R.B. Braithwaite, "The Role of Values in Scientific Inference," in Kyburg and Nagel, *Induction,* Middletown, Wesleyan University Press, pp. 180-193, 1963.
14. Leonard J. Savage, *The Foundations of Statistical Inference,* New York, John Wiley and Sons, 1962.
15. Alan Birnbaum, "Intrinsic Confidence Methods," *Bulletin de l'institut International de Statistique,* p. 2, 1961.

APPENDIX

The reference classes among which we may choose are determined by the language we use, not by the nature of the world. We are thus led to specify a certain set of *expressions* — reference terms — which may appropriately occupy the second place in the metalinguistic relation Ran (a, B, C, K), just as subsets of the set of *statements* are referred to by K, and individual terms (names and definite descriptions) are referred to by a. That is: an appropriate substituend for 'a' denotes an individual term or a set naming term, an appropriate substituend for 'B' or 'C' denotes a set naming term; and an appropriate substituend for 'K' denotes a set of statements.

No problem appears to arise in connection with a; we may let a be any term of our language.

The problem that arises in connection with K is just the problem of induction and acceptance; it is the problem of determining the criteria — perhaps relative to a number of parameters — that serve to authorize the acceptance of empirical statements. It is not my purpose to examine such criteria here. We assume that they exist, or at least that in a given situation agreement can be reached somehow concerning what is to be regarded as background knowledge.

The set K of statements representing our body of knowledge may contain statements of the form $[H(\alpha, \beta) = p]$ or $[p_1 < H(\alpha, \beta) \le p_2]$ and the like. These are the statements representing our statistical knowledge. We stipulate — as a meaning postulate, perhaps — that if α is finite, $[H(\alpha, \beta)]$ denotes the ratio of the cardinality of the set denoted by $[\alpha \cap \beta]$ to the cardinality of the set denoted by α.

B and C, however, offer problems; we can refer to all sorts of classes in an ordinary language that are totally inappropriate as possible reference classes: for example, the class consisting of tosses of this coin landing heads union the unit set of the next toss. We must therefore somehow restrict the set of possible substituends for 'B' and perhaps 'C'.

Let us first consider a language containing a finite number of one place first order predicate expressions, a finite or denumerable number of O-place operation expressions (names), a two-place operation H taking pairs of sets into real numbers, and, for logical machinery, membership and identity.

Let $\Pi = \{$'P_0', 'P_1', $\ldots$, 'P_m'$\}$ denote the set of primitive predicates. We define $A\Pi$, the set of atomic predicates:

Definition. $A\Pi = \{\phi : \bigvee_{\psi} (\psi \in \Pi \wedge (\phi = \Psi \vee \phi = [\neg \psi]))\}$.

We define the set of conjunctive propositional functions, $CA\Pi$, as follows:

Definition. $CA\Pi = \{\phi : \bigvee_{n \in \omega} \bigvee_{\alpha_0, \ldots, \alpha_{n-1}} \bigvee_{\phi_0, \ldots, \phi_{n-1} \psi} (\bigwedge_{i < n} (\phi_i \in A\Pi \wedge$
$\alpha_i \in \text{Vbl}) \wedge \psi \in \text{LFmla} \wedge \phi = [\psi \wedge \phi_0 \alpha_0 \wedge \ldots \wedge \phi_{n-1} \alpha_{n-1})\}$.

The need for clause ψ will be apparent shortly. In powerful languages the condition that ψ contain no non-logical terms may not be sufficient to avoid sophisticated troubles, but we suppose the intent to be clear.

Let Q be the set of strings of quantifiers of our language (existential or universal), including the empty string.

The set or reference class expressions, R, may now be defined:

Definition. $R = \{\phi : \bigvee_{\psi} \bigvee_{n < \omega} \bigvee_{\alpha_0 \ldots \alpha_{n-1}} \bigvee_q (\psi \in CA\Pi \wedge q \in Q \wedge \bigwedge_{i < n} \alpha_i$ is free

in $q\psi \wedge \bigwedge_{\beta} (\beta$ is free in $q\psi \rightarrow \bigvee_{i < n} \beta = \alpha_i) \wedge \phi = [\{\langle \alpha_0, \ldots, \alpha_{n-1} \rangle : q\psi\}])\}$.

We observe that R satisfies the closure conditions mentioned in the text:

Theorem 1. $\alpha \in R \wedge \beta \in R \rightarrow \bigvee_{\gamma} (\gamma \in R \wedge \vdash [\gamma = \alpha \cap \beta])$

let $\quad \alpha = [\{\langle \alpha_0, \alpha_1, \ldots \alpha_{k-1} \rangle : q(\psi \wedge \phi_0 \alpha_0 \wedge \ldots \wedge \phi_{k-1} \alpha_{k-1})\}]$

let $\quad \beta = [\{\langle \alpha'_0, \alpha'_1, \ldots \alpha'_{j-1} \rangle : q'(\psi' \wedge \phi'_0 \alpha'_0 \wedge \ldots \wedge \phi'_{j-1} \alpha'_{j-1})\}]$

Case I: $j \neq k$ $\vdash [\alpha \cap \beta = 0]$

Since $\vdash "0 = \{x : P_1 x \wedge \neg P_1 x\}"$,

$\vdash [\alpha \cap \beta = \{x : P_1 x \wedge \neg P_1 x\}]$,

$\vdash$ where $"\{x : P_1 x \wedge \neg P_1 x\}" \epsilon R$

Case II: $j = k$

Rewrite the bound variables of β in such a way that the variables of q' are replaced by completely new variables foreign also to α, to obtain $q*$ and $\psi*$, and replace $\alpha'_0 \ldots \alpha'_{j-1}$ by $\alpha_0 \ldots \alpha_{k-1}$ respectively. Then

$\vdash [\alpha \cap \beta = \{\langle \alpha_0 \ldots \alpha_{k-1} \rangle : q(\psi \wedge \phi_0 \alpha_0 \wedge \ldots \wedge \phi_{k-1} \alpha_{k-1}\} \cap$

$$\{\langle \alpha'_0 \ldots \alpha'_{j-1} \rangle : q'(\psi' \wedge \phi'_0 \alpha'_0 \ldots \phi'_{j-1} \alpha'_{j-1})\}]$$

$\vdash [\alpha \cap \beta = \{\langle \alpha_0 \ldots \alpha_{k-1} \rangle : q(\psi \wedge \phi_0 \alpha_0 \wedge \ldots \wedge \phi_{k-1} \alpha_{k-1})$

$$\wedge \, q*\, (\psi* \wedge \phi'_0 \alpha_0 \ldots \wedge \phi'_{j-1} \alpha_{k-1})\}]$$

Since none of the variables of $q*$ are free in the left hand conjunct, and since conjunction is associative,

$\vdash [\alpha \cap \beta = \{\langle \alpha_0 \ldots \alpha_{k-1} \rangle$

$$: qq*\,((\psi \wedge \psi*) \wedge \phi_0 \alpha_0 \wedge \ldots \wedge \phi_{k-1} \alpha_{k-1} \wedge \phi'_0 \alpha_0 \wedge \ldots)\}]$$

But the right hand since of this identity is an expression belonging to R.

Theorem 2. $\alpha \epsilon R \wedge \beta \epsilon R \to \underset{\gamma}{V} (\gamma \epsilon R \wedge \vdash [\gamma = \alpha \times \beta])$

let $\alpha = [\{\langle \alpha_0 \ldots \alpha_{k-1} \rangle : q(\psi \wedge \phi_0 \alpha_0 \wedge \ldots \wedge \phi_{k-1} \alpha_{k-1})\}]$

let $\beta = [\{\langle \alpha'_0 \ldots \alpha'_{j-1} \rangle : q'(\psi' \wedge \phi'_0 \alpha'_0 \wedge \ldots \wedge \phi'_{j-1} \alpha'_{j-1})\}]$

Rewrite the variables of α and β to be all distinct.

$\vdash [\alpha \times \beta = \{\langle \beta_1, \beta_2 \rangle : \underset{\substack{\alpha_0 \ldots \alpha_{k-1} \\ \alpha'_0 \ldots \alpha'_{j-1}}}{V} qq'\,(\beta_1 = \langle \alpha_0 \ldots \alpha_{k-1} \rangle$

$$\wedge \beta_2 = \langle \alpha'_0 \ldots \alpha'_{j-1} \rangle \wedge \psi \wedge \psi' \wedge \phi_0 \alpha_0 \wedge \ldots \wedge \phi'_{j-1} \alpha'_{j-1})\}]$$

The right hand side is clearly a member of R.

Theorem 3. $\alpha \epsilon R \to \underset{n < \omega}{\wedge} \underset{\gamma}{V} (\gamma \epsilon R \wedge \vdash [\gamma = \alpha''])$

let $\alpha = [\{\langle \alpha_0 \ldots \alpha_{n-1} \rangle : q(\psi \wedge \phi_0 \alpha_0 \wedge \ldots \wedge \phi_{k-1} \alpha_{k-1})\}]$

Choose $n \cdot (k + r)$ new variables, where r is the number of bound variables in $q\psi$. Generate n expressions of the form

$$q_i(\psi_i \wedge \phi_0 \alpha_{i, 0} \wedge \ldots \wedge \phi_{k-1} \alpha_{i, k-1})$$

which have no variables in common.

$\vdash [\alpha'' = \{\langle \beta_0 \ldots \beta_{n-1} \rangle : q_0 \ldots q_{n-1} \underset{\alpha_{0, 0} \ldots \alpha_{n-1, k-1}}{V} \quad (\underset{i < n}{\wedge} \beta_i =$

$\langle \alpha_{i, 0}, \alpha_{i, 1}, \ldots \alpha_{i, k-1} \rangle \wedge \psi_0 \wedge \ldots \wedge \psi_{n-1} \wedge \phi_0 \alpha_{0, 0} \wedge \ldots \wedge \phi_{k-1} \alpha_{n-1, k-1})\}]$

The right hand side belongs to R.

Reference class expressions as just defined are the appropriate substituends for 'B' in 'Ran (a, B, C, K)'. We can be more generous regarding substituends for 'C'. Let us call the set of appropriate substituends for 'C' the set of Predicate Class Expressions, P, to be confused neither with the set of primitive predicates, nor with the general set of propositional functions.

Definition. $P = \{\phi : \underset{\psi_0 \ldots \psi_{k-1}}{\vee} \underset{q}{\vee} \underset{\alpha_0 \ldots \alpha_{n-1}}{\vee} (\underset{i<k}{\wedge} \psi_i \in CA \, \Pi \wedge q \in Q$

$\wedge \underset{i<n}{\wedge} \alpha_i$ is free in $q(\psi_0 \vee \ldots \vee \psi_{k-1}) \wedge \underset{\beta}{\wedge} (\beta$ is free in $q(\psi_0 \vee \ldots \vee \psi_{k-1}) \rightarrow$

$\underset{i<n}{\vee} \beta = \alpha_i) \wedge \phi = [\{\langle \alpha_0 \ldots \alpha_{n-1} \rangle : q(\psi_0 \vee \psi_1 \vee \ldots \vee \psi_{k-1})\}]]\}$

The set of functions referred to in the text is just the set of characteristic functions of the set of sets denoted by expressions in the set P. The domain of the functions is as required a set of sets. Conditions $IV - IX$ of the text become:

T-IV $R \subseteq P$

T-V $\alpha \in P \wedge \beta \in P \rightarrow \underset{\gamma}{\vee} (\gamma \in P \wedge \vdash [\gamma = \alpha \cap \beta])$

T-VI $\alpha \in P \wedge \beta \in P \rightarrow \underset{\gamma}{\vee} (\gamma \in P \wedge \vdash [\gamma = \alpha \times \beta])$

T-VII $\alpha \in P \rightarrow \underset{n<\omega \gamma}{\wedge} \vee (\gamma \in P \wedge \vdash [\gamma = \alpha^n])$

T-VIII $\alpha \in P \wedge \beta \in P \rightarrow \underset{\gamma}{\vee} (\gamma \in P \wedge \vdash [\gamma = \alpha \cup \beta])$

T-IX $\alpha \in P \wedge \beta \in P \rightarrow \underset{\gamma}{\vee} (\gamma \in P \wedge \vdash [\gamma \in \alpha - \beta])$

COMMENTS

I.D.J. Bross:

As some of you know, I take the rather disturbing view that the languages and sublanguages that we speak have a profound—I might even say controlling—influence on our habitual speech and behaviour. Moreover the more highly specialized sublanguage we use, the more closely we are prisoners of our own jargons. Extremely specialized artificial languages — such as those used nowadays in logic — lead to speech and behaviour patterns which, in terms of ordinary language, are extreme and even bizarre.

Let me illustrate my point with a sample from this text. We are told that (page 83), "When it comes to general scientific hypotheses then the purely pragmatic, decision theoretic, approach has nothing to offer us. The classical statistical testing procedures, derived from this viewpoint, are nevertheless employed in psychological, social, medical, etc. sciences".

The statistical jargons used by working statisticians and the scientific jargons used by working scientists are fact-limited languages—that is, the statements tend to be restricted to those which are in fairly good agreement with the facts. Many of the jargons spoken at this meeting belong to a different language family where there is no such factual limitation and where statements flatly contrary to the facts are habitually made. Let me make it plain that this propensity is an integral part of the language — this flagrant disregard of facts affects any native speaker of these languages so my remarks are never *ad hominem* but "to the languages".

In these fact-free jargons one can say that classical methods are derived from a decision-theoretic approach although, in point of historical fact, the latter came along a generation later in time. Again one can flatly assert that pragmatic approaches "have nothing to offer us" and at the same time state that they are used throughout the sciences for dealing with scientific hypotheses. It is consistent with the bizarre logic of these so-called logical languages that one can say in effect "these thousands of working scientists whose regular business is doing scientific research really know nothing at all about scientific hypotheses and how to deal with them". What makes this particularly bizarre is that most of the people who make this kind of statement have never actually carried out a scientific study in their lives and wouldn't know where to start.

These fact-free jargons facilitate the making of arrogant, pretentious, and factually false statements about science. This has been going on for years but it's time it stopped. There are two reasons why I feel impelled to speak out on this. First of all — time is running out on all of us; if between now and year 2000 we haven't licked some of our major problems, there won't be any more time left to do so. I don't see how we will solve these problems if our brightest people are off playing silly games. My second reason for hating this mythology of science is somewhat personal. Unlike most of my colleagues my primary mission is cancer research—not teaching statistical inference. For example, Dr. Slack, Dr. Blumenson, and I recently published a paper in *Cancer* entitled "Therapeutic Implications from a Mathematical Model Characterizing the Course of Breast Cancer". What lies behind this title? Well for the past fifty years women with breast cancer have been subjected to severe and completely useless trauma because of the myths about breast cancer. In other words, on top of the severe trauma from the breast cancer itself they've been hit by heavy doses of x-ray and cut up by radical surgery that did no good at all. Statistical inference isn't just a game to me—it's a means of putting a stop to these dangerous myths. This isn't an abstract logical question and neither is statistical inference. It doesn't help matters one bit to pretend otherwise. If you want to talk abstract logical or set-theoretic languages, go right ahead. Just don't claim to be talking about statistical inference.

V.P. Godambe:

In spite of my general admiration of Professor Kyburg's paper I think it displays a lack of statistical realism. For instance, none of Kyburg's philosophization seems to have any bearing on the most basic problem of statistical inference, namely, the one associated with sampling of a finite population. This problem is characterized in detail in two recent publications: (1) V. P. Godambe, 1969, "Some Aspects of Theoretical Developments in Survey-Sampling," *New Developments in Survey-Sampling*, Wiley-Interscience (2) V.P. Godambe, 1970, "Foundations of Survey-Sampling," *American Statistician*, Vol. 24, No. 1.

I.J. Good:

I think I would be able to understand Dr. Kyburg's theory of probability better if he would relate it to previous theories. Philosophers, mathematicians and scientists usually regard history as an aid to understanding.

One distinction between Dr. Kyburg's theory and the theories of Keynes, Koopman, and myself is that Dr. Kyburg talks about the probabilities of statements instead of propositions. I am not clear whether a proposition could be included under Dr. Kyburg's definition of a statement by defining his class C as the class of true propositions.

If Dr. Kyburg prefers to use subjective probabilities rather than *credibilities* (logical probabilities) then his position is closer to mine than to Keynes. Koopman was noncommittal on this point. Ramsey, de Finetti, and Savage did not emphasize inequalities.

Dr. Kyburg said that the principle of total evidence has no non-circular justification. I think this is an error and that a justification can be given in terms of the usual principle of rationality (Raiffa and Schlaifer: Lindley; and especially Good, *Brit. J. Philos. Sc.*, c. 1967). It is not quite obvious; at any rate, it was missed by 17 philosophers of science when A.J. Ayer raised the matter in c. 1958.

D.V. Lindley:

Consider the comparison of two simple hypotheses and the diagram, as in Kyburg's paper, in which the axes correspond to the two error probabilities. Any experiment will give a convex curve passing through (1,0) and (0,1) of admissible values of α and β, and the statistician has to select one point on this curve as the best in some sense. This will be done if the experiment is forgotten and all points (α,β) in the unit square are given a preference ordering — with (0,0) as the best — and the selection for a given experiment made amongst the subset of admissible values. In the economist's language, let us describe indifference curves in the (α,β)-plane.

Suppose we are indifferent between (α_1,β_1) and (α_2,β_2). Then, by an argument due to von Neumann, if one is indifferent between tea and coffee one won't mind if the choice is made by means of a (possibly bent) coin. Applying this here one is indifferent between all points on the segment joining (α_1,β_1) and (α_2,β_2). Consequently the indifference 'curves' are straight lines. A generalization of this argument involving (α_1,β_1) (α_2,β_2) and the origin shows that the lines must be parallel. Consequently it is only necessary to fix the slope of these lines and then to use the resulting system to determine the best choice of (α,β) for a given experiment. This is the Bayesian rule.

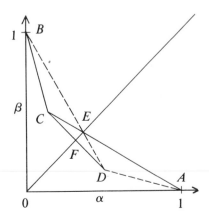

The minimax strategy always makes $\alpha = \beta$. Consider an experiment with admissible curve BCA in the diagram. According to minimax E is preferred to C. Consider a second experiment with admissible (dotted) curve BDA. Again, minimax prefers E to D. Now apply the 'coffee-tea' principle to C and D and reach F. Since E is preferred to both C and D it is preferred to F. This is ridiculous since $\alpha_F < \alpha_E$ and $\beta_F < \beta_E$. Hence the same goes for minimax.

(The above argument is due to L.J. Savage and myself, c. 1955).

REPLY

To Professor Bross:

Dr. Bross argues persuasively, not to say frighteningly, that statistical inference is a serious business, and that there is no time between now and the year two thousand for people to go off playing silly games. Speaking of traditional cancer 'therapy' he says: "Statistical inference isn't just a game to me—it's a means of putting a stop to these dangerous myths." I agree with him whole-heartedly. I think we are in a position now where we must learn from experience, that is, acquire scientific knowledge, faster, more accurately, and with less mistakes than ever before, or any one of a dozen different disasters will overtake us. It is precisely because I feel this way that I think the abstract logical questions and the logic games are so tremendously important.

Evidence for this can be found in Dr. Bross's own comments. At one point he argues as if there were a coherent, unified, and unquestionable body of doctrine regarding statistical inference which is well known to all scientists. Isn't it absurd, he suggests, to suppose that all these thousands of working scientists whose regular business is doing scientific research really know nothing at all about scientific hypotheses and how to deal with them? No more absurd, I think, than to suggest that scientists always agree on when a hypothesis is adequately supported by the evidence for it, or that they agree on whether a certain hypothesis has any evidential support. We all know perfectly well that the application of significance tests is often naive to the point of absurdity. How many pages of scientific journals are devoted, not to presenting new facts, but to arguing about the evidential support which someone else has claimed for a certain scientific hypotheses? Can we really believe that everything is as straightforward as ordinary language and Dr. Bross would have us believe?

We can find the answer easily enough in Dr. Bross's example. For the past fifty years, he says, women with breast cancer have been subjected to severe and completely useless trauma because of the myths about breast cancer. There may have been no evidence supporting these myths but those who believed them surely believed that they had evidential support. Those who first promulgated them surely thought that they had evidence supporting them. It has taken fifty years, apparently, to get this claim concerning evidential support straightened out. If Dr. Bross is correct, this is fifty years of misery that should never have happened. There should be no dispute about evidential support. We need only look at the hypotheses that scientists (surgeons and radiologists, for example) accept. But this is an instance where Dr. Bross does not believe his own account!

Dr. Bross believes that scientists, except of course when they believe things that are mythological, accept hypotheses only when the evidence for them is adequate, and that appropriate concepts of evidence are built into fact-limited languages. I believe he has given a clear-cut counterexample himself. One can argue, of course, about whether there are general concepts of languages that cut across all scientific disciplines, and about whether logical or set theoretic languages provide an appropriate framework for examining such concepts. I can't prove that that's the case, but I'm willing to gamble on it. I'm willing to bet that at least some of the fifty years of misery that Dr. Bross deplores could have been avoided by greater logical sophistication. I'm willing to bet—indeed I am betting, and the stakes are the lives of my children—that we'll be better off in the year two thousand if some of us concern ourselves with evidence and inference in the abstract than if we, all of us, spend all our time doing research within the fact-limited languages that happen to have been handed down to us by history. If it takes fifty years to overcome a groundless myth in such a language, we don't have the time.

To Professor Good:

If C is the class of true propositions then the probability of any proposition X is 1 or 0 according as $X \in C$ or $X \notin C$, and life is very simple.

Propositions have always seemed a little mystical to me, which is why I prefer to talk of statements. But there is no great issue here.

Like Keynes, I regard probability as a logical relation—that is, probabilities impose *constraints* on degrees of belief. These constraints are logically imposed by the evidence, that is, by the total body of knowledge relative to which the judgment is made. Unlike Keynes, I do not suppose that these constraints determine degrees of belief uniquely. There is some latitude within the constraints, and within the constraints the subjectivist view seems perfectly acceptable. There is not much that separates Dr. Good from me: just the existence of constraints of a logical character on degrees of belief.

I have seen various arguments purporting to establish the principle of total evidence, including several mentioned by Dr. Good, but I remain unconvinced that the principle of total evidence is not smuggled either into the premises of the argument or into the rules of inference. It is a peripheral point and I am ready to be convinced.

To Professor Lindley:

Of course if you have precisely determined indifference curves in this situation, it is the same as having a precise subjective probability. I do not find it easy to tell whether three points lie on the same indifference curve, and I think this is essentially the same as my confusion about very fine judgments (say to six decimal places) of my degrees of belief. I should regard any of a number of slopes of indifference curves as reasonably representing my preferences; and as in the straight Bayesian argument, I would stipulate that the slopes of these lines be bounded by the constraints imposed by known frequencies.

THE PROBABILISTIC EXPLICATION OF INFORMATION, EVIDENCE, SURPRISE, CAUSALITY, EXPLANATION, AND UTILITY

I. J. Good

Virginia Polytechnic Institute

My purpose in this paper is to review some of my life's work in the mathematics of philosophy, meaning the application of mathematics in the philosophy of science. Apart from the clarification that the mathematics of philosophy gives to philosophy, I have high hopes for its application in machine intelligence research, just as Boolean logic, a hundred years after its invention, became important in the design of computers. I think philosophy and technique are both important, but I do not intend to argue the case for the use of subjective probability on this occasion. I would just like to quote a remark by Henry E. Daniels (private communication, c. 1956) that *each statistician wants his own methods to be adopted by everybody*. (Also see the Appendix.)

I shall cover a variety of topics and will have to be too succinct for complete clarity. But I hope to give some impression of the results attained, and I shall refer to the original sources for fuller details.

Some unifying themes for this work are the simple concepts of weight of evidence and amount of information, and also what Carnap calls the *desideratum-explicatum* approach to the analysis of linguistic terms. The desiderata are extracted from normal linguistic usage of a term, and an explicatum that satisfies the desiderata constitutes a sharpened form of the term, likely to be of more use in scientific contexts than the original vaguely defined term. Of course, the explicatum is not necessarily uniquely determined by the desiderata, and this ambiguity gives us an opportunity of enriching the language.

The desideratum-explicatum approach has been used by several writers for arriving at the usual axioms of subjective probability and utility. In the present paper, I shall take the usual axioms of probability and utility for granted. By doing so, I do not imply that judgments are precisely representable by numbers; on the other hand, I think that all judgments are judgments of inequalities: see for example, Keynes (1921), Good (1950a). I hope that this point is well taken since I do not intend to repeat it; and I hope what I shall say later will not give the impression of being overprecise merely because symbols are used.

In order to clarify my presuppositions, it is necessary to make one or two comments concerning the principle of rationality, the recommendation to maximize expected utility. I call this *rationality of type 1* and, when allowance is also made for the cost of theorizing, *rationality of type 2* (1962).* Rationality of type 1 implies complete logical consistency with the axioms of rationality; but rationality of type 2 should be adopted in practice. For example, you should not normally and knowingly allow any blatant contradictions in your judgments and discernments when the axioms of rationality are taken into account; but an exception is reasonable if a decision is extremely urgent. In particular, apparently non-Bayesian methods are often acceptable to me. I think this compromise resolves all the important fundamental controversies in statistics, but we shall go on arguing because, being mortal, we are anxious to justify our existence, and for other reasons not mentioned in polite society.

The notion of rationality of type 2 is closely related to that of evolving probabilities. An evolving probability is one that changes in the light of reasoning alone, *without the intervention of new empirical information*. I shall return to this matter later.

People will say that the principle of rationality is inapplicable in a situation of conflict, and that then the theory of games and minimax solutions are more fundamental. I do not agree. I think the principle of rationality is still the overriding principle; but you should, of course, take into account your opinions concerning your opponent's (randomized) strategy (which opinions might be largely based on his past behavior) and whether his past behavior appears to have been based on yours. In particular, if you are convinced that he is playing a minimax strategy then, by von Neumann's theorem, you maximize your expected utility by also adopting a minimax strategy in a zero-sum game.

I mentioned that point in order to emphasize that the principle of rationality has no real exceptions, although pseudo-utilities can be used, as discussed later.

It is convenient to make a distinction between information and evidence, and I shall discuss information first. My approach is fairly closely related to that of Shannon, the main difference being that he was concerned with the average amount of information transmitted by a communication channel, whereas my approach is in terms of the amount of information concerning one proposition that is provided by another one.

Let E, F, G, and H be propositions, and let $I(H:E|G)$ denote the amount of information concerning H provided by E when G is given throughout. Let us consider the explication of $I(H:E|G)$ as a real number in terms of probabilities.

We write $E.F$ for the logical conjunction of E and F.

From six reasonable axioms (see 1966a) we can deduce that $I(H:E|G)$ must be a continuous increasing function of the *association factor* (Keynes'

---

*When a date is modestly given without a name, then the omitted name is "I. J. Good."

term)

$$P(H.E|G)/[P(H|G)P(E|G)] \text{ or } P(E|H.G)/P(E|G).$$

If we are concerned only with preserving the ordinal relations (inequalities) between amounts of information, then we might as well, by convention, select the logarithm of the association factor, to some base exceeding unity, since this choice leads to simple additive properties. I shall do this.

The analysis applies whether the probabilities are physical, logical or subjective, in which case we are talking about physical, logical, or subjective information respectively.

The main axiom assumed was that $I(H:E.F|G)$ is some function of $I(H:E|G)$ and $I(H:F|E.G)$.

In terms of communications, H can be interpreted as the hypothesis that a particular message was transmitted on a particular occasion, and E as the event that a particular message was received. Then, upon taking expectations with respect to both H and E we obtain the rate of transmission of information, as in Shannon's theory of communication.

A philosophical approach is not necessary for the theory of communication since the main theorems of that theory deal with efficient coding for transmission through a given communication channel; and these theorems can be proved without even giving a definition for the rate of transmission of information. Nevertheless, I think the theorems are easier to understand when such a definition is given.

I should now like to discuss *evidence*. This differs from information in the following respect. In legal and other circumstances, we talk about the evidence *for or against* some hypothesis, but we talk about the information *relevant to or concerning* a hypothesis. Thus the notion of evidence, as ordinarily used, makes almost explicit reference both to a hypothesis H and to its negation. On the other hand, *information* concerning a hypothesis H does not seem to refer primarily to the question of discriminating H from its negation. It seems linguistically appropriate to regard the *weight of evidence*, in favor of a hypothesis H, provided by evidence E, as identifiable with the degree to which the evidence corroborates H as against its negation $\bar{H}$. Note that I am distinguishing between evidence E and weight of evidence, just as in ordinary English.

Now Popper (1959) laid down nine compelling desiderata that corroboration C should satisfy. To these I (1969d, 1968b) made some minor modifications and also added the assumption that $C(H:E.F|G)$ is some function of $C(H:E|G)$ and $C(H:F|E.G)$. This assumption is reasonable although it is not as compelling as the other axioms. But, since it is possible to assume it and still find an *explicatum* that satisfies all the conditions, this explicatum can be expected to be the most convenient one to accept. Any explicatum that does not satisfy this assumption would come under suspicion. It seems reasonable to say that if an explicatum satisfies all the compelling axioms and also the reasonable ones, then it is better than an explicatum that satisfies only the compelling axioms.

The assumptions made in (1960d) did not lead to a unique explicatum so in (1968b) I added the further assumption that corroboration is objective in some circumstances, i.e. that if $C(H:E)$ depends only on $P(H)$, $P(E|H)$, and $P(E|\bar{H})$, then it depends only on $P(E|H)$ and $P(E|\bar{H})$. This approach led to the conclusion that the best explicatum for corroboration is some monotonic increasing function of weight of evidence W, defined (1950a) as the logarithm of the Bayes factor (or Bayes-Jeffreys-Turing factor) in favor of a hypothesis provided by evidence. Formally, the weight of evidence is

$$W(H{:}E|G) = \log F(H{:}E|G),$$

where the base of the logarithms merely determines the unit of measurement of weight of evidence, and the factor F is defined by

$$F(H{:}E|G) = P(E|H.G)/P(E|\bar{H}.G)$$
$$= O(H|E.G)/O(H|G),$$

where O denotes odds. (The odds corresponding to a probability p are defined as $p/(1-p)$.) It is because of the expression in terms of odds that Turing (1940, private communication) suggested the excellent name *factor in favor of a hypothesis*. Note that

$$W(H{:}E|G) = I(H{:}E|G) - I(\bar{H}{:}E|G)$$

and this gives especially clear expression to the previous remark that evidence is concerned with a comparison of a hypothesis with its negation.

Sometimes we wish to talk about the weight of evidence in favor of H as against, or as compared with, some other hypothesis H'. A convenient notation for this is $W(H/H'{:}E)$ and it is equal to $W(H{:}E|H \vee H')$, where the sans serif ($\vee$) denotes logical disjunction. A similar notation can be used for Bayes factors F and corroborations C. The colon can be read *provided by*.

The expression W for weight of evidence is more clearly appropriate in terms of ordinary English than is I for amount of information since, as M. S. Bartlett has pointed out in conversation (about 1951), it is not linguistically clear whether information relevant to H should ever be allowed to take negative values. It is not unreasonable of course, since information *can* be misinformation.

The calculation of the expressions $I(H{:}E|G)$ and $W(H{:}E|G)$ both usually depend on a Bayesian philosophy, especially the former. But W is non-Bayesian when both H and $\bar{H}$ are simple statistical hypotheses, since in this case the factor in favor of H is equal to the simple likelihood ratio. Note though that even in this case, the interpretation as a factor on the odds has more intuitive appeal to the non-statistician. It would be more appropriate to say that the name *likelihood ratio* is jargon than to say it of *weight of evidence* or *factor in favor of a hypothesis*.

Proofs of some coding theorems more general than Shannon's Fundamental Theorem can be expressed with much intuitive appeal in terms of weight of evidence, and it turns out that amount of information occurs in this theory merely as an approximation to weight of evidence. (See Good and Toulmin, 1968.)

An entertaining philosophical application of weight of evidence occurs in the discussion of Hempel's paradox of confirmation. In a nutshell, the paradox is that since a case of a hypothesis supports it, and since the assertion that all crows are black is logically equivalent to the assertion that all non-black objects are non-crows, it follows that the observation of a white shoe supports the hypothesis that all crows are black. This paradox might seem trivial but if left unresolved it would undermine the whole of statistical inference. I shall here merely mention some of the relevant references (1960b, 1961c, 1967a, 1968c, and Hempel, 1967).

Since I have just said that a paradox might undermine statistical inference, I should like to take this opportunity to correct an error I made (1966b, p. 382) when discussing Miller's paradox concerning the axioms of probability. Succinctly, his paradox can be expressed thus: for all propositions E we have
$$P(\bar{E}) = P(E \mid P(E) = P(\bar{E})) = P(E \mid P(E) = \tfrac{1}{2}) = \tfrac{1}{2}.$$
My attempted resolution was that $P(E) = P(\bar{E})$ is impossible unless $P(E) = \frac{1}{2}$ and no sensible theory of probability permits an impossible proposition to be *given**. But, as Miller has pointed out (private communication), this resolution is wrong since $P(E) = P(\bar{E})$ is not impossible unless it is *known* that $P(E) \neq \frac{1}{2}$. One moral is that it is dangerous to allow the propositions to the right of the vertical stroke to make explicit reference to probabilities, as pointed out in (1950a, p. 41) and Koopman (1940, p. 275). This paradox is more or less of the self-referring kind familiar in logic. But this resolution of the paradox will not work if the given information is interpreted in terms of long-run frequencies. Then (1969b) I think we can resolve the trouble by quoting more precisely the theorem in the theory of subjective or logical probability that lies behind the intuitive feeling that $x = P(E \mid P(E) = x)$. This theorem, which is a form of the law of large numbers is that, provided that the initial density of $P(E)$ is positive at x, then $x = P(E \mid s = x)$, where s denotes the limiting relative proportion of *successes* in an infinite sequence of *trials,* and where x is an assigned real number. Since $P(\bar{E})$ is not an assigned real number, Miller's paradox is not a threat to the axioms of subjective or logical probability, at least as I understand them.

An algebraic analogue of Miller's paradox was pointed out by Mackie (1966). It can be expressed succinctly thus:
$\frac{1}{2} =$ (The value of a if $a = \frac{1}{2}$) = (The value of a if $a = 1 - a$) $= 1 - a$. Here the resolution is that an expression involving a cannot legitimately be said to be a *value* of a, so that the last step in the argument is illogical.

Weight of evidence is especially appropriate for medical diagnosis and one example occurs in connection with the analysis of the relationship between lung cancer on the one hand and smoking, morning cough, and degrees of rurality or urbanity on the other.

It turned out in this example that degree of urbanity (rural, urban, highly urban) gave a weight of evidence that could be added to that derived from the evidence concerning coughing and smoking, but that coughing and smoking could not be treated as additive. In other words there was evidential interaction between coughing and smoking in relation to lung cancer, but there was not important interaction between urbanity on the one hand and coughing and smoking on the other. In this statement, the interaction can be defined as
$$W_1(H:E,F) = W(H:E.F) - W(H:E) - W(H:F) = I(E:F \mid H) - I(E:F \mid \bar{H}).$$
Second-order interactions are naturally defined by the expression
$$W_2(H:E,F,G) = W(H:E.F.G) - W(H:F.G) - W(H:G.E) - W(H:E.F) + W(H:E) + W(H:F) + W(H:G)$$
as in (1960c) and Card and Good (1969). It will be useful for diagnostic purposes if it turns out in many instances that interactions of the second order can be ignored.

---

*This *is* permitted in Popper's theory, which is sensible, but the above argument is unaffected.

In my work, I put much stress on the notion of weight of evidence. This enthusiasm is shared in a forthcoming book by Myron Tribus (1969), and has also been much used, though in somewhat different manners, by Jeffreys (1939/61) and Kullback (1959). Entropy enthusiasts are more numerous and include E. T. Jaynes, Jerome Rothstein, and S. Watanabe. In my terminology, Kullback's book would have been called *Weight of Evidence and Statistics*.

The next topic I should like to discuss is the concept of *surprise*. The British economist G.L.S. Shackle has argued (1954/57) that in business decisions the notion of *potential surprise* is more meaningful or at least more often used than that of subjective probability. He avoided trying to express the notion of surprise in terms of probability. Nevertheless, a connection has been proposed by Weaver (1948) for circumstances where the number of outcomes is descrete. Weaver's surprise index is

$$\lambda_1 = \mathcal{E}(p^*)/p,$$

where p is the probability of the event that actually occurs, and $\mathcal{E}(p^*)$ is the expected probability of all the events that might have occurred. This expectation equals the sum of the squares of all these probabilities, otherwise known as the *repeat rate* (Turing's terminology, see Good 1953).

There is something arbitrary about the definition of λ_1 since it can be generalized while continuing to have its main features. In fact (1954/57, 1956b) let

$$\lambda_u = [\mathcal{E}(p^{*u})]^{1/u}/p \ (u>0),$$

$$\lambda_0 = \exp\{\mathcal{E}(\log p^*) - \log p\} = G.E.\ (p^*)/p,$$

where $G.E.$ means *geometric expectation*, and let

$$\Lambda_u = \log \lambda_u \ (u \geqslant 0).$$

We may call Λ_u a *logarithmic surprise index*. It is additive if the results of several experiments are combined into a single experiment. The best measure of surprise is Λ_0 because its expectation is zero before observations are taken. It was used by Bartlett (1952) in connection with "odd bits of information," and not in relation to the concept of surprise. Note that the notion of negative surprise is intuitively entirely reasonable. ("Only to be expected.")

Almost the same mathematics can be applied to the problem of paying a consultant in order to encourage him to make good probability estimates (Good 1952; McCarthy 1956; Marshak 1959). The expected fee involves the ubiquitous entropy expression.

The use of a surprise index as the summary of an experiment or observation evades Jeffreys's well-known objection to the use of tail area probabilities, but all the same it might be better to use the tail area associated with the value of a surprise index. This would come to the same thing as the use of the distribution of the likelihood or likelihood density (Good 1956b). The possibility of using a surprise index itself (not the associated tail area) as a substitute for a tail area probability was suggested by G. C. Wall (private communication, about 1954).

Over a long series of observations, the expectation of Λ_0 remains zero, but its standard deviation tends to infinity. Hence in the long run, we shall sometimes be very surprised and sometimes very bored, and we shall oscillate between these two extremes. A long stretch of any given stationary time series will usually be either very surprising or very much *only to be expected* when taken as a whole. In the course of its development, it will oscillate between these extremes. This paints a reasonable picture of most history, although it is not easy to see history as a stationary time process.

Since surprise can be measured in terms of subjective probability, we can use our judgments of potential surprise in order to check the consistency of and extend our bodies of beliefs. Thus we can add substantially to our repertoire of possible judgments, if Shackle's opinion mentioned above is correct.

My next topic is *probabilistic causality*. This I found more difficult to express quantitatively than any of the other concepts in this paper, and my first attempt (Good 1959) was unsatisfactory, as was pointed out by L. J. Savage and E.M.L. Beale (private communications). I believe my second attempt (Good 1961b) was satisfactory. The main question was how to define $Q(E:F)$: the tendency of an event F to cause an event E, or the causal support for E provided by F. Note that this is distinct from Norbert Wiener's work on the causal relationship between two time series (Wiener, 1956).

After trying a large number of different approaches, I decided that the best explicatum for $Q(E:F)$ was $W(\bar{F}:\bar{E}|H.U)$, the weight of evidence against F if E does not occur, given the laws of nature, H and the state, U of the universe immediately before F started. A slight weakness in the explicatum is that it has to be assumed that a cause always precedes its effects. This would rule out precognition and some speculative ideas in modern physics concerned with backward time, unless time itself is more complicated than we suppose.

If precognition becomes scientifically accepted then it might be only a matter of definition whether we were to say that an effect can precede its cause or whether we would say instead that time has a complicated structure. Current linguistic usage does not resolve this problem and we cannot predict with certainty how language would develop, although someone with precognitive powers might be able to do so!

I have recently found that there is a relationship between Q and one of the measures of association for 2 by 2 contingency tables. Sheps (1959) points out that in a medical context, the ratio of survival rates might be used and that it is of course not a function of the ratio of mortality rates. In my terminology this is the same as saying that the weight of evidence against F if E does not occur is not a function of the weight of evidence in favour of F if E does occur. The ratio of survival rates is also mentioned by Greenberg (1969) in an article on smoking and lung cancer, with reference to Sheps. These authors do not try to make a definite choice between the various measures of association, but point out that the appropriate measure depends on the application. This is consistent with the fact that in ordinary English, causation is not the same thing as evidence.

Another example of Q in a classical statistical context is in linear regression. Consider the linear multivariate normal model $y = N(B\mathbf{x}, C)$, where C is the covariance matrix of the errors. Suppose that the "treatment" $\mathbf{x}$ is changed to $x + \xi$ and applied to *essentially the same physical system*. This last condition is necessary in order to avoid mixing up causality with spurious correlation. Then the degree Q to which this change of treatment tends to produce a change in the expected value of y is equal to $\frac{1}{2}\xi'B'C^{-1}B\xi$ or $\beta^2\xi^2/(2\sigma^2)$ in the univariate case. Note that this is invariant under a linear transformation of the model, as it should be. Also it is additive for two independent causes.

It is also possible to explicate the causal strength S and the causal resistance R of causal nets, somewhat as if they were electrical networks. Apart from trivial transformations, I found that I was forced to take

$$R = -\log(1 - e^{-S}), \, S = -\log(1 - e^{-R}).$$

For a single causal link, $S(E\!:\!F) = Q(E\!:\!F)$. The resistance of a Markov chain is the sum of the resistances of its links, and the strength of a bundle of parallel independent causal chains is the sum of the strengths. An example of such a bundle is a firing squad, in which the event F is the command to fire, and the event E is that the victim is hit. An extension can be made to arbitrary causal nets, and *the extent to which an event E is actually caused* by an event F can then be defined as the limit of the strengths as the mesh of the net becomes smaller. (I have here quoted from 1961a.) It can be seen by means of a simple example that the extent to which E was caused by F cannot be identified with the *tendency* of F to cause E.

I believe that the philosophical analysis of *strict* causality, itself by no means trivial, can be inferred as a limiting case of this treatment of *probabilistic* causality.

Next consider *explanation*. Here again Popper (1959) suggested some desiderata for the degree $D(H\!:\!E)$ to which a hypothesis explains an event E and again I (1960d, 1968b) made some modifications and appended the assumption that $D(H\!:\!E.F)$ is some function of $D(H\!:\!E)$ and $D(H\!:\!F|E)$. This led to the conclusion that D must be some increasing function of $I(H\!:\!E)$, and we might as well identify it with $I(H\!:\!E)$ itself. In this way we enjoy the simple additive property,

$$I(H\!:\!E.F) = I(H\!:\!E) + I(H\!:\!F|E).$$

But one of the desiderata assumed here was arguable, namely that $D(H.K\!:\!E) = D(H\!:\!E)$ when K has nothing to do with H and E. Do we or do we not wish to claim that the explanatory power of a hypothesis H, for explaining E, is unaffected if the hypothesis is cluttered up with irrelevancies? It seems to me that we cannot answer yes or no here, but that it is better to say there are two kinds of explanatory power, depending on whether clutter is relevant or not. If it is irrelevant, let us refer to explanatory power in the *weak sense*, but if clutter is regarded as a disadvantage then we refer to explanatory power in the *strong sense*. Denoting the latter by D', I made the assumption that $D'(H.K\!:\!E.F)$ is some function of $D'(H\!:\!E)$ and $D'(K\!:\!F)$ when H and E have nothing to do with K and F. This led to the explicatum

$$D'(H\!:\!E) = I(H\!:\!E) - \gamma I(H) = (1 - \gamma)I(H\!:\!E) - \gamma I(H|E),$$

where γ is a constant strictly between 0 and 1. (Here $I(H)$ and $I(H|E)$ are abbreviations for $I(H\!:\!H)$ and $I(H\!:\!H|E)$.) The value of γ depends on the relative weights we wish to give to explanation and to the avoidance of clutter. Taking $\gamma = 0$ gives rise to weak-sense explanation, and taking $\gamma = 1$ would assign maximum explanatory power to the simplest possible hypothesis namely an obvious tautology. This would clearly be too extreme and perhaps $\gamma = \frac{1}{2}$ is adequate.

The above formula for strong explanatory power D' can be used for selecting a hypothesis for the explanation of given observations E; in fact the hypothesis of maximum strong explanatory power is in my opinion the one to be preferred. But if we wish to make predictions about future observations, I think the right thing to do is to try to maximize the expected strong explanatory power,

$$\sum_i P(E_i)D'(H:E_i),$$

where E_i runs through the possible mutually exclusive future observations. This expectation is equal to

$$\sum_i P(E_i) \log P(E_i|H) + \gamma \log P(H) - \sum_i P(E_i) \log P(E_i)$$

which can be written

$$\mathcal{E} D'(H:E) = -P(H) \, \text{ent}(E|H) - \gamma I(H) + \text{ent}(E),$$

where *ent* denotes entropy.

The maximization of $\mathcal{E} D'(H:E)$ can be interpreted in two ways, either as maximization with respect to H for selecting the most useful hypothesis; or, when H is fixed, as a maximization by choice of experimental design. In the latter case, the principle comes to the same as the maximization of the expected amount of information in an experiment, as proposed by Cronbach (1953), Lindley (1956), and myself (1956a). Lindley's paper developed the idea in considerable detail. Thus the principle of maximizing the expected strong explanatory power gives unity to inference and experimental design when the aim is uncluttered explanation.

If an event E is already known to be true, and we envisage the possibility of future mutually exclusive observations $F_1, F_2, \ldots$, then we should replace E_i in the above formula by $E.F_i$, and a natural notation for the expected strong explanatory power is

$$\mathcal{E} D'(H:E.\mathbf{F}).$$

When there are no future observations that concern us, then **F** is absent and we return to the simple form mentioned first, where we are concerned only with the explanation of an event E.

As an example where we are concerned only with explaining a given observation, consider the estimation of a binomial parameter p from a sample of size n in which there are r successes, and assume a uniform initial distribution for p. Then the hypothesis of greatest strong explanatory power gives an interval of values of p defined approximately by the inequality

$$(p - r/n)^2 \le r(n-r) \, \xi/n^3,$$

where $\xi = \xi(\gamma)$ is the unique positive solution of

$$\xi = (1-\gamma)e^{\xi^2/2} \int_0^\xi e^{-x^2/2} dx$$

$$[\ \xi\,(0) = 0,\ \xi\,(1/4) = 0.92,\ \ \xi\,(1/2) = 1.40,\ \ \xi\,(3/4) = 1.94,\ \ \xi\,(1) = \infty.\ ]$$

Note that $\gamma = 0$, which corresponds to weak explanatory power, gives maximum-likelihood point estimation, whereas values of γ that are very close to 1 put so much emphasis on a high initial probability of the hypothesis that they give too wide an interval for the estimated parameter.

If two hypotheses imply the observed events, then Occam's razor recommends the choice of the simpler hypothesis. This is a special case of the principle of maximizing the strong explanatory power provided that we assume that the simpler of two hypotheses has the larger initial probability. I am referring here of course to logical or subjective probability and not to physical probability.

J. Agassi (private communication, 1960) pointed out an objection to the assumption that the simpler of two hypotheses is the one of higher probability. His example referred to Fresnel's laws of optics and Maxwell's equations of electromagnetism. Maxwell's equations seem simpler and yet they imply Fresnel's laws, so that Fresnel's laws must be more probable than Maxwell's equations. My reply to this is that the *evolving* (or unfolding, developing, shifting, sliding, unsteady, vagrant) probability of Fresnel's equations suddenly increased when Maxwell pointed out how they could be derived. I infer from this that the explication or explanation does not merely involve the use of subjective probabilities, but must make use even of evolving probabilities, which I suppose are yet more controversial. I do not think they *should* be controversial: for example, they are used all the time in a complicated game of so-called perfect information such as chess. (The *evolving* information is imperfect.) They also occur in research in pure mathematics (1950a, p. 49). The probabilities that occur in Polya's thesis on plausible reasoning (Polya 1954) are clearly evolving probabilities. Also in non-mathematical problems, I do not believe we ever finish all possible relevant thinking, so perhaps all subjective probabilities are evolving ones in practice.

There are two things that make Occam's razor insufficient for the selection of hypotheses, even if strict utilities cannot be estimated. In the first place, we often have to choose between two hypotheses one of which is more complicated but makes the observed events more probable. This difficulty is intended to be met by the definition of strong explanatory power. Secondly, one of the hypotheses might imply other results not yet observed and thus have greater predictive value. This is intended to be met by the idea of taking the expected value of the strong explanatory power as described above. All these remarks are intended to apply to general scientific theories and not merely to statistical hypotheses.

My final topic is *utility,* especially in relation to the utility of a distribution (1960a, 1968d, 1969a). Statisticians have been giving their customers estimates of distributions of random variables for a long time, and they hardly ever stop to consider what the loss in utility is if the estimate is not accurate. In order to find out something about this, let us denote by $U(G|F)$ the utility of asserting that a random variable $\mathbf{x}$ has the distribution G when the true distribution is F. Let us suppose that $U(G|F)$ is some form of generalized expectation of $v(\mathbf{x},\mathbf{y})$, where $v(\mathbf{x},\mathbf{y})$ denotes the utility of asserting that the value of the random variable is $\mathbf{y}$ when it is really $\mathbf{x}$. We naturally assume that $v(\mathbf{x}, \mathbf{y}) \le v(\mathbf{x}, \mathbf{x})$. Compelling desiderata are

(a) if a constant is added to v then the same constant is added to U;
(b) additivity for mutually irrelevant vectors

$$U(GG^*|FF^*) = U(G|F) + U(G^*|F^*);$$

(c) invariance under non-singular transformations of $\mathbf{x}$: $U(G|F)$ is unchanged if a non-singular transformation $\mathbf{x} = \psi(\mathbf{x'})$, $\mathbf{y} = \psi(\mathbf{y'})$ is made, subject to the obvious desideratum that the transformed form of v is

$$v(\psi(\mathbf{x'}), \psi(\mathbf{y'}).$$

A doubly infinite system of functionals satisfying these desiderata is

$$U_\alpha^\beta(G|F) = \frac{1}{\beta} \log \int dF(\mathbf{x}) \left[\int e^{\alpha v(\mathbf{x}, \mathbf{y})} dG(\mathbf{y}) \right]^{\beta/\alpha}$$

where $0 \le \alpha \le \infty$, $0 \le \beta \le \infty$. (Another system of solutions can be obtained by interchanging F and G.) When $\beta \to 0$ we obtain

$$U_\alpha(G|F) = \frac{1}{\alpha} \int dF(\mathbf{x}) \log \int e^{\alpha v(\mathbf{x}, \mathbf{y})} dG(\mathbf{y}).$$

If it is required further that (d) $U(F|F) \ge U(G|F)$ for all F and G we must let $\alpha \to \infty$ and when G has a density function g we obtain *up to a linear transformation* (irrelevant for utility measures)

$$U_\infty^\beta(G|F) = U_\infty(G|F) = \int \log \{g(\mathbf{x}) |\Delta(\mathbf{x})|^{-1/2}\} dF(\mathbf{x}) \tag{1}$$

where

$$\Delta(\mathbf{x}) = \left\{ \frac{\partial^2 v(x, y)}{\partial y_j \partial y_k} \bigg|_{\mathbf{y} = \mathbf{x}} \right\} j, k = 1, 2, \ldots$$

(The solutions with F and G interchanged cannot satisfy condition [d]).

If also F has a density function f.

$$U_\infty(F|F) = \int f(\mathbf{x}) \log \{f(\mathbf{x}) |\Delta(\mathbf{x})|^{-1/2}\} d\mathbf{x}. \tag{2}$$

This can be regarded as minus an *invariantized* entropy and equation (1) as an invariantized *cross-entropy*.

When there is *quadratic loss*, that is, when v is a quadratic form, the factor involving $\Delta(\mathbf{x})$ is constant and can be ignored and equation (1) reduces to an ordinary cross-entropy which was called by Kerridge (1961) the *inaccuracy* of G when F is true. This reduction occurs also if v is any twice differentiable function of a quadratic form.

The formulae could be used in the design of experiments and in the summarizations of their results. Another interpretation can be given if there is a density function $f_0(\mathbf{x})$ proportional to $|\Delta(\mathbf{x})|^{1/2}$, that is, if $\int |\Delta(\mathbf{x})|^{1/2} d\mathbf{x}$ converges. Then $U_\infty(F|F)$ can be minimized by taking F to have the density function f_0, so that, with this utility measure, f_0 is the *least favorable* initial density. It is thus the minimax initial density (Wald, 1950). The principle of selecting the least favorable initial distribution may be called *the principle of least utility*. For discrete $\mathbf{x}$ it leads to the principle of maximum entropy (Jaynes, 1957) if $v(\mathbf{x}, \mathbf{x})$ is constant, and for continuous $\mathbf{x}$ if $|\Delta(\mathbf{x})|$ is constant. Otherwise, for continuous $\mathbf{x}$ it leads to a principle of maximum invariantized entropy. I think it is very interesting to see that the principle of maximum entropy can be regarded as a minimax procedure; and also to see how it should apparently be applied for continuous distributions: that U_∞ reduces to negentropy when the loss function is quadratic or a function of a quadratic.

Suppose that $\mathbf{x}$ is the parameter in the distribution $T(\mathbf{z}|\mathbf{x})$ of another random vector $\mathbf{z}$, the density function being $t(\mathbf{z}|\mathbf{x})$. In the Jeffreys-Perks invariance theory (Jeffreys, 1946; Perks, 1947) the initial density is taken as the square root of the determinant of the information matrix

$$\left\{-\int t(\mathbf{z}|\mathbf{x})\frac{\partial^2 \log t(\mathbf{z}|\mathbf{x})}{\partial \mathbf{x}_j \partial \mathbf{x}_k} d\mathbf{z}\right\} j, k = 1, 2, \ldots$$

(strictly Perks was concerned only with the one-dimensional case). This comes to the same as using the principle of least utility with v defined in terms of the expected weight of evidence

$$v(\mathbf{x},\mathbf{x}) - v(\mathbf{x},\mathbf{y}) = \int \log \frac{dT(\mathbf{z}|\mathbf{x})}{dT(\mathbf{z}|\mathbf{y})} \, dT(\mathbf{z}|\mathbf{x})$$

for distinguishing the true from the assumed distribution of $\mathbf{z}$. This formula follows at once from equation (1), quite *irrespective of the minimax interpretation*, for we must have by definition

$$v(\mathbf{x},\mathbf{y}) = U_\infty \ (T(\mathbf{z}|\mathbf{y})|\, T(\mathbf{z}|\mathbf{x}))$$

if the only use of $\mathbf{x}$ is to serve as a parameter in the distribution of $\mathbf{z}$. Hence $\Delta(\mathbf{x})$ is equal to Fisher's information matrix. Here v and Δ are defined for the *random variable* $\mathbf{x}$ and should not be confused with the corresponding functions for $\mathbf{z}$.

Thus we see that Harold Jeffreys's invariant density can be derived from a minimax procedure provided that utility differences are identified with weight of evidence. The disadvantage of Jeffreys's brilliant suggestion can therefore be attributed to that of minimax procedures in general. (See also 1967b.)

The maximization of entropy is a very reasonable method for the estimation of probabilities in contingency tables, and in Markov chains, since it leads to hypotheses of generalized independence that are satisfactory to the intuition of statisticians. But in a problem such as medical diagnosis it is reasonable, when acquiring information, to try to *minimize* the entropy of a set of mutually exclusive diseases. In a medical diagnostic search tree, one is involved both with the estimation of probabilities and with the acquisition of new information. Hence a reasonable procedure is to try to minimax the entropy in the sense of the theory of games (1968a; Card and Good 1969: rival formulae are given in these references).

Very closely related to and somewhat more general than the principle of maximum entropy is a principle of minimum discriminability in which expected weights of evidence are used in place of the entropy (Kullback 1959). The formula for U_∞ gives some support for this. A satisfying property of minimum discriminability was shown in (1966b): if we have a chain of hypotheses $H_1, H_2, \ldots, H_n$, concerning various probabilities, where the hypotheses satisfy increasing sets of linear constraints, and if we introduce additional constraints, and determine the next hypothesis H_{n+1} by minimum discriminability from any one of the earlier hypotheses, then we always arrive at the same hypothesis H_{n+1} This is by no means obvious, but the proof is not difficult.

When we aim to maximize the expectation of any expression, that expression can be regarded as a pseudo-utility or quasi-utility, whether it be entropy, weight of evidence, or something else. As a historical matter, my first introduction to the use of expected weight of evidence as a measure of the value of a statistical investigation was in 1940 when working with Turing on a war-time project in which he invented sequential analysis. The idea later occurred in Wald's work. Expected weight of evidence also occurs in my definition of a decision (1964), but I shall not discuss that here.

It is useful to use pseudo-utilities when true utilities are difficult to estimate, as they often are, especially in problems of inference. The use of weight of evidence as a pseudo-utility is especially appropriate when we are trying to decide whether a hypothesis or its negation is true. If the negation of the hypothesis is sufficiently vague, or if we are not sure which of several hypotheses we are really interested in, then the entropy serves as a reasonable pseudo-utility. Moreover, owing to an additive property of entropy, the principles of maximizing and minimizing entropy are consistent when applied to a pair of completely independent problems. For a further discussion of this point see (1968a).

There seems to be a constant interplay between the ideas of entropy and expected weight of evidence, or *dientropy* as it might be called since it refers to two distributions. Even in statistical mechanics, it seems that expected weight of evidence is a useful concept, as conjectured in (1950b) and demonstrated by Koopman (1969) for non-equilibrium problems.

My conclusion is that the mathematics of the philosophy of inference is a useful and interesting pursuit.

References

1. Bartlett, M.S., "The Statistical Significance of Odd Bits of Information," *Biometrika*, 39, 328-337, 1952.
2. Card, W.I. and Good, I.J., "A Mathematical Theory of the Diagnostic Process," 1969.
3. Cronbach, L.J., "A Consideration of Information Theory and Utility Theory as as Tools for Psychometric Problems," Technical Report, College of Education, University of Illinois, Urbana, 1953.

4. Good, I.J., *Probability and the Weighing of Evidence,* London, Griffin; New York, Hafners, 1950a.
5. Good, I.J., Contribution to the discussion of a paper by E. C. Cherry in *Symposium on Information Theory, Report of Proceedings,* London, Ministry of Supply, 167-168, 1950b.
6. Good, I.J., "Rational Decisions," *Journal of the Royal Statistical Society, B,* 14, 107-114, 1952.
7. Good, I.J., "On the Population Frequencies of Species and the Estimation of Population Parameters," *Biometrika,* 40, 237-264, 1953.
8. Good, I.J., "Mathematical Tools," Chapter 3 of *Uncertainty and Business Decisions* (see Shackle, 1954/57), 1954/57.
9. Good, I.J., "Some Terminology and Notation in Information Theory," *Proceedings of the Institution of Electrical Engineers, Part C (3),* 103, 200-204, 1956a or Monograph 155R, 1955.
10. Good, I.J., "The Surprise Index for the Multivariate Normal Distribution," *Annals Mathematical Statistics,* 27, 1130-1135, 1956b, unnumbered page, 1957.
11. Good, I.J., "A Theory of Causality," *Brit. J. Philos. Sc.,* 9, 307-310, 1959.
12. Good, I.J., Contribution to the discussion of a paper by E.M.L. Beale, "Confidence Regions in Non-Linear Estimation," *J. Roy. Statist. Soc., B,* 22. 79-82, 1960a.
13. Good, I.J., "The Paradox of Confirmation," *Brit. J. Philos. Sc.,* 11, 145-148, 1960b.
14. Good, I.J., "Effective Sampling Rates for Signal Detection: Or Can the Gaussian Model be Salvaged?" *Information and Control,* 3, 116-140, 1960c.
15. Good, I.J., "Weight of Evidence, Corroboration, Explanatory Power, Information, and the Utility of Experiments," *J. Roy. Statist. Soc., B,* 22, 319-331, (30, 1968, 203), 1960d.
16. Good, I.J., "Weight of Evidence, Causality, and False-Alarm Probabilities," *Fourth London Symposium on Information Theory,* edited by E.C. Cherry, London, Butterworths, 125-136, 1961a.
17. Good, I.J., "A Causal Calculus," *Brit. J. Philos. Sc.,* 11, 305-318, 1961b; 12, 43-51, 1961; 13, 88, 1962.
18. Good, I.J., "The Paradox of Confirmation, II," *Brit. J. Philos. Sc.,* 12, 63-64, 1961c.
19. Good, I.J., "How Rational Should a Manager Be?," *Management Science,* 8, 383-393, 1962. Reprinted with numerous minor improvements, in *Executive Readings in Management Science,* edited by Martin K. Starr, New York, Macmillan, 88-98, 1965.
20. Good, I.J., "Measurements of Decisions," *New Perspectives in Organization Research,* Proceedings of a Conference on Organization Research, Pittsburgh, Carnegie Institute of Technology, June, 1962, edited by W.W. Cooper, H.J. Leavitt, and M.W. Shelly II, New York, London, Sydney, Wiley, 391-404, 1964.
21. Good, I.J., "A Derivation of the Probabilistic Explication of Information," *J. Roy. Statist. Soc., B,* 28, 578-581, 1966a.
22. Good, I.J., "How to Estimate Probabilities," *J. Inst. Maths. Applics.,* 2, 364-383, 1966b.
23. Good, I.J., "The White Shoe is a Red Herring," *Brit. J. Philos. Sc.,* 17, 322, 1967a.
24. Good, I.J., "A Bayesian Significance Test for Multinomial Distributions," *J. Roy. Statist. Soc., B,* 29, 399-431, (with discussion), 1967b.
25. Good, I.J., "Some Statistical Methods in Machine Intelligence Research," *Journal of the Virginia Academy of Science,* 19, 101-110, 1968a. Reprinted with some improvements in *Math. Biosciences.*
26. Good, I.J., "Corroboration, Explanation, Evolving Probability, Simplicity and a a Sharpened Razor," *Brit. J. Philos. Sc.,* 19, 123-143, 1968b.
27. Good, I.J., "The White Shoe *qua* Herring is Pink," *Brit. J. Philos. Sc.,* 19, 156-157, 1968c.

28. Good, I.J., "Utility of a Distribution," *Nature,* 219, 1392, 1968d.
29. Good, I.J., "What is the Use of a Distribution?" in *Second International Symposium on Multivariate Analysis,* edited by P.R. Krishnaiah, New York, Academic Press, 183-203, 1969a.
30. Good, I.J., "A Suggested Resolution of Miller's Paradox," submitted to *Brit. J. Philos. Sc.,* 1969b.
31. Good, I.J., and Toulmin, G.H., "Coding Theorems and Weight of Evidence," *J. Inst. Math. Applics.,* 4, 94-105, 1968.
31 A. Greenberg, B.G., "Problems of Statistical Inference in Health with Special Reference to the Cigarette Smoking and Lung Cancer Controversy," *J. Amer. Stat. Assoc.,* 64, 739-758, 1969.
32. Hempel, C.G., "The White Shoe: No Red Herring," *Brit. J. Philos. Sc.,* 18, 239-240, 1967.
33. Jaynes, E.T. "Information Theory and Statistical Mechanics," *Phys. Rev.,* 106, 620-630, 1957.
34. Jeffreys, Harold, *Theory of Probability,* Oxford, University Press, 1939/61.
35. Jeffreys, Harold, "An Invariant Form for the Prior Probability in Estimation Problems," *Proceedings of the Royal Society,* A, 186, 453-461, 1946.
36. Kerridge, D.F., "Inaccuracy and Inference," *J. Roy. Statist. Soc.,* B, 23, 184-194, 1961.
37. Keynes, J.M., *A Treatise on Probability,* London, Macmillan, 1921.
38. Koopman, B.O., "The Axioms and Algebra of Intuitive Probability," *Annals of Math.,* 41, 269-292, 1940.
39. Koopman, B.O., "Relaxed Motion in Irreversible Molecular Statistics," *Advances in Chemical Physics,* 1969, to appear.
40. Kullback, S., *Information Theory and Statistics,* New York, Wiley; London, Chapman and Hall, 1959.
41. Lindley, D.V., "On the Measure of the Information Provided by an Experiment," *Ann. Math. Statist.,* 27, 986-1005, 1956.
42. Mackie, J.L., "Miller's So-called Paradox of Information," *Brit. J. Philos. Sc.,* 17, 144-147, 1966.
43. Marshak. J., "Remarks on the Economics of Information," in *Contributions to Scientific Research in Management,* Berkeley, University of California Press, 79-98, 1959.
44. McCarthy, John, "Measure of the Value of Information," *Proceedings of the National Academy of Science,* 42, 654-655, 1956.
45. Perks, W., "Some Observations on Inverse Probability Including a New Indifference Rule," *J. Inst. Actuar. Students' Soc.,* 73, 285-334, (with discussion), 1947.
46. Polya, G., *Mathematics and Plausible Reasoning,* Princeton University Press, Two Volumes, 1954.
47. Popper, K.R., *The Logic of Scientific Discovery,* London, Hutchinson; New York, Basic Books, 1959.
48. Shackle, G.L.S., "Expectation in Economics," in *Uncertainty and Business Decisions,* edited by C.F. Carter, G.P. Meredith, and G.L.S. Shackle, Chapter IX, *Liverpool, University Press,* 1954/57.
49. Tribus, M., *Rational Descriptions, Decisions and Designs,* New York, Pergamon, 1969.
50. Wald, A., *Statistical Decision Functions,* New York, Wiley, 1950.
51. Weaver, W., "Probability, Rarity, Interest and Surprise," *Scientific Monthly,* 67, 390-392, 1948.
52. Wiener, Norbert, "The Theory of Prediction," Chapter 8 of *Modern Mathematics for the Engineer,* edited by E.G. Beckenbach, New York, McGraw-Hill, 1956.

APPENDIX

Twenty-Seven Principles of Rationality

In the body of my paper for this symposium I originally decided not to argue the case for the use of subjective probability since I have expressed my philosophy of probability, statistics, and (generally) rationality on so many occasions in the past. But after reading the other papers I see that *enlightenment* is still required. So, in this appendix I give a succinct list of 27 priggish principles. I have said and stressed nearly all of them before, many in my 1950 book, but have not brought so many of them together in one short list. As Laplace might have said, they are *au fond le bon sens,* but they cannot be entirely reduced to a calculus.

1. Physical probabilities probably *exist* (I differ from de Finetti and L. J. Savage here) but they can be *measured* only with the help of subjective probabilities. There are several kinds of probability (Good, 1959, 1966. The latter paper contains a dendroidal categorization.).

2. A familiar set of axioms of subjective probability are to be used. Kolmogorov's axiom (complete additivity) is convenient rather than essential. The axioms should be combined with rules of application and less formal suggestions for aiding the judgment. Some of the suggestions depend on theorems such as the laws of large numbers which makes a frequency definition of probability unnecessary. (It is unnecessary and insufficient.)

3. In principle these axioms should be used in conjunction with inequality judgments and therefore they often lead only to inequality discernments. The axioms can themselves be formulated as inequalities but it is easier to incorporate the inequalities in the rules of application. In other words most subjective probabilities are regarded as belonging only to some interval of values the end points of which may be called the lower and upper probabilities. (Keynes, 1921; Koopman, 1940; Good, 1950, 1962a; C. A. B. Smith, 1961.)

4. The principle of rationality is the recommendation to maximize expected utility.

5. The input and output to the abstract theories of probability and rationality are judgments of inequalities of probabilities, odds, Bayesian factors (ratios of final to initial odds), log-factors or weights of evidence, information, surprise indices, utilities, *and any other functions of probabilities and utilities.* (For example, Good, 1954.) It is often convenient to forget about the inequalities for the sake of simplicity and to use precise estimates (see Principle 6).

6. When the expected time and effort taken to think and do calculations is allowed for in the costs, then one is using the principle of *rationality of type ·II.* This is more important than the ordinary principle of rationality, but is seldom mentioned because it contains a veiled threat to conventional logic by incorporating a time element. It often justifies *ad hoc* procedures such as confidence methods and this helps to decrease controversy.

7. The purposes of the theories of probability and rationality are to enlarge bodies of beliefs and to check them for consistency, and thus to improve

the objectivity of subjective judgments. This process can never be* completed even in principle, in virtue of Gödel's theorem concerning consistency. Hence the type II principle of rationality is a logical necessity.

8. For clarity in your own thinking, and especially for purposes of communication, it is important to state what judgments you have used and which parts of your argument depend on which judgments. The advantage of likelihood is its mathematical independence of initial distributions (priors), and similarly the advantage of weight of evidence is its mathematical independence of the initial odds of the null hypothesis. The subjectivist states his judgments whereas the objectivist sweeps them under the carpet by calling assumptions *knowledge,* and he basks in the glorious objectivity of science.

9. The vagueness of a probability judgment is defined either as the difference between the upper and lower probabilities or else as the difference between the upper and lower log-odds (Good, 1962b). I conjecture that the vagueness of a judgment is strongly correlated with its variation from one judge to another. This could be tested.

10. The distinction between type I and type II rationality is very similar to the distinction between the standard form of subjective probabilities and what I call *evolving or sliding probabilities.* These are probabilities that are currently judged and they can change in the light of thinking only, without change of empirical evidence. The fact that probabilities change when *empirical* evidence changes is almost too elementary a point to be worth mentioning in this distinguished assembly, although it was overlooked by the knight R. A. Fisher in his fiducial argument. More precisely he talked about the probabilities of certain events or propositions without using the ordinary notation of the vertical stroke or any corresponding notation, and thus fell into a fallacy that he was too proud to withdraw. Evolving probabilities are essential for the refutation of Popper's views on simplicity (see Good, 1968).

11. My theories of probability and rationality are theories of consistency only, that is, consistency between judgments and the basic axioms and rules of application of the axioms. Of course, these are usually judgments about the objective world. In particular it is incorrect to suppose that it is necessary to inject an initial probability distribution from which you are to infer a final probability distribution. It is just as legitimate logically to assume a final distribution and to infer from it an initial distribution. (For example, Good, 1959, pp. 35, 81.) To modify an aphorism quoted by Dr. Geisser elsewhere, "Ye priors shall be known by their posteriors".

12. This brings me to the *device of imaginary results* (Good, 1950) which is the recommendation that you can derive information about an initial distribution by an imaginary (*Gedanken*) experiment. Then you can make discernments about the final distribution after a real experiment. A recent writer attributed the device of imaginary results to Laplace, having deleted my name from his first draft, after finding this attribution made by another recent writer whom he cannot now recall!

---

*The words "known to be" should be inserted here to cope with Dr. Barnard's comment.

13. *The Bayes/non-Bayes compromise* (Good, 1957, p. 863; and also many more references in Good, 1965 under the index entry "Compromises"). Briefly: use Bayesian methods to produce statistics, then look at their tail-area probabilities and try to relate these to Bayes factors. A good example of both the device of imaginary results and of the Bayes/non-Bayes compromise was given in Good (1967). I there found that Bayesian significance tests in multiparameter situations seem to be much more sensitive to the assumed initial distribution than Bayesian estimation is.

14. The weakness of Bayesian methods for significance testing is also a strength, since by trying out your assumed initial distribution on problems of significance testing, you can derive much better initial distributions and these can then be used for problems of estimation. This improves the Bayesian methods of estimation!

15. Compromises between subjective probabilities and credibilities are also desirable because standard priors might be more general-purpose than non-standard ones. In fact it is mentally healthy to think of your subjective probabilities as estimates of credibilities (Good, 1952, 108). Credibilities are an ideal that we cannot reach.

16. The need to compromise between simplicity of hypotheses and the degree to which they explain the facts was discussed in some detail in Good (1968), and the name I gave for the appropriate and formally precise compromise was "a Sharpened Razor". Occam (actually his eminent teacher John Duns Scotus) in effect emphasized simplicity alone, without reference to *degrees* of explaining the facts.

17. The relative probabilities of two hypotheses are more relevant to science than the probabilities of hypotheses *tout court*. (Good, 1950, pp. 60, 83-84.)

18. The objectivist or his customer reaches precise results by throwing away evidence; for example (a) when he keeps his eyes averted from the precise choice of random numbers by using a Statistician's Stooge; (b) when his customer uses confidence intervals for betting purposes, which is legitimate *provided that he regards the confidence statement as the entire summary of the evidence.*

19. If the objectivist is prepared to bet, then we can work backwards to infer constraints on his implicit prior beliefs. These constraints are of course usually vague, but we might use precise values in accordance with type II rationality.

20. When you don't trust your estimate of the initial probability of a hypothesis you can still use the Bayes factor or a tail-area probability to help you decide whether to do more experimenting (Good, 1950, p. 70).

21. Many statistical techniques are legitimate and useful but we should not knowingly be inconsistent. The Bayesian flavour vanishes when a probability is judged merely to lie in the interval (0,1), but this hardly ever happens.

22. A hierarchy of probability distributions, corresponding in a physical model to populations, superpopulations, etc., can be helpful to the judgment even when these superpopulations are not physical. I call these *distributions of types I, II, III,...*" (partly in order to be noncommital

about whether they are physical) but it is seldom necessary to go beyond the third type. (Good, 1952; 1965; 1967.)

23. Many compromises are possible, for example, one might use the generalizations of the likelihood ratio mentioned by Good (1960), page 80.

24. *Quasi- or pseudo-utilities.* When your judgments of utilities are other wise too wide it can be useful to try to maximize the expectation of something else that is of value, known as a quasi-utility or pseudo-utility. Examples are (a) weight of evidence, when trying to discriminate between two hypotheses; (b) information in Fisher's sense when estimating parameters; (c) information in Shannon's sense when searching among a set of hypotheses; (d) strong explanatory power when explanation is the main aim: this includes example (c); (e) and (ea) tendency to cause (or a measure of its error) if effectiveness of treatment (or its measurement) is the aim; (f) f(error) in estimation proablems, where $f(x)$ depends on the application and might, for example, reasonably be $1 - e^{-\lambda x^2}$ (or might be taken as x^2 for simplicity); (g) financial profit when other aims are too intangible. In any case the costs in money and effort have to be allowed for.

25. The time to make a decision is largely determined by urgency and by the current rate of acquisition of information, evolving or otherwise. For example, consider chess timed by a clock.

26. In logic, the probability of a hypothesis does not depend on whether it was typed accidentally by a monkey, or whether an experimenter pretends he has a train to catch when he stops a sequential experiment. But in practice we do allow for the degree of respect we have for the ability and knowledge of the person who propounds a hypothesis.

27. All scientific hypotheses are numerological but some are more numerological than others. Hence a subjectivistic analysis of numerological laws is relevant to the philosophy of induction (Good, 1969).

I have not gone systematically through my writings to make sure that the above list is complete. I fact there are, for example, a few more principles listed in Good (1956, 1962c). But I believe the present list is a useful summary of my position.

References

1. Good, I. J., *Probability and the Weighing of Evidence,* London, Charles Griffin; New York, Hafners, 1950.
2. Good, I. J., "Rational Decisions," *J. Roy. Statist. Soc., B,* 14, 107-114, 1952.
3. Good, I. J., Contribution to the discussion on R. D. Clark's paper "The Concept of Probability" at the Institute of Actuaries, 26th October, 1953, *J. Inst. Actuaries,* 80, 19-20, 1954.
4. Good, I. J., "Which Comes First, Probability or Statistics?" *J. Inst. Actuaries,* 82, 249-255, 1956.
5. Good, I. J., "Saddle-point Methods for the Multinomial Distribution," *Annals Math. Statist.,* 28, 861-881, 1957.
6. Good, I. J., "Kinds of Probability," *Science,* 129, 443-447, 1959; (Italian translation by Fulvia de Finetti in *L'Industria,* 1959.) Reprinted in *Readings in Applied Statistics,* edited by William S. Peters, Prentice Hall, 28-37, 1969.
7. Good, I. J., Contribution to the discussion of a paper by E. M. L. Beale, "Confidence Regions in Non-Linear Estimation," *J. Roy. Statist. Soc. B,* 22, 79-82, 1960.

8. Good, I. J., "Subjective Probability as the Measure of a Non-Measurable Set," *Logic, Methodology, and Philosophy of Science: Proc. of the 1960 International Congress,* Stanford, 319-329, 1962a.
9. Good, I. J., "How Rational should a Manager Be?" *Management Science*, 8, 383-393, 1962b; Reprinted with numerous minor improvements, in *Executive Readings in Management Science,* edited by Martin K. Starr, New York, Macmillan, 88-98, 1965.
10. Good, I. J., Contribution to the discussion of a paper by Charles Stein, "Confidence Sets for the Mean of a Multivariate Distribution," *J. Roy. Statist. Soc. B,* 24, 289-291, 1962c.
11. Good, I. J., *The Estimation of Probabilities: An Essay on Modern Bayesian Methods,* M.I.T. Press, 1965; Paperback edition, 1968.
12. Good, I. J., "How to Estimate Probabilities," *J. Inst. Maths. Applics.* 2, 364-383, 1966.
13. Good, I. J., "A Bayesian Significance Test for Multinomial Distributions," *J. Roy. Statist. Soc. B*, 29, 399-431, with discussion, 1967.
14. Good, I. J., "Corroboration, Explanation, Evolving Probability, Simplicity, and a Sharpened Razor," *Brit. J. Philos. Sc.,* 19, 123-143, 1968.
15. Good, I. J., "A Subjective Evaluation of Bode's Law and an 'Objective' Test for Approximate Numerical Rationality," *J. Amer. Statist. Assn.,* 64, 23-66, with discussion, 1969.
16. Keynes, J. M., *A Treatise on Probability,* London, 1921.
17. Koopman, B. O., "The Axioms and Algebra of Intuitive Probability," *Annals of Math.*, 41, 269-292, 1940.
18. Smith, C. A. B., "Consistency in Statistical Inference and Decision," *J. Roy. Statist. Soc. B*, 23, 1-37, including discussion, 1961.

COMMENTS

G. A. Barnard:

I must protest against the statement in priggish principle 10, that Fisher accepted the fiducial argument merely because he did not, on that occasion, use the usual notation for conditional probability. It is inconceivable that someone who, like Fisher, had the perception to isolate the concept of *sufficiency*, which makes essential use of conditional probability, should commit such an elementary and obvious blunder in a matter central to his argument. If anything, the confusion would seem to be in Dr. Good's mind, in his apparent insistence that Pr $[x|\theta]$ (the probability of x when θ is known) must always be interpreted as a conditional probability, whereas it can easily be shown that such a view leads to contradictions (see Barnard, Jenkins and Winsten, 1962, near end of paper).

Further, priggish principle 7 seems to be false. Gödel's theorem suggests that the process in question cannot be *known* to be completed, not that it cannot ever be completed. I make this point because I feel it to illustrate again a weakness in Professor Good's general approach, that he often fails to make the distinction, vital in the present context, between a proposition being true, and the proposition being *known* to be true.

Irwin D. J. Bross:

The speakers of highly specialized languages often have the task of translating from their specialized language into something closer to ordinary, everyday English. The further that the technical language or jargon has evolved from the mother tongue, the more difficult it is to make an adequate translation. At the same time it is the speakers of abstract, esoteric, or highly theoretic languages who most need to translate their jargons into ordinary language. As a consequence, the speakers of abstract languages have evolved various linguistic devices for translation. One such device is what Carnap calls "the *desideratum-explicatum* approach to the analysis of linguistic terms". All of these devices set up a correspondence between some word or phrase in ordinary English—call it "*O*"—and an expression in the specialized language —call it "*X*". In effect, the device leads to an equation of the form "*O = X*". This equation is then used for translating into and out of the jargon.

Carnap's choice of a Latin name, *desideratum-explicatum,* tells us something important. It tells us that the schema is based on the classic notions about language that have come down to us from the Greek and Roman grammarians. But we are living in the 1970's. Even though few philosophers seem to realize it, a modern science of language has come of age in the past generation. This science is often called *descriptive* or *structural* linguistics but I will refer to it as *modern linguistics*. With the advent of the new science of language, there is no more excuse for promulgating archaic beliefs about language than there is for using Ptolemaic notions about the solar system or Aristotelian concepts of biological species. The *desideratum-explicatum* schema is 2000 years out of date. It is in no way improved by expressing it, as Carnap does, in the current notation of symbolic logic.

One of the first things that the modern linguists learned when they began to study natural language with their new scientific techniques was that most of the existing dictionaries were seriously inadequate. In translating from one language to another the dictionary-makers had relied on naive, intuitive devices that were based on ethnocentric notions about language. The resulting translations were confusing, misleading, and often outright wrong. The trouble was that the intuitive equations of "*O = X*" were more of a reflection of the prejudices and predelictions of the dictionary-maker than of the real structures of the two languages.

With this in mind let us consider the dictionary offered by Dr. Good in "The probabilistic explication of Information, Evidence, Surprise, Causality, Explanation, and Utility". Since the list begins with the word *information,* let's consider it first. Dr. Good favors a particular *explicatum, "X"* but he is almost alone in this preference. Most mathematicians seem to prefer a different *explicatum.* However Dr. Good wants to call this other *explicatum* "weight of evidence". He would like to rename Kullback's book *Weight of Evidence and Statistics* (page 113). This is the sort of quibbling that can be expected when a linguistic device is based on subjective impressions rather than upon the observed performance of native speakers. Note the consequences of this quibbling. Instead of "sharpening the terminology" as was claimed, what has actually been done is to blur the distinction between two terms, *information* and *weight of evidence*, which are not likely to be confused in ordinary language.

This kind of blurring occurs constantly but I will give just one more example. Dr. Good uses the initial letters of *degree of explanation* and *information* as the symbols for these quantities. In the next quotation I will substitute the names for the symbols. From the *desideratum-explicatum* approach Dr. Good was "led to the conclusion that the degree of explanation was some increasing function of the information and we might as well identify it with the information itself" (page 115). Again, in ordinary usage the *degree of explanation* is not the same thing as *information* and a failure of the schema to make this distinction should have been a warning that something was wrong with the approach.

What is the trouble here? Clearly, it is this reliance on the highly subjective impressions of a native speaker in setting up the *desiderata* and on his *post hoc* rationalizations for the justification of the *explicatum*. For example, at one point we are told (page 113) that "the notion of negative surprise is intuitively entirely reasonable". At another point (page 111) we are told that "The expression *W* for weight of evidence is clearly more appropriate in terms of ordinary English than is *I* for amount of information...since it is not linguistically clear whether information relevant to *H* should ever be allowed to take negative values. It is not unreasonable, of course, since information *can* be misinformation." These are clearly *post hoc* rationalizations of the unsatisfactory results that are produced by the *desideratum-explicatum* schema.

There is no time to go into the many other examples of mistranslation so let me get to the gist of my message: false translation equations are doubly dangerous. They lead to a situation where statements which are true in the specialized language are false when they are translated into ordinary language. The speaker is deluded—he believes he is telling the truth when he is actually making false statements. The hearer is misled—he may believe the false statement because of the sincerity of the speaker. Since both the speaker and the hearer are fooled, the consequences of the false translations can be especially devastating.

I have time for only one example of how statements with some plausibility in a specialized language lead to absurdities when mistranslated into ordinary language (page 111). In reference to Hempel's paradox that "the observation of a white shoe supports the hypothesis that all crows are black" the following assertion is made:

> "This paradox might seem trivial but if left unresolved it would undermine the whole of statistical inference."

> If *statistical inference* is taken in its ordinary sense the statement is just plain silly. Nothing in actual statistical inference is derived from or requires any support from the abstract languages of symbolic logic where this paradox occurs. Of course if *statistical inference* is equated to certain mathematical schemas that rely on logical languages, then the paradox may constitute a threat to these schemas.

V. P. Godambe:

I admire the considerable amount of flexibility in Professor Good's approach to statistical inference. I would appreciate his telling what the implications of his approach are for what I consider to be the most realistic problem of statistical inference; namely, the inference problem associated with the sampling of a finite population. This problem is characterized in detail in two recent publications: (1) V. P. Godambe, 1969, "Some Aspects of Theoretical Developments in Survey-Sampling," *New Developments in Survey-Sampling,* Wiley-Interscience; (2) V. P. Godambe, 1970, "Foundations of Survey-Sampling," *American Statistician,* Vol. 24, No. 1.

H. E. Kyburg:

Personalistic or subjectivistic principles of probability impose certain conditions on a body of beliefs. Ideally, your body of beliefs should be coherent. Professor Good recognizes that this is an unattainable ideal, and hence not a necessary condition for rationality in the sense in which we might argue that people ought to be rational, that is, for Good's rationality of type II. I should like to argue that it is not sufficient, either, in view of the fact that coherence is not time or evidence dependent. Let us suppose that you have a certain body of beliefs B_t at time t; we may characterize this body of beliefs as a set of propositions, together with a (relatively) coherent probability function P_t whose domain is this set of propositions, and whose value for any proposition is reflected by the least odds at which you would bet on the truth of that proposition. During the interval of time between t and t', you acquire certain evidence E by observation or experience. At t' you will now have a certain body of beliefs $B_{t'}$, which may again be characterized as a set of propositions, together with a (relatively) coherent probability function $P_{t'}$. The Bayesian claim is that for any proposition X, $P_{t'}(X) = P_t(X/E)$. But there seems to be nothing in the subjectivistic doctrine to support this claim. The only constraint on $P_{t'}$ is that it be relatively coherent. We may also impose the constraint that $P_{t'}(E) = 1$, but even this will not support the Bayesian claim except for propositions X entailed by E. As Good himself points out, we can start with the function $P_{t'}$, and argue backwards to come to conclusions about what the belief function should have been at t. Suppose, for example, that $P_{t'}(Y) \neq P_t(Y/E)$; in order to maintain the appropriateness of $P_{t'}$ (supposing that it is coherent) this merely requires a retroactive adjustment of $P_t(Y \cdot E)$ and $P_t(E)$. We say, "Aha, I guess my belief function at t wasn't really P_t, but P^*_t," where P^*_t is coherent, and, among other things, $P^*_t(E) \cdot P_{t'}(Y) = P^*_t(Y \cdot E)$.

The personalistic principles are therefore essentially static principles, and none of Professor Good's twenty-seven principles seem to escape this static character. Inference, however, concerns changes of state; for inference, we need kinematic principles, principles which relate the belief function at one time with the belief function at another, which will relate $P_{t'}$ to P_t. Bayes's theorem is not such a principle, in the first place because it can be worked both ways, and in the second place because it is unclear why, if $P_{t'}$ is relatively coherent at t', P_t should have any bearing on it at all.

There is one kinematic principle, which seems to be implicit in the subjectivist development, though Professor Good does not state it explicitly. This is the principle of conservatism: Don't change P_t; if that was your belief function at t you're stuck with it, whatever the consequences. Although this principle seems to be implicit in much of what Professor Good writes, his doctrine of evolving probabilities – the doctrine that mere reflection can lead us to change our probability function – seems to be at variance with it. The conservative principle does not strike me as a good one but without it, it is hard to see how a subjectivist can throw any light on the kinematical problem of how our beliefs ought to change with the acquisition of evidence.

REPLY

To Professor Bross:

Carnap's use of the expression *desideratum-explicatum* does not imply that the technique is derived from the Greek and Roman grammarians; in fact I think it is a rather modern idea, at least in its mathematical usage. One might as well attribute cybernetics to the ancients. But my main reply to Dr. Bross is that I was not aiming at a description of ordinary English, but rather at the extraction of some of the wisdom from ordinary English for use in scientific and statistical contexts. Please re-read the third paragraph of my paper. This makes most of Dr. Bross's comments irrelevant. The use in a precise technical sense of words previously used in natural language is familiar and hardly misleading; for example, "force", "work", "energy", "heat", "velocity". The desideratum-explicatum approach has been effectively used previously in a mathematical manner; for example, in order to "demonstrate" the axioms of subjective probability, by Serge Bernstein (1917), R. T. Cox (1946), L. J. Savage (1954) and others. Also Popper (1959), who is certainly a modern, had gone some way with the method for explicating "corroboration" and "explanation". I have taken this work to a point where I believe it has begun to pay dividends.

I pointed out that the explication $I(H:E)$ is not as satisfying for amount of information as $W(H:E)$ is for weight of evidence. Dr. Bross has seized on this admission in order to attack the entire desideratum-explicatum approach. Clearly he is assuming here again that my main object was to contribute to ordinary linguistics, so no further reply is necessary.

When I said "the notion of negative surprise is intuitively entirely reasonable" I was quoting from another of my papers where I used the description "only to be expected". I have now put this in the text of my lecture since apparently the point was not otherwise clear enough.

I am convinced that $W(H:E)$ is the best possible quantitative explication for weight of evidence, already of use in medical and other contexts. (See, for example, M. Tribus, *Rational Description, Decisions and Designs,* 1969.) Its expectation is highly relevant to problems of discrimination and I cannot understand why Dr. Bross wishes to blur the distinction between its expectation and the expectation of amount of information. I think the title of Dr. Kullback's book was somewhat misleading since the book has nothing to do with the coding theorems of Shannon, but more with applications of expected weight of evidence. These coding theorems are the main point of Shannon's work. Their value is like the principle of conservation of energy in discouraging people from attempting the impossible. One important communication device that was apparently inspired by Shannon's work was invented by Wozencraft and Herstein. Shannon's work, like General Relativity, is scientifically important even if it has not yet had much direct application. The applications of $I(H:E)$ and $W(H:E)$ and of their expectations may be more important than those of the coding theorems. My paper contains references from which various applications by various people can be traced.

Certainly there have been too many publications on information theory. Shannon himself made the point in an invited editorial entitled "The Bandwagon" in an *IEEE* periodical.

I think Dr. Bross's sample of the use of the word "information" must be biased, because most writers, when using the word in a technical sense in statistics or communication theory, mean either Fisher's sense or Shannon's or Shannon's without expectations (as in my paper). Perhaps it will be of some interest to mention some of the history of the distinction between weight of evidence and information. I was a referee of Dr. Kullback's book and I recommended it for publication although I thought he had blurred the distinction between the two concepts. Independently of myself, the expression "weight of evidence" was proposed effectively for the sense used by me by Charles Peirce (1878) and Minsky and Selfridge (1961). Dr. Bross should know of these references because I mentioned them in a paper in *J. Am. Stat. Assoc.,* 64, 1969, page 26, and he attacked that paper vigorously. The expression "amount of information" is used in my sense, again independently, by Wiener, Fano, Carnap and Bar-Hillel, Stanford Goldman, and others. Its expectation is what Shannon called "amount of information" and is used in many hundreds of papers on information theory. Shannon was not concerned with the expected weight of evidence, a concept that to my knowledge was first used in unpublished work by Turing (1940) and later by me in my 1950 book which was written before Shannon's work on communication theory appeared (apart from an insert made in proof on page 75 in which I mentioned his work). It was only in 1967 that it was shown by George H. Toulmin and myself that coding theorems could be very conveniently derived by making use of expected weight of evidence instead of expected amounts of information.

Dr. Bross next objects to the identification of explanatory power (in the weak sense) with information. I thought it was clear from my paper that the identification was with $I(H:E)$ and I do not know whether Dr. Bross objects to this or whether he is repeating his objection to the explication of "information" here. But my main point in the discussion of explanation was the suggestion of trying to maximize strong explanatory power or its expectation. I hope people will judge this part of my work by trying out some consequences of this recommendation instead of splitting hairs and chopping up my logic.

Regarding Hempel's paradox, I still maintain that logical paradoxes ought to be resolved. The resolution of Hempel's paradox shows that *a "case" of a hypothesis does not necessarily support it.* The reason is given in the listed references. The result *seems* paradoxical because we mentally confuse $W(H:\text{Black Crow})$ with $W(H:\text{Black}|\text{Crow})$, where H asserts that all crows are black. The latter expression *is* necessarily positive. The symbolism helps to elucidate the matter.

It is true that much of science and mathematics does not depend on its foundations, but this is a symposium on the foundations and moreover the foundations often do turn out to be important. The history of science is bestrewn with quarrels about the importance of rigor, and most of these quarrels were rather pointless because there is room in the world for both theoreticians and practitioners. Even trivial-sounding paradoxes have led to important advances; for example, Russell's paradox about the class of all classes that are not members of themselves led him to the theory of types, and this led to an emphasis on the distinction between a language and a metalanguage and hence to Gödel's work (although Hilbert's influence was perhaps greater than Russell's). This, as Quine points out (*Scientific American,* September 1964) led to the branch of mathematics known as *proof theory.* Again, most experimental chemists in the 1930's despised the efforts via quantum mechanics to contribute to chemistry but these efforts were crowned with success in the next decade (M. C. Longuet-Higgins, *The Times Literary Supplement,* May 8, 1970). In fact the continual tension and interplay of logic and experiment, and the illumination and heat that it has generated, is one of the main themes of the history of science and of its sociology.

To Professor Godambe:

In reply to Dr. Godambe's question about his two stimulating papers, I must first agree that survey sampling is a very important part of statistics. Perhaps it has been down-graded in theoretical statistics courses because it was thought to be easy and uninteresting, or obversely because the interesting problems were thought to be too difficult to face.

I agree further that much sampling is from finite populations, but I think rationality of type 2 would usually justify the assumption that the finite population is a random sample from a hypothetical infinite superpopulation. This I think leads to simpler though more advanced mathematics.

In the finite model, every hypothesis concerning the precise constitution of the finite population has the same likelihood if it permits the sample to be a possible one, *assuming that all the x's in the population are unequal.* But a reasonable rounding-off of the x's will usually make many of them equal. The complication arising from the arbitrariness of the rounding off can be avoided by making use of the hypothetical infinite superpopulation, and then the likelihood takes a very familiar and simple form. An analogy is that if we take a multinomial sample $(n_1, n_2, \ldots, n_t)$ $(\Sigma n_i = N)$ from a t-category multinomial population with category probabilities $(p_1, p_2, \ldots, p_t)$ and we compute $X^2 = \sum [(n_i - Np_i)^2/Np_i]$, then X^2 tells us all the n_i's if no non-trivial rational linear combination of $1/p_1, \ldots, 1/p_t$ vanishes and if X^2 is known exactly. The use of continuous methods in discrete problems is too convenient to sacrifice, as in hydrodynamics.

When a Bayesian approach is adopted, the price that has to be paid for using the hypothetical infinite superpopulation is that we are forced to make use of a measure in function space. In what follows I shall for the most part assume the superpopulation model but some of my remarks could be adapted to the finite model.

133

Dr. Godambe stated in the two papers mentioned that the mean of the sample is intuitively the most appropriate point estimate of the mean of a finite population. I do not agree that this is always true. For example, if the sample values were all powers of 2 and were all distinct, and if the sample size were known to be much smaller than the population size, most statisticians would not have this intuition. A more intuitive result is that the *median* of the sample is the most appropriate point estimate of the median of the finite population. If we make no assumptions about the population we *cannot* estimate its mean unless the sample size is an appreciable fraction of the population size! In practice there must always be a non-zero (but perhaps negligible) probability that there is an extreme outlier in the population that has not occurred in the sample. Hence it is safer to estimate the median than the mean. It is sensible to estimate the mean when the initial probability of an extreme outlier is very small. The point seldom *seems* to arise because *most statisticians are Bayesians without knowing it.*

The estimation of the distribution, or adequately the density function, is the ideal, although usually all we need is a typical value and a measure of spread such as the median and a few percentiles. For greater theoretical completeness I shall consider the estimation of the density function, and my comment on the paper by Orear and Cassel becomes relevant. But that discussion referred only to homogeneous (i.e. permutable = exchangeable) and one-dimensional populations and I have to extend it to more general problems of survey sampling here. (In my opinion a judgment of permutability is always provisional and approximate: see *Synthese,* 20, 1969, 16-24.) My discussion will show that I agree with Dr. Godambe in recognizing the relevance of general statistical theory to sample surveys.

Let the expression *surveyed property* refer only to the x's, the random scalar or vector whose distribution is sought. In addition, *individuals* (units) have other features. Some of the features might be taken as *strata* but some might be regarded as irrelevant either initially or later. Any feature is a value taken within a *facet* of features, where I use the word, "facet", as in the literature of information retrieval.

A judgment that a feature is irrelevant, should not be held dogmatically: a sensible man does not *really* assume that two probabilities are precisely equal even when he *says so* for brevity (unless the equality can be proved from the axioms). (See Priggish Principle No. 3.) A pilot sample should be taken to test such judgments and, if rejected, the corresponding facet will become the basis for a new stratification in the next sample. (Likewise strata can be lumped together in the light of experience.) Accordingly survey samples should be sequential in the sense that at least one pilot sample should be taken, unless there are good economic reasons for not doing so.

The fact that an initial guess of approximate irrelevance can become overruled by a large sample is well exemplified by Turing's suggestion (private communication, 1940) that the "frequencies of the frequencies" become relevant in multinomial sampling for a very large number of classes when the sample is large, as in the species sampling problem; see *Biometrika,* 40, 1953, 237-264 and 43, 1956, 45-63. Dr. Godambe makes essentially the same point when he refers to the empirical Bayes method in his paper in the *American Statistician,* and this is no coincidence since Turing's idea was an anticipation of the empirical Bayes method in a special case (essentially the binomial case). There is no reason why the Bayesian or the non-Bayesian should disagree with the empirical Bayes method when it is used with judgment: it attempts to make use of known information that is often disregarded. The Bayesian too makes use of all information *in theory*, and this includes half-remembered information as well as what is remembered. This is one reason why judgment is so essential. (By "in theory" I mean when using rationality of type 1.)

In a sample survey, let $\mathbf{x}$ denote the surveyed property and ξ the relevant features or concomitant variables (for which stratification is desirable). Then the problem we face is the estimation of the density functions $f(\mathbf{x}|\xi)$. For each value of ξ there is a density function to estimate but we might expect some relationship between these different density functions. For example, $\mathbf{x}$ might denote height and ξ sex. Then we would expect the means of the two distributions to be different but the variances to be roughly equal, and we might put this assumption into a precise model.

There are various kinds of Bayesians, and some of them would assume a density ϕ (f) in function space for the density functions f. (ϕ is here a *functional*: its argument is a *function*, not the *value* of a function.) They would compute the likelihood function

$$\Pi_i f(\mathbf{x}_i|\xi_i)$$

where the subscript runs through the labels of the sampled individuals. (The likelihood function is easier to apprehend for this "superpopulation" model than for the finite-population model.) Their final density in function space would be

$$\Phi(f)\,\Pi_i f(\mathbf{x}_i|\xi_i).$$

They would then try to select the f that maximizes this final density. "In theory" (again by Principle 3) I would prefer to use a *class* of initial densities, say ϕ_k (f) for a set of values of k, and would hope to maximize

$$\Phi_k\,(f)\,\Pi_i f(\mathbf{x}_i|\xi_i)$$

by selection of f and k. (*Type 2 maximum likelihood*; see *J. Roy. Stat. Soc., B*, 1967, 399-431.) This is similar to Dr. Godambe's use of a class of "priors", here adapted to the superpopulation model. In my terminology it is not the *labels* that are relevant but the concomitant *features*, but I think this is what Dr. Godambe really has in mind.

If we are sure that a facet is relevant, and if our sample is large enough with respect to each value (stratum) of that facet, then we can safely break the sampling problem into independent problems and ignore the fact that the density functions for each stratum have some interrelationships. This is because with a large enough sample the choice of the initial distribution (prior) is of little importance. This situation often occurs if there is only one facet. If there is still only one facet, but we are not sure whether it is relevant, even after taking our sample, then I should be interested in trying to maximize

$$\Phi(f_r)\,\Pi_t f_r\,(x_{rt})\,\Pi_s\,\{p_{rs}\Pi_t f_r\,(x_{st}) + (1-p_{rs})\,\textstyle\int_g \Phi(g)\,\Pi_t\,g\,(x_{st})\}$$

by choice of the density function f_r, where r refers to the rth stratum of the facet, x_{rt} to the values (for the various individuals) of the surveyed property in the rth stratum, p_{rs} the initial probability that the strata r and s can be lumped together, and

$$\Phi(f) = \frac{\exp\left[-A \int \frac{[f'(x)]^2}{f(x)} dx\right]}{\int_f \exp\left[-A \int \frac{[f'(x)]^2}{f(x)} dx\right]}$$

corresponding to the penalties for roughness in my comments on the paper by Orear and Cassel. $\int_f$ denotes an integration over function space and resembles a Feynman "integral over all paths". (To avoid divergences we must at first restrict f to a finite range, and later take limits.) The above formula is intended to give the relative final probability of any choice for f_r. I should prefer to give extra weight to normal and other simply defined f's. In practice I should test for normality by means of a significance test, Bayesian or otherwise. In other words the measure in function space should not itself be as smooth as the above formula suggests although it is designed to induce smoothness in the f's. The measure in function space in principle should have approximate singularities at places corresponding to simply defined f's.

In view of the numerical difficulties in the above method, it would be worth experimenting with the more *ad hoc* formula

$$\sum_t \log f_r(x_{rt}) + \sum_{s,t}^{s \neq r} q_{rs} \log f_r(x_{st}) - A \int \frac{[f_r'(x)]^2}{f_r(x)} dx$$

where q_{rs} is an estimate of the final probability that strata r and s can be lumped together. For the model with finite populations it is even more difficult to allow for assumptions of normality, mixtures of normality, smoothness, simplicity, etc.

Sometimes we can decide fairly definitely whether any two strata should be lumped together, but the intermediate situation mentioned above would often occur when we were dealing with many facets (a vector facet, a compound facet) because it would then not be practicable to make the sample large enough to separate the "vector strata" decisively. This difficulty with many strata is highly analogous to one that occurs when estimating probabilities for the cells of a multidimensional contingency table. It is an example of the estimation of the probabilities of events that have never occurred: these are the *only* problems of probability estimation in practice, and it appears otherwise only because we continually make judgments of approximate irrelevance.

Dr. Godambe stated that the simplest estimator of a population mean is unbiased and of minimum variance only if the individual labels are ignored. In my terminology it is the specific concomitant features of the individuals that have to be ignored. If they are not ignored then virtually *nothing* can be deduced from the sample without Bayesian assumptions. This point comes up in connection with all applications of randomized designs and is not tied to sample surveys (see Priggish Principle No. 18; *J. Inst. Actuar.*, 82, 1956, page 255; *Information Theory: London Symp. Sep. 1955*, ed. Colin Cherry, London,

1956, page 13; *British J. Philos. Sc.*, 9, 1958, page 252; and *Amer. Statist.*, October 1969, page 44). The entire purpose of randomization is to avoid the use of difficult and controversial Bayesian judgments, and the aim can be achieved only with the help of our friend the Statistician's Stooge. We ask the Stooge to withhold information in order that we can achieve objectivity. I think Fisher was never explicit on this point.

Another place where Dr. Godambe and I think alike is in connection with his definition (in *New Developments in Survey Sampling*, ed. N. L. Johnson and Harry Smith, Jr., Interscience and Wiley, 1969, page 46) of a "Bayesian sufficient statistic". This is similar to the "efficacious statistic" in *J. Am. Stat. Assoc.*, 53, 1958, pp. 803 and 805-6.

Consider the survey sampling problem when the survey property is a vector. Leaving problems of stratification aside we wish to estimate a multidimensional density function. Non-Bayesian discussions of this problem have been given (for example, V. K. Murthy in *Multivariate Analysis*, ed. P. R. Krishnaiah, 1966, 43-56). A Bayesian approach which I think will be better for effectively small samples, can be based on a generalization of my discussion of the paper by Orear and Cassel. *Effectively* small samples are usual in multidimensional problems (and this raises questions about the meaning of "statistical phenomena"). We need a formula for the penalty for roughness that should be subtracted from the log-likelihood.

If the hypersurface representing the density function is compared with its shift by an infinitesimal vector ϵ, the expected weight of evidence in favor of no shift can be seen to be $\frac{1}{2} \epsilon' F \epsilon$ natural bans, where F is Fisher's information matrix. The penalty for roughness can then naturally be taken as proportional either to tr F or to log det F by analogy with mean curvature and the logarithm of Gaussian curvature, in fact the trace of any analytic function of F should be considered. (The definition of the smoothness of a hypersurface naturally has as much ambiguity as the definition of its curvature.) The constant of proportionality depends on how much weight you wish to give to the smoothness. The two specific forms of penalty mentioned are equivalent to the assumption of (improper) initial densities in density-function space of the forms $\exp(-A \, \text{tr} \, F)$ and a power of det F, both of which appropriately factorize when f does. They are invariant under orthogonal transformations of $\mathbf{x}$ and they both can be made completely invariant by replacing F by the expectation of the mixed tensor $-l_{\mu}^{\nu} = -\Sigma_a \, g^{\nu a} \, l_{\mu a}$ where

$$-l_{\mu\nu} = \sum_{r,s} \frac{1}{2} g^{rs} \left(\frac{\partial g_{\mu s}}{\partial x_{\nu}} + \frac{\partial g_{\nu s}}{\partial x_{\mu}} - \frac{\partial g_{\mu\nu}}{\partial x_s} \right) \frac{\partial l}{\partial x_r} - \frac{\partial^2 l}{\partial x_{\mu} \, \partial x_{\nu}}$$

where

$$l(x) = \log f(x) - \frac{1}{2} \log \det \{g_{\mu\nu}\},$$

$$\{g_{\mu\nu}\} = -\frac{\partial^2 v(\mathbf{x}, \mathbf{y})}{\partial y_{\mu} \, \partial y_{\nu}} \bigg|_{\mathbf{y} = \mathbf{x}} = \{g^{\mu\nu}\}^{-1}$$

and $v(\mathbf{x}, \mathbf{y})$ is the utility of asserting that a property has the value $\mathbf{y}$ when the true value is $\mathbf{x}$. But for reasons omitted here I believe that a better form for the penalty is

$$A\mathcal{E}\sum_{\mu,\nu}\frac{\partial l}{\partial x_\mu}\frac{\partial l}{\partial x_\nu}g^{\mu\nu}.$$

If v is a function of a quadratic in $\mathbf{x}-\mathbf{y}$ then $g_{\mu\nu}$ is independent of $\mathbf{x}$ Our space is then so to speak euclidian and the expectation of $\{-l_{\mu\nu}\}$ reduces to F. If v cannot be judged at all it means you have no motivation for the sample survey! The tensorial generalization of Fisher's information matrix should be compared with the invariantized form of entropy [*Nature,* 219, 1968, page 1392 (where a minus sign was omitted in error in the definition of the fundamental tensor); *Multivariate Analysis II,* ed. P. R. Krishnaiah, 1969, 183-203]. A penalty for roughness could be taken proportional to the invariantized negentropy (compare *Annals Math. Statist.,* 34, 1963, page 931; *The Estimation of Probabilities,* 1965, page 76) and this should be considered, but I think the above-mentioned penalties will turn out to be preferable because they force smoothness more effectively; the negentropy seems more appropriate for contingency tables.

About the most general problem of survey sampling is to estimate the densities $f(\mathbf{x}|\xi)$ where the surveyed property $\mathbf{x}$ is a vector and also the relevant concomitant features ξ form a vector compound stratum, that is, we have a cross-stratification. We might then be able to get somewhere by combining the ideas already suggested, but I believe, as a practical matter, that at present we might usually be forced to assume multinormal densities, perhaps after suitable transformations. The parameters of the multinormal densities might sometimes be estimable by regression methods, using the strata as concomitant variables. This does not sound Bayesian, but in my philosophy I ask only that the axioms of subjective probability should not be knowingly contradicted (see Principles 6 and 21).

To Professor Barnard:

Dr. Barnard is correct but misleading when he says that Priggish Principle number 7 is *false,* although I am grateful to him for pointing out a minor *slip,* and I have inserted the words "known to be" in proof. In this corrected form the principle still establishes the logical necessity of using type II rationality in practice. We can never prove our body of beliefs to be consistent so we have to judge when to stop thinking in each application of the theory of probability or of rationality. To quote Pascal (*Pensées,* 1670; English translation by W. F. Trotter, London, Dent, priggish thought number 272), though out of context, "There is nothing so conformable to reason as this disavowal of reason".

The original form of principle number 7 could also have been corrected by deleting the words "concerning consistency" following "Gödel's theorem". That is, we can always enlarge a body of beliefs by appending to it a true but unprovable "Gödel proposition".

I agree that if we do not examine the fiducial argument carefully, it seems almost inconceivable that Fisher should have made the error which he did in fact make. It is because (i) it seemed so unlikely that a man of his stature should *persist* in the error, and (ii) because, as he modestly says (*Statistical Methods and Scientific Inference,* 1956, page 54), his 1930 "explanation left a good deal to be desired", that so many people assumed for so long that the argument was correct. They lacked the *daring* to question it.

Of course Fisher was very familiar with conditional probability but he avoided the use of the usual notation and I can find no example of it in his collected papers. It is a tribute to a good notation when a man of such stature blunders through not using it. I shall try to *pinpoint* the error in the example treated by Fisher, *l.c.,* page 54.

He stated correctly that the inequality

$$\theta > \frac{T}{2n} \chi^2_{2n} (P) \qquad (26) \text{ (Fisher's number)}$$

(where T is a random variable) is verified with frequency P. In the standard notation he meant

$$\text{Prob} \{\theta > \frac{T}{2n} \chi^2_{2n} (P) | \theta\} = P,$$

and from this it *does* follow that

$$\text{Prob}\{\theta > \frac{T}{2n} \chi^2_{2n} (P)\} = P,$$

since the left side is equal to

$$\int \text{Prob} \{\theta > \frac{T}{2n} \chi^2_{2n} (P) | \theta\} \, dF(\theta) = P \int dF(\theta) = P,$$

where F is the cumulative distribution function of θ. It is now a very tempting mistake to assume "by symmetry" that we have further

$$\text{Prob} \{\theta > \frac{T}{2n} \chi^2_{2n} (P) | T\} = P.$$

But this time the argument breaks down because

$$\int \text{Prob}\{\theta > \frac{T}{2n} \chi^2_{2n} (P) | T\} dG(X),$$

where G is the distribution function of T, reduces to

$$\text{Prob}\{\theta > \frac{T}{2n} \chi^2_{2n} (P) | T\}$$

only under the assumption that this expression is mathematically independent of T. Thus the fiducial argument, in this example, *assumes implicitly that the credibility distribution of* θ / T, *when T is given, is mathematically independent of T.* Some subjectivists might accept this assumption, as a formulation of scale invariance, but it should have been made explicitly.

Fisher says that his inequality (26) is first proved for a given value of the parameter θ and then says "It has, however, been proved for all values of θ, and so is applicable to the enlarged reference set of all pairs of values (T, θ) obtained from all values of θ". Here again he assumes that knowing T has no effect but he does not notice that it is an assumption.

I think a decisive objection to the fiducial argument is that a final fiducial probability, based on two observations, can depend on the order in which these observations are taken into account. This is shown in *The Estimation of Probabilities,* 1965, Appendix A, which gives a modified form of an argument of Lindley's. This modified argument is not open to an objection involving "recognizable subsets", and as far as I know there has not yet been a reply to it.

Dr. Barnard tries to save the fiducial argument by stating that $P(x|\theta)$ cannot be always interpreted as a conditional probability and that such an interpretation leads to contradictions. For me this is a conditional probability *by definition,* but I agree that a different definition of conditional probability has unfortunately been used and it does lead to the paradox pointed out by Barnard, Jenkins, and Winsten (*J. Roy. Statist. Soc., A*, 1962, page 351). If I know the bivariate density function of x and θ, then I can infer the probability density of x given that, correct to a zillion places of decimals, $\theta = 0$, but I cannot deduce the probability density of x conditional on $\theta = 0$ exactly because I cannot divide by zero. If θ is transformed to a new variable, the new variable will not in general be known to a fixed accuracy, so no paradox will arise in my system. This is essentially the same resolution of the paradox that is given by Barnard *et al*, except that I disapprove of the definition of conditional probability density as used by Feller and some others. (It is of course very unusual for Feller to make a mistake, but still.) In my system the conditional probability given that $\theta = 0$ *exactly* can be assumed to be unique in the abstract theory, but *its value cannot be deduced from the bivariate density.* This is also in accordance with the approach of Loève (1955), Chapter 7.

In my system (*Probability and the Weighing of Evidence,* 1950, Chapter 3) I often require the *given* proposition to have positive probability in order that simple theorems should be applicable, and the above paradox does not worry me. I agree that people might make mistakes by using a bad definition of conditional probability density, but I do not agree that *this* mistake was made by Fisher or me. Fisher's inequality (26) can be interpreted as conditional on a value of θ being given to a zillion places of decimals: there is no need to interpret it in terms of a bad definition of conditional density.

I hope my attack on the fiducial argument will not be taken as an attack on Fisher's contributions taken as a whole. He was a great man but was only too human.

That I am fully aware of the distinction between a proposition's being true and being *known* to be true is borne out by my comment (*Synthese* 20, 1969, page 22) "But I do not see how de Finetti would, without undue complexity, express the statement that 'really the trials are physically independent although we do not *know* that they are' ". But, like Fisher, the Pope, Karl Marx, and Father Christmas, I am fallible. Sometimes I assume $7 \times 8 = 72$ but I do not *insist* upon it.

To Professor Kyburg:

In my reply to Dr. Kyburg's comments I must repeat that the 27 principles do not exhaust my philosophy, nor even what I have published in the past. In *Probability and the Weighing of Evidence* (1950), page 49, I pointed out that, in the interests of realism, the axiom that $P(E|H) = P(F|H)$ and $P(H|E) = P(H|F)$, when E and F are logically equivalent, should be qualified by "If you have seen that E and F are equivalent". It seems to me that this does introduce a dynamic element that Dr. Kyburg requests. It precludes you from being forced to say "Aha, I guess my belief function at t wasn't really P_t . . .". The modified axiom should be a part of Principle 6, but I left it out for brevity.

On a point of notation I would not write $P_t(X)$ for $P_t(X|E)$, where E denotes all current knowledge, although this would be acceptable to me if the symbol P were changed to some other symbol, say PP, to denote a probability given "all you know", the *comprehensive* probability as it were.

I think there might be another misunderstanding, because when using rationality of type II it is not possible to insist on coherence, but perhaps by "relatively coherent" Dr. Kyburg means "coherent as far as you have seen (up to time t)".

A semantic point that can cause confusion is that, when *empirical* evidence changes, our beliefs might change or not change depending on what is meant: that is, if the evidence changes from E to F, then we can say that $PP(A)$ changes, whereas $P(A|E)$ and $P(A|F)$ might remain what they were before. But in practice we might not have computed $P(A|F)$ at time t; it is only after the evidence becomes F that we might be *interested* in the value of $P(A|F)$.

If my writings sometimes appear to be inconsistent, one reason is that, in the interests of simplicity, I often write as if I were accepting rationality of type I, together with precise probabilities. My justification for writing this way, when I really think type II rationality to be more important, is that precise statements are too lengthy and are unintelligible. It is better to be succinct and slightly misleading than garrulous, precise, and unintelligible: there is an "uncertainty principle" here. Since Dr. Kyburg did not explicitly mention any of my inconsistencies I cannot answer him in more detail.

LIMITING PROPERTIES OF LIKELIHOODS AND INFERENCE

Jaroslav Hájek
Tallahassee

Introduction

This paper supports to some extent the idea that the likelihood principle is applicable within a wide area. The sphere of applicability does not depend on the form of the sample space but on the form of likelihood functions (likelihoods, for short). If the likelihoods are exactly normal, then being given a likelihood function we may draw inference ignoring the statistical model in the background. This has been pointed out in Hájek (1967) and is resumed in Section 2. The main purpose of the present paper is to investigate how far the background model may be ignored, if the likelihoods are only approximately normal. The investigation will be done in the framework of asymptotic theory, whose basic features the paper discusses.

Although my initial aim was to explore only principles and to illustrate the points by examples, I could not resist answering some technical problems, which occurred to me in the course of the writing. Thus the paper contains some new propositions, which are fully stated, but given without proofs. The proofs will be published elsewhere.

My negative attitude towards the likelihood principle, if advocated as a *general* principle, will not be extended here, because the paper is long enough already. The last section contains, however, a few remarks directed against a strict use of the likelihood principle even in cases where it is asymptotically justifiable.

1
Notation

The distribution law of a random variable Y, if θ is the true parameter and value, will be denoted by $L(Y|\theta)$. The normal distribution with expectation vector a and covariance matrix Γ will be denoted by $N(a, \Gamma)$. The fact that a

142

probability measure P has probability density p with respect to some dominating measure μ will be denoted by $dP = pd\mu$. A randomization random variable, usually uniformly distributed over $(0, 1)$, will be denoted by U. If $A - \Gamma$ is a positive semidefinite matrix, we shall write $A \geq \Gamma$. Given a one-dimensional density f, f' will denote its derivative, F the corresponding distribution function, and F^{-1} will be the inverse of F. For $x = (v_1, \ldots, v_k)$, $\|x\|$ will denote the usual norm $(\Sigma_{i=1}^{k} v_i^2)^{1/2}$. The k-dimensional Euclidean space will be denoted by R^k. $E_{n\theta}$ will denote the expectation referring to the probability distribution $P_{n\theta}$. For $x = (v_1, \ldots, v_k)$ and $y = (u_1, \ldots, u_k)$, $x'v$ will denote the scalar product $\Sigma_{i=1}^{k} v_i u_i$, and xy' will denote the matrix $\{v_i u_j\}_{i,j=1}^{n}$.

2

Exactly Normal Likelihoods

Let $\{p_\theta, \theta \in R^k\}$ be a family of densities relative to a σ-finite measure on an arbitrary sample space (X, A). We shall say that the family has *exactly normal likelihood functions*, if there exist a positive definite $(k \times k)$-matrix Γ, a function $c(x)$ and a vector statistic $T(x) = [T_1(x), \ldots, T_k(x)]$ such that for all $x \in X$

$$p_\theta(x) = c(x) \exp\left[-\frac{1}{2}(T(x) - \theta)'\Gamma(T(x) - \theta)\right].$$

The statistic T will be called the *centering statistic*.

The following conclusions can be made for experiments with exactly normal likelihoods:

Proposition 1.

1. $T(x)$ is sufficient.
2. $L(T|\theta) = N(\theta, \Gamma^{-1})$.
3. In any decision problem to any randomized decision function $d(x, U)$ there exists another decision function $d^*(T(x), U)$ providing the same risks.
4. If for some estimate $\hat{\theta}$ the relation
$$E[(\hat{\theta} - \theta)(\hat{\theta} - \theta)'|\theta] \leq \Gamma^{-1} \text{ for all } \theta \in R^k$$
holds, then $\hat{\theta} = T$.
5. The distribution of T given θ may be established from any single likelihood function.
6. For a uniform prior we have $L(\theta|x) = N(T(x), \Gamma^{-1})$.

Proof:

Property 1 is obvious. Property 2 has been proved for $k = 1$ in Hájek (1967) on the basis of completeness of the normal family; the generalization to $k > 1$ presents no new problems. Property 3 follows from 1. Property 4 follows from the proof of admissibility of the usual estimator for $k = 1$. If 4 did not hold, then for some $(c_1, \ldots, c_k) \in R^k$ the estimate $\Sigma_{i=1}^{k} c_i T_i$ would not be admissible for $\Sigma_{i=1}^{k} c_i \theta_i$ under usual mean square deviation risk. Property 5 simply means that any likelihood provides the information about Γ. Property 6 is obvious.

The main topic of the present paper is to discuss asymptotic versions of properties 1–6, if the likelihoods are normal only in an asymptotic sense.

3
Asymptotically Normal Likelihoods

Consider a sequence of families of densities $\{p_n(x_n, \theta), \theta \in \Theta\}$ relative σ-finite measures μ_n on arbitrary sample spaces (X_n, A_n), $n = 1, 2, \dots$. Assume that Θ is an open subset of R^k. Represent the densities in the form

$$p_n(x_n, \theta) = c_n(x_n) \exp \left\{ -\frac{1}{2} n (T_n - \theta)' \Gamma_\theta (T_n - \theta) + Z_n(x_n, \theta) \right\}$$

where $T_n(x_n) = [T_{n1}(x_n), \dots, T_{nk}(x_n)]$ are some *centering* statistics, Γ_θ are $(k \times k)$-matrices, and $Z_n(x_n, \theta)$ are *remainders*.

By saying that the likelihoods are asymptotically normal, we shall understand that the following assumptions are satisfied:

Assumptions A.

The following holds uniformly on every compact subset K of Θ:

A1. T_n is a square root consistent estimate of θ.

A2. The remainders $Z_n(x_n, \theta)$ are continuous in θ for every x_n, and, for every $a > 0$,

$$\max_t \, [|Z_n(x_n, t)|: \sqrt{n}|t - \theta| \leq a] \to 0 \text{ in } P_{n\theta}\text{-probability}$$

where $dP_{n\theta} = p_n(x_n, \theta) \, d\mu_n$.

A3. In $\Gamma_\theta = \{\gamma_{ij}(\theta)\}_{i,j=1}^k$ the functions $\gamma_{ij}(\theta)$ are continuous in θ and $\det \Gamma_\theta \geq \epsilon_K > 0$.

If uniformity holds over all Θ, we shall talk about *Assumptions A**.

The spirit of the above assumptions is the same as of LeCam's (1960) definition of differentially asymptotically normal families: we do not restrict ourselves to n independent replications of a basic experiment and we also do not derive the existence of statistics T_n from some more elementary assumptions. It may be proved, in a non-trivial way, that Assumptions A are somewhat stronger than LeCam's (1960) conditions DN1-DN7 (see page 57, loc. cit.), which are entailed, if we put $\Delta_n(\theta) = \sqrt{n}(T_n - \theta)$ and $\mu(\theta) = 0$. Among others it means that $\sqrt{n}(T_n - \theta)$ is asymptotically normal with parameters $(0, \Gamma_\theta^{-1})$ which will be formulated as a proposition in Section 5. On the other hand, Assumptions A are considerably simpler and, besides uniformities on compact subsets, the only part which is stronger than DN1-DN7 is A2, where we assume the convergence of maximas instead of random variables $Z_n(x_n, t_n)$ for $\sqrt{n}(t_n - \theta) \to h$.

To summarize: Assumptions DN1-DN7 are supremely economical, but drawing consequences from them is a very delicate matter; Assumptions A provide a simpler theory and have the advantage that no explicit assumption of normality of any random variables is made.

144

4
Asymptotic Sufficiency

Following LeCam, we shall say that the centering statistics T_n are asymptotically sufficient on compacts for the families $\{p_n(\cdot, \theta), \theta \in \Theta\}$, if they are sufficient for some families $\{q_n(\cdot, \theta), \theta \in \Theta\}$ such that for any compact $K \subset \Theta$

$$\sup_{\theta \in K} \int |p_n(x_n, \theta) - q_n(x_n, \theta)| \, d\mu_n(x_n) \to 0 \text{ as } n \to \infty. \tag{4.1}$$

If (4.1) holds for K replaced by Θ, we shall say that T_n is asymptotically sufficient.

Asymptotic sufficiency of T_n under Assumptions A may be easily proved by the method of LeCam (1956, p. 142). Choose a sequence of positive numbers $b_n \to \infty$ and $\epsilon_n \to 0$ and put

$$\chi_n(x_n) = 1 \text{ if } \max_t \{|Z_n(x_n, t)| : \sqrt{n} | T_n(x_n) - t| < 2b_n\} < \epsilon_n \tag{4.2}$$
$$= 0 \text{ otherwise, or if the above set of } t\text{-values is empty}$$

and

$$v_n(T_n - \theta) = 1 \text{ if } \sqrt{n} | T_n - \theta| < b_n \tag{4.3}$$
$$= 0 \text{ otherwise.}$$

It follows easily from Assumptions A that ϵ_n and b_n may be chosen so that we have

$$P_{n\theta}(\chi_n(x_n)v_n(T_n(x_n) - \theta) = 1) \to 1 \tag{4.4}$$

uniformly on any compact $K \subset \Theta$. Then we put

$$q_n(x_n, \theta) = k_n(\theta)c_n(x_n)\chi_n(x_n)v_n(T_n - \theta)\exp\{-\tfrac{1}{2}n(T_n - \theta)'\Gamma_\theta(T_n - \theta)\} \tag{4.5}$$

where the functions $k_n(\theta)$ are introduced so as to obtain

$$\int q_n(x_n, \theta)d\mu_n = 1, \theta \in \Theta$$

Proposition 2.

Under Assumptions A the densities $q_n(x_n, \theta)$ of (4.5) satisfy (4.1) for a choice of numbers ϵ_n and b_n in (4.2) and (4.3). Furthermore, then

$$\sup_{\theta \in K} |k_n(\theta) - 1| \to 0 \text{ as } n \to \infty \tag{4.6}$$

for any compact $K \subset \Theta$.

Since T_n is evidently sufficient for $\{q_n(x_n, \theta), \theta \in \Theta\}$, it is asymptotically sufficient on compacts for $\{p_n(x_n, \theta), \theta \in \Theta\}$.

In view of the presence of the functions $k_n(\theta)$ and $v_n(T_n - \theta)$ the likelihoods of the densities $q_n(x_n, \theta)$ are only approximately normal. There are, however, a few cases when the likelihoods in $q_n(x_n, \theta)$ can be made exactly normal. To obtain such a result, we must obviously have $\Theta = R^k$.

Example 1. Put $x_n = (v_1, \ldots, v_n) \in R^n, \Theta = R$ and

$$p_n(x_n, \theta) = \prod_{i=1}^{n} f(v_i - \theta) \tag{4.7}$$

and assume that f has a finite Fisher's information I_f. Introducing $\phi(u) = -f'(F^{-1}(u))/f(F^{-1}(u)), 0 < u < 1$, it means that

$$I_f = \int_0^1 \phi^2(u)\,du < \infty. \tag{4.8}$$

Then Assumptions A* are satisfied and there exist statistics $c_n^*(x_n)$ and $T_n(x_n)$ such that the densities

$$q_n(x_n, \theta) = c_n^*(x_n)\exp\{-\tfrac{1}{2}n(T_n - \theta)^2 I_f\} \tag{4.9}$$

satisfy

$$\sup_{-\infty < \theta < \infty} \int \ldots \int |q_n(x_n, \theta) - p_n(x_n, \theta)|\,dv_1 \ldots dv_n \to 0. \tag{4.10}$$

In fact the integrals may be made constant in θ.

The centering statistic T_n may be obtained as *restricted* maximum likelihood estimates over the regions $\{\theta: |\theta - \hat{\theta}_n| \leq 1\}$ where $\hat{\theta}_n$ is some consistent location invariant estimate (see LeCam, 1968). If

$$\int_{-\infty}^{\infty} f(x) \log f(x)\,dx > -\infty, \tag{4.11}$$

then T_n may be the maximum likelihood estimate over the whole space $-\infty < \theta < \infty$. If (4.11) does not hold, the behaviour of the unrestricted maximum likelihood estimate is not known. Assuming that $\phi = \phi_1 - \phi_2$ where both ϕ_1 and ϕ_2 are non-decreasing and square integrable, which covers all reasonable cases, then we may take for T_n any estimate defined by

$$T_n = \tilde{\theta} - I_f^{-1} n^{-1} \sum_{i=1}^{n} \frac{f'(v_i - \tilde{\theta})}{f(v_i - \tilde{\theta})}, \tag{4.12}$$

where $\tilde{\theta}$ is an arbitrary location invariant and square root consistent estimate. For example, we may take for θ a properly shifted sample α-quantile for any α such that $f(F^{-1}(\alpha)) > 0$.

Since the likelihoods are *exactly* normal under q_n of (4.9), we have from Proposition 1, part 2,

$$L(\sqrt{n}(T_n - \theta)|q_{n_\theta}) = N(0, I_f^{-1}). \tag{4.13}$$

146

In view of (4.10), it entails

$$L(\sqrt{n}\,(T_n-\theta)\,|p_{n\theta}) \to N(0, I_f^{-1}). \tag{4.14}$$

The cases when the approximating family $\{q_n\}$ can be made to have exactly normal likelihoods are relatively rare. In some instances we may get the result for a suitably transformed parameter.

Example 2. Let us consider the sequence of gamma densities:

$$p_n(x, \theta) = \theta^{-n}[\Gamma(n)]^{-1}\, x^{n-1} \exp\,(-x/\theta),\, x > 0,\, \theta > 0. \tag{4.15}$$

Transforming the parameter by

$$\tau = \log\theta \tag{4.16}$$

we can approximate the family by

$$q_n(x, \theta) = c_n(x)\, \exp\,[-n(\log x - \tau)^2] \tag{4.17}$$

which has exactly normal likelihoods in terms of τ. We again have

$$\sup_{0 < \theta < \infty} \int_0^\infty |p_n(x, \theta) - q_n(x, \theta)|\,dx \to 0 \text{ as } n \to \infty. \tag{4.18}$$

The reader recognizes in (4.16) the well-known variance-stabilizing transform. However, not every transform of this type provides an approximation family with exactly normal likelihoods. In the binomial family and Poisson families, for example, the transformed parameter space does not coincide with R. In the normal correlation family again appropriate group structural properties are missing.

There are situations where the parameter space as well as its dimension may depend on n. This may be illustrated by the following.

Example 3. Fix some $b > 0$ and put

$$\Theta_n = \{\theta_n = (d_1, \ldots, d_n) \in R^n: \sum_{i=1}^n d_i = 0,\, \sum_{i=1}^n d_i^2 \le b\}. \tag{4.19}$$

Then take a density f with finite Fisher's information and, denoting $x_n = (v_1, \ldots, v_n) \in R^n$, consider the families

$$p_n(x_n, \theta_n) = \prod_{i=1}^n f(v_i - d_i),\, \theta_n \in \Theta_n. \tag{4.20}$$

Introducing the vector of ranks $R_n = (R_{n1}, \ldots, R_{nn})$ where $R_{ni}(x_n)$ denotes the number of coordinates of x_n that are smaller or equal to v_i, it may be derived ([4], VII 1.2) that the families may be approximated, uniformly on Θ_n, by families

$$q_n(x_n, \theta_n) = c_n(x_n)\, B_n(\theta_n)\, \exp\,\{-\frac{1}{2}\sum_{i=1}^n [d_i - I_f^{-1} a_n(R_{ni})]^2\}$$

where $a_n(i) = E\phi(U_n^{(i)})$, with $\phi(u) = -f'(F^{-1}(u))/f(F^{-1}(u))$ and $U_n^{(i)}$ denoting the i-th order statistic in a sample of size n from the uniform distribution over $(0, 1)$.

Of course, since the approximation of p_n by q_n holds uniformly on θ_n, it holds uniformly on any subsets of Θ_n that may be appropriate for applications. Since the vector of ranks R_n is evidently sufficient for the family $\{q_n(x_n, \theta_n), \theta_n \in \Theta_n\}$, it follows that it is asymptotically sufficient for $\{p_n(x_n, \theta_n), \theta_n \in \Theta_n\}$.

5
Asymptotic Normality of Centering Statistics

It is possible to prove the following.

Proposition 3.

Under Assumptions A

$$L(\sqrt{n}(T_n - \theta)|\theta) \to N(0, \Gamma^{-1})$$

holds uniformly on compacts $K \subset \Theta$.

The proposition shows that asymptotic normality may be proved from square root consistency and certain properties of likelihood functions. The usefulness of this fact is exhibited by the following.

Example 4. Let $X_n = (Y_1, \ldots, Y_n)$ be a normal stationary autoregressive sequence of order one with zero expectations. Putting $\theta = (a, b)$ and $\theta = \{(a, b) : a > 0, |b| < 1\}$, we obtain the following family of densities:

$$p_n(x_n, \theta) = (2\pi)^{-\frac{1}{2}n} a^n (1 - b^2)^{\frac{1}{2}}$$

$$\exp\left\{-\frac{1}{2}a^2(b - \hat{b}_n)^2 \sum_{i=2}^{n-1} y_i^2 - \frac{1}{2}a^2\left(\sum_{i=1}^{n} y_i^2 - \hat{b}^2 \sum_{i=2}^{n-1} y_i^2\right)\right\} \qquad (5.1)$$

where

$$\hat{b}_n = \frac{\displaystyle\sum_{i=1}^{n-1} y_i y_{i+1}}{\displaystyle\sum_{i=2}^{n-1} y_i^2} \qquad \text{and } x_n = (y_1, \ldots, y_n). \qquad (5.2)$$

Now we may prove without difficulty that $\hat{b}$ is a square root consistent estimate of b uniformly on compacts $K \subset \Theta$, and $\frac{1}{n}\sum_{i=2}^{n-1} y_i^2$ is a square root consistent estimate of $a^{-2}(1 - b^2)^{-1}$. Consequently, we may write

$$p_n(x_n, \theta) = c_n(x_n) \, a^n \exp \left\{ -\frac{1}{2} n (b - \hat{b}_n)^2 / (1 - b^2) - \frac{1}{2} a^2 (1 - \hat{b}_n{}^2) \right.$$

$$\left. \sum_{i=1}^{n} y_i^2 + Z_n(x_n, \theta) \right\}. \quad (5.3)$$

Fixing a, we can see that Assumptions A are satisfied for $T_n = \hat{b}_n$ and $\Gamma_b = (1 - b^2)^{-1}$. Consequently whatever may be a, $\hat{b}_n$ is asymptotically normal $(b, n^{-1}(1 - b^2))$.

6

Asymptotic Testing of Contiguous Alternatives

Fix a simple hypothesis θ_o and introduce local coordinates λ as follows:

$$\theta = \theta_o + n^{-1/2} B \lambda, \quad (6.1)$$

where B is a matrix satisfying

$$B'B = \Gamma_{\theta_o}{}^{-1}. \quad (6.2)$$

Now choose a bounded subset L of R^k and consider alternatives K_n consisting of θ-points that correspond to $\lambda \in L$:

$$K_n = \theta_o + n^{-1/2} BL, \quad n = 1, 2, \ldots. \quad (6.3)$$

We shall call a sequence of critical functions Φ_n asymptotically most powerful (in the maximin sense) on level α for testing θ_o against K_n, if

$$\limsup_{n \to \infty} \int \Phi_n(x_n) p_n(x_n, \theta_o) d\mu_n \leq \alpha \quad (6.4)$$

and the minimum power

$$\beta_n = \inf_{\theta \in K_n} \int \phi_n(x_n) p_n(x_n, \theta) d\mu_n \quad (6.5)$$

has the largest possible limit.

For families satisfying Assumptions A the solution may be given in three steps:
1. We prove that Φ_n may depend on x_n through T_n.
2. We suggest a critical function which gives the best solution for testing $N(0, I)$ against the family $\{N(\lambda, I), \lambda \in L\}$, where I denotes the identity convariance matrix.
3. We prove that there is no better solution.

Step 1 depends on sufficiency of T_n. Given any sequence of critical functions Φ_n let us put

$$\Phi_n^*(t) = E[\Phi(x_n) \mid T_n = t, q_{n\theta} \text{ holds}]. \quad (6.6)$$

Since T_n is sufficient for $\{q_{n\theta}\}$, $\phi_n^*(t)$ will be independent of θ. From (6.6) it follows that

$$\int \Phi_n^*(T_n(x_n)) q_n(x_n, \theta) d\mu_n = \int \Phi_n(x_n) q_n(x_n, \theta) d\mu_n, \quad \theta \in \Theta.$$

This combined with (4.1) entails that

$$\int [\Phi_n(x_n) - \Phi^*(T_n(x_n))] p_n(x_n, \theta) d\mu_n \to 0 \tag{6.7}$$

uniformly on compact subsets of Θ. Since there obviously is a compact set containing all the K_n's, the first step is by (6.7) proved.

Step 2 is based on the fact that

$$L(n^{1/2} B^{-1}(T_n - \theta_0) \mid \theta) \to N(\lambda, I) \tag{6.8}$$

uniformly for $\|\lambda\| < M < \infty$, if λ is related to θ, and B to Γ_{θ_0}, by (6.1) and (6.2). Statement (6.8) follows easily from Proposition 3 and continuity of Γ_θ. Now if $\Phi^0(t)$ is the best critical function for testing $N(0, I)$ against $\{N(\lambda, I) \lambda \in L\}$, we obtain from (6.8)

$$\sup_{\|\lambda\| < M} | \int \phi^0[n^{\frac{1}{2}} B^{-1}(T_n - \theta_0)] dP_{n\theta} - \int \phi^0(t) (2\pi)^{-\frac{1}{2}k}$$

$$\exp[-\frac{1}{2} \sum_{i=1}^{k} (t_i - \lambda_i)^2] dt_1 \ldots dt_k | \to 0 \tag{6.9}$$

provided that $\Phi^0(t)$ is continuous almost everywhere (Lebesgue measure). This will be true for all reasonable alternatives L. However, even a completely general Φ^0 would be tractable, since it may be proved that there exist random variables Y_n of the form

$$Y_n = h_n(T_n, U) \tag{6.10}$$

where h_n are some functions, and U has uniform distribution over $(0, 1)$ independently of T_n, such that $L(Y_n \mid P_{n\theta})$ converges to $N(\lambda, I)$ in variation uniformly for $\|\lambda\| < M$. Then (6.9) holds with $\phi^0[n^{-1/2} B^{-1}(T_n - \theta_0)]$ replaced by $\int_0^1 \phi^0[h_n(T_n, u)] du$. Now, if

$$E[\Phi^0(T) \mid N(0, I)] \leq \alpha, \quad \inf_{\lambda \in L} E[\Phi^0(T) \mid N(\lambda, I)] = \beta, \tag{6.11}$$

then (6.5) entails

$$\limsup_{n \to \infty} \int \Phi^0[n^{1/2} B^{-1}(T_n - \theta_0)] dP_{n\theta_0} \leq \alpha \tag{6.12}$$

$$\lim_{n \to \infty} [\inf_{\theta \in K_n} \int \Phi^0[n^{1/2} B^{-1}(T_n - \theta_0)] dP_{n\theta_0}] = \beta. \tag{6.13}$$

150

Example 5. Choose $c > 0$ and put $L = \{\lambda: \|\lambda\|^2 = c\}$. The corresponding set in the Θ space will be, in view of (6.1) and (6.2)

$$K_n = \{\theta: n(\theta - \theta_0)'\Gamma_{\theta_0}(\theta - \theta_0) = c\}. \tag{6.14}$$

Now it is well-known that the best critical function Φ^0 for testing $N(0, I)$ against $\{N(\lambda, I), \lambda \in L\}$ equals 1 if

$$\|t\|^2 = \sum_{i=1}^{n} t_i^2 \geq k_\alpha \tag{6.15}$$

and equals 0 otherwise. Since the Lebesgue measure of the set $\|t\|^2 = k_\alpha$, on which Φ^0 is discontinuous, equals 0, the above theory applies and the test which rejects θ_0 in favor of K_n if and only if

$$\|n^{-1/2}B^{-1}(T_n - \theta_0)\|^2 = n(T_n - \theta_0)'\Gamma_{\theta_0}(T_n - \theta_0) \geq k_\alpha \tag{6.16}$$

has the same asymptotic significance level and asymptotic minimal power as Φ^0 in testing $N(0, I)$ against $\{N(\lambda, I), \|\lambda\|^2 = c\}$.

In order to complete step 3 it does not suffice to know that the distribution of T_n is asymptotically normal. We shall illustrate that by the following.

Example 6. Let ξ_n be a binomial random variable with parameters n and $p = 1/2$. Consider two distributions of T_n defined by means of ξ_n as follows:

$$P_{n0}(T_n = k) = P(\xi_n = k) \tag{6.17}$$

$$P_{n1}(T_n = k + [\sqrt{n}] + \tfrac{1}{2}) = P(\xi_n = k) \tag{6.18}$$

$$\left.\right\} \, k = 0, 1, \ldots, n.$$

where $[\sqrt{n}]$ denotes the integral part of $\sqrt{n}$. Then, obviously, for every n the distributions P_{n0} and P_{n1} are disjoint and may be distinguished with no error (i.e. $\alpha = 0, \beta = 1$) by any critical function such that

$$\Phi(k) = 0, \Phi(k + \tfrac{1}{2}) = 1, \text{ if } k \text{ is an integer.} \tag{6.19}$$

On the other hand the limiting distributions

$$L[n^{-1/2}(T_n - \tfrac{1}{2}n) \,|\, P_{n0}] \to N(0, \tfrac{1}{4}) \tag{6.20}$$

$$L[n^{-1/2}(T_n - \tfrac{1}{2}n) \,|\, P_{n1}] \to N(1, \tfrac{1}{4}) \tag{6.21}$$

lack this property and the sum of errors of both kinds cannot be smaller than a certain positive number.

Fortunately, in our case we know in addition that the likelihoods are approximately normal. From this we are able to derive, in a nontrivial way, that for every sequence of test functions $\Phi_n(x_n)$ there exist a subsequence Φ_{n_j} and a test function $\psi(t)$ in R^k such that

$$\int \Phi_{n_j}(x_n) p_{n_j}(x_n, \theta) d\mu_n \to \int \psi(t)(2\pi)^{-\frac{k}{2}} \exp[-\frac{1}{2}\sum_{i=1}^{k}(t_i - \lambda_i)^2] dt_1 \ldots dt_k \quad (6.22)$$

uniformly for $\|\lambda\| < M$.

All three steps yield the following.

Proposition 4.
Under Assumptions A the best test for θ^0 against K_n of (6.3), where L is bounded, has in the limit the same power as the best test for testing $N(0, I)$ against $\{N(\lambda, I), \lambda \in L\}$.

Historical notes. Asymptotic hypothesis testing theory was founded by A. Wald (1943) in his very long and difficult paper. He considered n independent replications of a basic experiment and used assumptions that entail Assumptions A* with T_n being the maximum likelihood estimate. Using the *paving* method he proved that for any event $A_n \in \mathscr{A}_n$ there exist a Borel subset $B = B(A_n)$ of R^k such that

$$|P_{n\theta}(A_n) - \int_{B(A_n)} (2\pi)^{-\frac{k}{2}}|\det \Gamma_\theta|^{1/2} \exp[-\frac{1}{2}n(t-\theta)'\Gamma_\theta(t-\theta)]dt_1 \ldots dt_k| < \epsilon$$

holds for $n > n_\epsilon$ uniformly in $A_n \in A_n$ and $\theta \in \Theta$. From this Proposition 4 easily follows.

The most important subsequent development is due to L. LeCam (1956 and 1960). In his (1956)-paper he relaxed Wald's conditions and replaced the maximum likelihood estimate by estimates obtainable from square-root consistent estimates by *improving* as exemplified by (4.12) above. He proved asymptotic sufficiency of centering statistics and then he established the existence of a function $\phi_n : R^k \to R^k$, such that $\sup_t |\phi_n(t) - t| \to 0$ as $n \to \infty$ and

$$|P_{n_v}(T_n \in B) - \int_{\phi_n^{-1}(B)} (2\pi)^{-\frac{k}{2}}|\det \Gamma_\theta|^{1/2} \exp[-\frac{1}{2}n(t-\theta)'\Gamma_\theta(t-\theta)]dt_1 \ldots dt_k| < \epsilon$$

for $n > n_\epsilon$ uniformly for Borel subsets B of R^k and θ belonging to compact subsets of K. This also yields Proposition 3 immediately. The point transforms ϕ_n established by LeCam are more appropriate for estimating problems than the set transforms used by Wald. In his (1960)-paper LeCam elaborated consequences of the conditions of differential asymptotic normality DN1-DN7 mentioned in Section 3. Proposition 4 could also be derived under these extremely economical conditions.

In Hájek-Šidák (1967) asymptotic theory is developed for rank tests and an alternative approach to Proposition 4 is used.

In all four above sources not only simple hypotheses are considered, but also composite ones.

We shall conclude by explaining why we called the alternatives K_n contiguous in the title. Actually, Proposition 3 entails easily, by a LeCam lemma (Hájek-Šidák 1967, VI. 1.2), that the sequences $\{P_{n\theta_n}\}$, $\theta_n \in K_n$ are contiguous to the sequence $\{P_{n\theta_0}\}$. Contiguity implies that convergence in $P_{n\theta_0}$-probability entails the convergence in $P_{n\theta_n}$-probability. Thus for $\alpha = 0$ the power of the best test converges to 0, for example.

7

Asymptotic Estimation

Having obtained an asymptotic result we usually are not able to tell how far it applies to particular cases with finite n. In other words, given $\epsilon > 0$ we are not able to decide whether some n of interest is or is not larger than n_ϵ whose existence the limit theorem ensures. Theorems providing a satisfactory answer to this question would be much more modest and much less elegant. Consequently, in applications we are guided by two epistemologically very different sources of knowledge: on one side, we have limit theorems giving some hope, but no assurance for practical sample sizes; on the other side, we dispose with some numerical experience, which we extend to cases that seem to us to be *similar*.

Especially misinformative can be those limit results that are not uniform. Then the limit may exhibit some features that are not even approximately true for any finite n.

Example 7. In Hájek-Šidák (1967) a test is constructed that is asymptotically optimal for testing regression in location whatever may be the underlying density with finite Fisher's information. The convergence of the power, however, is not uniform with respect to the choice of density. Thus no matter that the limiting power is the best possible one for all the densities, for any finite n, the actual power will be more than poor for most densities under consideration. A theorem, where convergence is replaced by uniform convergence, would pertain to much more modest classes of densities.

Example 8. Super efficient estimates produced by L. J. Hodges (see LeCam 1953, p. 280) have their shocking properties only in the limit. For any finite n they behave quite poorly for some parameter values. These values, however, depend on n and disappear in the limit.

Fisher's program in the asymptotic theory of estimation was to prove that the best possible limiting distribution of an estimate is (under regularity conditions) $N(\theta, n^{-1}\Gamma_\theta)$ and that this optimum result is reached only by estimates that are asymptotically equivalent to the maximum likelihood estimate. If we interpret this conjecture in the sense of generally nonuniform convergence we come to paradoxes of Hodges' examples, and the conjecture is easily disproved. However, to come closer to what Fisher might have had in mind we must introduce some uniformity considerations. The requirements either impose some restrictions on the class of estimates that are allowed to compete (see Wolfowitz 1965 or Hájek 1970, e.g.) or they give a stricter meaning to the words *best possible* (see Huber 1966). Wolfowitz, while defending the first approach, argues that the limiting distribution of any *operational* estimate should be estimable. We may ask, however, why we should exclude from competition estimates whose limiting distribution is equal to or better than some estimable distribution? On the other hand, in Huber's approach we do not exclude any estimates, and only require that favourable behaviour of one parameter point should not be depreciated by poor behaviour at neighboring points.

The next proposition has been obtained by a different method than used by Huber (1966).

Proposition 5.

Let Θ be an open subset of the line, i.e. $k = 1$. Consider a nondecreasing function $l(y), y \geq 0$, introduce

$$r(\sigma^2) = (2\pi)^{-1/2}\sigma^{-1}\int_{-\infty}^{\infty} l(|y|)\exp(-\frac{1}{2}y^2/\sigma^2)dy \qquad (7.1)$$

and assume that $r(\sigma^2) < \infty$ for any $\sigma^2 > 0$. Fix $\theta \in \Theta$.

Then, under Assumptions A, any sequence of estimates $\hat{\theta}_n$ either satisfies

$$\liminf_{n \to \infty} E_{n\theta} l(\sqrt{n}|\hat{\theta}_n - \theta|) \geq r(\Gamma_\theta^{-1}) \qquad (7.2)$$

or there exists a sequence of parameter points θ_n such that for some constant $M < \infty$

$$\sqrt{n}|\theta_n - \theta| < M \qquad (7.3)$$

holds and

$$\limsup_{n \to \infty} E_{n\theta_n} l(\sqrt{n}|\hat{\theta}_n - \theta_n|) > r(\Gamma_\theta^{-1}). \qquad (7.4)$$

Example 9. For $l(y) = y^2$ we obtain that either the Cramér-Rao bound is not exceeded, i.e.

$$\liminf_{n \to \infty} E_{n\theta}[n(\hat{\theta}_n - \theta)^2] \geq \Gamma_\theta^{-1} \qquad (7.5)$$

or it is exceeded by some positive number, independent of n, in the neighborhood of θ for some n's of arbitrary magnitude. Putting $l_b(y) = \min(y^2, b)$, we obtain either

$$\lim_{b \to \infty}\liminf_{n \to \infty} E_{n\theta} l_b(\sqrt{n}|\hat{\theta}_n - \theta|) \geq \Gamma_\theta^{-1} \qquad (7.6)$$

or uniformly sharp inequalities for a subsequence of a sequence θ_n satisfying (7.3).

Also, if $L[\sqrt{n}\,(\hat{\theta}_n - \theta_n)\,|\,\theta_n] \to G$ for any sequence $\theta_n \to \theta$, then

$$\int y^2 dG(y) \geq \Gamma_\theta^{-1}$$

and, generally,

$$\int l(|y|)\,dG(y) \geq r(\Gamma_\theta^{-1}).$$

Example 10. Another interesting loss function is

$$\lambda_a(y) = \begin{matrix} 0 \text{ if } 0 \leq y \leq a \\[4pt] 1 \text{ if } a < y < \infty. \end{matrix}$$

If we apply Proposition 5 to this function, we obtain that either

$$\liminf_{n \to \infty} P_{n_\theta}\,(\sqrt{n}\,|\hat{\theta}_n - \theta| \geq a) \geq 2\Phi(-a\Gamma_\theta^{1/2}),$$

where $\Phi(x) = (2\pi)^{-1/2}\int_{-\infty}^{x}\exp(-\tfrac{1}{2}y^2)\,dy$, or that there is a uniformly sharp inequality for a subsequence of a sequence satisfying (7.3).

In order to obtain analogues of Proposition 5 for a multidimensional parameter space, we must consider the loss function also multidimensional (i.e. vector-valued or matrix-valued).

Example 11. If $\theta \in \Theta \subset R^k, k$ arbitrary, then either

$$\liminf_{n \to \infty} E_{n\theta}(\hat{\theta}_n - \theta)(\hat{\theta}_n - \theta)' \geq \Gamma_\theta^{-1}$$

or there exist a vector $c \in R^k$, a constant $M < \infty$, and a sequence θ_n such that $\sqrt{n}\,\|\theta_n - \theta\| < M$, for which

$$\limsup_{n \to \infty} E_{n\theta_n}[c'(\hat{\theta}_n - \theta_n)'(\hat{\theta}_n - \theta_n)c] > c'\Gamma_\theta^{-1}c.$$

As is known from nonasymptotic consideration (Stein 1956) Proposition 5 (with $|\hat{\theta}_n-\theta|$ replaced by $\|\theta_n-\theta\|$) is not correct for $k > 2$. It may be correct for $k = 2$, but this has not been yet proved. The explanation of failure for $k > 2$ may be found in the fact that estimating a multidimensional parameter with a one-dimensional loss function includes as a special case repeated estimation of a one-dimensional parameter that varies from one experiment to another. If k is sufficiently large we may successfully estimate and utilize prior distribution of this parameter, if interested in minimizing the average risk.

Proposition 5 allows us to say that Fisher's program has been essentially sound.

If we know the family $\{p_n(x_n, \theta)\}$ and a centering statistic we can estimate Γ_θ by simply substituting T_n for θ. This gives a consistent estimate, uniformly on compacts. If we are, however, adherents of the likelihood principle, we must seek for estimation methods that are independent of our knowledge of $\{p_n(x_n, \theta)\}$. One possibility is to estimate Γ_θ by

$$\left\{ -\frac{1}{n} \frac{\partial^2}{\partial \theta_i \partial \theta_j} \log p_n(x_n, \theta) \bigg|_{\theta = T_n \theta_n} \right\}_{i, j=1}^{k}$$

where θ_n is the maximum likelihood estimate.

To prove consistency of this estimate would require conditions much stronger than our Assumptions A (namely in the point A2), but these alternative conditions are often satisfied in practical models. A more general method is given in the following.

Proposition 6.

Under Assumptions A there exists a sequence $b_n \to \infty$ such that the matrix

$$c_n(x_n) \quad \underset{\sqrt{n}\|t - T_n\| < b_n}{\int \ldots \int} (t - T_n) \, (t - T_n)' p_n(x_n, t) dt_1 \ldots dt_k$$

where $t = (t_1, \ldots, t_k)$ and $c_n(x_n)$ is defined by

$$c_n(x_n) \quad \underset{\sqrt{n}\|t - T_n\| < b_n}{\int \ldots \int} p_n(x_n, t) dt_1 \ldots dt_k = 1,$$

is a consistent estimate of Γ_θ^{-1}, uniformly on compact subsets of Θ.

The above proposition is not exactly in the likelihood principle spirit, since the speed of convergence of $b_n \to \infty$ and the choice of T_n depend on the sequence of families $\{p_n(x_n, \theta)\}$. However, it shows that methods based on the knowledge of likelihood functions only, with ignorance of the distribution family in the background, may be often successful.

9
Bayes Estimates

Consistently using the likelihood principle, we also have to define the centering statistics as functionals on the likelihoods, without reference to the family of probability densities in the background. One way is to take for the centering statistic the maximum likelihood estimate. Of course, satisfaction of Assumptions A is not sufficient for proving even consistency of the maximum likelihood estimate. However, if we know that the ML estimate is square-root consistent, uniformly on compacts, then it is easy to show that it has all properties of a centering statistic under A.

Another possibility is to take for the centering statistic conditional expectation of θ given X_n with respect to some prior distribution, possibly depending on X_n.

Proposition 7.

Under Assumptions A, there exists a sequence $b_n \to \infty$ such that

$$T_n^* = c_n(x_n) \quad \int \ldots \int_{\sqrt{n}\|t-T_n\| \leq b_n} t \, p_n(x_n, t) \, dt_1 \ldots dt_k \qquad (9.1)$$

is also a centering statistic. Consequently, $L[\sqrt{n}(T_n^* - \theta)|\theta] \to N(0, \Gamma_\theta^{-1})$ as $n \to \infty$, uniformly on compacts.

Of course, T_n^* will be a proper likelihood principle estimate if T_n has the same property and b_n is defined independently of the underlying family of distributions. Both T_n and b_n may be removed from consideration, if the integration in (9.1) could be extended over all Θ. A partial result of this kind is given in the following.

Example 12. Let $f(x)$ possess a finite Fisher's information I_f and let there exist $\alpha > 0$ such that

$$\sup_{-\infty < x < \infty} f(x)|x|^\alpha < \infty. \qquad (9.2)$$

Then, for the family considered in Example 1, the statistics

$$T_n^* = [\int_{-\infty}^\infty (\prod_{i=1}^n f(v_i - t)) \, dt]^{-1} \int_{-\infty}^\infty t \prod_{i=1}^n f(v_i - t) \, dt \qquad (9.3)$$

are well-defined for $n > 1 + 1/\alpha$ and possess the properties of centering statistics. Consequently, $L(\sqrt{n}(T_n^* - \theta)|\theta) \to N(0, I_f^{-1})$ uniformly for $\theta \in R$.

For the Cauchy density, for example, (9.2) is satisfied for any $0 < \alpha < 1$. For more general results see P. Bickel (1969).

10

The Local Theory and the Large Deviations Theory

The theory exposed in preceding sections has the local character. Its results do not depend on the behaviour of likelihoods at joint points (θ, x_n) such that $|T_n(x_n) - \theta| > \epsilon$, whatever positive ϵ may be. For this reason we were able to truncate the likelihoods as severely as was done in (4.5).

The local type of asymptotic theory is supported by the following fact: any two families of densities satisfying Assumptions A such that their likelihoods coincide over regions $|T_n(x_n) - \theta| < b_n/\sqrt{n}$, $b_n \to \infty$, are asymptotically undistinguishable in the sense that any test designed to separate them has the sum of errors of both kinds tending to 1. Therefore, it is desirable not to let the procedures depend on features that cannot be empirically tested.

However reasonable this consideration may sound, it has been recently shown that *tails* of the likelihoods are very well worth studying. This is being done under the name of large deviation theory. The parts of the structure of a sequence of density families $\{p_n(x_n, \theta)\}$ studied by each of these two theories are essentially independent: without changing the local results we can vary the tails of the likelihoods so as to obtain any large deviation results we wish, and the opposite also is true. Practically, however, for well-behaved families there is a smooth transition from large deviation results to local results. Combining both theories we may hope to get better fit of the asymptotic results to finite sample sizes.

It is very urgent for the applications of large deviation results to know the model precisely. The same is true to a lesser extent with respect to local theory. If we are vague about the model we can be quite confused by the mess of mutually inconsistent recommendations. Theoretical answers to this kind of difficulty are: robustness and uniform optimality.

11
Concluding Remarks

The above discussion shows that not much harm may be caused, at least for sufficiently large n, if we base our inferences on certain functionals of likelihoods that are independent of the family of densities (models, for short) in the background. On the other hand, it is also true that in order to prove it, we would need much stronger assumptions than our Assumptions A. Furthermore, it is notoriously known that maximum likelihood estimates and Bayes estimates are often very difficult to obtain explicitly. This is well illustrated by Example 4, where the maximum likelihood estimate for $\theta = (a, b)$ is a difficult function of observations. On the other hand, utilizing the knowledge of the model, we may modify the likelihoods [by neglecting the term $Z_n(x_n, \theta)$ in (5.3), e.g.] so as to obtain very simple asymptotically sufficient statistics as $\hat{b}_n$ of (5.2). In Example 3 again the analysis of the model provides asymptotic sufficiency of the vector of ranks—a kind of information, which is not apparent from the likelihood function. Also, if the likelihood functions are very wild, we can replace the observations X by a statistic $T(X)$ for which the likelihoods are smoothened enough to be a basis for inference. To do that, however, we again need to know the model.

For these and similar reasons, I believe that strict adherence to the likelihood principle would complicate and impoverish the statistical theory.

References

1. Bickel, P. J. and Yahav, J. A., "Some Contributions to the Asymptotic Theory of Bayes Solutions," *Z. Wahrscheinlichkeitstheorie verw. Geb.*, 11, 257-276, 1969.
2. Hájek, J., "On basic Concepts of Statistics," *Proceedings of the 5th Berkeley Symposium on Math. Statist. and Probability*, Vol. 1, 139-162, University of California Press, 1968.
3. Hájek, J., "A Characterisation of Limiting Distributions of Regular Estimates," *Z. Wahrscheinlichkeitstheorie verw. Geb*, To be published, 1970.

4. Hájek, J. and Šidák, Z., *Theory of Rank Tests,* Academic Press, London, 1967.
5. Huber, P., "Strict Efficiency Excludes Super Efficiency," Unpublished, Abstract in *Annals of Mathematical Statistics,* 37, p. 1425, 1966.
6. LeCam, L., "On the Asymptotic Theory of Estimation and Testing Hypotheses," *Proc. 3rd Berkeley Symposium on Math. Statist. and Probability,* Vol. I University of California Press, 129-156, 1956.
7. LeCam, L., "Locally Asymptotically Normal Families of Distributions," *University of California Publications in Statistics,* Vol. 3, No. 2, 37-98, 1960.
8. LeCam, L., "On the Assumptions Used to Prove Asymptotic Normality of Maximum Likelihood Estimates," Preprint, 1969.
9. Stein, C., "Inadmissibility of the Usual Estimator for the Mean of a Multivariate Normal Distribution," *Proc. 3rd Berkeley Symposium on Math. Statist. and Probability,* University of California Press, 197-206, 1956.
10. Wald, A., "Tests of Statistical Hypotheses Concerning Several Parameters When the Number of Observations is Large," *Trans. Amer. Math. Soc.,* 54, 426-482, 1943.
11. Wolfowitz, J., "Asymptotic Efficiency of the Maximum Likelihood Estimator," *Teorie, Vyerojatn. Primen.,* 10, 267-281, 1965.

COMMENTS

V. P. Godambe:

Though I do not share the view often expressed that, asymptotically, answers to the questions are independent of the questions asked, I believe that a considerable portion of the asymptotics in statistics has no strong relevance to the central problem of statistical inference. This is particularly borne out by the recent literature on non-parametric statistics which many times consists of meaningless investigations of the asymptotic properties of some arbitrarily chosen estimators or test criteria (Refer to papers in *Ann. Math. Statist.* during the last few years). I am glad to say that a happy exception to my foregoing remarks is Professor Hájek's work in this field.

To me it looks as though asymptotic investigation is necessitated occasionally by too narrow formulation of statistical concepts. For instance an optimum property of maximum likelihood estimator obtained by carrying the criterion of unbiassedness and minimum variance to its logical ends (mostly ignored by statisticians) is: Let X be any abstract sample-space, Ω the parameter space (real interval) and for every $x \in X$ and $\theta \in \Omega$, $f(x \mid \theta)$ the frequency function. Then it can be very easily shown that under mild regularity conditions, of all the real functions $g(x, \theta)$ on $X \times \Omega$ such that $E(g \mid \theta) = 0$ the variance of $g / E\left(\dfrac{\partial g}{\partial \theta} \,\middle|\, \theta\right)$ is minimized for every $\theta \in \Omega$, for $g = \dfrac{\partial \log f(x \mid \theta)}{\partial \theta}$ (Godambe, 1960, *Ann. Math. Statist.*). Surely this property of m.l. estimator which has its appeal for physicists (see for instance Janossy's book, 1963, on statistics) and for anyone who has a feeling for statistical concepts has no appeal whatsoever to many mathematical statisticians who take on mathematical formalism of statistics without its conceptual statistical content. [Professor Barnard here added that the property of m.l. estimation referred to by Godambe was an important one which deserved more publicity than it had received.]

C. R. Rao:

Le Cam-Hájek formulation of the problem providing an asymptotically sufficient statistic seems to involve undue assumptions and the method of finding such a statistic also seems to be somewhat involved. An alternative approach which I have advocated (see for instance the book, *Linear Statistical Inference and its Applications*) is as follows. If $L_n(x_n, \theta)$ denotes the log likelihood based on sample size n, then it is well known that

$$\frac{d}{d\theta} L_n(x_n, \theta) = S_n$$

has very important properties, especially when n is large, in drawing inference on θ. It is shown that under mild regularity conditions s_n is asymptotically equivalent to $T_n - \theta$ uniformly in compact intervals of θ, where T_n is the maximum likelihood estimator. This shows that for purposes of inference T_n is as good as the sample in large samples. The argument can be extended to many parameters and also to the case of dependent observations.

H. Rubin:

Professor Hájek's approach is inadequate for the decision-theoretic viewpoint. For example,

$$\text{let } T_n(x) = \begin{cases} x_1 \text{ if } x_1 \text{ is in the largest } n^{1/2} \text{ observations,} \\ \bar{x} \text{ otherwise.} \end{cases}$$

If the loss ρ is approximately squared error for small deviations and monotone, the risk of $\bar{x}$ is $\dfrac{\sigma^2}{n}$ but that of T_n is greater than $n^{-1/2} E(\rho(x_1 - \theta) \mid x_1$ is in the largest $n^{1/2}$ observations) which is asymptotically infinitely greater. However T_n satisfies Hájek's conditions.

Professor Hájek's discussion of large deviations, brief though it is, conveys the impression that all deviations of the type he has not considered can be called large. *Large* deviations is a technical term in the literature, as are *excessive*, and *moderate*. The appropriate term for this whole class is *extraordinary*.

REPLY

To Professor Rubin:

Proposition 5 gives a lower bound and generally does not claim that T_n satisfies it. If $l(y) = C \mid y \mid^\alpha$ then $l(\sqrt{n} \mid T_n - \theta \mid) = n^{\frac{1}{2}\alpha} l(\mid T_n - \theta \mid)$, otherwise there of course may be a difference. The former choice of passing to the limit makes the theory more feasible.

To Professor Godambe:

As is well-known many asymptotic results have been tested numerically

160

with great success, namely in the nonparametric field. As I stressed in my paper, in general, we do not have a more satisfactory substitute for them. Professor Godambe's suggestion how to prove "optimality" of the maximum likelihood estimate for any finite n is for me not convincing enough. This estimate fails even to be admissible in many instances, for example, in estimating the location when the density is not normal.

To Professor Rao

I welcome Professor Rao's comment. The following is an attempt to clarify the relation between the two approaches.

On page 285 of his book, *Linear Statistical Inference*, Professor Rao introduces the concept of asymptotic efficiency of a sequence of estimates. Such estimates T_n must be square-root consistent and satisfy

$$\sqrt{n} \left| T_n - \theta - \beta(\theta) \frac{1}{n} \frac{\partial}{\partial \theta} \log p_n(x_n, \theta) \right| \to 0 \qquad (*)$$

in probability. This condition for T_n is very close to what in the language of my paper would be described as "the densities have asymptotically normal likelihoods with T_n serving for the centering statistic." Actually, if log $p_n(x_n, \theta)$ is differentiable, we have

$$p_n(x_n, \theta) = p_n(x_n, T_n) \exp \left\{ -\frac{1}{2} n(T_n - \theta)^2 \Gamma_\theta + \int_{T_n}^{\theta} \left[\frac{\partial}{\partial t} \log p_n(x_n, t) - n(T_n - t) \Gamma_t \right] dt \right\}$$

so that conditions A of my paper correspond roughly to (*) with

$$c_n(x_n) = p_n(x_n, T_n), \quad \Gamma_\theta = [\beta(\theta)]^{-1}$$

and

$$Z_n(x_n, \theta) = \int_{T_n}^{\theta} \left[\frac{\partial}{\partial t} \log p_n(x_n, t) - n(T_n - t) \Gamma_t \right] dt.$$

Now since $|T_n - \theta|$ is of order $1/\sqrt{n}$, the smallness of (*) and of $Z_n(x_n, \theta)$ (see A2) are related conditions, even if neither implies the other. In (*) one stresses the closeness of $T_n - \theta$ to $\frac{1}{n} [\partial/\partial \theta] \log p_n(x_n, \theta)$, in assumptions A, I prefer to stress the approximately normal form of likelihoods, because just this fact is crucial for deriving propositions of my paper. I do not see that assumptions A would be much more "undue" than the assumption that there exists an estimate satisfying (*). I also do not see how (*) could be used to prove more refined theorems corresponding to propositions 4 and 5 of my paper. Let me also remind again that under (*) we derive asymptotic normality of T_n via asymptotic normality of $[\partial/\partial \theta] \log p_n(x_n, \theta)$, whereas in proposition 2 of my paper a different way not using the central limit theorem is used.

161

One source of misunderstanding may be the fact that the theory of my paper starts at the point where the theory in Professor Rao's book ends. Professor Rao, introducing traditional differentiability and uniformity conditions (page 299 of his book) proves that the maximum likelihood estimator, if consistent, satisfies (*). In my paper, assuming that there exists an estimate T_n satisfying assumptions A (which are a variant of (*)), we prove certain properties that are even not considered in Rao's book. True enough, the proofs of these properties are too difficult to be embodied in a textbook, but the results themselves could and should be included easily, and only these properties justify introduction of such terms as asymptotically efficient or sufficient.

Not being concerned with the construction of T_n from more basic assumptions, I do not see why Professor Rao mentions that methods of finding T_n are "involved." If he means that the maximum likelihood estimates are always easiest to find, there are plenty of examples showing the opposite. Let me refer to Example 4 of my paper in that connection.

A NOTE ON BASU'S EXAMPLES OF ANOMALOUS ANCILLARY STATISTICS

G. A. Barnard and D. A. Sprott
University of Essex & University of Waterloo

G. A. Barnard and D. A. Sprott
University of Essex & University of Waterloo

1

Introduction

Basu (1964) has given an interesting and comprehensive review of Fisher's use of ancillary statistics in the theory of estimation. In it he gives three examples of possible anomalies or inconsistences that can arise in the use of ancillary statistics. The first two of these are examples of non-uniqueness; it was concluded there existed, in these cases, no unique way of recovering all of the information via ancillary statistics. His third example was to exhibit a unique maximal ancillary statistic that purportedly leads to "a rather curious and totally unacceptable 'reference set'. Accordingly, Basu states that a further specification is necessary to delimit the appropriate use of ancillary statistics. For this he proposes the concept of *performable experiments.*

We propose to show that if proper attention is paid to the likelihood function as the primary inference from an experiment, and the role of the ancillary statistic in describing its *shape*, the difficulties raised by Basu disappear. We do not claim to present a general theory showing that all possible difficulties can be dealt with in this way; the general question must be regarded as remaining open.

2

Ancillary Statistics and Likelihood: Position and Shape

We take the usual formulation of an experiment as represented by a sample space S to which the observed data X belong, a parameter space Ω to which the parameter value θ belongs, and a function $f: S \times \Omega \to R$ representing the probability (density) function of X in S (relative to an appropriate measure μ). Then, for given X, any function

$$g(X)f(X,\cdot)\colon \Omega \to R$$

is a representative of the observed value of the likelihood function statistic $\mathscr{L}(X)$. The values of the statistic $\mathscr{L}(X)$ are thus equivalence classes of functions from Ω to R under the relation that two functions $h(\theta)$, $k(\theta)$ are equivalent if and only if their ratio $h(\theta)/k(\theta)$ is constant in Ω. It is customary to represent $\mathscr{L}(X)$ by the relative likelihood function

$$R(\cdot) = C(X)f(X,\cdot)$$

where
$$1/C(X) = \sup_{\alpha \in \Omega} f(X, \alpha).$$

It may happen that $f(X, \cdot)$ is unbounded and for this or some other reason another normalisation may be convenient.

In emphasizing the importance of the likelihood function as the primary inference from an experiment and in recommending the examination of its *shape* and other related mathematical characteristics, we do not wish to imply that in any particular problem the likelihood is the *only* item to be considered or that all other characteristics and structures of the problem should be ignored. We do not believe there is a 1-1 correspondence between inferences and likelihoods; the approach in this paper is not to be taken as support for the universal applicability of the likelihood principle.

Similarly, the above mathematical formulation of an experiment does not deny nor minimize the importance of the antecedent problem of the objective testing of the validity of the model $f(X, \cdot)$. We merely restrict ourselves to the aspects of inference related to the above formulation in order to deal with the specific questions about ancillary statistics, raised by Basu.

The observed value $\mathscr{L}(X)$ of the likelihood function may be thought of, in a way, as representing a *distribution* over the parameter space Ω. Indeed, if a Bayesian model of inference with a prior distribution uniform over Ω were in a particular case appropriate, then the likelihood function, normalised to integrate to unity over Ω, would actually be the posterior distribution of θ in the light of the observed value X. If we think of $\mathscr{L}(X)$ in this way*, then for varying X we have a family of distributions parameterised by X. If, loosely speaking, we can find two functions $T(X)$, $A(X)$, such that $T(X)$ represents a *position parameter* for the family while $A(X)$ represents a shape and/or scale parameter for the family, and such that $\mathscr{L}(X)$ is uniquely determined by $T(X)$ and $A(X)$:

$$\mathscr{L}(X) = \Lambda[T(X), A(X)],$$

---

*It is important to the understanding of the relationship between probability and likelihood that the first is a set function while the second is a point function. Our reference to the possibility of regarding the likelihood function as *distribution* should not be interpreted as implying any weakening of this distinction. It can make just as good sense to speak of a position or a scale parameter for a family of point functions as for a family of set functions.

then $T(X)$ will give us an estimate of θ whose accuracy is specified by the *ancillary statistic A(X)*.

Before proceeding to a more precise or more general formulation it will be best to consider a simple example, and then proceed to consider Basu's examples. After that we return to the general statement with more precision.

Example 1. Consider a pair of observations (x_1, x_2) from a Cauchy distribution. The likelihood is proportional to

$$\frac{1}{[1+(x_1-\theta)^2][1+(x_2-\theta)^2]}.$$

The position can be measured by $\bar{x} = (x_1 + x_2)/2$, the centre of symmetry of the likelihood. The shape is then measured by the ancillary statistic $a = (x_1 - x_2)/2$; $(\bar{x}, a)$ together are equivalent to (x_1, x_2) and determine the likelihood function. It is intuitively obvious that $\bar{x}$ measures position of the likelihood or size of θ. That a measures shape can be seen by noting that if $|a| \geqslant 1$, the likelihood is unimodal and symmetric about $\bar{x}$; if $|a| > 1$, the likelihood is bimodal and symmetric at $\bar{x}$ (the local minimum) with distance $2\sqrt{a^2-1}$ between the modes. Thus a describes everything about the likelihood but its position; $\bar{x}$ gives the position of the likelihood and hence estimates θ, but contains no information about shape.

In this example the concepts of position and shape can be given mathematical precision because we have a group of transformations characterising the problem. In the Cauchy distribution above, the fact that $\bar{x}$ measures position θ and nothing else can be seen because shifts in position of the distribution, $\theta \rightarrow \theta + K$ imply similar shifts in the position of the likelihood, $\bar{x} \rightarrow \bar{x} + K$. But a remains fixed. Thus $\bar{x}$ contains all the information about position θ; a shift in position does not alter shape as reflected by a remaining constant. The separation into position and shape is therefore complete and mathematically well defined. The totality of possible likelihoods can be separated into mutually exclusive subsets of likelihoods of the same shape; likelihoods of a given shape forming such a subset can be obtained from one another by location transformations.

3

Basu's Examples

Example 2. We consider Basu's second example first. This was that of a biased die with probability distribution

score X	1	2	3	4	5	6
probability	$\dfrac{1-\theta}{12}$	$\dfrac{2-\theta}{12}$	$\dfrac{3-\theta}{12}$	$\dfrac{1+\theta}{12}$	$\dfrac{2+\theta}{12}$	$\dfrac{3+\theta}{12}$.

Basu restricted the range to $0 \leqslant \theta \leqslant 1$. However the distribution is defined in the range $-1 \leqslant \theta \leqslant 1$; more symmetry is retained by using the whole parameter space $(-1, 1)$ without altering the logic of the argument. It is merely necessary to replace the maximum likelihood estimate $T = 0, 0, 0,$ 1, 1, 1 corresponding to the observation $X = 1, 2, 3, 4, 5, 6$ (appropriate to $0 \leqslant \theta$) by $T = -1, -1, -1, 1, 1, 1$. The six ancillary statistics given by Basu are:

X	1	2	3	4	5	6
$Y_1(X)$	0	1	2	0	1	2
$Y_2(X)$	0	1	2	0	2	1
$Y_3(X)$	0	1	2	1	0	2
$Y_4(X)$	0	1	2	2	0	1
$Y_5(X)$	0	1	2	1	2	0
$Y_6(X)$	0	1	2	2	1	0

The effect of each of these ancillaries in describing the likelihood function can be seen by plotting the six possible likelihoods. The maximum of the likelihoods occur, as recorded above, at the maximum likelihood estimates $-1, -1, -1, 1, 1, 1$ and are $\frac{2}{12}, \frac{3}{12}, \frac{4}{12}, \frac{2}{12}, \frac{3}{12}, \frac{4}{12}$ respectively. Standardizing the likelihood with respect to these maxima, the six possible relative likelihood functions (that is, all standardized to have a maximum of unity) are

X	1	2	3	4	5	6
$R(X)$	$\frac{1-\theta}{2}$	$\frac{2-\theta}{3}$	$\frac{3-\theta}{4}$	$\frac{1+\theta}{2}$	$\frac{2+\theta}{3}$	$\frac{3+\theta}{4}$.

These are plotted in Figure 1 and are numbered according to the observation $X = 1, 2$, etc., that gives rise to it. From this graph the effect of each of the ancillaries $Y_1, Y_2, \ldots$ can be visualized. For instance, Y_2 separates the six possible likelihoods into three groups of two each: (R_1, R_4), (R_2, R_6), and (R_3, R_5). The groupings produced by the other Y_i can similarly be obtained.

By considering the purpose of an ancillary statistic to describe the shape of the likelihood, the uniquely appropriate ancillary can be seen. The groupings produced by $Y_2, Y_3, \ldots, Y_6$ are completely artificial. For instance. the shapes of the likelihood (R_2, R_6) and of (R_3, R_5), pairs which are produced by Y_2, have no similarity. From the graph it would appear that the uniquely appropriate pairing in respect of shape is (R_1, R_4), (R_2, R_5), and (R_3, R_6) as yielded by Y_1. R_1 has a shape much more similar to R_4 than to any other R. They both have the same value of $\alpha = |\theta|$ such that $R(\alpha) = 0$. That is, the absolute values of the θ intercepts of the likelihoods R_1 and R_4 are the same.

The same is true of the pairs (R_2, R_5) and (R_3, R_6). Thus the absolute value of the θ intercept is an ancillary equivalent to Y_1. The position is specified by their maximum; the shape is specified by the fact they are mirror images. This can be made mathematically precise by noting that the distribution is invariant under the group of reflections $\theta \to -\theta$. Under this group $X_1 \leftrightarrow X_4, X_2 \leftrightarrow X_5, X_3 \leftrightarrow X_6$. The same effect is produced on the likelihoods: R_1 and R_4 are interchanged as are R_2 and R_5, R_3 and R_6. Thus, under the groups of reflections the pairing produced by Y_1 remains invariant, the likelihoods in any one pair being obtained from one another by the reflections $\theta \to -\theta$. The structure therefore of position and shape is essentially the same as that described above for the Cauchy distribution.

166

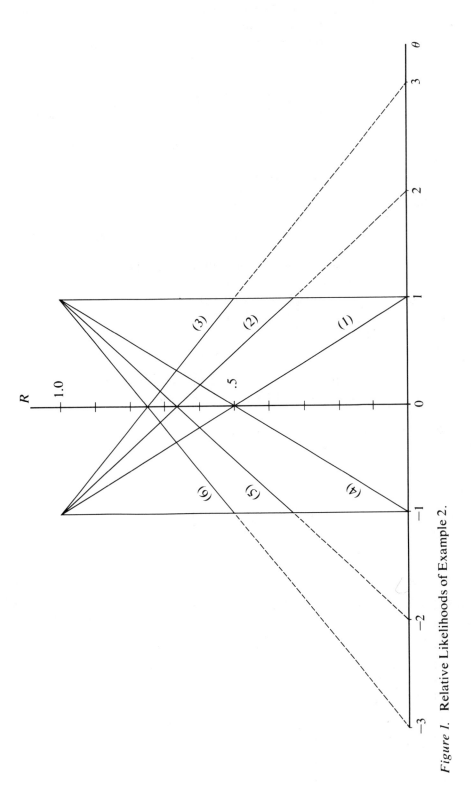

Figure 1. Relative Likelihoods of Example 2.

167

From the considerations of position and shape therefore there is a unique way of recovering the ancillary information. It is the ancillary statistic Y_1, or equivalently the absolute value of the θ intercept of the likelihood. This remains invariant under the group $\theta \to -\theta$ thus serving to specify the shape.

Example 3. We next consider Basu's third example.
It is that of a random variable X with a uniform distribution in the range $\theta \leqslant X < \theta + 1$. The integer part $[X]$ of X can be considered an estimate of θ; the fractional part $\Phi(X)$ of X is an ancillary statistic having a uniform distribution between $(0,1)$. Basu states that $\Phi(X)$ leads to a "curious and unacceptable" reference set as follows.

Since

$$\theta = [\theta] + \Phi(\theta) \leqslant X = [X] + \Phi(X) < [\theta] + 1 + \Phi(\theta) = \theta + 1,$$

it follows because of the definition of $\Phi(\theta)$ and $\Phi(X)$ that

$$[X] = [\theta] \text{ if } \Phi(X) \geqslant \Phi(\theta)$$
$$[X] = [\theta] + 1 \text{ if } \Phi(X) < \Phi(\theta).$$

The conditional distribution of X given $\Phi(X)$ is therefore

$$\Pr\{X = [\theta] + \Phi(X)\} = \Pr[\Phi(X) \geqslant \Phi(\theta)]$$
$$\Pr\{X = [\theta] + 1 + \Phi(X) = \Pr[\Phi(X) \leqslant \Phi(\theta)]$$

so that as soon as θ is known, the numerical value of X is known with probability one.

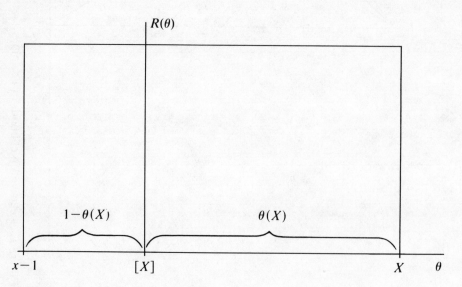

Figure 2. A Representative Relative Likelihood of Example 3.

168

There is, however, nothing curious or anomalous about that. The likelihood with a given $\Phi(X)$ is shown in Figure 2. This likelihood is divided at the point $[X]$ in the ratio $[1-\Phi(X)]\colon \Phi(X)$. $[X]$ therefore measures position of the likelihood and $\Phi(X)$ its shape in the sense of the proportion into which $[X]$ divides the likelihood. The reference set determined by a fixed $\Phi(X)$ is the subset of all uniform unit width likelihoods divided at an integer into the same proportion as that of the observed likelihood $1-\Phi(X)\colon\Phi(X)$. There are, of course many other equivalent ways of describing the above likelihood function. In fact, any interior point could be given along with the ratio into which it divides the likelihood. All of these seem unnecessarily complicated, the simplest way being merely to give one of the end points.

Example 4. Finally, we consider Basu's first example. This concerns a sample of n independent pairs (X_i, Y_i) from a bivariate normal distribution with zero means, unit variances, and unknown correlation Θ. Basu points out that $X=(X_1,\ldots,X_n)$ and $Y=(Y_1,\ldots,Y_n)$ are both ancillary statistics as are also $S_x=\Sigma X_i^2, S_y=\Sigma Y_i^2$, so that there is non-uniqueness.

However, the likelihood function is proportional to

$$(1-\theta)^{-\frac{n}{2}}\exp\left[-\frac{1}{2(1-\theta^2)}(U-2V\theta)\right]$$

where

$$U=\Sigma(X_i^2+Y_i^2),\quad V=\Sigma X_i Y_i.$$

If, therefore, we were to represent the family of possible likelihood functions in terms of *position* and *shape* parameters, these parameters would have to be functions of U and V. For otherwise we would have two identical likelihood functions having different positions or different shapes. However, although most of the information about θ is contained in the value of V, and one is therefore tempted to think of V as a position parameter, the complementary statistic U is not ancillary in that its distribution involves θ^2. Thus it appears impossible to find functions of U and V only which can be thought of as representing position and shape in this case and no ancillary statistic of the type we are requiring appears to exist.

4
Definition of Position and Shape

We now return to general considerations. As is always the case with problems of statistical inference we are confronted with logical notions, here referred to as *position* and *shape*, which ought *so far as possible* to be expressed in mathematical terms. It is doubtful whether a complete solution to this problem of expression in mathematical terms can be given. But a partial solution is available when there is a group G of transformations involved in the problem in the following way. Referring to the beginning of Section 2 we require that a group G acting on S and a corresponding group H acting on

Ω should both exist with the element h in H corresponding to the element g of G and such that for any measurable $B \subset S$

$$\int_{g(B)} f(\cdot, h\theta) \, d\mu \, [g(\cdot)] = \int_{B} f(\cdot, \theta) \, d\mu(\cdot).$$

In this case, if a minimal sufficient statistic of the form (T, A) exists, with T taking values in Ω and such that

$$A[g(\cdot)] = A(\cdot)$$

while $T[G(\cdot)]$ acts transitively on $T(S)$, then T may be regarded as specifying the *position* of $\mathscr{L}(X)$ while A specifies its *shape*. All possible likelihoods can then be separated into mutually exclusive subsets of likelihoods all of the same shape determined by the fixed A. All likelihoods in a given subset can be generated from a likelihood of the same shape by the group of transformations.

If the distribution of A depends on another parameter α for which (T, A) is also minimal sufficient, then A is sufficient for α in the absence of knowledge of θ (Barnard, 1963). Then A determines the shape of the likelihood of θ for a given α and also the *marginal* likelihood of α (Fraser, 1968; Kalbfleisch and Sprott, 1970).

5
Summary

Basu gave three examples that have received considerable attention as demonstrating possible anomalies of ancillary statistics which in turn would lead to difficulties for the theory of conditional inference. This note indicates how these apparent anomalies can be dealt with.

References

1. Barnard, G. A., "Some Logical Aspects of the Fiducial Argument," *Journal of the Royal Statistical Society, B,* 25, 111-114, 1963.
2. Basu, D., "Recovery of Ancillary Information," *Sankhya, A,* 26, 3-16, 1964.
3. Fraser, D. A. S., *The Structure of Inference,* New York, John Wiley and Sons, 1968.
4. Kalbfleisch, J. D. and Sprott, D. A., "Application of Likelihood Methods to Problems Involving Large Numbers of Parameters," *J. Roy. Statist. Soc., B,* to be published, 1970.

COMMENTS

D. Basu:

I thank you very much for granting me nearly equal time to discuss Professor Barnard and Professor Sprott's very interesting and thought-provoking work on ancillary information. However, I must admit that I do not understand what Barnard and Sprott plan to do with ancillary statistics and that I have not, as yet, any feeling for their classification of likelihood functions by their 'position' and 'shape'. Also I do not understand what group of transformations is doing what to achieve this classification. I had similar trouble when I first attempted a serious study of Fisher's writings on the subject — elusive notions like information and its recovery, reference sets, intrinsic accuracy, sufficient estimate, ancillary statistics and red herrings like the Problem of the Nile. It took me the greater part of seven years to make sense of what Fisher was trying to say and so, to-day, I do not feel unduly depressed if I do not understand the 'position' and 'shape' and the 'classification' of likelihood functions. When everything falls into its place, when I am able to rehabilitate myself with the new ideas, maybe, I shall construct a few more counter examples and be thankful to Barnard and Sprott for giving me an opportunity to write yet another paper!

Since no mention was made to-day of what Fisher meant by ancillary statistics and how he wanted to recover ancillary information, let us see if we can agree on that. For the sake of simplicity, let us consider the point estimation problem with a single unknown parameter θ and sample x. Let $T = T(x)$ be the maximum likelihood estimate of θ. If T is a sufficient statistic then, according to Fisher, there is no loss of information. If T is not sufficient, then how to recover the information lost? Fisher would then look for an 'ancillary complement' to T, that is a statistic $A = A(x)$ whose sampling distribution does not involve the parameter θ and which together with T jointly constitute the minimal sufficient statistic. In this situation the ancillary A determines the reference set — the set of all possible sample points x' such that $A(x') = A(x)$. Fisher would then assess the performance characteristics of T by referring it to the above reference set, that is, from the conditional sampling distribution of T given A. For instance, if Fisher needs to record the variance of T then his appropriate variance function is not the unconditional variance $V(T|\theta)$ but the conditional variance $V(T|A, \theta)$. This is the conditionality principle in the simplest case and this is what Fisher described as the process of recovery of information. Let us not go into the logic of the above conditionality principle but rather examine the examples referred to by Professor Sprott in the above context.

Example 4 was constructed to demonstrate the simple fact that two ancillary statistics $A_1=(x_1,x_2,...,x_n)$ and $A_2=(y_1,y_2,...,y_n)$ may jointly be sufficient and indeed be equal to the whole sample $(x_1, y_1), (x_2, y_2),..., (x_n,y_n)$. In example 3 the minimal sufficient statistic is x and the likelihood function $L(\theta|x)$ is flat over the interval $(x-1, x]$. We may look at the integer part $[x]$ of x as a maximum likelihood estimate of θ. The fractional part $\phi(x)=x-[x]$ is the ancillary complement to $[x]$. Six years ago when I wrote the paper on ancillary statistics, the fact that the conditional distribution of the sample x, given the ancillary $\phi(x)$, is degenerate at the point $[\theta]+\phi(x)$ if $\phi(\theta) \leq \phi(x)$ or at the point $[\theta]+1+\phi(x)$ if $\phi(\theta) > \phi(x)$ appeared to me to be a very curious and isolated phenomenon. Since then I have realized that this kind of a phenomenon is typical of every non-sequential sample survey situation. In my survey sampling paper (to be presented to you tomorrow) I have discussed in some details this phenomenon of a flat likelihood function and the existence of an ancillary statistic that leads to a degenerate reference set — namely, the sample itself. This phenomenon bothers me no more as I have come to the realization that in every statistical problem there is only one reference set with only one point in it, namely, the sample point.

I am sure that Fisher would have found the reference set produced by $\phi(x)$ in example 3 very disturbing and would have had a lot of trouble accommodating his theory to the particular case of this example. Given $\phi(x)$, the conditional distribution of $[x]$ (the ML estimator of θ) is concentrated at the point $[\theta]$ or at the point $[\theta]+1$ depending on whether $\phi(\theta) \leq \phi(x)$ or $\phi(\theta) > \phi(x)$. Unconditionally, $[x]$ is an unbiased estimator of θ, but conditionally

$$E\{[x]|\phi(x),\theta\} = \begin{cases} [\theta] & \text{if } \phi(\theta) \leq \phi(x) \\ [\theta]+1 & \text{if } \phi(\theta) > \phi(x). \end{cases}$$

Unconditionally, $V\{[x]|\theta\} = \phi(\theta)(1-\phi(\theta))$, but conditionally the variance of $[x]$ is zero! The conditional mean square error of $[x]$ is $\{\phi(\theta)\}^2$ or $\{1-\phi(\theta)\}^2$ according as $\phi(\theta) \leq \phi(x)$ or $\phi(\theta) > \phi(x)$. One may endlessly study the conditional performance characteristics of $[x]$, but what for? The whole business of recovery of information in the context of the present example makes no sense to me at all. The entire information about θ supplied by the sample x is contained in the likelihood function

$$L(\theta|x) = \begin{cases} 1 & \text{if } x-1 < \theta \leq x \\ 0 & \text{otherwise.} \end{cases}$$

How this information is to be utilized by the statistician depends on his prior information about θ.

It must be obvious to you now that I believe in the likelihood principle. And I always thought that so does Professor Barnard. Indeed, I used to think of Professor Barnard as the principal exponent of the likelihood principle. To-day, I am, therefore, greatly puzzled to find that Barnard is joining with Sprott in looking at not only the likelihood function that is provided by the data, but at the family of all possible likelihood functions — one for each point x in the sample space X. For instance, in example 2 there are six such functions (only one of which is the real one). The graphs of these six functions are all linear in shape and they all hover over the same θ interval $[-1, 1]$. Yet we find Barnard and Sprott trying to classify these six linear segments according to their 'shape' and 'position'. But why such a classification? Why do we need this, and then, what are we going to do with this? What has ancillary statistics got to do with all these?

I must admit that at the moment I am a very confused man. But I am sure that there must be an element of truth in what Barnard and Sprott are trying to do and that in due course the truth will dawn on me. A major source of my confusion is the way a group structure is exploited in example 2. Mr. Chairman, please permit me to register my feelings of helplessness and despair at the manner this whole business of a group structure has seized the whole attention of a sizable part of the statistical community. Yesterday, when I was listening to Professor Fraser, my mind plunged into a helpless state of utter confusion. The standard medium of communication has apparently broken down. To-day, Professor Sprott told us how he will begin with example 2 by extending the parameter space to the wider interval $[-1, 1]$. This enabled him to have a group structure. What if this extension is disallowed? Will his theory then fall into pieces? What if I restate my example in the following manner? The variable X now takes the eight values $1, 2, \ldots, 8$ with probabilities

$$\frac{1-\theta}{24}, \frac{2-\theta}{24}, \frac{3-\theta}{24}, \frac{1+\theta}{24}, \frac{2+\theta}{24}, \frac{3+\theta}{24}, \frac{\theta}{2}, \frac{1-\theta}{2},$$

where θ has necessarily to lie between 0 and 1. As before there are many ancillaries. However, I cannot think of a non-trivial transformation that leaves the problem invariant. How then are we going to tackle the problem of arriving at a unique ancillary? And the question remains: What are we going to do with an ancillary statistic?

D. R. Cox:

In some ways it may help to consider not one trial from Basu's multinomial example, but n trials. The unconditional argument uses a multinomial likelihood. The authors' recommended ancillary treats the data as in effect 3 independent binomial experiments; $n_1 + n_4$ trials with n_1 successes, probability of success $\frac{1}{2} - \frac{1}{2}\theta$, etc. How are these to be analyzed? One way is to test $\theta = \theta_0$ by a locally most powerful test using the efficient score, the derivative of the combined log likelihood, whose exact distribution is in principle available; the efficient score itself is unaffected by conditioning but its distribution is affected. The authors' arguments make it plausible that their particular choice of ancillaries partitions the overall distribution into components as distinct as possible in some sense. If this is indeed true, it provides a resolution of the non-uniqueness independent of group arguments.

O. Kempthorne:

I find this paper most interesting. The examples of Basu are very challenging. I confess that I do not find the present paper highly informative. It is interesting, however, at a highly intuitive level. My own interest in the matter is in connection with tests of significance for which a population of repetitions is crucial. This is the aspect, incidentally, that the orthodox Neyman-Pearson school seems unable to understand or refuses to admit as relevant. I believe that school is totally wrong insofar as data analysis is directed to the "evidential" content of data.

Some weeks ago, I examined the likelihood function for a "haphazard" sample for a Basu example, and I was comforted a bit to note that the likelihood function is quite pleasing.

I was struck by the aptness of the comments of Dr. D.R. Cox on the present paper. As usual, he is "bang on", I believe. His point is related, I think, to the question of tests of significance and inversion thereof, that I mention above.

Dr. Cox's discussion, moreover, exhibits the divergence between those who advocate the likelihood principle, whatever that is, and those who are interested in properties of the likelihood function under repetition — what I refer to in my essay, as a possible property of a mapping rule.

R. L. Prentice:

I would like to state in Professor Fraser's absence what I think he thinks par-

ticularly with regard to Professor Williams' comments. Professor Williams*
has started with the response space X and *defined* what he calls an error variable
E in terms of X and parameter value θ and has indicated that subsequent reductions *as if* we had a structural model may be useful in resolving some
uniqueness problems with ancillaries. Such arguments are not enough to solve
uniqueness problems as indicated, for instance, by Professor Lindley in his
review of Fraser's book; that is, by factoring a bivariate normal into upper and
lower triangular components. Fraser, however, illustrated clearly yesterday that
these were not to be the conditions under which a structural analysis applied,
that is, I think he meant what he said when he indicated that an error system
is to be something objective and physically identifiable in an experiment, something that is present without regard to the observed response. When thought of
in this way the uniqueness problems disappear and the various reductions in
the analysis are for quite different reasons than in the case discussed by Professor Williams and I think construing the structural analysis in this way contributes to the confusion between structural and fiducial inference.

C. R. Rao:

ON BASU'S QUESTION: I would like to give a mathematical formulation
of what Fisher calls recovery of information through ancillary statistics. Let
(T,A) be a sufficient statistic, for θ where A is ancillary. If $I_T(\theta)$ is the information on θ contained in T and $I(\theta)$, that in the whole sample, then we
have the relationship $I(\theta) \geq I_T(\theta)$. Then it is easy to verify

$$\int I_{T|A}(\theta) \, dF(A) = I(\theta)$$

where $I_{T|A}(\theta)$ is the information on θ in the conditional distribution, of T
given A. This relationship implies that in the long run, no information is lost
by using the conditional distribution of T given A in drawing inference on θ.

J.S. Williams:

The three examples of ancillary statistics proposed by Dr. Basu are particularly
instructive in the search for conditioning principles because they lack a property I regard as essential for conditional inference when pivotal variables exist.
Very briefly, if $X = X(\theta,E)$ where X is observable, E is pivotal, and inferences
are made on the parameter θ, then inferences associated with Bayes, fiducial
and structural arguments of the form "$\theta = \hat{\theta}$ given $X = x$" are equivalent to
inferences of the form "$E = \tilde{E}$ given $X = x$" where $x = X(\hat{\theta}, \tilde{E})$. Operating characteristics of these and most standard forms of inference are, for fixed θ, determined from the sampling distribution of E. Now if inferences are conditioned
on an ancillary outcome $A(X) = a$, one should require that $A(X) = a$ translate into a known constraint on the sample space of E, that is, $A(X) = B(E)$
where the function B is not indexed by θ. Only then will the appropriate con-
ditional sampling distribution of E not be indexed by θ; and will $E|A(X) = a$,
be pivotal.

In the structural model, if we use the notation in Dr. Fraser's recent book,
$X = \theta E$, and $A(X) = D(X)$ is the orbit reference point defined by $D(X) =$

---

*Professor Williams's comments to follow

174

$[X]^{-1}X = [\theta E]^{-1}\theta E = [E]^{-1}E = B(E)$. Clearly the ancillaries have the desired property. In Dr. Basu's example of a random variable uniformly distributed on $[\theta, \theta+1]$, $\theta \in R^1$, the ancillary $X-[X] = \theta + E - [\theta+E]$ regarded as a function of E is indexed by θ.

Professor Rao remarked on the recovery of lost information in the exhaustive estimation scheme. I want to point out two extensions of this idea which may be of practical importance.

If (T, A) is a minimal sufficient statistic, but A is not ancillary, then the lost information cannot be recovered. However if θ is nonidentifiable in the conditional distribution $F_\theta(A|T)$ of A, no *useable* information can be extracted from A once T is observed. In effect T is a sufficient statistic. On the other hand if θ is nonidentifiable in the marginal distribution $F_\theta(A)$ of A, all the *useable* information about θ is contained in $(T|A)$, and in effect A is ancillary.

When pinned down, I will have to refer to Koopmans and Reiersöl for a definition of a nonidentifiable parameter, and surely this will have to be broadened somewhat to implement any search for effectively sufficient and ancillary statistics.

REPLY

We are grateful to the other discussants for their comments which seem to us to carry matters forward in an area where final solutions are not yet available, but we hope they will forgive us if we concentrate our reply on Dr. Basu, since his original paper lies at the centre of the issues with which we have tried to deal.

In his discussion, Dr. Basu has repeated much of what he originally said in his 1964 paper. Our paper dealt with the salient aspects of this, and so there is no need to repeat here what was said in our paper. Accordingly we shall concentrate on the points raised by Dr. Basu that were not dealt with previously.

We did not discuss the use to which ancillaries are put, as this seemed irrelevant to the main issue raised by Dr. Basu's original paper, namely, the possible anomalies (non-uniqueness, etc.) of ancillary statistics as illustrated by his examples. However, once these are dealt with, there would appear to be no difficulties in their use. Thus, in Example 2, the appropriate ancillary statistic could be used in the manner outlined by Professor Cox in his discussion of our paper or indeed as described by Dr. Basu himself in 1964. In this regard, however, we did not advocate the use of $\phi(x)$ as an ancillary in Example 3. We agree with Dr. Basu that the entire information about θ supplied by the sample x is contained in the likelihood function. Indeed we specifically stated that such a description of the likelihood in terms of an ancillary $\phi(x)$ is "unnecessarily complicated, the simplest way being merely to give one of the end points". We were only pointing out that the existence of the ancillary $\phi(x)$ did not imperil the theory of ancillary statistics or conditional inference. This is also a good example where it is completely uninformative to evaluate the performance of the estimate T by its variance (either conditional on A or otherwise).

The existence of group structure is not essential to the problem. The relevant requirement is that such symmetries and invariance properties as are possessed by the problem must (in the absence of other overriding considerations) be preserved in the inference. Such symmetries may be expressed in structures other than groups, for example, combination of groups, embeddings in a group, or sets that are completely unrelated to groups. The simplest structure would appear to be that of a group, and as this was sufficient to deal with Dr. Basu's examples, nothing more general was considered. However, his modification of Example 2 serves to illustrate a slight extension beyond the simple case of group invariance, and Section 7 could be slightly modified to allow for this.

In this modification, the experiment is represented by a sample space $S = S_1 \cup S_2$ and a parameter space $\Omega = \Omega_1 \times \Omega_2$. Groups G_1 and G_2 act on S_1 and S_2 respectively, and the corresponding groups H_1 and H_2 act on Ω_1 and Ω_2. In the modified Example 2 $S_1 = \{1, 2, 3, 4, 5, 6\}$, $\Omega_1 = \{-1 \le \theta \le 1\}$, $S_2 = \{7, 8\}$, $\Omega_2 = \{0 \le \theta \le 1\}$. H_1 and G_1 are as in the original Example 2 and G_2 is the group of transformations $\{7 \rightarrow 8, 8 \rightarrow 7\}$ and H_2 is $\{\theta \rightarrow 1 - \theta\}$. The unique ancillary statistic left invariant under the structure is $Y = (0, 1, 2, 0, 1, 2, 3, 3)$, that is, the pairing of possible likelihoods is

$$\left(\frac{1-\theta}{24}, \frac{1+\theta}{24}\right), \left(\frac{2-\theta}{24}, \frac{2+\theta}{24}\right), \left(\frac{3-\theta}{24}, \frac{3+\theta}{24}\right), \left(\frac{\theta}{2}, \frac{1-\theta}{2}\right).$$

Dr. Basu refers to one of us as the main advocate of the likelihood principle (LP). Perhaps we should clarify the position here. LP refers only to problems for which the model consists of a sample space S, with measure μ, a parameter space Ω, and a family of probability functions $f : S \times \Omega \rightarrow R^+$ such that for all θ, $\int_S f d\mu = 1$. For two such problems, $\{S, \mu, \Omega, f\}$ and $\{S', \mu', \Omega, f'\}$, LP asserts that if $x \in S$ and $x' \in S'$ and $f(x, \theta)/f'(x', \theta)$ is independent of θ, then the inference from x must be the same as the inference from x'.

We may distinguish three forms of LP:
1. Strongly restricted LP: LP applicable only if $\{S, \mu, \Omega, f\} = \{S', \mu', \Omega, f'\}$. This is equivalent to the sufficiency principle.
2. Weakly restricted LP: LP applicable (a) whenever $\{S, \mu, \Omega, f\} = \{S', \mu', \Omega, f'\}$ and (b) when $\{S, \mu, \Omega, f\} \ne \{S', \mu', \Omega, f'\}$ but there are no structural features of $\{S, \mu, \Omega, f\}$ (such as group structure) which have inferential relevance and which are not present in $\{S', \mu', \Omega, f'\}$.
3. Unrestricted LP: LP applicable to all situations which can be modelled as above —
4. Totally unrestricted LP: as 3, but, further, all inferential problems are describable in terms of the model given.

As we understand the situation, almost everyone would accept 1, while full Bayesians would accept 4. GB's own position is now, and has been since 1947, 2.

SOME ASPECTS OF STATISTICAL INFERENCE IN PROBLEMS OF SAMPLING FROM FINITE POPULATIONS

C. Radhakrishna Rao
Indian Statistical Institute

Summary

The problem of survey sampling of a finite population is stated in a somewhat wider context than what is generally considered. It is shown that statistical analysis of survey data highlights the inherent difficulties in evolving suitable theories of statistical inference.

The concept of information provided by randomization is introduced and its role discussed in obtaining estimates of population mean or total by observing only a subset of units of the population. The mean of values from distinct units in SRSWR (Simple Random Sampling with Replacement) is shown to be a maximum likelihood estimate of the population mean considering the distribution of suitable sample statistics induced by randomization in the choice of sample units.

Finally, optimum properties of the sample mean in SRSWOR and the Horvitz-Thompson estimator in pps sampling are established under the superpopulation model generated by permutations of values (or some quantities based on them) attached to the distinct units.

1
Introduction

A number of papers have appeared in recent times purporting to examine the logical foundations of statistical inference based on a sample from a finite population. They have created the impression that the statistical analysis of such data known as survey type poses new problems not encountered in other areas of application and that there is a need for developing new methods. Unfortunately, the same situation prevails in all other areas and considerable literature in statistics is devoted to an examination of the foundations of statistical methodology. The recent controversies associated with survey sampling have once again brought to the forefront the difficulties involved in evolving general theories of statistical inference, and we are indebted to Godambe for the current interest in the problem. His first paper written as early as in 1955 threw doubts on claims made about properties of certain estimators and he has been since then seeking to obtain more satisfactory estimators and trying to interest others.

I may mention that in statistical methodology, the existence of uniformly optimum procedures (such as UMV unbiased estimator, uniformly most powerful critical region for testing a hypothesis) is a rare exception rather than a rule. That is the reason why *adhoc* criteria are introduced to restrict the class of procedures in which an optimum may be sought. It is not surprising that the same situation obtains in sampling from a finite population. However, it presents some further complications which do not seem to exist in sampling from infinite or hypothetical populations. In sampling from a finite population, there is what is called a *frame,* which in a simple case may be a list of units which are individually distinguishable and which constitute the population to be surveyed. We are interested in estimating the total value of a particular characteristic defined on the units in the population on the basis of observations on a subset of the units. The sample then consists of observed pairs $(y_1, x_1),\ldots, (y_n, x_n)$ where y stands for the distinguishing label of a unit (such as its number in the frame) and x for the value of a characteristic on the unit y. The question raised is whether the knowledge of $y_1,\ldots, y_n$ provides information, in addition to the values observed, in estimating the total for the population. Investigations by Godambe and others show that some additional information possibly exists but its utilization seems to be difficult and is, therefore, a problem for research.

2

Problems of Survey Sampling

I shall try to present the problems of sampling from a finite population in a slightly wider setup which is applicable to many situations familiar to statisticians.

Let us consider N-distinct groups which we may number as $1,\ldots,N$ and which constitute a population. Let X be a r.v. which has the d.f. $F_i\,(x, \theta_i)$ in the ith group depending on an unknown parameter θ_i, $i = 1,\ldots, N$. (In general both θ_i and X could be vector valued, but for simplicity we assume that they are scalars). We are interested in estimating a characteristic of the population specified by a function

$$g(\theta_1, \ldots, \theta_N) \tag{2.1}$$

of the unknown parameters, by sampling a subset of the N groups. Typical functions in which statisticians have been interested are

$$\theta = (\theta_1 + \ldots + \theta_N)/N, \tag{2.2}$$

$$\sigma^2 = \Sigma(\theta_i - \theta)^2/N, \tag{2.3}$$

and more generally weighted functions such as

$$\theta = \pi_1 \theta_1 + \ldots + \pi_N \theta_N, \tag{2.4}$$

$$\sigma^2 = \Sigma \pi_i (\theta_i - \theta)^2. \tag{2.5}$$

It may be recognised that the functions (2.3) and (2.5) are described in statistical literature as variance between populations as different from variance within populations, and the problem is known as estimation of variance components.

In the sample survey situation, $F_i(x, \theta_i)$ is considered to be degenerate with probability 1 for the value $x = \theta_i$ may not be observable without error.

A slightly extended version of the problem arises in situations where the groups do not exist as distinct entities so that prior selection of a group and then sampling for the r.v. X within it is not possible. However, when the value of the r.v. X is observed from the composite population it may be possible to know the group to which it belongs. The problem is again that of estimating functions of the type (2.1) - (2.5) where $\theta_1, \ldots, \theta_n$ are parameters associated with the N groups.

Such a situation arises in an anthropometric survey where we can only sample individuals in a given region, and along with physical measurements on a chosen individual we can also ascertain the caste (group) to which he belongs.

To consider a simpler example let us consider a 100-faced die with 0 or 1 written on each face. In addition each face carries one of 20 colours and each colour is represented on exactly five faces. Each face has the same probability of occurrence in a single toss, and when a face turns up we note the colour as well as the number on it, which may be 0 or 1. The problem is to estimate the proportion of zero faces by tossing the die, say ten times. The answer is simple, viz, the observed proportion of zeros in 10 tosses, if we did not observe the colour of the face in addition to the value on it. Do the observations on colour enable us to get a more efficient estimator of the unknown proportion when nothing is specified about the association between the colours and values on the faces?

These two types of situations occur in practice, where we can first choose a subset of the groups and sample for the r.v. in each group, and where we sample directly for the r.v. in the composite population and ascertain the group to which it belongs. In theory, at least when we know the proportions of individuals belonging to different groups in the composite population, the two situations are similar.

We shall now examine the statistical methodology for estimating unknown functions of the parameters or for testing statements concerning them.

3

Information Provided by a Single Observation

To make the discussion simple let us suppose that the r.v. X has the density $f_i(x, \theta_i)$ in the ith group. We denote by Y the r.v. which takes values $1, \ldots, N$

representing the N groups, in which case the r.v. we are observing is the pair (Y, X). Let us consider the sampling procedure of first drawing a group giving probability π for the ith group and observing the r.v. X. Then the probability density for the pair $(Y = i, X = x)$, or the likelihood of $(\theta_1, \ldots, \theta_N)$ given (i, x) is

$$L(\theta; i, x) = \pi_i f_i(x, \theta_i) \qquad (3.1)$$

where $\theta = (\theta_1, \ldots, \theta_N)$. If $I_i(\theta_i)$ denotes the information on θ_i contained in the distribution of X in the ith group, then the overall information contained in X arising from the likelihood (3.1) may be shown to be

$$I(\theta; Y, X) = \Sigma \pi_i I_i(\theta_i). \qquad (3.2)$$

We note that the likelihood (3.1) based on the observed pair (i, x) is completely uninformative on parameters other than θ_i and tells us that when the ith unit is drawn we have good information only on θ_i, the parameter of the ith group. The expression (3.2) for information, however, contains all the parameters and it only tells us that in the long run, as we repeatedly sample, we will have information on all the unknown parameters.

Indeed the likelihood (3.1) is relevant for drawing any inference on the particular parameter θ_i. In addition, we enquire whether the knowledge that the ith group has been obtained by a process of randomisation over a given set of groups enables us to draw any inference on other parameters (that is, those relating to the unobserved groups).

To indicate such a possibility let us consider a statistic T which is function of the r.v., (Y, X) such that the likelihood based on an observed value $T = t$.

$$L(\theta, t) \qquad (3.3)$$

contains all the parameters or a subset of parameters of interest, and thus enables us to compare the odds for different sets of parameters. If the information associated with (3.3) is denoted by $I(\theta; T)$, then for a wide class of definitions of information, it can be shown that

$$I(\theta; T) \leq I(\theta; Y, X) \qquad (3.4)$$

and the equality holds iff T is sufficient; that is the average information on θ in T is generally less than that in (Y, X).

Thus, if the choice of a likelihood is to be made for inference purposes on the basis of maximising information, one has to prefer the likelihood (3.1) to (3.3). But information as computed in (3.2) is what we expect in the *long run* and has no relevance to a *particular realisation*. This clearly points out the pitfalls involved in getting ourselves tied to particular principles such as basing inference on likelihood alone.

On the other hand, the choice of likelihood can be made to depend on the information in the observed sample itself and not on the average over all possible samples. One definition of information which has some relevance in statistical inference is the log ratio of likelihoods for two alternative values

180

of the parameters θ and ϕ. In the problem under consideration we have the information

$$\log\frac{f_i(x, \theta)}{f_i(x, \phi_i)} \qquad (3.5)$$

based on (i, x) and the information

$$\log\frac{L(\theta, t)}{L(\phi, t)} \qquad (3.6)$$

based on a statistic t. While (i,x) is informative about the parameter in the ith group only, the statistic t, when the expression (3.6) involves all the components of θ, distributes information on all the parameters although the total quantity of information utilized may be less.

Let the probability density of $T = h(Y,X)$ in the ith group be $g_i(t, \theta_i)$. Then the total probability density of T (due to randomisation over the groups) is

$$\Sigma\pi_i g_i(t, \theta_i) \qquad (3.7)$$

which is the likelihood based on the statistic t and which may involve all the components of θ or at least a subset. We shall call (3.7) *the likelihood of θ based on t and the corresponding information $I(\theta ; T)$ as provided by randomisation.*

In a situation like the one we are considering where the full likelihood does not satisfy our purpose, we may have to depend on a statistic which for every observed value supplies information (however poor it may be) on parameters of interest. In choosing a statistic T for this purpose we may be guided by $I(\theta ;T)$, the information due to randomisation. Unfortunately, no unique choice of T which maximises $I(\theta ;T)$ may be possible unless some further restrictions are placed on the class of statistics to be considered.

In our discussion we have considered a single observation (Y,X) but the situation is the same when we have any sample design which selects a subset of the groups and the r.v. X is observed in each of the chosen subsets. Then we will have a sample

$$(i_1, x_{i_1}) \ldots (i_s, x_{i_s}) \qquad (3.8)$$

and we need the choice of a statistic (which may be a vector) depending on $i_1,\ldots,i_s$ and $x_{i_1},\ldots,x_{i_s}$ for writing the likelihood and drawing inference on unknown parameters.

Let us examine a similar situation in the analysis of data obtained by throwing the 100 faced die described in Section 2. Besides 0 or 1, each face carries one of twenty colours and exactly five faces have the same colour (in a more general situation the frequency distribution of colours over the faces may not be known.). We throw the die 10 times and each time we record the colour and the number on the face that comes up. The object of the experiment is to estimate the proportion of faces with the value 0, when nothing is known about the association between the colour and the number on each face.

If observations on colours have not been made, then a good estimate of the unknown proportion is $r/10$, where r is the number observed 0's. How can we use the information provided by the colours?

Let θ_i be the proportion of zeros on the faces with the colour i. Then the parameter under estimation is

$$(\theta_1 + \ldots + \theta_{20})/20. \tag{3.9}$$

We observe that at most ten colours can be represented in ten tosses and the likelihood based on the observed data (colours and numbers on the ten faces that turned up) involves only the parameters associated with observed colours. The likelihood principle allows us to say something only about the proportion of zeros on faces with colours represented in our sample. This is somewhat discomforting and the advocates of likelihood may say that one has to keep on throwing the die till all colours are represented and the Bayesians may make more unreasonable demands. All these may be desirable but, should it not be possible to estimate the parametric function (3.9) based on ten tosses only? Even if one ensures that all colours are represented, there may be other unobserved characteristics in which the faces of the die may differ and all possible distinct categories may not be represented in the sample.

In such a situation it seems to be necessary to consider the likelihood which is based on a function of the observations and which involves all the known parameters or at least the subset of parameters or parametric functions in which we are interested.

In the example of the die, the knowledge that all colours are equally represented would enable us to obtain a better estimator than $r/10$, but this cannot be achieved by considering the original likelihood. One may have to start with the likelihood based on $(r_i, s_i), i=1,\ldots,k$ where r_i and s_i are the numbers of zeros and unities observed with the same colour (but ignoring what particular k colours have been observed).

Another illuminating example is provided by Fisher's analysis of Randomised Block experiments. We may recall that the varieties are assigned at random to plots in a block. The observed data from an experiment can be provided in the form y_{ijk}, where i stands for the variety, j for the block number and k for the plot number in the jth block. The likelihood based on y_{ijk} is again not helpful. However, *valid analysis* can be carried out by ignoring k and considering the distribution of y_{ij}.

We shall consider some examples in the next section.

4
Simple Random Sampling (SRS) with Replacement

Let us suppose that each group has an equal chance of being chosen and we choose n groups independently (allowing repetitions) and from each group observe a value of X. Let us represent the sample by

$$(i_1, x_{i_1}), \ldots, (i_n, x_{i_n}). \tag{4.1}$$

What kind of inferences can we draw from such a sample? There are several possibilities all of which are valid.

(a) Knowing that we have observations from the groups $i_1,...,i_n$, some of which may be the same, we may confine our attention to the parameters $\theta_{i_1},...,\theta_{i_n}$ only and draw inferences about them. The appropriate likelihood for this purpose is

$$\frac{1}{N^n}\prod_{r=1}^{n} f_{i_r}(x_{i_r}, \theta_{i_r}).\qquad(4.2)$$

In estimating the parameters $\theta_{i_1},...,\theta_{i_n}$ or testing hypotheses about them, the fact that the groups $i_1,...,i_n$ are the result of a random choice does not play any role. We can behave as if we had only these n groups and we took observations from them.

In the sample survey situation where in each group the r.v. X takes a single value with probability one, the likelihood (4.2) only tells us that the parameters attached to the observed groups are known without error and nothing about parameters of the unobserved groups.

(b) How does random choice of the groups enable us to draw inference of a wider nature than that considered in (a)?

To avoid complications due to notation, let us consider a sample of size 3 in which say the jth unit has *occurred* twice giving two observations x_1, x_2 on X and the kth unit has occurred once giving the observation x_3. The probability density is

$$\frac{1}{N^3} f_j(x_1, \theta_j) f_j(x_2, \theta_j) f_k(x_3, \theta_k).\qquad(4.3)$$

Let us consider the statistic

$$[(x_1, x_2), x_3]\qquad(4.4)$$

where the information on the groups that occurred in the sample is supressed, and (x_1, x_2) indicates that they are observed from the same group. The probability density of (4.4), obtained from (4.3) by summing up over different possible indices j,k, is

$$\frac{1}{N^3}\sum_{j \neq k}\sum f_j(x_1, \theta_j) f_j(x_2, \theta_j) f_k(x_3, \theta_k).\qquad(4.5)$$

The expression (4.5) considered as the likelihood based on $[(x_1, x_2), x_3]$ involves all the unknown parameters and can therefore constitute a suitable starting point for drawing inference on all the parameters. We are not claiming at the moment that the statistic $[(x_1,x_2), x_3]$ is the best choice for this purpose, but only observe that (4.4) is a candidate for extracting *information due to randomisation*.

In the sample survey situation, the expression (4.5) reduces to u/N^3 where $u = ab$, a being the number of distinct groups having the value x_1, and b

the number with the value x_3. To obtain maximum likelihood estimate of the N parameters we have to maximise u. The maximum is attained when

$$\left[\frac{N}{2}\right] \quad \text{groups have the value } x_1$$

$$\left[\frac{N}{2}\right] + 1 \quad \text{groups have the value } x_3 \tag{4.6}$$

or

$$\left[\frac{N}{2}\right] + 1 \quad \text{groups have the value } x_1$$

$$\left[\frac{N}{2}\right] \quad \text{groups have the value } x_3 \tag{4.7}$$

An estimate of the total value of the groups, if needed, is $\left[\dfrac{N}{2}\right](x_1 + x_3) + x_3$ or $\left[\dfrac{N}{2}\right](x_1 + x_3) + x_1$ depending on the choice (4.6) or (4.7) of the maximum likelihood solution. We may choose the average of these two estimates

$$N\left(\frac{x_1 + x_3}{2}\right) \tag{4.8}$$

as an estimate of the total. It may be seen that the estimate is Nx_1 if only one distinct unit is observed and $N(x_1 + x_2 + x_3)/3$ if three distinct units are observed. This is the familiar estimator based on *distinct units in SRSWR*.

It is easy to generalise the method to a sample of any size and obtain the estimate of the total as N times the average of the values of the distinct units observed in the sample.

The above discussion differs somewhat from the arguments used by Hartley and J.N.K. Rao (1968, 1969) and Royall (1967) in arriving at the estimator (4.8). First, no initial reduction of the sample into distinct units is made. The principle of sufficiency is not invoked anywhere in the argument. It is automatically taken care of in our discussion which uses the knowledge that some units are repeated in our sample. Second, some of the concepts vital to the *new theory* of Hartley and Rao such as discretisation of values attached to the units and scale-load estimator are not used.

(c) As mentioned earlier, the statistic $[(x_1, x_2), x_3]$ is only one possible candidate. As an alternative let us consider

$$[(j, x_1), (j, x_2), x_3] \tag{4.9}$$

where we suppress only the label attached to x_3. The likelihood associated with (4.9) is

$$\frac{1}{N^3} f_j(x_1, \theta)\, f_j(x_2, \theta_j) \sum_{k \neq j} f_k(x_3, \theta_k) \tag{4.10}$$

184

which also provides a valid likelihood for purposes of inference on all the parameters.

In the sample survey situation the maximum likelihood estimates of the parameters derived from (4.10) are

$$\theta_j = x_1 \text{ and } \theta_k = x_3, \text{ for all values of } k \text{ except } k = j, \tag{4.11}$$

giving the estimate of the total as

$$x_1 + (N-1)x_3. \tag{4.12}$$

In a more general situation, if $x_1, \ldots, x_n$ are the values observed on distinct units, we may construct an estimate of the type

$$x_1 + \ldots + x_\mu + (N - \mu)\frac{x_{\mu+1} + \ldots + x_n}{n - \mu}. \tag{4.13}$$

The estimate (4.13) is of considerable interest in the following situation.

The *quality* of information provided by randomisation depends on how close the distributions in the different groups are, or in other words how homogeneous the different groups are. This is the reason why a sample survey practitioner takes the trouble of dividing his population into strata within which the different units (groups) are as homogeneous as possible.

Suppose that we have a sample in which some of the observations are rather large compared to the others (outliers in some sense). In such a case, we may consider our sample as an element of the set of samples in which the outliers are always observed and the others are randomly chosen from the rest of the groups. In computing the variance of (4.13), we allow for variations in $x_{u+1}, \ldots, x_n$ only. In such a case the precision of the estimator (4.13) may be considerably higher than that of

$$N(x_1 + \ldots + x_n)/n. \tag{4.14}$$

Some may object to the procedure I am suggesting and contend that the design of the sample survey was defective and that proper stratification of the units should have been made earlier, or the decision to always include some of the units should have been taken before drawing the sample. No doubt the comments are in order. Probably proper judgment was not exercised by the survey practitioner, or he was ignorant of the wide disparity between units. But I submit that the object of statistical analysis is to learn from given data and make appropriate inference and not blindly to apply given decision rules.

5
A Justification for the Sample Mean in SRS

We shall consider SRSWOR for simplicity of notation and treatment.

Let us suppose that there are N counters in a bag, N_1 of which carry the number 0 and N_2 the number 1. Further let the counters with value zero be numbered from 1 to N_1 and the counters with value unity from $N_1 + 1$ to N, so that all the counters are labelled. It is clear that any permutation of the labels within the sets of counters carrying the same number does not alter the character of the population in the bag. Now suppose that we draw n random numbers from 1 to N and we obtain n_1 numbers from 1 to N_1 and n_2 numbers from N_1+1 to N. Our sample will consist of n_1 zeros and n_2 unities, although the actual counters drawn may depend on a particular permutation of numbers within the sets 1 to N_1 and $N_1 + 1$ to N. Thus two practitioners using the same random numbers but with different permutations (which keep the population same in respect of labels and values) will both have n_1 zeros and n_2 unities although the actual counters may be different (for instance they may differ in some other unobserved aspects such as colour or some defects on it). In such a case it is reasonable to assert that the estimate should depend only on the observed number of 0's and 1's; that is, the two practitioners should provide the same estimate of the unknown proportion of counters with zero value.

Consider an estimator $T = f(n_1, n_2)$ and impose the condition of unbiasedness; that is,

$$\sum_{n_1 + n_2 = n} \sum f(n_1, n_2) P(n_1, n_2 \mid N_1, N_2) = \frac{N_1}{N} \qquad (5.1)$$

for all N_1, N_2 such that $N_1 + N_2 = N$, where $P(n_1, n_2 \mid N_1, N_2)$ denotes the probability of observing n_1 zeros and n_2 unities for given N_1, N_2. Since n_1 is a complete sufficient statistic (with respect to $P(n_1, n_2 \mid N_1, N_2)$, for varying N_1) it follows that the only solution of $f(n_1, n_2)$ satisfying (5.1) is

$$f(n_1, n_2) = n_1/n \qquad (5.2)$$

which is the familiar estimator. Thus n_1/n is the UMV unbiased estimator in the class of estimators considered.

We may now extend the same argument to the situation where there are m types of counters carrying distinct numbers, N_1 of them carrying the value α_1, N_2 the value α_2 etc. We may label these counters from 1 to N but observe that any permutation of labels within the set of counters having the same number leaves the population invariant. In such a case, the estimator should depend only on the observed vector $(n_1, \ldots, n_m)$ where n_i is the number of counters drawn with the number α_i. We note that, with respect to $P(n_1, \ldots, n_m \mid N_1, \ldots, N_m)$ for variations in $N_1, \ldots, N_m$, the statistic $(n_1, \ldots, n_m)$ is sufficient and complete. Consequently there is a unique function of $n_1, \ldots, n_m$ whose expectation is a given function of $N_1, \ldots, N_m$. Thus $\Sigma \alpha_i n_i/n$ is the UMV unbiased estimator of $\Sigma \alpha_i N_i/N$ in the class of estimators considered. These considerations seem to provide a justification for the sample mean in the general case. Hartley and J. N. K. Rao (1968, 1969) and Royall (1967) justify the sample mean starting from a different premises.

Kempthorne (1969) has given another justification for the sample mean by showing that it has average minimum variance, for permutations of values attached to the units, in the general class of linear unbiased and translation invariant estimators. The condition of translation invariance, however, seems to be redundant as shown below.

To simplify the notation, let us consider without loss of generality samples of size 3. If the i, j, k, units appear in the sample, then the general linear estimator of the population total is of the form

$$T = e_{ijk} X_i + e_{jki} X_j + e_{kij} X_k \qquad (5.3)$$

with the restrictions

$$\sum_{j,k} e_{ijk} = \binom{N}{3}, \quad i = 1, \ldots, N \qquad (5.4)$$

to ensure unbiasedness. Now

$$E(T^2) = \frac{1}{\binom{N}{3}} \sum_{i,j,k} (e_{ijk} X_i + e_{jki} X_j + e_{kij} X_k)^2. \qquad (5.5)$$

The average value of $E(T^2)$ over the $N!$ permutations of $X_1, \ldots, X_n$ is found as

$$\frac{1}{\binom{N}{3}} \sum_{i,j,k} (\alpha e_{ijk}^2 + \alpha e_{jki}^2 + \alpha e_{kij}^2 + 2\beta e_{ijk} e_{jki} + 2\beta e_{ijk} e_{kij} + 2\beta e_{jki} e_{kij}) \qquad (5.6)$$

where α and β are suitably defined.

Introducing N Lagrangian multipliers $\lambda_1, \ldots, \lambda_N$ for the restrictions (5.4), the equations minimising the expression (5.6) are

$$\alpha\, e_{ijk} + \beta\, e_{jki} + \beta\, e_{kij} = \lambda_i$$

$$\beta\, e_{ijk} + \alpha\, e_{jki} + \beta\, e_{kij} = \lambda_j \qquad (5.7)$$

$$\beta\, e_{ijk} + \beta\, e_{jki} + \alpha\, e_{kij} = \lambda_k$$

for $i \neq j \neq k$. The solution of (5.7) is of the form

$$e_{ijk} = a\, \lambda_i + b\, \lambda_j + b\, \lambda_k$$

$$e_{jki} = b\, \lambda_i + a\, \lambda_j + b\, \lambda_k \qquad (5.8)$$

$$e_{kij} = b\, \lambda_i + a\, \lambda_j + b\, \lambda_k$$

for suitable values of a and b. Summing up the first equation of (5.8) over all possible values of j, k, we obtain

$$\binom{N}{3} = a \binom{N-1}{2} \lambda_i + b(N-2) \sum_{j \neq i} \lambda_j \qquad (5.9)$$

$$i = 1, \ldots, N$$

which shows that $\lambda_1 = \ldots = \lambda_N$ and consequently $e_{ijk} = e_{jik} = e_{kij} = e_{i'j'k'}$ etc., which lead to the sample mean.

One may question the wisdom of minimising the average variance over the permutations of the values of the units $(X_1,...,X_N)$. But it could be argued that if the coefficients in (5.3) depend only on the labels and not on the knowledge of the values $X_1,...,X_N$ it is only natural to minimise the average loss incurred in using the estimator (5.3) by different practitioners who may have the same values for the units but choose to label them differently.

6

Sampling with Unequal Probabilities

We shall consider a population of N distinct units labelled by numbers $1, \ldots, N$ and denote by p_{ijk} the probability of observing the units with the labels i, j, k. The observed values on these units are denoted by X_i, X_j, X_k. Let us consider a statistic of the type

$$T = e_{ijk} X_i + e_{jki} X_j + e_{kij} X_k \tag{6.1}$$

as in (5.3). If T is unbiased for the parametric function $(X_1 + \ldots + X_n)$ then we have the conditions

$$\sum_{j,k} p_{ijk} e_{ijk} = 1, \quad i = 1, \ldots, N. \tag{6.2}$$

As in (5.5), we can compute

$$E(T^2) = \sum_{i,j,k} p_{ijk} (e_{ijk}^2 X_i^2 + e_{jki}^2 X_j^2 + e_{kij}^2 X_k^2 + 2 e_{ijk} e_{jki} X_i X_j$$

$$+ 2 e_{ijk} e_{kij} X_i X_k + 2 e_{jki} e_{kij} X_j X_k). \tag{6.3}$$

In the case of SRS, we took the average of $E(T^2)$ over permutations of $X_1, \ldots, X_k$ which appeared as a natural class of populations to which the estimator T is applicable. But in the case of unequal probability sampling, *presumably* large values of X_i are given large probabilities of inclusion. In such a case it seems natural to consider the set of parameter values $(p_1 y_{i_1}, \ldots, p_N y_{i_N})$ obtained by permutations of $y_1, \ldots, y_N$, keeping $p_1, \ldots, p_N$ fixed, where $y_i = X_i/p_i$ and $p_i = \sum_{r,k} p_{ijk}$. Taking the average of (6.3) over the set of parameters considered we obtain an expression of the form

$$R(T^2) = \sum p_{ijk} \left\{ \alpha(p_i^2 e_{ijk}^2 + p_j^2 e_{jki}^2 + p_k^2 e_{kij}^2) \right.$$

$$\left. + 2\beta(p_i p_j e_{ijk} e_{jki} + p_i p_k e_{ijk} e_{kij} + p_j p_k e_{jki} e_{kij}) \right\}. \tag{6.4}$$

Let us minimise (6.4) subject to the conditions (6.2).

Introducing Lagrangian multipliers $\lambda_1, \ldots, \lambda_N$, the minimising equations are

$$p_{ijk}(\alpha p_i^2 e_{ijk} + \beta p_i p_j e_{jki} + \beta p_i p_k e_{kij}) = \lambda_i p_{ijk}$$

$$p_{ijk}(\beta p_i p_j e_{ijk} + \alpha p_j^2 e_{jki} + \beta p_j p_k e_{kij}) = \lambda_j p_{ijk} \tag{6.5}$$

$$p_{ijk}(\beta p_i p_k e_{ijk} + \beta p_j p_k e_{jki} + \alpha p_k^2 e_{kij}) = \lambda_k p_{ijk}$$

for $i \neq j \neq k$. We may assume that $p_{ijk} \neq 0$, in which case we can solve the equations (6.5) and obtain

$$p_i e_{ijk} = a \frac{\lambda_i}{p_i} + b \frac{\lambda_j}{p_j} + b \frac{\lambda_k}{p_k} \qquad (6.6)$$

for suitable values of a, b not depending on i, j, k. Multiplying both sides of (6.6) by p_{ijk} and summing over j, k we have

$$1 = a \frac{\lambda_i}{p_i} + b \sum_{j,k} \frac{\lambda_j}{p_j p_i} p_{ijk} + b \sum_{j,k} \frac{\lambda_k}{p_k p_i} p_{ijk} \qquad (6.7)$$

$$i = 1, \ldots, N.$$

We find that

$$\frac{\lambda_1}{p_1} = \frac{\lambda_2}{p_2} = \ldots = \frac{\lambda_N}{p_N} = (a + 2b)^{-1} \qquad (6.8)$$

is a possible solution. Using this solution in (6.6)

$$p_1 e_{ijk} = p_j e_{jki} = p_k e_{kij} = 1 \qquad (6.9)$$

which give the Horvitz-Thompson estimator

$$T = \frac{X_1}{p_1} + \frac{X_2}{p_2} + \frac{X_3}{p_3} \qquad (6.10)$$

as the best choice. We have considered sampling of three units only but the argument holds good for the general case.

It may be noted that averaging over a different set of the parameters might give a different answer. As such the result obtained can be considered only as an additional property of the H-T estimator, which seems to be worthy of putting on record.

7

Concluding Remarks

The object of the paper has been to pose some more problems, in addition to those already known in sampling from finite populations, to bring to light the difficulties involved in evolving suitable theories of statistical inference. It points out the limitations of a monolithic structure for statistical inference. In his treatise on *Statistical Methods and Scientific Inference*, R. A. Fisher advocates several principles of statistical inference each appropriate to a given situation. The prime place given to likelihood in his famous theory of estimation is taken away by his equally famous principle of randomisation in design of experiments. For a discussion on statistical inference and the need for a variety of statistical tools reference may be made to Rao (1965, Chapters 7 and 8).

I have not discussed the important problem of design of sample surveys, but confined my attention to statistical analysis of data obtained from a given survey utilising the information provided by the data alone. This is a definite and a challenging problem. Of course, if auxiliary information is available, it should be used in the statistical analysis; however, it would be more profitable to use it in designing the survey as well. I have not commented on Bayesian methods in survey data analysis. I hope to examine these questions in greater detail in a forthcoming paper.

References

1. Fisher, R.A., *Statistical Methods and Scientific Inference,* Second Edition, Oliver and Boyd, 1959.
2. Godambe, V.P., "A Unified Theory of Sampling from Finite Populations," *Journal of the Royal Statistical Society, B,* 17, 268-278, 1955.
3. Hartley, H.O. and Rao, J.N.K., "A New Estimation Theory for Sample Surveys," *Biometrika,* 55, 547-557, 1968.
4. Hartley, H.O. and Rao, J.N.K., "A new Estimation Theory for Sample Surveys --- II," *New Developments in Survey Sampling,* University of North Carolina Symposium, 147-169, 1969.
5. Kempthorne, Oscar, "Some Remarks on Statistical Inference," *New Developments in Survey Sampling,* University of North Carolina Symposium, 671-692, 1969.
6. Rao, C. Radhakrishna, *Linear Statistical Inference and its Applications,* New York, John Wiley and Sons, 1965.
7. Royall, Richard M., "An Old Approach to Finite Population Sampling Theory," *Technical Report 420,* J.H.U., 1967.

COMMENTS

D. Basu:

Please permit me to make a statement to be followed by a question. Let θ be the unknown state of nature and suppose there are k experiments

$$E_1, E_2, \ldots, E_k$$

each of which may be performed to gain a meaningful quantum of information on θ. The scientist allots probabilities

$$\Pi_1, \Pi_2, \ldots, \Pi_k \quad (\Sigma \Pi_i = 1)$$

(the Π_i's do not depend on θ) to the k experiments and thereby selects an experiment say E_s. He then performs the experiment E_s thereby generating the data x. It seems axiomatic to me that at the time of analyzing the data x, the scientist should refer the data to the experiment E_s that he has actually performed and forget about the other experiments that he might have performed. It follows that the selection probabilities $\Pi_1, \Pi_2, \ldots, \Pi_k$ have nothing to do with the analysis of the data. I do hope that Professor Rao will agree with me.

Professor Rao's descriptions of a typical survey operation is essentially the same as above. A subset s of population units is selected by a sampling scheme that is unrelated to the state of nature $\theta = (x_1, x_2, \ldots, x_N)$ of all the population values. The selection of s is a miniscule part of the usually large survey operation. The major part of the survey operation may be characterized as the experiment E_s which is the act of actually surveying the units in s. The experiment E_s produces the data. The nature and extent of the information that is contained in the data naturally depend on the prior knowledge about θ that the scientist has.

My question to Professor Rao is the following:

Having performed the experiment E_s should the scientist relate the data so generated to the experiment E_s actually performed or should he also speculate about all the other experiments that might have been performed? In other words, is the sampling design really relevant at the analysis stage?

I. J. Good:

The problem of the die with 100 faces can be regarded as a problem of probability estimation from a contingency table. More general problems are concerned with multidimensional contingency tables. I have considered such problems from a Bayesian point of view in at least two places (*J. Roy Statist. Soc., B,* c. 1955; and *The Estimation of Probabilities,* MIT Press, 1965). I have not settled many of the problems, but would like to emphasize that I was especially concerned with cells of the tables containing no entries: the *probabilities of events that have never occurred.*

O. Kempthorne:

While we should give credit, I believe, to Dr. Godambe and others for raising the basic question of the logic of inference from a random sample of a finite population, I believe that the main direction of most of the work, including that of Dr. Rao is not strongly relevant. The provision of a "best" point estimate is a pure decision problem. Suppose, indeed, that we know that the UMVU estimate of a parameter is 5.7 in our particular case. As an ordinary (hopefully intelligent) citizen I ask, How reliable is the result? Could the parameter be as high as 20? or as low as 1? There is uniformity in the failure to approach this type of question, which I regard as *the* basic question of inference. The reduction of inference to decision, and, in particular, to point estimation is, I believe, one of the basic errors of much mathematical statistics of the past 30 years.

There have been a variety of discussions of the likelihood function for the case under hand, which must be about as confusing to an outsider as they could possibly be. Dr. Rao has added to the volume. I can only express my opinion that he has not lessened the confusion. But words mean different things to different people and perhaps his set of words is compelling to some. I find the whole discussion by everyone wholly confusing because likelihood equals probability of actual data (apart from a multiplicative constant), and obviously the randomness in the sampling process does not depend *at all* on the values $x_1, x_2, \ldots, x_N$ of the units. The probability of a sample does not depend on these, and *hence* the likelihood does not. If one adopts what I regard as sleight of hand, one obtains the notion that the likelihood function in the space R^N for the parameters $x_1, x_2, \ldots, x_N$ is equal to $1 | \binom{N}{n}$ for the observed units having x's equal to the known values, with *no* dependence on the x-values of unobserved units. This renders any discussion of the maximum of the likelihood null and void, I believe. This, also, is the graveyard of the likelihood principle, because if any statistician claims that there is not partial evidence on the set of unobserved x-values, he should, I believe, visit a psychiatrist or leave statistics. I do not find Dr. Rao's discussion convincing.

Dr. Rao discusses briefly the case of the randomized block design. The logic here is obscure but is not, I believe, based on likelihood ideas. I discussed in my Fisher lecture the peculiar (to me) bivalence of R. A. Fisher in this respect. In my opinion the basis of inference in the experiment is the test of significance (as discussed in my *J. Am. Stat. Ass.*, paper of 1955, for example, and in my book). Cornfield has written on the matter and shown how the incorporation of a prior leads to a conclusion. But, *of course*, a prior with no data gives a conclusion, so we should not, I think, be highly comforted by this.

I am very curious on what notion of information Dr. Rao would use. If there are in fact distributions around the θ_i as in equation (2.1) I have no difficulties. I tend to the view that the notion of information is not viable in this problem. To this I would add the difficulty of the notion of information except in what one may term highly regular cases, an area in which Dr. Rao himself gives a good description in his book.

I am interested by Dr. Rao's proof with regard to the average minimum variance of the sample mean. I was apparently mistaken in requiring translation in variance. I shall examine the matter at leisure. This area has been plagued with erroneous proofs. It is interesting to note that the work of Horvitz and Thompson was based very much on the obscurity of *linear* in Neyman's theorem, which was thought for years to be well established.

W. J. Hall:

(This was not given in the discussion but only communicated in writing for publication in the *Proceedings*.)

My first reactions to a central idea of this paper was: (1) How is it sensible to suppress part of the data in order to learn something about parameters that can't be learned from the full data? (2) How is it possible that the likelihood for (X, Y) should have fewer parameters than the likelihood for X alone or for a statistic $T = t(X, Y)$? I don't know the answers but wish to record two brief comments.

The first is roughly analogous to a common method of dealing with nuisance parameters. If two parameters appear in the distribution of the full data and we are interested only in one of them, we may reduce the data (from (X, Y) to T, say) to a statistic whose distribution depends only on the parameter of interest. So why not do the analogous thing to gain information on additional parameters?

Secondly, the likelihood function of (X, Y) is a function of *all* parameters — it simply is constant along certain coordinate directions. No new parameters are introduced by reducing to T, it's just that the constancies need not persist. These simple observations helped me appreciate this interesting paper.

J. A. Hartigan:

Why can't the likelihood method be saved in your example by the following approach to parametrization? Let there be two populations, means μ_1 and μ_2, variance one. Let a population be selected at random, and the observation x taken. Then

$$\text{Likelihood} = f(x \mid \mu_1)$$

and only information about μ_1 is available, as you say. But, parametrize $\underset{\sim}{\mu} = (\mu_1, \mu_2)$ where the order (μ_1, μ_2) is not specified. The probability of $x \mid \underset{\sim}{\mu}$ is still known because of randomization, and in fact

$$\text{Likelihood} = \tfrac{1}{2} f(x \mid \mu_1) + \tfrac{1}{2} f(x \mid \mu_2)$$

and information, about μ_1 and μ_2 both, is available. Generally with n populations equally likely to be selected, the parameter $\underset{\sim}{\mu} = (\mu_1, \ldots, \mu_n)$, with order of μ_i ignored, should be used. Perhaps the implication is the likelihood man does use randomization, since it allows him to choose different parametrizations, and thereby different likelihood functions.

It is not true that in the Bayesian approach only prior information will be available about non-observed populations. The prior will be stated in such a form that information about one population applies to others.

For example consider two normal populations with means μ_1, μ_2. Let

$$\text{Prior } (\mu_1, \mu_2) \propto \exp\left[-\frac{1}{2}\left(\frac{\mu_1 - \mu_2}{\sigma}\right)^2\right].$$

On one observation x, from the population μ_1 the

$$\text{Likelihood} \propto \exp\left[-\tfrac{1}{2}(x - \mu_1)^2\right],$$

and the posterior density of (μ_1, μ_2) is

$$\exp -\tfrac{1}{2}\left[(x - \mu_1)^2\left(\frac{1}{\sigma^2} + 1\right) - 2\frac{(x - \mu_1)(x - \mu_2)}{\sigma^2} + \frac{(x - \mu_2)^2}{\sigma^2}\right].$$

Thus $\text{corr}(\mu_1, \mu_2 \mid x) = 1/(1 + \sigma^2)^{1/2}$. $\text{Var}[\mu_1 \mid x] = 1$ (*no loss*). $\text{Var}[\mu_2 \mid x] = [1 + \sigma^2]$ (*we don't know μ_2 so well*).

J. C. Koop:

I must congratulate Dr. Rao on his ingenious attempt in formulating likelihood functions and the associated information functions which cover situations both in classical estimation theory and in estimation theory for sample surveys. Together with likelihood he has used the principles of sufficiency and completeness to derive a U.M.V. estimator in simple random sampling without replacement.

Now it is possible to construct a class of linear estimators where the co-efficients attached to the counters depend on the type of counter (which is a manifestation of the identity of the unit) and the distinct sample. This is somewhat in line with his thinking that "the average information on θ in T is generally less than that in (Y,X)". Some possible members belonging to this class and their mean square errors (M.S.E.'s) will be considered in the following:

Example. Consider a bag containing $N=5$ counters where there are only $m=2$ types of counters. $N_1=3$ counters carry the value α_1 and $N_2=2$ counters carry the value α_2. We want to estimate

$$\tfrac{3}{5}\alpha_1 + \tfrac{2}{5}\alpha_2$$

on the basis of a simple random sample (drawn without replacement) of size $n=2$. Clearly there are only 10 possible samples, six composed of counters (α_1, α_2), three of (α_1, α_1) and finally only one of (α_2, α_2). Thus the three possible U.M.V. estimators are

$$\tfrac{1}{2}(\alpha_1 + \alpha_2), \quad \alpha_1 \text{ and } \alpha_2.$$

The probabilities of realizing the three distinct samples, and therefore also the corresponding estimators, are respectively $\frac{6}{10}$, $\frac{3}{10}$ and $\frac{1}{10}$, so that the variance of the U.M.V. estimator (say) $\bar{y}$ is

$$V(\bar{y}) = \frac{6}{10}\left\{\frac{\alpha_1 + \alpha_2}{2} - \frac{3\alpha_1 + 2\alpha_2}{5}\right\}^2 + \frac{3}{10}\left\{\alpha_1 - \frac{3\alpha_1 + 2\alpha_2}{5}\right\}^2 + \frac{1}{10}\left\{\alpha_2 - \frac{3\alpha_1 + 2\alpha_2}{5}\right\}^2$$

$$= \frac{9}{100}(\alpha_1 - \alpha_2)^2.$$

The members of the alternative class of estimators, constructed in the way stated in the foregoing account, and their M.S.E.'s (given in the last line of the table) are as follows:

Possible distinct sample	Estimators				
	$\hat{y}$	$\hat{y}_1$	$\hat{y}_2$	$\hat{y}_3$	$\hat{y}_4$
(α_1, α_2)	$\frac{3}{4}\alpha_1 + \frac{1}{4}\alpha_2$	$\frac{2}{3}\alpha_1 + \frac{1}{3}\alpha_2$	$\frac{3}{5}\alpha_1 + \frac{2}{5}\alpha_2$	$\frac{7}{10}\alpha_1 + \frac{3}{10}\alpha_2$	$\frac{9}{10}\alpha_1 + \frac{1}{10}\alpha_2$
(α_1)	α_1	α_1	α_1	α_1	α_1
(α_2)	α_2	α_2	α_2	α_2	α_2
M.S.E.	$\frac{31}{400}(\alpha_1 - \alpha_2)^2$	$\frac{13}{150}(\alpha_1 - \alpha_2)^2$	$\frac{21}{250}(\alpha_1 - \alpha_2)^2$	$\frac{9}{100}(\alpha_1 - \alpha_2)^2$	$\frac{77}{500}(\alpha_1 - \alpha_2)^2$

It may be noted that the mean square errors of $\hat{y}$, $\hat{y}_1$ and $\hat{y}_2$ are smaller than the variance of the U.M.V. estimator $\bar{y}$. Reasoning in this sense, $\hat{y}_3$ is as good as $\bar{y}$, but $\hat{y}_4$ is worse. Certainly the class in question contains an infinity of estimators some of which are better and some of which are worse than the U.M.V. estimator. Indeed if one looks hard at the table, it becomes evident that the "U.M.V." estimator itself, which in this case is the sample mean, could have been constructed as a member of this class. In general it is not difficult to show that there are no best linear estimators in this class, a result in line with Godambe's 1955 work.

Hartley and J. Rao (*Biometrika*, 1968, 55, page 547) introduced the notion of scale load estimators and justified the sample mean reasoning through maximum likelihood, sufficiency and completeness. Royall's justification (*J. Am. Statist. Assoc.*, 1968, 63, page 1269) was through Hacking's notion of support, which is a version of the principle of maximum likelihood. In the light of the above counter examples the relevance of all these concepts in estimation theory for finite populations comes into question.

H. Rubin:

Let me consider a "practical" problem somewhat similar to Prof. Rao's die problem. Suppose it is desired to estimate the number of blonds in the student body (10,000) at the University of Waterloo. The investigator is provided with a directory whose entries include name, age, sex, marital status, home address, class, university address, and telephone number, and can select any 200 students to observe (we assume hair dye is not used). Let us also assume that the loss is squared error.

What would a Bayesian do? He would start out with the prior probability that each of the sets of students is the set of all blonds, choose the 200 students so that the expected conditional variance of the number of blonds given the observations on the 200 individuals is smallest, observe these 200, and report the conditional mean. However, a computer to contain the prior information would require a memory of $2^{10000} > 10^{3000}$ words. Unfortunately, the current estimate of the number of particles in the universe is considerably smaller than 10^{100}. Therefore, it is necessary to use an optimality of type II approach.

There is considerable information available which I now treat from my personal viewpoint. If the student's name is Mr. Singh or Miss Singh, no appreciable risk is encountered by assuming this individual is dark. I would also be willing to ignore, but with some reluctance, any individual named Mrs. Singh if her home address is in India, but not if her address were Waterloo. There is much more information — I believe Mr. Black is considerably less likely to be blond than Mr. Anderson. What I would do would be to divide the individuals into strata, to decide on an approximate joint prior distribution of intrastratum exchangeable distributions, use this to decide on the number of students to observe from each stratum by random sampling and what estimation procedure to use, and then try to improve my estimate by using a substratified random sample.

D. A. Sprott:

It is not possible that the supposed contradiction between randomization and likelihood entailed by suppressing the subscript k in y_{ijk}, as well as the different likelihoods considered in this paper might be a result of a failure to specify more precisely the reference set to which the observed result is supposed to belong, along with the recognizable and relevant subsets, as discussed at great length by Fisher? Even in sampling a finite population it seems necessary to consider the hypothetical reference set of all such similar populations, for example, all urns containing 10 balls coloured red and black and with distinguishing marks such as the numbers from 1 to 10. If the observer has no knowledge of the method of numbering the balls, then although the ith ball drawn may be numbered 1, and this certainly is in the recognizable subset of all balls numbered 1 in the hypothetical reference set, this subset is not *relevant* since his knowledge concerning this subset is the same as that of the whole set. The numbers in that case must be ignored and the likelihood is informative.

If all characteristics of an item are considered to define recognizable and relevant subsets, the likelihood becomes uninformative, the only element in any subset being the observed item itself. Variation between the two extremes can occur where there is some knowledge concerning numbering and colouring of the balls which may define relevant subsets and so alter the likelihood accordingly. It seems that the various different likelihoods in Dr. Rao's paper might be considered to have arisen in this way, and in that case there is no difference in this regard between sampling from a finite population and the more classical problem of sampling from an infinite hypothetical population considered by Fisher.

M. E. Thompson:

Dr. Rao's result in Section 6 of his paper is more generally true. In particular, the requirement that the unbiassed estimator be linear, that is, of the form (6.1), is not necessary.

Following generally the notation of the paper by Godambe and Thompson in these *Proceedings*, we denote by

$$\underline{x} = (x_1, \ldots, x_i, \ldots, x_N)$$

the *population vector*, that is, x_i is the value (real) associated with the individual i of the population, $i = 1, \ldots, N$. The domain of the values of $\underline{x}$ is assumed to be the Euclidean N-space R_N. Again a *sampling design* is denoted by (S, p) where S is the set of all possible subsets of $(1, \ldots, i, \ldots, N)$ and p is the probability function defined on S. An element of S will be denoted by s and $\nu(s)$ will denote the number of individuals $i \in s$. We will be concerned with sampling designs satisfying the following two conditions:

$$[\nu(s) \neq n] \rightarrow [p(s) = 0] \qquad \text{for } s \in S$$

and

$$\Sigma_{S(i)} p(s) = \pi_i, \qquad i = 1, \ldots, N$$

where π_i (> 0), $i = 1, \ldots, N$ and n are some *given* numbers and $S(i)$ is the subset of S consisting of those elements s containing i. An *estimator* is a real function $e(s, \underline{x})$ on $S \times R_N$ which depends on $\underline{x}$ only through those x_i for which

$i \in s$. Naturally the estimator $e(s, \underline{x})$ can also be denoted by $e(s, x_i: i \in s)$. For a given sampling design (S, p), the estimator e is said to be *unbiased* for the population total

$$\Sigma_1^N x_i = \tau(\underline{x}), \qquad \underline{x} \in R_N,$$

for

$$\Sigma_s e(s, x_i: i \in s) \, p(s) = \tau(\underline{x}), \qquad \underline{x} \in R_{N'}.$$

Now let ξ be a probability measure on R_N, such that

$$\int x_i^2 d\xi < \infty \qquad i = 1, 2, \dots, N.$$

For any unbiased estimator $e(s, x_i: i \in s)$ of $\tau = \sum_{i=1}^N x_i$ let

$$V_\xi(e) = \int \left[\sum_{s \in S} p(s) \, (e(s, x_i: i \in s) - \tau)^2 \right] d\xi.$$

The following theorem includes the result that among unbiassed estimators of τ the Horvitz-Thompson estimator has minimum average variance over permutations of the 'appropriate' population co-ordinates y_i.

Theorem. Let $y_i = x_i/\pi_i$, where $\pi_i = \Sigma_{S(i)} p(s)$. If ξ is such that the y_i have an exchangeable joint distribution, then

$$V_\xi(e) > V_\xi(\bar{e}),$$

where

$$e(s, x_i: i \in s) = \sum_{i \in s} (x_i/\pi_i).$$

Proof: Let

$$h(s, x_i: i \in s) = e(s, x_i: i \in s) - \bar{e}(s, x_i: i \in s).$$

Then

$$V_\xi(e) = V_\xi(\bar{e}) + V_\xi(h) + 2 \, \mathrm{Cov}_\xi(\bar{e}, h),$$

where

$$\mathrm{Cov}_\xi(\bar{e}, h) = \int (\sum_s \bar{e}(s, x_i: i \in s) \, h(s, x_i: i \in s) \, p(s)) \, d\xi.$$

To prove the theorem, then, it will suffice to show that $\mathrm{Cov}_\xi(\bar{e}, h)$ vanishes. Since $\sum_{s \in S} p(s) \, h(s, x_i: i \in s) = 0$ for all $\underline{x} = (x_1, \dots, x_N) \in R_N$, this follows by induction from the following lemmas.

Lemma 1. Let η be an exchangeable probability measure on R_N. For any n, $2 < n < N$, let σ denote the generic subset of $n-1$ elements of $\{1, \dots, N\}$. Suppose that for any function $g(\sigma, y_i; i \in \sigma)$ the equation

$$\sum_\sigma g(\sigma, y_i: i \in \sigma) = 0 \quad \text{for all } \underline{y} = (y_1, \dots, y_N) \in R_N \tag{1}$$

implies

$$\int \sum_\sigma (\sum_{i \in \sigma} y_i) g(\sigma, y_i: i \in \sigma) \, d\eta = 0. \tag{2}$$

Then for any function $f(s, y_i: i \in s)$, where s runs through n-element sets in S,

$$\sum_s f(s, y_i: i \in s) = 0 \qquad \text{for all } \underline{y} \in R_N$$

implies

$$\int \sum_s (\sum_{i \in s} y_i) f(s, y_i : i \in s)\, d\eta = 0. \tag{4}$$

Proof: Setting $y_j = 0$ for all $j \notin s$ in (3) yields the form

$$f(s, y_i : i \in s) = \sum_{\sigma \subset s} f_{s,\sigma}(\sigma, y_i : i \in \sigma) \tag{5}$$

(where $\sigma \subset s$ means "σ contained in s"); and (3) becomes

$$\sum_\sigma g(\sigma, y_i : i \in \sigma) = 0 \tag{6}$$

where $g(\sigma, y_i : i \in \sigma) = \sum_{s \supset \sigma} f_{s,\sigma}(\sigma, y_i : i \in \sigma)$, the notation $s \supset \sigma$ meaning "s containing σ". Now

$$
\begin{aligned}
\sum_s (\sum_{i \in s} y_i) f(s, y_i : i \in s) &= \sum_s (\sum_{i \in s} y_i) \cdot \sum_{\sigma \subset s} f_{s,\sigma}(\sigma, y_i : i \in \sigma) \\
&= \sum_\sigma \sum_{s \supset \sigma} (\sum_{i \in s} y_i) f_{s,\sigma}(\sigma, y_i : i \in \sigma) \\
&= \sum_\sigma \sum_{s \supset \sigma} (\sum_{i \in \sigma} y_i) f_{s,\sigma}(\sigma, y_i : i \in \sigma) \\
&\quad + \sum_\sigma \sum_{k \notin \sigma} y_k f_{\sigma \cup \{k\}, \sigma}(\sigma, y_i : i \in \sigma).
\end{aligned}
$$

Thus by (2),

$$
\begin{aligned}
\int \sum_s (\sum_{i \in s} y_i) f(s, y_i : i \in s)\, d\eta &= \int \sum_\sigma \sum_{k \notin \sigma} y_k f_{\sigma \cup \{k\}, \sigma}(\sigma, y_i : i \in \sigma)\, d\eta \\
&= \frac{1}{N-n+1} \sum_\sigma \int (\sum_{i \notin \sigma} y_i) \sum_{s \supset \sigma} f_{s,\sigma}(\sigma, y_i : i \in \sigma)\, d\eta \\
&= -\frac{1}{N-n+1} \int \sum_\sigma (\sum_{i \in \sigma} y_i) g(\sigma, y_i : i \in \sigma)\, d\eta = 0.
\end{aligned}
$$

Lemma 2. If η is an exchangeable probability measure on R_N and if

$$\sum_{i=1}^N g_i(y_i) = 0 \quad \text{for all } y \in R_N, \tag{7}$$

then

$$\int \sum_{i=1}^N y_i g_i(y_i)\, d\eta = 0. \tag{8}$$

Proof: Equation (7) implies that $g_i(y_i)$ is some constant g_i, where $\sum_{i=1}^N g_i = 0$. Hence

$$\int \sum_{i=1}^N y_i g_i(y_i)\, d\eta = \sum_{i=1}^N g_i \int y_i\, d\eta = 0.$$

REPLY

1. I shall first deal with the comments made by Dr. Basu, Dr. Hall and Dr. Kempthorne on section 2 of my paper which stresses the role of (or information due to) randomization. I must make it clear that no new information is created by randomization but it enables us to extract the desired information using likelihood and the traditional concepts of estimation and testing of hypothesis. To explain this, I must emphasize what I have already stated in my paper.

We shall denote the observed set of labels by L, the values on observed units by V and the full sample of labels and associated units by $L + V$.

(a) The likelihood based on $L + V$ (which we may write as $l(L+V)$) is constant over the possible values of unobserved units, and therefore the *likelihood approach* does not help us in inferring about unobserved units from the sample. It is in this sense that $l(L+V)$ does not involve or is uninformative about the unknown parameters. This need not imply that the sample itself is uninformative. [I am glad that Dr. Kempthorne agrees with me in this respect although he uses the unfortunate expression that the information is of a *partial* nature]. It is not difficult to demonstrate that the sample contains the desired information. For instance, if we draw a unit at random and observe its value, we are generating a *random variable* whose expectation is equal to the parameter we want to estimate, viz., the population mean (μ say). This implies that observations on such a random variable must carry information on μ. However, we may disagree on methods of estimating μ.

(b) What is wrong with $l(L+V)$? Indeed $l(L+V)$ becomes effective if we have some information on the association between L and V. Such a dependence may be specified by a stochastic relationship between the two or through Bayesian approach by considering a prior distribution for values of the sampling units. [If independent priors are assumed for values of different sampling units, the posterior distribution for the unobserved values is same as the prior, but this difficulty does not arise if one considers the type of priors suggested by Dr. Hartigan, although I do not see any practical relevance of such priors]. Thus, it appears that the likelihood, $l(L+V)$, is somewhat *underspecified* to take full advantage of the observations on the labels, unless the nature of dependence between L and V is known. Similar situations exist in applications of multivariate analysis (see advanced statistical methods in Biometric Research by the author).

(c) Can we draw inferences on the population mean (or more specifically on the values of unobserved units) when the relationship between L and V is unspecified or independently of any relationship between the two? The answer seems to be, yes.

One method is to consider a suitable statistic $T = f(L, V)$ which provides a likelihood which is not constant over the possible sets of parameter values and thus enables us to use the likelihood approach. For writing the likelihood, the probability distribution of the statistic T has to be computed. This can be done if samples are drawn subject to know probabilities. Thus randomization plays a key role in our approach.

For instance, if we consider $T=V$(that is, ignoring the labels) we may obtain a likelihood satisfying our criterion, and inference could be based on such a likelihood. Nobody will deny that the procedure is not valid, for otherwise it would mean that no inference on the unknown population mean is possible if information on labels is not recorded. We can legitimately maintain that when L is known, the analysis based on $T=V$ alone *may* not be fully efficient.

Our choice between $l(L+V)$ and $l(T)$ is guided by the fact that the former focuses attention on parameters of the observed units while the latter gives information on all the parameters, although such information may be of a lower quality. If in a given problem we need to extract information on all parameters by the likelihood approach, we may have to choose $l(T)$. [At this stage the question raised by Dr. Basu on mixture of experiments can be answered. If the likelihood based on the knowledge of the *particular* chosen experiment and the resulting observations is informative on desired unknown parameters (in the sense described above) there is no need to reduce the sample to a statistic such as considering only the observations and ignoring the knowledge that a particular experiment is chosen. If each experiment provides information on the same parameter, I do not see any reason for reducing the sample. The trouble arises when different experiments refer to different parameters and inference is needed on a function of the parameters by conducting only a subset of the experiments.]

(d) If it is recognized that there is a need to reduce the sample to obtain a suitable likelihood, can we lay down some criteria to avoid over reduction (that is to provide optimum reduction). This is a problem to which I addressed myself but succeeded in giving only a partial answer.

I have shown that the knowledge of the labels can be used in various ways, such as in the type of reduction of the sample in equation (3.4) and in post stratification on the basis of outlying observations as in equation (3.9). The result obtained seemed to be reasonable and the approach is less dogmatic and less restrictive than that of Royall and of Hartley and Rao. I do not subscribe to the view that the labels are uninformative but for application of statistical methods based on proper likelihood, it may be necessary to reduce the sample to begin with. I have shown that there may be several ways of reducing the sample without completely ignoring the information on the labels. Indeed, there will be other ways of using the labels which are still to be explored.

2. I am surprised at Dr. Kempthorne's remark that the problem of randomized block design is different from that of sample survey. He seems to be guided or misguided by the testing aspect emphasized by Fisher. But if we consider *estimation* of differences between treatments (varieties) which is more important, then the problem is exactly the same as that of sample survey. The labels are identification numbers of plots and an estimate of the total yield of a variety in a given area is to be obtained by observing yields on a subset of plots.

I am equally surprised about Dr. Kempthorne's remarks on the precision of an estimate. I believe it is the general practice to provide an estimate of the sampling error of an estimate, which gives some idea of the precision. This is not the place to discuss the role of estimation in statistical inference, but Dr. Kempthorne, as a (practicing) statistician ought to be aware that an individual point estimate one makes may be an intermediate step in data analysis and not the final result.

3. Dr. Good referred to his paper on estimation of probabilities in a contingency table. Everyone is aware that this is possible from a Bayesian point of view, but I am inquiring whether something can be done when realistic priors are not available, as is usually the case.

4. Dr. Koop has not correctly applied the principles I have laid down. He ignores the knowledge that there are 3 counters with the same value and 2 counters with another but the same value in applying the method of maximum likelihood. My method would lead to what Dr. Koop designates as $\hat{p}^2$ which seems to be a reasonable estimator.

5. Dr. Sprott raised the interesting question whether Fisher's principle of "recognizable and relevant subsets" can be invoked to resolve the difficulty such as the one encountered in my example of the 100 faced die. I think it is possible to do so by interpreting relevance in a broad sense. For instance in the die example, we can observe only a subset of observed colors? It is recognizable but it is relevant iff the *nature* of the difference between the proportions of zeros on faces with this subset of colors and with all colors is specified, or in other words if the relationship between colors and values on faces is unknown, the observed subset of colors may not be relevant for estimating the overall proportion of zeros.

To show what statements of interest can be made from observed data, let us suppose that 10 tosses avoiding repetition of faces resulted in the following:

	zero	one
Violet	5	0
Indigo	2	1
Blue	1	1

What do we learn from the data? We may provide valid estimates for the proportions of zero on faces with violet, or indigo or blue color, on faces with any one of the observed colors, or on all faces. Each will have its own precision.

	Violet	Indigo	Blue	Three	All Colors
Estimate of the proportion of zeros	1	2/3	1/2	8/10	8/10
Estimate of variance (precision)	0	1/27	3/32	8/1400	16/1100

We note that the estimate 8/10 is the same for the three observed colors as well as for all the colors, but the precision depends on how we consider it. The wider interpretation of the estimate 8/10 is obtained by treating the observed colors as not relevant for estimating the proportion of faces with value zero.

6. Dr. Godambe mentioned that he and Dr. Thompson proved the type of optimality of H-T estimator established in section 5 of my paper by considering a wider class of unbiased estimators. My object in proving such a result was to remove an unnecessary assumption made by Dr. Kempthorne in a previous publication and to extend the results to pps sampling. It is nice to know that the optimality established is true for the entire class of unbiased estimators.

However, it is enough to prove optimality when the variance is averaged over the permutations of values $Y_i = X_i / \pi_i$, for this could imply that the result is true for any symmetric distribution of Y_i. A simple proof can be given to prove the desired optimality by showing that

$$\sideset{}{'}\sum_s \sum \left(\sum_{i \in s} Y_i \right) h_s (Y_i, i \in s) p(s) = 0$$

where $\sum'$ denotes summation over permutations or Y_i and $h_s(Y_i, i \in s)$ is a function whose expectation is zero.

7. I agree with the remarks made by Dr. Rubin, but the problem of the die referred to in my paper is somewhat different.

202

AN ESSAY ON THE LOGICAL FOUNDATIONS OF SURVEY SAMPLING, PART ONE*

D. Basu

The University of New Mexico and Indian Statistical Institute

1

An Idealization of the Survey Set-up

It is a mathematical necessity that we idealize the real state of affairs and come up with a set of concepts that are simple enough to be incorporated in a mathematical theory. We have only to be careful that the process of idealization does not distort beyond recognition the basic features of a survey set-up, which we list as follows:

(a) There exists a population—a finite collection $\wp$ of distinguishable objects. The members of $\wp$ are called the (sampling) units. [Outside of survey theory the term population is often used in a rather loose sense. For instance, we often talk of the infinite population of all the heads and tails that may be obtained by repeatedly tossing a particular coin. Again, in performing a Latin-square agricultural experiment the actual yield from a particular plot is conceived of as a sample from a conceptual population of yields from that plot. It is needless to mention that such populations are not real. The existence of a down-to-Earth finite population is a principal characteristic of the survey set-up.]

(b) There exists a sampling frame of reference. By this we mean that the units in $\wp$ are not only distinguishable pairwise, but are also *observable* individually; that is, there exists a list (frame of reference) of the units in $\wp$ and it is within the powers of the surveyor to pre-select any particular unit from the list and then observe its characteristics. Let us assume that the units in $\wp$ are listed as

$$1, 2, 3, \ldots, N,$$

*Research supported in part by NSF Grant GP-9001.

where N is finite and is known to the surveyor. [We are thus excluding from our survey theory such populations as, for example, the insects of a particular species in a particular area or the set of all color-blind adult males in a particular country. Such populations as above can, of course, be the subject matter of a valid statistical inquiry but the absence of a sampling frame makes it impossible for such populations to be *surveyed* in the sense we understand the term survey here.]

(c) Corresponding to each unit $j \in \mathscr{P}$ there exists an unknown quantity Y_j in which the surveyor is interested. The unknown Y_j can be made known by observing (surveying) the unit j. The unknown state of nature is the vector quantity

$$\theta = (Y_1, Y_2, \ldots, Y_N).$$

However, the surveyor's primary interest is in some characteristic (parameter)

$$\tau = \tau(\theta)$$

of the state of nature θ. [Typically, the Y_j's are vector quantities themselves and the surveyor is seeking information about a multiplicity of τ's. However, for the sake of pinpointing our attention to the basic questions that are raised here, we restrict ourselves to the simple case where the Y_j's are scalar quantities (real numbers) and $\tau = \Sigma Y_j$.]

(d) The surveyor has prior knowledge K about the state of nature θ. This knowledge K is a multi-dimensional complex entity and is largely of a qualitative and speculative nature. We consider here the situation where K has at least the following two well-defined components. The surveyor *knows* the set Ω of all the *possible* values of the state of nature θ and, for each unit j, $(j=1,2,\ldots,N)$, he has access to a record of some known auxiliary characteristic A_j of j. [Typically, each A_j is a vector quantity. However, in our examples we shall take the A_j's to be real numbers.] The set Ω and the vector

$$\alpha = (A_1, A_2, \ldots, A_N)$$

are the principal measurable components of the surveyor's prior knowledge K. Let us denote the residual past of the knowledge by R and write

$$K = (\Omega, \alpha, R).$$

(e) The purpose of a survey is to gain further knowledge (beyond what we have described as K) about the state of nature θ and, therefore, about the parameter of interest $\tau = \tau(\theta)$. Since the surveyor is supposed to know the set Ω of all the possible values of θ, he knows the set $\mathscr{T}$ of all the possible values of τ. Initially, the surveyor's *ignorance* about τ is, therefore, *spread* over the set $\mathscr{T}$. [Later on, we shall quantify this initial *spread of ignorance* as a prior probability distribution.] In theory, the surveyor can dispel this ignorance and gain complete knowledge by making a total survey (complete enumeration) of $\mathscr{P}$. If he observes the Y-characteristic of every unit $j(j=1,2,...,N)$, then he knows the actual value of $\theta=(Y_1,Y_2,...,Y_N)$ and, therefore, that of $\tau(\theta)$. We are, however, considering the case where a total survey is impracticable. By a *survey* of the population $\mathscr{P}$ we mean the selection of a (usually small) subset

$$u=(u_1,u_2,...,u_n)$$

of units from $\mathscr{P}$ and then observing the corresponding Y-values

$$y=(Y_{u_1},Y_{u_2},...,Y_{u_n})$$

of units in the subset u.

(f) We make the simplifying assumption that there are no *non-response* and *observation* errors; that is, the surveyor is able to observe every unit that is in the subset u, and when he observes a particular unit j, he finds the true value of the hitherto unknown Y_j without any error.

(g) The surveyor's blueprint for the survey is usually a very complicated affair. The survey plan must take care of myriads of details. However, in this article we idealize away most of these details and consider only two facets of the survey project, namely, the *sampling plan* and the *fieldwork*. The sampling plan is the part of the project that yields the subset u of $\mathscr{P}$ and fieldwork generates the observations y on members of u. The data (sample) generated by the survey is

$$x=(u,y).$$

For reasons that will be made clear later, it is important to distinguish between the two parts u and y of the data x.

(h) Let $\mathscr{S}$ stand for the sampling plan of the surveyor. The plan (when set in motion) produces a subset u of the population $\mathscr{P}=(1,2,...,N)$. We write $u=(u_1,u_2,...,u_n)$, where $u_1<u_2<...<u_n$ are members of $\mathscr{P}$. The fieldwork generates the vector $y=(Y_{u_1},Y_{u_2},...,Y_{u_n})$ which we often write as $(y_1,y_2,...,y_n)$. [Occasionally, we shall consider sampling plans that introduce a natural selection order among the units that are selected. For such plans it is more appropriate to think of u, not as a subset of $\mathscr{P}$, but as a finite sequence of elements $u_1,u_2,...,u_n$ drawn from $\mathscr{P}$ in that selection order. In rare instances, the sampling plan may allow the possibility of a particular unit appearing repeatedly in the sequence $(u_1,u_2,...,u_n)$. From the description of the sampling plan $\mathscr{S}$ it will usually be clear if we intend to treat u as a set or a sequence. In either case, we can think of u as a vector $(u_1,u_2,...,u_n)$ and y as the corresponding observation vector $(y_1,y_2,...,y_n)$ where $y_i=Y_{u_i}$.]

(i) *Summary*: Our idealized survey set-up consists of the following:
   (i) A finite population $\mathscr{P}$ whose members are listed in a sampling frame as $1,2,...,N$. Availability of each $j \in \mathscr{P}$ for observation.
   (ii) The unknown state of nature $\theta=(Y_1,Y_2,...,Y_N)$ and the parameter of interest $\tau=\tau(\theta)$.

(iii) The prior knowledge $K = (\Omega, \alpha, R)$.

(iv) Absence of non-response and observation errors.

(v) Choice of a sampling plan $\mathcal{S}$ as part of the survey design.

(vi) Putting the sampling plan $\mathcal{S}$ and the fieldwork into operation, thus arriving at the data (sample)

$$x = \{u = (u_1, u_2, \ldots, u_n), \quad y = (y_1, y_2, \ldots, y_n)\}$$

where $u_i \in \mathscr{P}$ and $y_i = Y_{u_i} (i = 1, 2, \ldots, n)$.

(vii) Making a *proper* use of the data x in conjunction with the prior knowledge K to arrive at a *reasonable judgment* (or decision) related to the parameter τ.

The operational parts of the survey are its design (v), the actual survey (vi) and the data analysis (vii). In this article we are concerned only with the design and the analysis of a survey.

2

Probability in Survey Theory

We posed the survey set-up as a classic problem of inductive inference – a problem of inferring about the whole from observations on only a part. The basic questions are: Which part does one observe? Does the part (actually observed) tell us anything about the whole? and, then the main question, Exactly what does it tell? Let us now examine how probability enters into the picture.

There are three different ways in which probability theory finds its way into the mathematical theory of survey sampling. First, there is the time honored way through a probabilistic model for observation errors. Indeed, this is how probability theory first infiltrated the sacred domain of science. When we observe the Y-value Y_j of unit j, there is bound to be some observation error. In current survey theory we classify this kind of error as *non-sampling* error. In this article we have idealized away this kind of probability by assuming that there exists no observation error. We have deliberately taken this simplistic view of the survey set-up. The idea is to concentrate our attention on the other two sources of probability.

In current survey theory, the main source of probability is randomization, which is an artificial introduction (through the use of random number tables) of randomness in the sampling plan $\mathcal{S}$. Randomization makes it possible for the surveyor to consider the set (or sequence) u, and therefore the data $x = (u, y)$, as random elements. With an element of randomization incorporated in the sampling plan $\mathcal{S}$, the surveyor can consider the space U of all the possible values of the random element u and then the probability distribution p_θ of u over U. [For sampling plans usually discussed in survey textbooks, the probability distribution p_u of u is uniquely determined (by the plan) and is, therefore, independent of the state of nature θ. In part two of this article we shall take a broader view of the subject and also consider plans for which the probability distribution of u involves θ.] Now, let X be the space of all the possible values of the data (sample) $x = (u, y)$ of which we have already recognized (thanks to randomization) the part u to be a random element. The space X is our sample space. Let P_θ be the probability distribution of x over the sample space X. If $T = T(x)$ is an estimate of τ, then (prior to sampling and fieldwork) we can consider T to be a random variable and speculate about its sampling distribution and its average performance characteristics (as an estimator of τ) in an hypothetical sequence of repeated experimentations. This decision-theoretic approach is not possible unless we regard randomization as the source of probability in survey theory. From the point of view of a frequency-probabilist, there cannot be a statistical theory of surveys without some kind of randomization in the plan $\mathcal{f}$.

Apart from observation errors and randomization, the only other way that probability can sneak into the argument is through a mathematical formalization of what we have described before as the residual part R of the prior knowledge $K = (\Omega, \alpha, R)$. This is the way of a subjective (Bayesian) probabilist. The formalization of R as a prior probability distribution of θ over Ω makes sense only to those who interpret the probability of an event, not as the long range relative frequency of occurrence of the event (in an hypothetical sequence of repetitions of an experiment), but as a formal quantification of the illusive (but nevertheless very real) phenomenon of *personal belief* in the truth of the event. According to a Bayesian, probability is a mathematical theory of belief and it is with this kind of a probability theory that one should seek to develop the guidelines for inductive behavior in the presence of uncertainty. The purpose of this essay is not to examine the logical basis of Bayesian probability nor to describe how one may arrive at the actual qualification of R into a prior probability distribution of θ over Ω. [Of late, a great deal has been written on the subject. See, for instance, I. R. Savage's delightfully written new book, *Statistics: Uncertainty and Behavior.*]

Can the two kinds of probability co-exist in our survey theory? This is what we propose to find out.

3
Non-Sequential Sampling Plans and Unbiased Estimation

By a non-sequential sampling plan we mean a plan that involves no fieldwork. If the sampling plan $\mathcal{f}$ is non-sequential, then the surveyor can (in theory) make the selection of the set (or sequence) u of population units right in his office and then send his field investigators to the units selected in u and thus obtain the observation part y of the data $x = (u, y)$. A great majority of survey theoreticians have so far restricted themselves to non-sequential plans that involve an element of randomization in it. In this section we consider such plans only. The essence of non-sequentialness of a plan $\mathcal{f}$ is that the probability distribution of u does not involve the state of nature θ. Thus, the sampling plan where we continue to draw a unit at a time with equal probabilities and with replacements until we get ν distinct units is a non-sequential plan.

Given $u = (u_1, u_2, ..., u_n)$, the observation part $y = (y_1, y_2, ..., y_n)$, where $y_i = Y_{u_i} (i = 1, 2, \ldots, n)$, is obtained through the fieldwork and is uniquely determined by the state of nature $\theta = (Y_1, Y_2, \ldots, Y_N)$. The conditional probability distribution of y given u is degenerate, the point of degeneration depending on θ. That is, for all y, u and θ

$$\text{Prob}(y \mid u, \theta) = 0 \text{ or } 1. \tag{3.1}$$

[We are taking the liberty of using the symbols u, y, x and θ both as variables and as particular values of the variables.]

For each sampling plan $\mathcal{S}$ we have the space U of all the possible values of u. The probability distribution p of u over U is θ-free, that is, is uniquely defined by the plan $\mathcal{S}$. The probability distribution p is clearly discrete. There is no loss of generality in assuming that $p(u) > 0$ for all $u \in U$. If the non-sequential plan is *purposive* (that is, the plan involves no randomization) then U is a single-point set and the distribution of u is degenerate at that point.

Let X be the sample space, the set of all possible samples (data) $x = (u, y)$ where u is generated by the plan $\mathcal{S}$ and y by the fieldwork. For each $\theta \in \Omega$, we have a probability distribution P_θ over X. Whatever the plan $\mathcal{S}$, the probability distribution P_θ is necessarily discrete. We write $P_\theta(x)$ or $P_\theta(u, y)$ for the probability of arriving at the data $x = (u, y)$ when θ is the true value of the state of nature. Clearly,

$$P_\theta(x) = P_\theta(u, y) = p(u) \text{ Prob } (y \mid u, \theta). \tag{3.2}$$

The surveyor takes a peep at the unknown $\theta = (Y_1, Y_2, \ldots, Y_N)$ through the sample $x = (u, y)$. Prior to the survey, the surveyor's ignorance about θ was spread over the space Ω. Once the data x is at hand, the surveyor has exact information about some coordinates of the vector θ. These are the co-ordinates that correspond to the distinct units that are in u. The data x rules out some points in Ω as clearly inadmissible. Let Ω_x be the subset of values of θ that are consistent with the data x. In other words, $\theta \in \Omega_x$ if $P_\theta(x) > 0$; that is, it is possible to arrive at the data x when θ is the true value of the state of nature. The subset Ω_x of Ω is well-defined for every $x \in X$. Without any loss of generality we may assume that no Ω_x is vacuous. From (3.1) and (3.2) it follows that the likelihood function $L(\theta)$ is given by the formula

$$L(\theta) = P_\theta(x) = \begin{cases} p(u) & \text{for all } \theta \in \Omega_x \\ 0 & \text{otherwise.} \end{cases} \tag{3.3}$$

In other words, whatever the data x, the likelihood function $L(\theta)$ is flat (a positive constant) over the set Ω_x and is zero outside Ω_x. This remark holds true for sequential plans also (Basu, 1969). The importance of the remark will be made clear later on.

A major part of survey theory is concerned with unbiased estimation. A statistic is a characteristic of the sample x. An estimator $T = T(x)$ is a statistic that is well-defined for all $x \in X$ and is used for estimating a parameter $\tau = \tau(\theta)$. By an unbiased estimator of $\tau(\theta)$ we mean an estimator T that satisfies the identity

$$E(T \mid \theta) = \sum_x T(x) P_\theta(x) \equiv \tau(\theta), \text{ for all } \theta \in \Omega. \tag{3.4}$$

Let $w(t, \theta)$ be the loss function. That is, $w(t, \theta)$ stands for the surveyor's assessment of the magnitude of error that he commits when he estimates the parameter $\tau = \tau(\theta)$ by the number t. We assume that

$$w(t, \theta) \geq 0 \text{ for all } t \text{ and } \theta,$$

the sign of equality holding only when $t = \tau(\theta)$. The risk function $r_T(\theta)$ associated with the loss function w and the estimator T is then defined as the expected loss

$$r_T(\theta) = E[w(T, \theta) \mid \theta] = \sum_x w(T(x), \theta)P_\theta(x). \tag{3.5}$$

[If the reader is not familiar with the decision-theoretic jargons of loss and risk, he may restrict himself to the particular case where $w(t, \theta)$ is the squared error $(t - \tau(\theta))^2$ and the risk function $r_T(\theta)$ is the variance $V(T \mid \theta)$ of the unbiased estimator T.] The following theorem proves the non-existence of a uniformly minimum risk (variance) unbiased estimator of τ.

Theorem. Given an unbiased estimator T of τ and an arbitrary (but fixed) point $\theta_0 \in \Omega$, we can always find an unbiased estimator T_0 (of τ) such that $r_{T_0}(\theta_0) = 0$, that is, T_0 has zero risk at θ_0.

Proof: We find it convenient to write $T(u, y)$ and $P_\theta(u, y)$ for $T(x)$ and $P_\theta(x)$ respectively. It has been noted earlier that the conditional distribution of y given u is degenerate at a point that depends on θ. Let $y_0 = y_0(u)$ be the point of degeneration of y, for given u, when $\theta = \theta_0$. Consider the statistic

$$T_0 = T(u, y) - T(u, y_0) + \tau(\theta_0). \tag{3.6}$$

The statistic $T(u, y_0)$ is a function of u alone and so its probability distribution, and therefore its expectation are θ-free. Indeed,

$$E[T(u, y_0)] = \sum_u T(u, y_0)p(u)$$

$$= \sum_{u,y} T(u, y)P_{\theta_0}(u, y)$$

$$= E[T(u, y) \mid \theta_0]$$

$$= \tau(\theta_0).$$

Thus, the statistic T_0 as defined in (3.6) is an unbiased estimator of τ. Now, when $\theta = \theta_0$, the statistics $T(u, y)$ and $T(u, y_0)$ are equal with probability one, and so $T_0 = \tau(\theta_0)$ with probability one. This proves the assertion that $r_{T_0}(\theta_0) = 0$.

The impossibility of the existence of a uniformly minimum risk unbiased estimator follows at once. For, if such an estimator T exists then $r_T(\theta)$ must be zero for all $\theta \in \Omega$. That is, whatever the value of the state of nature θ it should be possible to estimate $\tau(\theta)$ without any error (loss) at all. Unless the sampling plan $\mathcal{S}$ is equivalent to a total survey of the population, such a T clearly cannot exist for a parameter τ that depends on all the coordinates of θ. The following two examples will clarify the theorem further.

Example 1. Consider the case of a simple random sample of size one from the population $\mathcal{P} = (1, 2, ..., N)$. The sample is (u, y) where u has a uniform probability distribution over the N integers $1, 2, ..., N$ and $y = Y_u$. Let the population mean

$$\overline{Y} = \frac{1}{N}(Y_1 + ... + Y_N)$$

be the parameter to be estimated. Clearly, y is an unbiased estimator of $\overline{Y}$. Let $\theta_0 = (a_1, a_2, ..., a_N)$ and $\bar{a} = (\Sigma a_j)/N$. The statistic

$$T_0 = y - a_u + \bar{a} \tag{3.7}$$

is an unbiased estimator of $\overline{Y}$ with zero risk (variance) when $\theta = \theta_0$. The variance of y is $\Sigma(Y_j - \overline{Y})^2/N$ and that of T_0 is $\Sigma(Z_j - \overline{Z})^2/N$ where $Z_j = Y_j - a_j$ $(j = 1, 2, ..., N)$.

Example 2. Let $\mathcal{S}$ be an arbitrary non-sequential sampling plan that allots a positive selection probability to each population unit. That is, the probability Π_j that the unit j appears in the set (or sequence) u is positive for each $j(j = 1, 2, ..., N)$. Since $\mathcal{S}$ is non-sequential, the vector

$$\Pi = (\Pi_1, \Pi_2, ..., \Pi_N)$$

is θ-free. Let Y be the population total ΣY_j. A particular unbiased estimator of Y that has lately attracted a great deal of attention is the so-called Horvitz-Thompson (HT) estimator (relative to the plan $\mathcal{S}$). The HT-estimator is defined as follows. Let $u_1 < u_2 < ... < u_\nu$ be the distinct population units that appear in u and let $\hat{y} = (y_1, y_2, ..., y_\nu)$ be the corresponding observation vector. Let $p_i = \Pi u_i$ $(i = 1, 2, ..., \nu)$. The HT-estimator H is then defined as

$$H = \frac{y_1}{p_1} + ... + \frac{y_\nu}{p_\nu}. \tag{3.8}$$

That H is an unbiased estimator of Y will be clear when we rewrite (3.8) in a different form. Let $E_j (j = 1, 2, ..., N)$ stand for the event that the plan $\mathcal{S}$ selects unit j, and let I_j be the indicator of the event E_j. That is, $I_j = 1$ or 0 according as unit j appears in u or not. It is now easy to check that

$$H = \sum_{j=1}^{N} \Pi_j^{-1} I_j Y_j. \tag{3.9}$$

That H is an unbiased estimator of Y follows at once from the fact that

$$\begin{aligned} E(I_j) &= \text{Prob}(E_j) \\ &= \Pi_j \quad (j = 1, 2, ..., N). \end{aligned}$$

Now, let $\theta_0=(a_1, a_2, \ldots, a_N)$ be a point in Ω that is selected by the surveyor (prior to the survey) and let H_0 be defined as

$$H_0 = H - \Sigma \, \Pi_j^{-1} I_j a_j + \Sigma a_j. \tag{3.10}$$

It is now clear that H_0 is an unbiased estimator of Y and that $V(H_0 \mid \theta)=0$ when $\theta = \theta_0$. Since the variances of H and H_0 are continuous functions of θ, it follows that

$$V(H_0 \mid \theta) < V(H \mid \theta) \tag{3.11}$$

for all θ in a certain neighborhood Ω_0 of the point θ_0. If the surveyor has the prior knowledge that the true value of θ lies in Ω_0, then the modified Horvitz-Thompson estimator H_0 is uniformly better than H. The estimator H_0 will look a little more reasonable if we rewrite it as

$$H_0 = [\Sigma \, \Pi_j^{-1} I_j(Y_j - a_j)] + \Sigma a_j. \tag{3.12}$$

If (3.9) is a reasonable estimator of $Y = \Sigma Y_j$, then the variable part of the right hand side of (3.12) is an equally reasonable estimator of

$$\Sigma(Y_j - a_j) = Y - \Sigma a_j.$$

The strategy of a surveyor who advocates the use of (3.12) [in preference to that of (3.9)] as an estimator of $Y = \Sigma Y_j$ is quite clear. Instead of defining the state of nature as

$$\theta = (Y_1, Y_2, \ldots, Y_N)$$

he is defining it as

$$\theta' = (Y_1 - a_1, Y_2 - a_2, \ldots, Y_N - a_N).$$

Suppose the surveyor has enough prior information about the state of nature, so that by a proper choice of the vector $(a_1, a_2, \ldots, a_N)$ he can make the co-ordinates of θ' much less variable than that of θ. He is then in a better position to estimate the total of the coordinates of θ' than one who is working with θ. Consider the situation where the surveyor knows in advance that the jth coordinate Y_j of θ lies in a small interval around the number a_j ($j=1, 2, \ldots, N$). In such a situation the surveyor ought to shift the origin of measurement (for θ) to the point $(a_1, a_2, \ldots, a_N)$ and represent the state of nature as

$$\theta' = (Y_1 - a_1, Y_2 - a_2, \ldots, Y_N - a_N).$$

If the numbers $a_1, a_2, \ldots, a_N$ has a large dispersion, then shifting the origin of measurement to $(a_1, a_2, \ldots, a_N)$ will cut down the variability in the co-ordinates of the state of nature to a large extent. The effect will be similar to what is usually achieved by stratification.

At this point one may very well raise the questions: Why must the surveyor choose his knowledge vector $(a_1, a_2, \ldots, a_N)$ before the survey?

Is it not more reasonable for him to wait until he has the survey data at hand and then take advantage of the additional knowledge gained thereby? Once the data is at hand, the surveyor knows the exact values of the surveyed coordinates of θ. The natural post-survey choice of a_j for any surveyed j is, therefore, Y_j. For a non-surveyed j, the surveyor's best estimate a_j of the unknown Y_j would still be of a speculative nature. If in formula (3.12) we allow the surveyor to insert a post-survey specification of the vector $(a_1, a_2, \ldots, a_N)$, then the first part of the right hand side of (3.12) will vanish and the estimator will look like

$$H_* = \Sigma \, a_j$$

$$= (Y_{u_1} + Y_{u_2} + \ldots + Y_{u_\nu}) + \sum_{j \notin u} a_j \qquad (3.13)$$

$$= S + S^*,$$

where S is the sum total of the Y-values of the distinct surveyed units and S^* is the surveyor's post-survey estimate of the total Y-values of the non-surveyed units.

A decision-theorist will surely object to our derivation of formula (3.13) as naive and incompetent. He will point out that we have violated a sacred cannon of inductive behavior, namely *never select the decision rule after looking at the data*. He will also point out that S^* in (3.13) is, as yet, not well-defined (as a function on the sample space X), and he will reject H_* (as an estimator of Y) with the final remark that the whole thing stinks of Bayesianism!

Nevertheless, the fact remains that formula (3.13) points to the very heart of the matter of estimating the population total Y. A survey leads to a complete specification of a part of the population total. This part is the sample total S as defined in (3.13). At the end of the survey the remainder part Y-S is still unknown to the surveyor. If the surveyor insists on putting down T as an estimate of Y, then he is in effect saying that he has reason to believe that T-S is close to Y-S. And then he should give a reasonable justification for his belief. Of course we can write any estimate T in the form

$$T = S + S^*$$

where S is the sample total and $S^* = T - S$. But then, for some T, the part S^* (of T) would appear quite preposterous as an estimate of the unknown part $Y - S$ of Y. The following two examples will make clear the point that we are driving at.

Example 3. The circus owner is planning to ship his 50 adult elephants and so he needs a rough estimate of the total weight of the elephants. As weighing an elephant is a cumbersome process, the owner wants to estimate the total weight by weighing just one elephant. Which elephant should he weigh? So the owner looks back on his records and discovers a list of the elephants' weights taken 3 years ago. He finds that 3 years ago Sambo the middle-sized elephant was the average (in weight) elephant in his herd. He checks with the elephant trainer who reassures him (the owner) that Sambo may still be considered to be the average elephant in the herd. Therefore, the owner plans to weigh Sambo and take $50\,y$ (where y is the present weight of Sambo) as an estimate of the total weight $Y = Y_1 + \ldots + Y_{50}$ of the 50 elephants. But the circus statistician is horrified when he learns of the owner's purposive samplings plan. "How can you get an unbiased estimate of Y this way?" protests the statistician. So, together they work out a compromise sampling plan. With the help of a table of random numbers they devise a plan that allots a selection probability of $99/100$ to Sambo and equal selection probabilities of $1/4900$ to each of the other 49 elephants. Naturally, Sambo is selected and the owner is happy. "How are you going to estimate Y?", asks the statistician. "Why? The estimate ought to be $50y$ of course," says the owner. "Oh! No! That cannot possibly be right," says the statistician, "I recently read an article in the *Annals of Mathematical Statistics* where it is proved that the Horvitz-Thompson estimator is the unique hyperadmissible estimator in the class of all generalized polynomial unbiased estimators." "What is the Horvitz-Thompson estimate in this case?" asks the owner, duly impressed. "Since the selection probability for Sambo in our plan was $99/100$," says the statistician, "the proper estimate of Y is $100y/99$ and not $50y$." "And, how would you have estimated Y," inquires the incredulous owner, "if our sampling plan made us select, say, the big elephant Jumbo?" "According to what I understand of the Horvitz-Thompson estimation method," says the unhappy statistician, "the proper estimate of Y would then have been $4900y$, where y is Jumbo's weight." That is how the statistician lost his circus job (and perhaps became a teacher of statistics!)

Example 4. Sampling with unequal probabilities has been recommended in situations that are less frivolous than the one considered in the previous example but the recommended unbiased estimators for such plans sometimes look hardly less ridiculous than the one just considered. Let us consider the so-called pps (probability proportional to size) plans about which so many research papers have been written in the past 20 years. A pps sampling plan is usually recommended in the following kind of situation. Suppose for each population unit j we have a record of an auxiliary characteristic A_j (the size of j). Also suppose that each A_j is a positive number and that the surveyor has good reason to believe that the ratios

$$\Lambda_j = Y_j/A_j \quad (j = 1, 2, \ldots, N) \tag{3.14}$$

are nearly equal to each other. In this situation it is often recommended that the surveyor adopts the following without replacement pps.

Sampling plan. Let $A = \Sigma A_j$ and $P_j = A_j/A$ $(j = 1, 2, \ldots, N)$. Choose a unit (say, u_1) from the population $\wp = (1, 2, \ldots, N)$ following a plan that allots a selection probability P_j to unit j $(j = 1, 2, \ldots, N)$. The selected unit u_1 is then removed from the sampling frame and a second unit (say, u_2) is selected from the remaining $N-1$ units with probabilities proportional to their sizes (the auxiliary characters A_j). This process is repeated n times so that the surveyor ends up with n distinct units

$$u_1, u_2, \ldots, u_n$$

listed in their natural selection order. After the fieldwork the surveyor has the sample

$$x = \{(u_1, y_1), \ldots, (u_n, y_n)\}$$

where $y_i = Y_{u_i} (i = 1, 2, \ldots, n)$. Let us write p_i for $P_{u_i} (i = 1, 2, \ldots, n)$ and

$$x^* = \{(u_1, y_1), \ldots, (u_{n-1}, y_{n-1})\}$$

for the vector defined by the first $n-1$ coordinates of x. It is then easy to see that (see Desraj [6] Theorem 3.13)

$$E\left(\frac{y_n}{p_n}\middle|x^*\right) = \Sigma' \frac{Y_j}{P_j} \cdot \frac{P_j}{1 - p_1 - p_2 - \ldots - p_{n-1}}$$

$$= (\Sigma' Y_j)/(1 - p_1 - \ldots - p_{n-1}),$$

where the summation is carried over all j that are different from $u_1, u_2, \ldots, u_{n-1}$. Since $\Sigma' Y_j = Y - (y_1 + \ldots + y_{n-1})$ it follows at once that

$$E\left(y_1 + \ldots + y_{n-1} + \frac{y_n}{p_n}(1 - p_1 - \ldots - p_{n-1}) \middle| x^*\right) = Y. \tag{3.15}$$

Therefore, the unconditional expectation of the lefthand side of (3.15) is also Y. And so we have the so-called Desraj estimator

$$D = y_1 + \ldots + y_{n-1} + \frac{y_n}{p_n}(1 - p_1 - \ldots - p_{n-1}), \tag{3.16}$$

which is an unbiased estimator of Y. Writing S for the sample total $y_1 + \ldots + y_n$ we can rewrite (3.16) as

$$D = S + S^* \tag{3.17}$$

where

$$S^* = \frac{y_n}{p_n}(1 - p_1 - \ldots - p_n).$$

Let us examine the face-validity of S^* as an estimate of Y^*, the total Y-values of the unobserved population units.

Writing $A = \Sigma A_j$, $a_i = A_{u_i} (i = 1, 2, \ldots, n)$ and $A^* = A - a_1 - \ldots - a_n$ (the total A-value of the unobserved units), we have

$$S^* = \frac{y_n}{p_n}(1 - p_1 - \ldots - p_n) \tag{3.18}$$

$$= \frac{y_n}{a_n} A^* \quad (\text{since } p_i = \frac{a_i}{A}).$$

214

Clearly, S^* would be an exact estimate of Y^* if and only if

$$\frac{y_n}{a_n} = \frac{Y^*}{A^*} = \frac{\Sigma' Y_j}{\Sigma' A_j} \qquad (3.19)$$

(the summation is over the unobserved j's).

Now, if the surveyor claims that according to his belief (3.17) is a good estimate of Y, then that claim is equivalent to an assertion of belief in the near equality of the two ratios

$$\frac{y_n}{a_n} \text{ and } \frac{Y^*}{A^*}.$$

What can be the logical basis for such a belief? We started with the assumption that the surveyor has prior knowledge of near equality in the N ratios in (3.14). At the end of the survey, the surveyor has observed exactly n of these ratios and they are

$$\frac{y_1}{a_1}, \frac{y_2}{a_2}, \ldots, \frac{y_n}{a_n}. \qquad (3.20)$$

The surveyor is now in a position to check on his initial supposition that the ratios in (3.14) are nearly equal. Suppose he finds that the observed ratios in (3.20) are indeed nearly equal to each other. This will certainly add to the surveyor's conviction that the unobserved ratios Λ_j (where j is different from $u_1, u_2, \ldots, u_n$) are nearly equal to each other and that they lie within the range of variations of the observed ratios in (3.20). Now, Y^*/A^* is nothing but a weighted average of the unobserved ratios (the weights being the sizes of the corresponding units). It is then natural for the surveyor to estimate Y^*/A^* by some sort of an average of the observed ratios. For instance, he may choose to estimate Y^*/A^* by $(y_1 + \ldots + y_n)/(a_1 + \ldots + a_n)$. This would lead to the following modification of the Desraj estimate (3.17):

$$D_1 = S + \frac{y_1 + \ldots + y_n}{a_1 + \ldots + a_n} A^* \qquad (3.21)$$

$$= \frac{y_1 + \ldots + y_n}{a_1 + \ldots + a_n} A$$

(and this we recognize at once as the familiar ratio estimate). Alternatively, the surveyor may choose to estimate the ratio Y^*/A^* by the simple average

$$\frac{1}{n} \left(\frac{y_1}{a_1} + \ldots + \frac{y_n}{a_n} \right)$$

of the observed ratios. This will lead to another variation of the Desraj estimate, namely

$$D_2 = (y_1 + \ldots + y_n) + \frac{1}{n} \left(\frac{y_1}{a_1} + \ldots + \frac{y_n}{a_n} \right) A^*. \qquad (3.22)$$

What we are trying to say here is the simple fact that both (3.21) and (3.22) have much greater face validity as estimates of Y than the Desraj estimate (3.17). In the Desraj estimate we are trying to evaluate Y^*/A^* by the nth observed ratio $y_n \mid a_n$ and are taking no account of the other $n-1$ ratios. This is almost as preposterous as the estimate suggested by the circus statistician in the previous example. Suppose the surveyor finds that the n observed ratios $y_i \mid a_i$ $(i = 1, 2, \ldots, n)$ are nearly equal alright, but $y_n \mid a_n$ is the largest of them all. In this situation how can he have any faith in the Desraj estimate

$$D = S + \frac{y_n}{a_n} A^*$$

being nearly equal to Y? [Remember, the factor A^* will usually be a very large number.] Again, what does the surveyor do when he discovers that his initial supposition that the ratios Y_j/A_j $(j = 1, 2, \ldots, N)$ are nearly equal, was way off the mark? Will it not be ridiculous to use the Desraj estimate in this case? Here we are concerned not with the mathematical property of un-biasedness of an estimator but with the hard-to-define property of face validity of an estimate. An estimate T of the population total Y has little face validity if after we have written T in the form

$$T = S + S^*$$

we are hard put to find a reason why the part S^* should be a good estimate of Y^*.

4
The Label-Set and The Sample Core

We have noted elsewhere that, for a non-sequential sampling plan $\mathscr{S}$, the label part u of the data $x = (u, y)$ is an ancillary statistic; that is, the sampling distribution of the statistic u does not involve the state of nature θ. The sampling distribution of u is uniquely determined by the plan. It is therefore obvious that the label part of the data cannot, by itself, provide any informa-tion about θ. Knowing u, we only know the names (labels) of the population units that are selected for observation. [When u is a sequence, we also know the order and the frequency of appearance of each selected unit in u.] With a non-sequential plan $\mathscr{S}$, the knowledge of u alone cannot make the surveyor any wiser about θ. The surveyor may, and often does, incorporate his prior knowledge of the auxiliary characters $\alpha = (A_1, A_2, \ldots, A_N)$ in the plan $\mathscr{S}$. But this does not alter the situation a bit. The label part u of the data x will still be an ancillary statistic.

If the label part u is informationless, then can it be true that the observa-tion part y of the data $x = (u, y)$ contains all the available information about θ? A little reflection will make it abundantly clear that the answer must be an emphatic, no. A great deal of information will be lost if the label part of the data is suppressed. Without the knowledge of u, the surveyor cannot relate the components of the observation vector y to the population units and so he cannot make any use of the auxiliary characters $\alpha = (A_1, A_2, \ldots, A_N)$ and whatever other prior knowledge he may have about the relationship between θ and α.

Let us call a statistic $T = T(u, y)$ *label-free* if T is a function of y alone. So far, the only label-free estimator that we have come across is the estimator y of $\bar{Y}$ in Example 1. If in this case the surveyor has prior knowledge that the true value of θ lies in the vicinity of the point $\theta_u = (a_1, a_2, \ldots, a_N)$, then he would naturally prefer the estimator (3.7) as an unbiased estimator of $\bar{Y}$. The surveyor can arrive at an estimate like (3.7) only if he has access to the information contained in u. In survey literature, we find several attempts at justifying label-free estimates. But a reasonable case for a label-free estimate can be made only under the assumption of a near complete ignorance in the mind of the surveyor. But, in these days of extreme specialization, who is going to entrust an expensive survey operation in the hands of a very ignorant surveyor?! To remain in survey business, the surveyor has to carefully orient himself to each particular survey situation, gather a lot of auxiliary data $A_1, A_2, \ldots, A_N$ about the population units, and then make intelligent use of such data in the planning of the survey and in the analysis of the survey data. Considerations of label-free estimates are, therefore, of only an academic interest in survey theory.

Let us denote by $\hat{u}$ the set of distinct population units that are selected (for survey) by the sampling plan. The set $\hat{u}$ is a statistic — a characteristic of the sample $x = (u, y)$. We call $\hat{u}$ the label-set and find it convenient to think of $\hat{u}$ as a vector

$$\hat{u} = (\hat{u}_1, \hat{u}_2, \ldots, \hat{u}_\nu),$$

where $\hat{u}_1 < \hat{u}_2 < \ldots < \hat{u}_\nu$ are the ν distinct unit-labels that appear in u, arranged in an increasing order of their label values. The *observation-vector* $\hat{y}$ is then defined as

$$\hat{y} = (\hat{y}_1, \hat{y}_2, \ldots, \hat{y}_\nu),$$

where $\hat{y}_i = Y_{\hat{u}_i} \ (i = 1, 2, \ldots, \nu)$.

We denote the pair $(\hat{u}, \hat{y})$ by $\hat{x}$ and call it the *sample-core*. For each sample $x = (u, y)$ we have a well-defined sample-core $\hat{x} = (\hat{u}, \hat{y})$. In the literature the sample-core has been called by other fancy names like *order statistic* or *sampley*, etc. [It should be noted that, though we can think of $\hat{u}$ as a subset of $\mathscr{P}$, we cannot think of $\hat{y}$ as a set, because the values in $\hat{y}$ need not be all different. Even if it were possible to think of $\hat{y}$ as a set, it would not be fruitful to do so. For, if $\hat{u}$ and $\hat{y}$ are both conceived as sets, then we have no way to relate a member of $\hat{y}$ to the corresponding label in $\hat{u}$. This is the reason why we prefer to think of the label-set $\hat{u}$ as a vector and of $\hat{y}$ as the corresponding observation-vector.]

The sample core $\hat{x}$ is a statistic. The mapping $x \to \hat{x}$ is usually many-one. For instance, in the pps plan of Example 4, the number ν (of distinct units selected) is the same as n, but, for each value of the label-set $\hat{u}$, there are exactly $n!$ values of u (corresponding to the $n!$ different selection-orders in which the n units might have been selected). Here the mapping $x \to \hat{x}$ is $n!$ to 1.

In part two of this essay we shall establish the fact that the sample-core $\hat{x}$ is a sufficient statistic. This means that, given $\hat{x}$, the conditional distribution of the sample x is uniquely determined (does not involve the unobserved part of the state of nature θ). The widely accepted principle of sufficiency tells us that if T be a sufficient statistic then every reasonable estimator (of every parameter τ) ought to be a function of T. Following Fisher we may call an estimator H *insufficient* if H is not a function of the sufficient statistic T. The Desraj estimator (3.17) of the population total Y is then an insufficient estimate. If we rewrite the Desraj estimate as

$$D = S + \frac{y_n}{a_n} A^*$$

where S is the sample Y- total and A^* is the total A-values of the unobserved units, then it is clear that both S and A^* are functions of the sample-core $\hat{x} = (\hat{u}, \hat{y})$. [Indeed, S is a function of $\hat{y}$ and A^* is a function of $\hat{u}$.] However, $y_n|a_n$ (the ratio corresponding to the last unit drawn in the without replacement pps plan) is not a function of $\hat{x}$. Knowing $\hat{x}$, we only know that the ratio $y_n|a_n$ may have been any one of the n ratios

$$\hat{y}_i|\hat{a}_i \quad (i = 1, 2, \ldots, n)$$

where $\hat{y}_i = Y_{\hat{u}_i}$ and $a_i = A_{\hat{u}_i}$.

If we define λ_i $(i = 1, 2, \ldots, n)$ as the conditional probability of $u_n|a_n$ being equal to $\hat{y}_i|\hat{a}_i$, then the λ_i's are well-defined (θ-free) constants, $\Sigma\lambda_i = 1$ and

$$\bar{D} = E(D \mid \hat{x}) = S + \left[\sum_1^n \lambda_i \frac{\hat{y}_i}{\hat{a}_i} \right] A^*. \tag{4.1}$$

Since D is an unbiased estimator of $Y = \Sigma Y_j$, so also is the estimator $\bar{D}$. From the Rao-Blackwell theorem it follows that (if $n > 1$) the variance of $\bar{D}$ is uniformly smaller than that of D. The estimator $\bar{D}$ has been variously called in the literature, the *symmetrized* or the *un-ordered* Desraj estimator. In view of what we explained in the previous section, the symmetrized Desraj estimator (4.1) looks much better than the original Desraj estimator on the score of face-validity. However, the coefficients $\lambda_1, \lambda_2, \ldots, \lambda_n$ in (4.1) are much too complicated to make $\bar{D}$ an acceptable estimator of Y. The estimates (3.21) and (3.22) have about the same face-validity as that of (4.1) and are much simpler to compute. However, (4.1) scores over the other two estimates on the dubious criterion of unbiasedness!

The estimator (4.1) cannot be the only unbiased estimator of Y that is a function of the sample core $\hat{x}$. Consider the estimator

$$\frac{y_1}{p_1} = \frac{y_1}{a_1} A, \tag{4.2}$$

where y_1 is the Y-value of the first unit that was drawn (by the pps plan of Example 4) and a_1 is the corresponding A-value. Clearly, (4.2) is an *insufficient* unbiased estimator of Y. The symmetrized version of (4.2) will be

$$\left(\Sigma\mu_i \frac{\hat{y}_i}{\hat{a}_i} \right) A, \tag{4.3}$$

where

$$\mu_i = P\left(\frac{y_1}{a_1} = \frac{\hat{y}_i}{\hat{a}_i} \,\middle|\, \hat{x}\right) \quad (i = 1, 2, \ldots, n).$$

The estimator (4.3) is unbiased and is a function of $\hat{x}$.

Of late, quite a few papers have been written in which the main idea is the above described method of *un-ordering an ordered estimate*, that is, making use of the Rao-Blackwell theorem and the sufficiency of the sample core $\hat{x}$. Whatever the sampling plan $\mathcal{S}$ is, the sample-core is always sufficient. Indeed, the sample-core is (in general) the minimum (minimal) sufficient statistic. However, for a non-sequential sampling plan $\mathcal{S}$, the sufficient statistic $\hat{x}$ is never *complete*. By the *incompleteness* of $\hat{x}$ we mean the existence of non-trivial functions of $\hat{x}$ whose expectations are identically zero for all possible values of the state of nature θ. This is because (when the plan is non-sequential) the label-set $\hat{u}$ (which is a component of $\hat{x}$) is an ancillary statistic. For every parameter of interest $\tau(\theta)$, there will exist an infinity of unbiased estimators each of which is sufficient in the sense of Fisher (that is, is a function of the minimal sufficient statistic $\hat{x}$.)

5

Linear Estimation in Survey Sampling

During the past several years a great many research papers have been written dealing exclusively with the topic of linear estimation of the population mean $\overline{Y}$ or, equivalently, the population total Y. Some confusion has, however, been created by the term *linear*. An estimator is a function on the sample space X. Unless X is a linear space we cannot, therefore, talk of a linear estimator. In our formulation, X is the space of all samples $x = (u, y)$ and so X is not a linear space. How then are we to reconcile ourselves to the classical statement that, in the case of a simple random sampling plan, the sample mean is the best unbiased linear estimate of the population mean? We have the often quoted contrary assertion from Godambe that in no realistic sampling situation (whatever the plan $\mathcal{S}$) can there exist a best estimator in the class of linear unbiased estimators of the population mean. This section is devoted entirely to the notions of the so-called linear estimates.

Consider first the case of a simple random sampling plan in which a number n (the sample size) is chosen in advance and then a subset of n units is selected from the population $\mathcal{P}$ in such a manner that all the $\binom{N}{n}$ subsets of $\mathcal{P}$ with n elements are alloted equal selection probabilities. Let us suppose that the plan calls for a selection of the n sample units one by one without replacements and with equal probabilities, so that we can list the selected units in their natural selection order as $u_1, u_2, \ldots, u_n$. The label part of the data is then the sequence $u = (u_1, u_2, \ldots, u_n)$ and the observation part is the corresponding observation vector $y = (y_1, y_2, \ldots, y_n)$. Clearly, the y_i's are identically distributed (though not mutually independent) random variables with

$$E(y_i) = (\Sigma Y_j)/N = \overline{Y} \quad (i = 1, 2, \ldots, n).$$

219

Now, if the surveyor chooses to ignore the label part u of the data, then he can define a linear estimator of $\overline{Y}$ as a linear function

$$T = b_0 + b_1 y_1 + b_2 y_2 + \ldots + b_n y_n \qquad (5.1)$$

of the observation vector y, where the coefficients $b_0, b_1, \ldots, b_n$ are pre-selected constants. All estimators of the above kind are label-free estimators. Let L be the class of all unbiased estimators of $\overline{Y}$ that are of the type (5.1). In other words, L is the class of all estimators of the type (5.1) with

$$b_0 = 0 \text{ and } b_1 + \ldots + b_n = 1. \qquad (5.2)$$

The sample mean $\bar{y} = (\Sigma y_i)/n$ is a member of L. The classical assertion that we referred to before is to the effect that, in the class L, there exists a uniformly minimum variance estimator and that is the sample mean $\bar{y}$. This result is well-known and a fairly straightforward proof may be given for the particular case where we define variance as the mean square deviation from the mean. We, however, consider it appropriate to sketch a proof that ties in well with the general spirit of this article.

Consider the sample core $\hat{x} = (\hat{u}, \hat{y})$ where we write the label-set $\hat{u}$ as a sequence $(\hat{u}_1, \ldots, \hat{u}_n)$ with $\hat{u}_1 < \hat{u}_2 < \ldots < \hat{u}_n$ and look upon $\hat{y}$ as the corresponding observation vector $(\hat{y}_1, \ldots, \hat{y}_n)$. Note that the mapping $x \to \hat{x}$ is $n!$ to 1 and that the vector $\hat{y}$ is obtained from the vector y by rearranging its coordinates in an increasing order of their corresponding unit labels. Now, given $\hat{x}$, the conditional distribution of x is equally distributed over the $n!$ possible values of x and so it follows that

$$E(y_i|\hat{x}) = (\Sigma \hat{y}_i)/n = \bar{y} \quad (i = 1, 2, \ldots, n). \qquad (5.3)$$

Thus, if $T = \Sigma a_i y_i$, with $\Sigma a_i = 1$, is any member of L then from (5.3) it follows that

$$E(T|\hat{x}) = \Sigma (a_i \bar{y}) = \bar{y}. \qquad (5.4)$$

And so from the Rao-Blackwell theorem it follows that $\bar{y}$ is better than T [and this is irrespective of the loss function $w(t, \theta)$ (see §3) as long as $w(t, \theta)$ is convex (from below) in t for each fixed value of θ]. Observe that, in the class L, the sample mean

$$\bar{y} = (\Sigma y_i)/n = (\Sigma \hat{y}_i)/n$$

is the only one that is a function of the sample core $\hat{x}$. Every other member of L is insufficient in the sense explained in the earlier section. And so it is no wonder that $\bar{y}$ beats every other member of L in its performance characteristics. The class $L - \{\bar{y}\}$ is certainly not worth any consideration at all.

At this stage one may ask: Why not consider the class of all linear functions of the vector $\hat{y} = (\hat{y}_1, \ldots, \hat{y}_n)$? The snag is that the variables $\hat{y}_1, \ldots, \hat{y}_n$ have very complicated distributions and their expectations are not easy to obtain. For instance, the variable $\hat{y}_1$ can take only the values $Y_1, Y_2, \ldots, Y_{N-n+1}$ and its expectation is a complicated linear function of these $N - n + 1$

values. It is, therefore, not easy to characterize the class of unbiased estimators of $\overline{Y}$ that are linear functions of the vector $\hat{y}$. In any case, our representation of $\hat{y}$ as a vector is a rather artificial one and it is difficult to see why we should consider linear functions of the vector $\hat{y}$.

Let us look at the problem from another angle. True, the sample space X is not linear, but the parameter space Ω of all the possible values of the state of nature $\theta = (Y_1, \ldots, Y_N)$ is a part of the N-dimensional linear space R_N. A linear function on Ω is a function of the type

$$B_0 + B_1 Y_1 + \ldots + B_N Y_N \qquad (5.5)$$

where $B_0, B_1, \ldots, B_N$ are constants. But (5.5) is a linear function of the parameter θ and cannot be conceived of as a statistic. Consider, however, a modification of (5.5) where we replace the coefficient B_j by the variable $B_j I_j$ where I_j is the indicator of the event E_j that the unit j is selected by the sampling plan $\mathcal{S}(j = 1, 2, \ldots, N)$. For each set of coefficients $B_0, B_1, \ldots, B_N$ we then have a sort of a linear function [see formula (3.9)]

$$T = B_0 + \sum_1^N B_j I_j Y_j \qquad (5.6)$$

on Ω, where the coefficients $B_j i_j (j = 1, 2, \ldots, N)$ are random variables. The indicator I_j is a function of the label-set $\hat{u}$ $[I_j(\hat{u}) = 1$ or 0 according as j is a member of $\hat{u}$ or not]. It is easy to recognize T as a statistic – indeed as a function of the sample core $\hat{x} = (\hat{u}, \hat{y})$. Only observe that we may rewrite (5.6) as

$$T = B_0 + \sum_1^n b_i \hat{y}_i \qquad (5.7)$$

where $b_i = B_{\hat{u}_i}$ $(i = 1, 2, \ldots, n)$. Let us repeat once again that T is not a linear function on the sample space X, but that we may stretch our imagination a little bit to conceive of T as a random linear function on Ω with coefficients that are determined by the label-set $\hat{u}$. If T is defined as in (5.6) then

$$E(T) = B_0 + \Sigma B_j \Pi_j Y_j \qquad (5.8)$$

where $\Pi_j = E(I_j) = P(E_j)$. And so T is an unbiased estimator of $\overline{Y}$ if and only if [we are assuming that each $\Pi_j > 0$ and that Ω does not lie in a subspace (of R_N) of dimension lower than N]

$$B_0 = 0 \text{ and } B_j = (N\Pi_j)^{-1}(j = 1, 2, \ldots, N). \qquad (5.9)$$

If we define a linear estimator as in (5.6), then it follows that the Horvitz-Thompson estimator [see (3.8) and (3.9)] is the only unbiased linear estimator of $\overline{Y}$. Following Godambe we, therefore, take one step further and define the class of linear estimators in the following manner:

Definition. Let $\beta_0, \beta_1, \ldots, \beta_N$ be well-defined functions of the label-set $\hat{u}$. By a generalized linear estimator T we mean a statistic that may be represented as

$$T = \beta_0 + \sum_1^N \beta_j I_j Y_j. \qquad (5.10)$$

Note that the β_j's and I_j's are functions of $\hat{u}$ and that it is only the observed Y_j's that really enter into the definition of T. We may rewrite T in the alternative form

$$T = \beta_0 + \sum_1^n \beta_{a_i} \hat{y}_i \qquad (5.11)$$

and thus recognize it as a function of the sample core $\hat{x}$.

The generalized linear estimator T [as defined in (5.10)] is an unbiased estimator of $\overline{Y}$ if and only if

$$E(\beta_0) = 0 \text{ and } E(\beta_j I_j) = N^{-1} \text{ for all } j. \qquad (5.12)$$

Let us denote by $\mathcal{L}$ the class of generalized linear unbiased estimators of $\overline{Y}$. If each $\Pi_j > 0$, then $\mathcal{L}$ is never vacuous, for we have already recognized the Horvitz-Thompson estimator

$$H = \Sigma(N\Pi_j)^{-1}I_jY_j \qquad (5.13)$$

as a member of $\mathcal{L}$. If $\theta_0 = (a_1, a_2, \ldots, a_N)$ be a fixed point in Ω and we define H_0 as

$$H_0 = \Sigma(N\Pi_j)^{-1}I_j(Y_j - a_j) + (\Sigma a_j)/N \qquad (5.14)$$

then H_0 is a member of $\mathcal{L}$ and has zero risk (variance) when $\theta = \theta_0$. It follows that in the class $\mathcal{L}$ of generalized linear unbiased estimators of $\overline{Y}$ there cannot exist a best (uniformly minimum risk) estimator. This then is the celebrated Godambe assertion that we referred to in the opening paragraph of this section.

If we go back to the case of simple random sampling and compare the two classes L and $\mathcal{L}$ [defined in (5.2) and (5.12) respectively] then we shall observe that the two classes have precisely one member in common, namely the sample mean

$$\bar{y} = \left(\sum_1^n y_i\right)/n = \sum_1^N (n^{-1}I_jY_j).$$

The Godambe class $\mathcal{L}$ of generalized linear estimators is not an extension of the class L. The two classes L and $\mathcal{L}$ are essentially different in character and scope. Thus, the classical assertion that the sample mean is the best linear unbiased estimate of the population mean and Godambe's denial that no such best linear unbiased estimate can ever exist are both true (each rather trivially) in their separate contexts.

Following Hanurav, we may extend the Godambe class of linear estimators by defining a linear estimate as

$$T^* = \beta_0^* + \Sigma \beta_j^* I_j Y_j \qquad (5.15)$$

where $\beta_0^*, \beta_1^*, \ldots, \beta_N^*$ are well-defined functions of u—the label part of the data $x = (u, y)$. The only difference between (5.10) and (5.15) is that in the former the β's are functions of $\hat{u}$, whereas in the latter, the β^*'s are functions of u. Once we remember that the I_j's are functions of the label-set $\hat{u}$, it follows at once that

$$E(T^* \mid \hat{x}) = \beta_0 + \Sigma \beta_j I_j Y_j \qquad (5.16)$$

where

$$\beta_j = E(\beta_j^* \mid \hat{x}) = E(\beta_j^* \mid \hat{u})$$

is a function of the label-set $\hat{u}$ ($j = 0, 1, 2, \ldots, N$). Thus, the conditional expectation of each T^*, given the sufficient statistic (sample-core) $\hat{x}$, is a T as defined in (5.10). From the Rao-Blackwell theorem it then follows that for each estimator of type (5.15) we can find an estimator of type (5.10) with a performance characteristic that is at least as good as (uniformly) that of the former. From the decision theoretic point of view the extension of the class (5.10) by the class (5.15) is, therefore, sort of vacuous.

6
Homogeneity, Necessary Bestness and Hyper-Admissibility

During the past few years, altogether much too much has been written on the subject of linear estimates of the population total Y. The original sin was that of Horvitz and Thompson who in 1952 sought to give a classification of linear estimates of Y. The tremendous paper-writing pressure of the past decade has taken care of the rest. For a plan $\mathcal{S}$ that requires that the n sample units be drawn one at a time, without replacements, and with equal or unequal probabilities, Horvitz and Thompson called an estimator T to be of T_1-type if T be of the form (5.1) with $b_0 = 0$, where $b_1, b_2, \ldots, b_n$ are pre-fixed constants and $y_1, y_2, \ldots, y_n$ are the n observed Y-values in their natural selection order. An estimator of the type (5.6) with $B_0 = 0$ (definable for an arbitrary plan $\mathcal{S}$) was classified as a T_2-type estimator. By a T_3-type estimator, Horvitz and Thompson meant an estimator $T = \beta S$, where S is the sample total and $\beta = \beta(\hat{u})$ is an arbitrary function of the label-set $\hat{u}$. That is, a T_3-type estimator is of the form (5.10) with $\beta_0 = 0$ and $\beta_1 = \beta_2 = \ldots = \beta_N$. Prabhu Ajgaonkar (1965) combined the features of the T_2 and T_3 type estimators to define his T_5-type (someone else must have defined the T_4-type!) estimators as estimators of the type (5.10) with

$$\beta_0 = 0 \text{ and } \beta_j = \beta B_j \quad (j = 1, 2, \ldots, N) \qquad (6.1)$$

where β is a function of $\hat{u}$ and $B_1, B_2, \ldots, B_N$ are pre-fixed constants. With the exception of the T_1-type estimators, all the other types are subclasses of the Godambe class of linear homogeneous estimators, that is, estimators of the type (5.10) with $\beta_0 = \beta_0(\hat{u}) \equiv 0$ for all values of $\hat{u}$. Let us denote the Godambe class of linear homogeneous unbiased estimators of Y by $\mathcal{L}_0$. The rest of this section is devoted to a study of the class $\mathcal{L}_0$.

The class $\mathcal{L}_0$ is the class of all estimators of the form

$$T = \Sigma \beta_j I_j Y_j, \qquad (6.2)$$

where β_j is a function of the label-set $\hat{u}$, I_j is the indicator of the event $j \in \hat{u}$ and

$$E(\beta_j I_j) = 1 \quad (j = 1, 2, \ldots, N). \qquad (6.3)$$

223

Let us count the degrees of freedom that we have in setting up an estimator in $\mathscr{L}_0$. Let U be the set of all the possible values (given a plan $\mathscr{S}$) of the label-set $\hat{u}$ and let U_j be the subset of those $\hat{u}$'s that include the unit j. [The event E_j that $j \in \hat{u}$ is then the same as $\hat{u} \in U_j$.] Let m_j be the number of members in the set U_j. We are assuming that no U_j is vacuous; that is, no E_j is an impossible event; that is, $\Pi_j = P(E_j) > 0$ for all j. Thus,

$$m = \Sigma m_j \geq N. \tag{6.4}$$

For defining a T in $\mathscr{L}_0$ we need to define the N functions $\beta_1, \beta_2, \ldots, \beta_N$ on U. Since $I_j = I_j(\hat{u}) = 0$ for all $\hat{u} \notin U_j$, it is clear that we really need to define β_j on the set U_j only $(j = 1, 2, \ldots, N)$. [The values of β_j outside the set U_j have no bearing on the statistic T as defined in (6.2).] Thus, we can think of each β_j as an m_j-dimensional vector. Now (6.3) is, in reality, a linear restriction on the m_j-dimensional vector β_j. We, therefore, have $m_j - 1$ degrees of freedom in our choice of the function (vector) β_j and so we have in all

$$\Sigma(m_j - 1) = m - N \tag{6.5}$$

degrees of freedom in our selection of a T in $\mathscr{L}_0$. We may visualize $\mathscr{L}_0$ as an $m - N$ dimensional surface (plane) in the m-dimensional Euclidean space R_m.

Let us stop for a moment to consider the extreme (and rather trivial) situation where $m = N$, that is, $m_j = 1$ for all j. This is the case of a unicluster (the terminology is Hanurav's) sampling plan, that is, a plan $\mathscr{S}$ that partitions the population $\mathscr{P}$ into a number of mutually exclusive and collectively exhaustive parts and then selects just one of these parts as the label-set $\hat{u}$. In this case we have no degree of freedom in the selection of a T; that is, the class $\mathscr{L}_0$ is a one point set consisting only of the Horvitz-Thompson estimator

$$T_0 = \Sigma \Pi_j^{-1} I_j Y_j. \tag{6.6}$$

Let us return to the non-trivial case where $m > N$. As we remarked before, a member T in $\mathscr{L}_0$ is then determined by our choice of $(\beta_1, \beta_2, \ldots, \beta_N)$ which we may look upon as an m-dimensional vector lying in an $m - N$ dimensional plane. The problem is to choose a T in $\mathscr{L}_0$ that has minimum variance. Now, if T be as in (6.2) then

$$V(T) = \sum_j V(\beta_j I_j) Y_j^2 + 2 \sum_{j<k} \text{Cov}(\beta_j I_j, \beta_k I_k) Y_j Y_k \tag{6.7}$$

which depends on the state of nature $\theta = (Y_1, Y_2, \ldots, Y_N)$. For each $\theta \in \Omega$, it is then clear that $V(T)$ is a (positive semi-definite) quadratic form in the m-dimensional vector $(\beta_1, \beta_2, \ldots, \beta_N)$. For each θ in Ω, there clearly exists a choice of the vector $(\beta_1, \beta_2, \ldots, \beta_N)$ that minimizes (6.7). Except in some very special situations (with Ω a very small set), there cannot exist a choice of $(\beta_1, \beta_2, \ldots, \beta_N)$ that will minimize (6.7) uniformly for all $\theta \in \Omega$. In the class $\mathscr{L}_0$ of all linear homogeneous unbiased estimators of Y there does not exist a uniformly minimum variance unbiased estimator (Godambe, 1955).

So the search was on for some other performance criterion that would uphold some estimator as the best in the class $\mathscr{L}_0$ (or in some other smaller or larger class). Of late two rather curious such criteria have been proposed for consideration. They are (a) Ajgaonkar's criterion of *necessary bestness* and (b) Hanurav's criterion of *hyper-admissibility*. Let us first consider necessary bestness, the curiouser of the two criteria.

"In order to choose a serviceable estimator from the practical point," writes Ajgaonkar (1965, p.638), "we propose the following criterion of the necessary best estimator."

Definition (Ajgaonkar). Between two unbiased estimators T and T' (of the population total Y) with variances

$$V(T) = \Sigma a_j Y_j^2 + 2 \sum_{j<k} a_{jk} Y_j Y_k$$

and

$$V(T') = \Sigma b_j Y_j^2 + 2 \sum_{j<k} b_{jk} Y_j Y_k$$

the estimator T is *necessary better* than T' if $a_j \leq b_j$ for all j. The estimator T (in the class C) is *necessary best in C* if it is necessary better than every other estimator in C.

From (6.7) and the above definition it then follows that the estimator $T = \Sigma \beta_j I_j Y_j$ is necessary best in the class $\mathcal{L}_0$ if and only if $V(\beta_j I_j)$ is uniformly minimum for all j. From the Schwarz inequality we have

$$V(\beta_j I_j) V(I_j) \geq [\mathrm{Cov}(\beta_j I_j, I_j)]^2 = [E(\beta_j I_j^2) - E(\beta_j I_j) E(I_j)]^2. \quad (6.8)$$

Since $I_j^2 = I_j$, $E(I_j) = \Pi_j$, $V(I_j) = \Pi_j(1 - \Pi_j)$ and $E(\beta_j I_j) = 1$ for all j, we at once have

$$V(\beta_j I_j) \geq (1 - \Pi_j)/\Pi_j \quad (j = 1, 2, \ldots, N). \quad (6.9)$$

The sign of equality holds for all j in (6.9) if we select

$$\beta_j = \Pi_j^{-1} \quad (j = 1, 2, \ldots, N),$$

that is, if T is the Horvitz-Thompson estimator. Thus, in $\mathcal{L}_0$ there exists a unique necessary best estimator and that is the Horvitz-Thompson estimator (5.22). [Ajgaonkar (1965) gave a very complicated looking proof of the necessary bestness of (6.6) in the subclass of T_5-type estimators as defined in (6.1), and for a particular class of sampling plans. The present proof is a simplification of a proof suggested by Hege (1967).]

But why necessary bestness? It is hard to figure out how Ajgaonkar stumbled across this curious name and definition. Let us hazard a guess. We begin with a most unrealistic assumption that the space Ω contains points of the type

$$(0, \ldots, 0, Y_j, 0, \ldots, 0),$$

that is, vectors with only one non-zero coordinate Y_j $(j = 1, 2, \ldots, N)$, and let

Ω_0 be the subset of all points of the above kind. For a typical $\theta \in \Omega_0$ the variance of $T = \Sigma \beta_j I_j Y_j$ is equal to

$$V(\beta_j I_j) Y_j^2 \text{ (for some } j \text{ and } Y_j).$$

Hence, if we restrict our attention to the subset Ω_0 of Ω, the necessary best estimator in $\mathscr{L}_0$ is also the uniformly minimum variance estimator. The Horvitz-Thompson estimator has uniformly minimum variance (in $\mathscr{L}_0$) over the subset Ω_0.

Let us now consider the hyper-admissibility thesis of Hanurav (1968). Hyper-admissibility as the name suggests, is a strengthening of the decision-theoretic notion of admissibility. In order not to draw the attention of the reader away from the present context, let us define admissibility in the narrow framework of unbiased point estimation (of the population total Y) with variance as the risk function. Let T_0 and T_1 be unbiased estimators of Y.

Definition. T_0 is *uniformly better* than T_1 if

$$V(T_0) \leq V(T_1) \quad \text{for all } \theta \in \Omega$$

with the strict sign of inequality holding for at least one $\theta \in \Omega$.

Let C be a class of unbiased estimators of Y. We tacitly assume that C is a convex class, that is, when T_0 and T_1 are both members of C then so also is $(T_0 + T_1)/2$. For instance, the class $\mathscr{L}_0$ is convex.

Definition. $T_0 \in C$ is *admissible* in C if there does not exist a $T_1 \in C$ that is uniformly better than T_0.

If T_0 is admissible in C, then for any alternative $T_1 \in C$ it must be true that T_1 is not uniformly better than T_0; that is, either

(a) $V(T_0) \equiv V(T_1)$ for all $\theta \in \Omega$, or

(b) $V(T_0) < V(T_1)$ for at least one $\theta \in \Omega$.

Now, in view of the admissibility of T_0 and the convexity of C, the alternative (a) is impossible. Suppose (a) holds. Consider the estimator

$$T_* = (T_0 + T_1)/2$$

and observe that

$$V(T_*) = \tfrac{1}{4}\{V(T_0) + V(T_1)\} + \tfrac{1}{2}\rho \sqrt{V(T_0)V(T_1)}$$

$$= V(T_0)(1+\rho)/2$$

$$\leq V(T_0),$$

where ρ is the correlation coefficient between T_0 and T_1. Since T_0 is admissible, it follows that $V(T_*) \equiv V(T_0)$ for all θ; that is, $\rho \equiv 1$ for all θ. Therefore, $T_0 = a + bT_1$. Since, T_0 and T_1 are both unbiased estimators of Y, it follows that $a = 0$ and $b = 1$. This contradicts the initial supposition that T_0 and T_1 are different estimators.

Thus, in our present context, we may redefine admissibility as

Definition (Hanurav). $T_0 \in C$ is admissible in C if, for any other $T_1 \in C$, it is true that

$$V(T_0) < V(T_1)$$

for at least one value of θ, say θ_{01}, in Ω. [The point θ_{01} will usually depend on T_0 and T_1.]

It is clear that the admissibility of an estimator T_0 depends on two things, namely, (a) the extent of the class C that T_0 is referred to and (b) the extent of the space Ω in which θ is supposed to lie. The smaller the class C and the larger the space Ω, the easier it is to establish the admissibility of a T_0 in C. A little while ago we noted that, in the class $\mathcal{L}_0$, the Horvitz-Thompson estimator (6.6) is the only one that has uniformly minimum variance over the set Ω_0 of all points θ with only one non-zero coordinate. If we are allowed to make the unrealistic assumption that $\Omega \supset \Omega_0$, then the admissibility of (6.6) in $\mathcal{L}_0$ follows at once. Godambe and Joshi (1965) proved the admissibility of (6.6) in the wider class of all unbiased estimators of Y, under the very unrealistic assumption that $\Omega = R_N$. As we have noted earlier [see (3.10) and (3.11)], the Horvitz-Thompson estimator is no longer admissible (even in the small class $\mathcal{L}$ of all linear unbiased estimators of Y) if it is known that Ω is a small neighborhood of a point $\theta_0 = (a_1, a_2, \ldots, a_N)$.

Hanurav sought to strengthen the notion of admissibility as follows. Following Godambe, he made the unrealistic assumption that $\Omega = R_N$ and then defined a *principal hyper-surface (phs)* of Ω as a linear subspace of all points $\theta = (Y_1, Y_2, \ldots, Y_N)$ with

$$Y_{j_1} = Y_{j_2} = \ldots = Y_{j_k} = 0$$

where $0 \le k < N$ and $(j_1, \ldots, j_k)$ is a subset of $(1, 2, \ldots, N)$. [The whole space Ω corresponds to the case $k = 0$. There are $2^N - 1$ phs's of Ω.] Let Ω^* be a typical phs in Ω. Let C be a class of unbiased estimators of Y.

Definition (Hanurav). $T_0 \in C$ is hyper-admissible in C if, for every phs $\Omega^* \in \Omega$, it is true that T_0 is admissible in C when we restrict θ to Ω^*.

It follows at once that the H-T estimator $T_0 = \Sigma \Pi_j^{-1} I_i Y_j$ is the unique hyper-admissible estimator in $\mathcal{L}_0$. Suppose $T = \Sigma \beta_j I_j Y_j$ is hyper-admissible in $\mathcal{L}_0$. Consider the phs Ω_j^* of all points θ with $Y_i = 0$ for all $i \neq j$. For a typical $\theta \in \Omega_j^*$

$$V(T) = V(\beta_j I_j) Y_j^2$$

227

and this [as we have noted in (6.9)] is greater than

$$V(T_0) = \Pi_j^{-1} (1 - \Pi_j) Y_j^2$$

unless $\beta_j = \beta_j(\hat{u}) = \Pi_j^{-1}$. Thus, the admissibility of T in each phs Ω_j^* implies that $T = T_0$. That T_0 is hyper-admissible, that is, is admissible on each phs, is equally trivial. Let Ω^* be a typical phs and let $T^* = \Sigma \beta_j^* I_j Y_j$ be a member of $\mathcal{L}_0$ such that

$$V(T^*) \leq V(T_0) \quad \text{for all } \theta \in \Omega^*. \tag{6.10}$$

For each one-dimensional phs $\Omega_j^* \in \Omega^*$, we must have the sign of equality in (6.10) for all $\theta \in \Omega_j^*$, and so it follows that $\beta_j^* = \Pi_j^{-1}$ for each j such that $\Omega_j^* \in \Omega^*$. Therefore, the sign of equality holds in (6.10) for all $\theta \in \Omega^*$. In other words, it is impossible to find an estimator T^* in $\mathcal{L}_0$ that is uniformly better than T_0 in the phs Ω^*; that is, T_0 is admissible (in the class $\mathcal{L}_0$) when we restrict θ to Ω^*.

In the context of the class $\mathcal{L}_0$, the twin criteria of *necessary bestness* and *hyper-admissibility* are mathematically equivalent. Before we proceed to examine the logical basis of the criterion of hyper-admissibility, let us point out a curious error committed by Hanurav (1968, p. 626). In his relation (3.2) Hanurav mistakenly asserts that T_0 is hyper-admissible (in C) if and only if, for every alternative $T_1 \in C$ and every phs $\Omega^* \subset \Omega$, we can find a point $\theta_{01} \in \Omega^*$ such that

$$V(T_0 | \theta = \theta_{01}) < V(T_1 | \theta = \theta_{01}). \tag{6.11}$$

We give an example to contradict the above assertion. Consider T_0 and T_1 where T_0 is as in (6.6) and

$$T_1 = \beta_1 I_1 Y_1 + \sum_{j=2}^{N} \Pi_j^{-1} I_j Y_j \quad \text{with } E(\beta_1 I_1) = 1.$$

In the phs Ω^* of all θ's with $Y_1 = 0$, it is clear that

$$V(T_0) \equiv V(T_1).$$

So in Ω^* we cannot find a point θ_{01} satisfying (6.11) and this in spite of T_0 being hyper-admissible in $\mathcal{L}_0$ and T_1 being an alternative member of $\mathcal{L}_0$.

The main result of Hanurav is to the effect that, for any nonunicluster sampling plan $\mathcal{L}$, the H-T estimator (6.6) is the unique hyper-admissible estimator in the class $\mathcal{M}^*$ of all polynomial unbiased estimators of Y. A quadratic estimator of Y is a statistic T of the form

$$T = \beta_0 + \Sigma \beta_j I_j Y_j + \Sigma \beta_{jk} I_{jk} Y_j Y_k \tag{6.12}$$

where $I_{jk} = I_j I_k$ is the indicator of the event that both j and k are in the label set $\hat{u}$, and the β's are functions of $\hat{u}$ with the (unbiasedness) conditions

$$E(\beta_0) = 0, \quad E(\beta_j I_j) \equiv 1, \quad E(\beta_{jk} I_j I_k) \equiv 0 \quad \text{for all } j \text{ and } k.$$

A polynomial estimator is similarly defined.

Now, let us examine the logical content of the hyper-admissibility criterion. Let $\mathscr{P}^*$ be an arbitrary but fixed subset (subpopulation) of the population $\mathscr{P}$ and let Y^* be the total Y-value of the units in $\mathscr{P}^*$; that is,

$$Y^* = \sum_{j \in \mathscr{P}^*} Y_j. \qquad (6.12)$$

Suppose, along with an estimate of Y, the surveyor also needs to estimate the parameter Y^*. Once the surveyor has decided upon an estimator $T = T(\hat{u}, \hat{y})$ for Y, he may choose to derive an estimate T^* for Y^* in the following manner. Recall that $\hat{y}$ is the vector $(\hat{y}_1, \hat{y}_2, \ldots, \hat{y}_\eta)$ of the (observed) Y-values of the μ distinct units $\hat{u}_1 < \hat{u}_2 < \ldots < \hat{u}_\eta$ in the label set $\hat{u}$. Define y^* as the vector

$$y^* = (y_1^*, y_2^*, \ldots, y_\eta^*),$$

where y_i^* is y_i or zero according as $\hat{u}_i$ is or is not a member of $\mathscr{P}^*$. In other words, we derive y^* by substituting by zeros those coordinates of the observation vector $\hat{y}$ that corresponds to units that are outside the sub-population $\mathscr{P}^*$. Now define

$$T^* = T(\hat{u}, y^*). \qquad (6.13)$$

If T is an unbiased estimator of Y, then it is almost a truism that T^* is an unbiased estimator of Y^*. If T is the linear homogenous estimator $\sum \beta_j I_j Y_j$, then T^* is the estimator $\sum^* \beta_j I_j Y_j$, where the summation $\sum^*$ extends over all j that belong to $\mathscr{P}^*$. In particular, if $\mathscr{P}^*$ is the single member subpopulation consisting of the unit j alone, then the H-T estimator $T_0 = \sum \Pi_j^{-1} I_j Y_j$ gives rise to the estimator

$$T_0^* = \Pi_j^{-1} I_j Y_j = \begin{cases} \Pi_j^{-1} Y_j & \text{if } j \text{ is surveyed} \\ 0 & \text{otherwise} \end{cases} \qquad (6.14)$$

for the parameter $Y^* = Y_j$.

The estimate (6.14) is similar to the one considered in example 3 of Section 3 and is, of course, utterly ridiculous. But in the makebelieve world of mathematicians, we are allowed to make any supposition. Let us pretend that when a surveyor estimates Y by T, he naturally commits himself to estimating each of the $2^N - 1$ subtotals Y^* by the corresponding derived estimate T^*. Given a class $C = \{T\}$ of estimators of Y, let us consider, for each subtotal Y^*, the class $C^* = \{T^*\}$ of derived estimators of Y^*. The estimator $T_0 \in C$ is hyper-admissible in C if, for each subtotal Y^*, the derived estimator T_0^* is admissible in C^*. According to Hanurav, if the sampling plan $\mathscr{S}$ is nonunicluster, then given any linear (or polynomial) unbiased estimator T that is different from the H-T estimator, he can always find another unbiased estimator T_1 and a subtotal Y^* such that the derived estimator T_1^* (for Y^*) is uniformly better than the derived estimator T^*.

Linear Invariance

We have idealized away many of the mathematically intractable features of the survey operation. But even with our oversimplified mathematical framework, the dimension N of the state of nature θ will usually run into several hundred thousands. It is clear that we are dealing with a most complex inference situation. A typical survey operation is an essentially non-repeatable, once in a lifetime affair. The surveyor, who is a specialist in the particular survey area, plans the survey, collects and analyzes the huge survey data and then arrives at his estimates of the various parameters of interest. Why does he need to consult a mathematician? How can the deductive processes of mathematics be of any use to the surveyor in his purely inductive inference making efforts? The author suspects that the answer lies in the general concensus among the scientific community that the mathematicians are the true watchdogs of rationalism. It may well be argued that this great reverence for mathematicians, this identification of rationalism with deduction, this over-eagerness to put every argument (be it in the realm of economics, psychology, survey theory, even philosophy) in the mold of pure deduction have done more harm than good to the general growth of knowledge. True, a good mathematician, having sharpened his mind with constant exercises in deductive reasoning, will often be able to comb out many a tangle created by unclear thinking on the part of the scientist. But new tangles are created by our over-eagerness to force a mathematical model for a situation that is essentially non-mathematical in nature. We close Part One of our essay with one more example of such a tangle in survey theory.

When the surveyor calls upon a decision-theorist (let us abbreviate the name to DT) to audit his survey work, the DT does not attempt to evaluate the thought process by which the surveyor arrived at his estimate T (for, say, the population total Y) from the data x. Indeed, the DT denies the very existence of a rational thought process that may lead us from the particular data x to the estimate T. [So far we have been freely using the two terms *estimate* and *estimator* and did not care to distinguish between them. But the whole controversy that is now raging in survey theory may be summarized as the difference between the estimate and the estimator. To the surveyor, the parameter Y is an unknown variable and the estimate T is a constant suggested by the data x at hand. The DT thinks of Y as an unknown constant and looks upon T as a random variable — a function on the sample space X.] As the DT cannot evaluate the estimate T, he proceeds to force an estimator out of the surveyor. For this he needs the sampling plan $\mathscr{S}$ to be randomized and, preferably, non-sequential. Once the DT has figured out the space X of all the possible data x (that the surveyor might have obtained from the survey), he would ask the surveyor to answer the impossible question of how he would have estimated Y for each x in X. If the function $T(x)$ is very complicated (as it would usually be) then that would be the end of the DT's audit. The estimator $T(x)$ better be simple enough so that the DT can evaluate the risk function — the average performance characteristics of the estimator. But before the risk function is evaluated, the DT would like to know the surveyor's loss function which again better be a simple one. As the DT cannot answer the question: How good (rational) is the estimate T?, he evades the issue and proceeds to answer what he thinks to be a nearly equivalent question: How good is the average performance characteristic of the estimator T?

Instead of looking at the average performance characteristics of T, the DT may try to evaluate the estimator T by examining it directly as a function on X. A criterion that is frequently used for such direct evaluation of the estimator T is the criterion of linear invariance. The DT tries to find out if the surveyor's estimate T of the population total depends in some way on the scale in which the population values (the state of nature θ) are measured. With a linear shift (change of origin and scale) in the measurement scale for the population values, the state of nature $\theta = (Y_1, Y_2, \ldots, Y_N)$ will be shifted to $\theta' = (a + bY_1, a + bY_2, \ldots, a + bY_n)$ and the parameter Y will be shifted to $Y' = Na + bY$. With the same shift in the measurement scale the data

$$x = \{(u_1, u_2, \ldots, u_n), (y_1, y_2, \ldots, y_n)\}$$

will appear as

$$x' = \{(u_1, u_2, \ldots, u_n), (a + by_1, \ldots, a + by_n)\}. \tag{7.1}$$

Since x and x' represent the same data (in two different scales) it is natural to require that they lead to the same estimate (in the two scales) of the population total. This leads us to the following

Definition. The estimator $T = T(x)$ is origin and scale invariant if

$$T(x') \equiv Na + bT(x) \tag{7.2}$$

for all x, a, and $b > 0$, where x' is defined as in (7.1). We call T scale invariant if the above identity holds with $a = 0$.

One reason why there is so much interest (see Section 6) in linear homogeneous estimators is that they are supposed to be scale invariant. [As we shall presently point out, the above supposition is true only under some qualifications.] The Horvitz-Thompson estimator $T_0 = \Sigma \Pi_j^{-1} I_j Y_j$ is clearly scale invariant. It will be origin invariant only if

$$\Sigma \Pi_j^{-1} I_j \equiv N \tag{7.3}$$

for all samples. Since the expected value of the left hand side is clearly equal to N, we may restate the identity (7.3) as $V[\Sigma \Pi_j^{-1} I_j] = 0$ or equivalently

$$\begin{aligned} N^2 &= E[\Sigma \Pi_j^{-1} I_j]^2 \\ &= \Sigma \Pi_j^{-1} + \sum_{j \ne k} [\Pi_{jk}/(\Pi_j \Pi_k)] \end{aligned} \tag{7.4}$$

where Π_{jk} is the probability that both j and k are in the sample. In the case of simple random sampling with sample size n, it is clear that $\Pi_j = n/N$ for all j and so the H-T estimator reduces to the simple origin and scale invariant estimator

$$G = (NS)/n \tag{7.5}$$

where S is the sample total.

231

So far mathematicians have generally avoided the non-homogeneous linear estimators of the type

$$\beta_0 + \Sigma \beta_j I_j Y_j \qquad (7.6)$$

in the mistaken belief that such estimators cannot possibly be scale invariant. It is tacitly assumed that any function $\beta = \beta(\hat{u})$ of the label-set $\hat{u}$ is necessarily *scale-free*; that is, the value of $\beta(\hat{u})$ depends only on $\hat{u}$ and not on the scale in which the population values are measured. That this need not be so is seen as follows. Suppose the surveyor defines β as

$$\beta(\hat{u}) = \Sigma I_j a_j \qquad (7.7)$$

where $\theta_0 = (a_1, a_2, \ldots, a_N)$ is a pre-selected fixed point in the space Ω. The function β is clearly scale invariant. That is, if the surveyor is told that, in the new measurement scale, each of the population values is to be multiplied by the scaling factor b, then he (the surveyor) will automatically represent the point θ_0 as $(ba_1, ba_2, \ldots, ba_n)$ and re-compute $\beta(\hat{u})$ as $b\beta(\hat{u})$. Let us look back on the modified H-T estimator

$$H_0 = \Sigma \Pi_j^{-1} (Y_j - a_j) + \Sigma a_j \qquad (7.8)$$

that we had considered earlier in (3.12), where $\theta_0 = (a_1, a_2, \ldots, a_N)$ is a pre-selected fixed point in Ω. A surveyor using (7.8) as his estimating formula for Y can never be accused of violating the canon of linear invariance. [We are not saying that H_0 is a respectable or a reasonable estimator of Y. We are only saying that, apart from being an unbiased estimator of Y with zero variance when $\theta = \theta_0$, the estimator H_0 is origin and scale invariant.] It has been repeatedly asserted by Godambe (see either of his 1968 papers) that, in the class of all estimators that are functions of the label-set $\hat{u}$ and the sample total S, the estimator $G = (NS)/n$ (where n is the number of units in $\hat{u}$) is the unique origin and scale invariant one. However, observe that if $\beta = \beta(\hat{u})$ is any scale-free function of $\hat{u}$ and

$$\beta_0(\hat{u}) = \Sigma a_j - \beta(\hat{u}) \Sigma I_j a_j ,$$

where $\theta_0 = (a_1, a_2, \ldots, a_N)$ is a fixed point in Ω, then the estimator

$$G_0 = \beta_0 + \beta S = \Sigma a_j + \beta \Sigma I_j (Y_j - a_j) \qquad (7.9)$$

is an origin and scale invariant function of $\hat{u}$ and S.

References

1. Ajgaonkar, S.G. Prabhu, "On a Class of Linear Estimates in Sampling with Varying Probabilities without Replacements," *Journal of the American Statistical Association,* 60, 637-642, 1965.
2. Basu, D., "On Sampling With and Without Replacements," *Sankhyā*, 20, 287-294, 1958.

3. Basu, D., "Recovery of Ancillary Information," *Sankhyā*, 26, 3-16, 1964.
4. Basu, D. and Ghosh, J.K., "Sufficient Statistics in Sampling from a Finite Universe," *Proceedings of the 36th Session of Int. Stat. Inst.*, 850-859, 1967.
5. Basu, D., "Role of the Sufficiency and Likelihood Principles in Sample Survey Theory," *Sankhyā*, 31, 441-454, 1969.
6. Desraj, *Sampling Theory*, McGraw-Hill, 1968.
7. Godambe, V.P., "A Unified Theory of Sampling from Finite Populations," *Journal of the Royal Statistical Society, B*, 17, 269-278, 1955.
8. Godambe, V.P., "An Admissible Estimate for Any Sampling Design," *Sankhyā*, 22, 285-288, 1960.
9. Godambe, V.P. and Joshi, V.M., "Admissibility and Bayes Estimation in Sampling Finite Populations – Part I, II, and III," *Annals of Mathematical Statistics*, 36, 1707-1742, 1965.
10. Godambe, V.P., "Contributions to the United Theory of Sampling," *Rev. Int. Stat. Inst.*, 33, 242-258, 1965.
11. Godambe, V.P., "A New Approach to Sampling From a Finite Universe, Part I and II," *J. Roy. Statist. Soc.*, 28, 310-328, 1966.
12. Godambe, V.P., "Bayesian Sufficiency in Survey-Sampling," *Ann. Inst. Stat. Math.* (Japan), 20, 363-373, 1968.
13. Godambe, V.P., "Some Aspects of the Theoretical Developments in Survey Sampling," *New Developments in Survey Sampling*, Wiley-Interscience, 27-53, 1968-69.
14. Hájek, J., "Optimum Strategy and Other Problems in Probability Sampling," *Casopis Pest. Math.*, 84, 387-423, 1959.
15. Hanurav, T.V., "On Horvitz and Thompson Estimator," *Sankhyā, A*, 24, 429-436, 1962.
16. Hanurav, T.V., "Hyper-Admissibility and Optimum Estimators for Sampling Finite Populations," *Ann. Math. Statist.*, 39, 621-642, 1968.
17. Hege, V.S., "An Optimum Property of the Horvitz-Thompson Estimate," *Journal of the American Statistical Association*, 62, 1013-1017, 1967.
18. Horvitz, D.G. and Thompson, D.J., "A Generalization of Sampling Without Replacements from a Finite Universe," *J. Am. Stat. Ass.*, 47, 663-685, 1952.
19. Joshi, V.M., "Admissibility of the Sample Mean as Estimate of the Mean of a Finite Population," *Ann. Math. Statist.*, 39, 606-620, 1968.
20. Midzuno, H., "On the Sampling System with Probability Proportionate to Sum of Sizes," *Ann. Inst. Stat. Math.* (Japan), 3, 99-107, 1952.
21. Murthy, M.N., "On Ordered and Unordered Estimators," *Sankhyā, A*, 20, 254-262, 1958.
22. Pathak, P.N., "Sufficiency in Sampling Theory," *Ann. Math. Statist.*, 35, 795-808, 1964.
23. Raiffa, H. and Schlaifer, R.O., *Applied Statistical Decision Theory*, Boston Division of Research, Graduate School of Business Administration, Harvard University, 1961.
24. Roy, J. and Chakravarti, I.M., "Estimating the Mean of a Finite Population," *Ann. Math. Statist.*, 31, 392-398, 1960.
25. Zacks, S., "Bayes Sequential Designs for Sampling Finite Populations," *J. Am. Stat. Ass.*, 64, 1969.

[D. Basu presented a very brief summary of his paper and then went on to make some additional remarks. The following is an abstract of these additional remarks.

233

The sample core $\hat{x} = (\hat{u}, \hat{y})$ is always a sufficient statistic. In general, it is minimal sufficient. The sufficiency principle tells us to ignore as irrelevant all details of the sample $x = (u, y)$ that are not contained in the sample core $\hat{x}$. The likelihood principle tells us much more. The (normalized) likelihood function is the indicator of the set Ω_x of all parameter points θ that are consistent with the sample x. The set Ω_x depends on the sample x only through the sample-core $\hat{x}$. Given $\hat{x}$, the set Ω_x has nothing to do with the sampling plan $\mathcal{J}$. And this is true even for sequential sampling plans. For one who believes in the likelihood principle (as all Bayesians do) the sampling plan is no longer relevant at the data analysis stage.

A major part of the current survey sampling theory was dismissed by D. Basu as totally irrelevant. He stressed the need for science oriented, down to earth data analysis. The survey problem was posed as a problem of extrapolation from the observed part of the population to the unobserved part. An analogy was drawn between the problem of estimating the population total ΣY_j and the classical problem of numerical integration. In the latter the problem is to 'estimate' the value of the integral $\int_a^b Y(u)\,du$ by 'surveying' the function $Y(u)$ at a number of 'selected points' $u_1, u_2, \ldots, u_n$. Which points to select and how many of them, are problems of 'design'. Which integration formula to use and how to assess the 'error' of estimation, are problems of 'analysis'. True, it is possible to set up a statistical theory of numerical integration by forcing an element of randomness in the choice of the points. But, how many numerical analysts will be willing to go along with such a theory?

The mere artifact of randomization cannot generate any information that is not there already. However, in survey practice, situations will occasionally arise where it will be necessary to insist upon a random sample. But this will be only to safeguard against some unknown biases. In no situation, is it possible to make any sense of unequal probability sampling.

The inner consistency of the Bayesian point of view is granted. However, the analysis of the survey data need not be Bayesian. Indeed, who can be a true Bayesian and live with thousands of parameters? According to the author, survey statistics is more an art than a science.]

COMMENTS

G. A. Barnard:

First, a point of detail. Dr. Basu suggested, in the unwritten part II of his paper, a method of estimation using an assistant and dividing the data into two parts, D_1 and D_2. He said that no estimate of error would be available. I simply want to point out that an error estimate could be obtained, in an obvious way, if Dr. Basu has a twin, Basu[1], and his assistant also has a twin, assistant[1]. Then the data are divided into four sets, $D_1, D_1', D_2,$ and D_2'.

Second, a general point. Dr. Basu and others here are concerned particularly with the problems which arise when it is necessary to make use of the additional information or prior knowledge α. In many sample survey situations this prior knowledge of individuals is negligible (at least for a large part of the population under discussion) and in this case the classical procedures, in particular the Horvitz-Thompson estimators, apply in a sensible manner. It is important that we should not appear in this conference to be casting doubt on procedures which experience has shown to be highly effective in many practical situations.

The problem of combining external information with that from the sample is in general difficult to solve. For instance, in ordinary (distribution-free) least squares theory, the additional information that one or more of the unknown parameters has a bounded range makes the usual *justifications* inapplicable and no general theory appears to be possible, though it is easy to see what we should do in some particular cases.

With Dr. Basu's elephants, a realistic procedure (on the data he has given) would seem to be to think of the measurement to be made on one elephant as providing some estimate of how the animals have put on weight, or lost it, during the past three years. The circus owner should be able to give good advice on how elephants grow, but in the absence of this it would seem plausible to assume that the heaviest elephant, being fully mature, will have gained nothing, and that the percentage growth will be a linear function of weight three years ago. It would then be wise to select an elephant somewhat lighter than Sambo for weighing. The estimation procedure is clear.

Evidently, as Dr. Basu suggests, no purely mathematical theory is ever likely to be able to account for an estimation procedure such as that suggested. But I do not think this implies that all mathematical theories are time wasting in this context. A judicious balance is necessary. In particular, as I have said, we should not throw overboard the classical theory, or the work of Godambe, Horvitz, Thompson and others, just because we can envisage situations where these results would clearly not be applicable.

V. P. Godambe:

Professor Basu has given a very interesting presentation of some ideas in survey sampling theory which many of us have been contemplating for some time. I find it difficult however to agree with him in one respect. The likelihood principle, which does not permit the use of the sampling distribution generated by randomization for inference purposes, is unacceptable to me in relation to survey sampling. It seems as though the likelihood principle has different implications for two intrinsically similar situations: for the coin tossing experiment the likelihood principle allows the use of binomial distributions while infering about the binomial parameter but if the experiment is replaced by one of the drawing balls from a bag containing black and white balls, the likelihood principle does not allow the use of corresponding binomial (or hypergeometric if sampling is without replacement) distribution to infer the unknown proportion of white balls in the bag.

Professor Basu's comments on HT-estimator (example of weighing elephant and so on) are humourous and I wonder if he wants us to take them at all seriously. The comments fail to take into account the fact that the inclusion probabilities involved in HT-estimator are inseparably tied to the prior knowledge represented or approximated by a *class* of (indeed a very very wide one)

235

prior distributions (Godambe, *J. Roy. Statist. Soc.,* 1955) on the parametric space. I believe the only way of making sense of sampling practice and theory is through studying the frequency properties implied by the distributions generated by different modes of randomization (that is, different sampling designs) of the estimators obtained on the basis of the considerations of prior knowledge; of course one should also study the implications of reversing the role of *frequency properties* and *prior knowledge* (reference: section 7, Godambe's and Thompson's Symposium paper).

At the end of his paper Professor Basu comments on "origin and scale invariant estimator" in my paper, "Bayesian Sufficiency in Survey-Sampling", *Ann. Inst. Stat. Math.,* 1968. My assertion about the uniqueness in the paper is certainly true. Basu's comments suggest a different type of invariance which is already discussed in our (Godambe and Thompson) symposium paper (section 3).

J. Hájek:

Professor Basu and myself both like the likelihood function connected with sampling from finite populations, but for opposite reasons. He likes it to support the likelihood principle in sample surveys, and I like it to discredit this principle by showing its consequences in the same area. We both are wrong, because the probabilities of selection of samples are in a vague sense dependent on the unknown parameter, because they depend on the same prior facts (prior means and expectations, etc.) that have influenced the values under issue. Consequently, we do not have exactly the situation assumed in applications of the likelihood and conditionality principles. Of course this dependence of parameter and sample strategy is hard to formalize mathematically. My recognition of this dependence is due to a discussion I had recently with Professor Rubin on the conditionality principle.

As to the Horvitz-Thompson estimate, its usefulness is increased in connection with ratio estimation. For example, if the probabilities of inclusion are π_i and we expect the Y_i's to be proportionate to A_i, then we should use the estimate

$$\left(\sum_{i=1}^{N} A_i \right) \frac{\sum_{i \, \epsilon \, s} Y_i / \pi_i}{\sum_{i \, \epsilon \, s} A_i / \pi_i},$$

which would save the statistician's circus job. This estimate is not unbiased but the bias is small, and the idea of unbiasedness is useful only to the extent that greatly biased estimates are poor no matter what other properties they have.

J.C. Koop:

Professor Basu's essay is very stimulating and sometimes also provocative.

Regarding Sambo, I find the choice of selection probability for him (equal to 99/100) rather unwise in the face of the existence of a list of elephants' weights taken three years ago in the owner's possession. Sambo, we are told, was a middle-sized elephant, and knowing the existence of Jumbo in the herd, it might have been wiser to choose the selection probabilities directly proportional to the respective weights of the elephants according to the available records. The reason being that if the elephants grew such that their present weights are directly proportional to their weights three years ago, then the variance of the estimate (equal to the selected elephant's weight divided by its selection probability), is zero. The circus statistician ought to have known better, and one should not be surprised that he was fired!

I am in complete agreement with him that the label of each unit in a sample (or in my terminology, the identity of a unit) cannot be discarded on the ground that it does not provide information. His discussion on this important point is very clear and can be read with profit.

However, I am somewhat surprised at his lack of appreciation for the basic ideas contained in Horvitz and Thompson's path-breaking paper of 1952 as evidenced by the following statement in section 6 of his paper: "During the past few years, altogether too much has been written on the subject of linear estimators of the population total Y. The original sin was that of Horvitz and Thompson who in 1952 sought to give a classification of linear estimates of Y. The tremendous paper-writing pressure of the past decade has taken care of the rest." These two writers constructed three linear estimators, each depending on one of the following three basic features of what I subsequently termed as the axioms sample formation in selecting units one at a time, namely, (i) the order of appearance of a unit in a sample, (ii) the presence or absence of a unit in the sample and (iii) the identity of the sample itself. Sample survey theorists have since benefited from their work. I for one felt in 1956 that the various types of estimators in the literature of that time needed classification and starting with these three features of sample formation, showed that $2^3 - 1 = 7$ types or classes of linear estimators, $T_1, T_2, \ldots, T_7$ were possible for one-stage sampling, three of which were those of Horvitz and Thompson. Godambe in his fundamental paper of 1955 found what I subsequently classified as the T_5-type of estimator, which should certainly not be attributed to Ajgaonkar, whose work began much later. In the process of this classification, it was found that an estimator given in the early pages of Sukhatme's text book of 1954 is of the T_4-type, that is, an estimator where the coefficients attached to the variate-values (observations in Basu's terminology) depended on the identity of the unit (label) and the order of appearance of the unit. Among other things, all this work was described in a thesis accepted by the North Carolina State University in 1957 and published in its *Institute of Statistics Mimeo Series* as No. 296, in 1961. Subsequently in 1963 I revised some of this work and amplified some of its ramifications in a paper in *Metrika*, Vol. 7(2) and (3).

One may ask what is the use of recognizing the three features of sample formation? In the context of the real world of sample surveys it must be said that they have physical meaning, which has some bearing on how an estimator may be constructed. Equally important, they point to the information supplied by the sample even before (field) observations on its members are made. In discussing Dr. C.R. Rao's excellent paper, I constructed a class of estimators where two of the features of sample formation were used, viz., (ii) which is equivalent to recognizing the identity of the distinct units (labels) and (iii) the identity of the sample itself, to show that an estimator of this class can have smaller M.S.E. than the U.M.V. estimator, derived through an appeal to the principles of maximum likelihood, sufficiency and completeness, carried over almost bodily from classical estimation theory, thus bringing into question the extent of relevance of these principles in estimation theory for sample surveys of a finite universe. (It must be stressed that this does not detract from Dr. Rao's valuable paper which I interpret as a probe to uncover the difficulties of the subject.)

R. Royall:

Although I agree with much of what is said in this paper, I must take exception to one fundamental point. In section 2 Professor Basu states that: "From the point of view of a frequency probabilist, there cannot be a statistical theory of surveys without some kind of randomization in the plan *S*."

"Apart from observation errors and randomization, the only other way that probability can sneak into the argument is through a mathematical formalization of what we have described before as the residual part *R* of the prior knowledge, $K = (\Omega, \alpha, R)$. This is the way of a subjective (Bayesian) probabilist. The formalization of *R* as a prior probability distribution of θ over Ω makes sense only to those who interpret the probability of an event, not as the long range relative frequency of occurrence of the event (in a hypothetical sequence of repetitions of an experiment), but as a formal quantification of the... phenomenon of *personal belief* in the truth of the event."

It seems frequently to be true that at some time before the values $y_1, y_2, \ldots, y_N$ are fixed it is natural and generally acceptable to consider these numbers as values, to be realized, of random variables $Y_1, Y_2, \ldots, Y_N$. For instance, these might be the numbers of babies born in each of the *N* hospitals in the state during the next month. What particular values will appear is uncertain, and this uncertainty can be described probabilistically. Although subjectivists would presumably accept these statements, in many finite populations such models are precisely as *objective* as those used everyday by frequentists. If such a model is appropriate before the *y*'s are realized, it seems to be equally appropriate after they are fixed but unobserved. If a fair coin is flipped, the probability that it will fall heads is one half; if the coin was flipped five minutes ago, but the outcome has not yet been observed, my statement that the probability of heads is one half is no less objective now than it was six minutes ago. The state of uncertainty is not transformed from objective to subjective by the single fact that the event which determines the outcome has already occurred.

It can be argued that since the event has already occurred, the outcome should be treated as a fixed but unknown constant (so that now the probability of heads is one if the fixed but unknown outcome *is* heads and otherwise is zero). Such an argument leads back to the conventional model but rests on an unduly restrictive notion of the scope of objective probability theory.

The probability of one half for heads arises from my failure to notice that the coin is slightly warped. It can be argued that all probability models for real phenomena are likewise conditioned on personal knowledge and should therefore be called subjective. Be that as it may, (i) many statisticians do not consider themselves to be subjectivists and (ii) *super-population* models are frequently as objective as any other probability models used in applied statistics. Since such models, in conjunction with non-Bayesian statistical tools, can be extremely useful in practice as well as in theory, it seems to me to be a mistake to insist that they are available only to subjective Bayesians without pointing out that in this context the term applies to essentially all practicing statisticians.

REPLY

Professor Koop and Professor Godambe seem to think that the real difficulty in the elephant problem lies in the 'unrealistic' sampling plan—a plan that is 'not related' to the background knowledge. I always thought that the real purpose of a sampling plan is to get a good representative sample. If the owner knows how to relate the present weight of the representative elephant Sambo to the total weight of his fifty elephants, then he ought to go ahead and select Sambo. Why does he need a randomized sampling plan? Professor Koop wants to allot larger selection probability to Jumbo, the large elephant. Does he really prefer to have Jumbo rather than Sambo in his sample? I think Professor Koop is actually indifferent as to which elephant he selects for weighing. He *knows* more about the circus elephants than the circus owner. He 'knows' that the 50 ratios of the present and past weights of the elephants are nearly equal. Therefore, he has made up his mind that the ratio estimate is a good one irrespective of which elephant is selected. But he is not prepared to go all the way with me and assert the goodness of the ratio estimate irrespective of the selection plan. Professor Koop needs to allot unequal selection probabilities (proportional to their known past weights) to the 50 elephants so that he can mystify his non-statistical customers with the assertion that his estimate is then an unbiased one. As a scientist he has been trained to make a show of objectivity. May I ask what Professor Koop would do if the elephant trainer informs him that Jumbo (the big elephant) is on hunger strike for the past 10 days? Should he not try to avoid selecting Jumbo? He should, because now he does not know how to relate the present weight of Jumbo to the total weight of the 50 elephants.

In survey literature, we often come across the term *representative sample*. But to my knowledge the term has never been properly defined. At one time it used to be generally believed that the simple random sampling plan yields a representative sample. However, the difficulty with this naive sampling plan was soon recognized and so surveyors turned to stratification and other devices (like ratio and regression estimation) to exploit their background information about a specific survey problem. It is not easy to understand how surveyors got messed up with the idea of unequal probability sampling. I think it started with the idea of making the ratio estimate look unbiased. Thus Lahiri devised his method of using the random number tables in such a manner that the probability of selecting a particular sample set of units is proportional to the total 'size' of the units. This plan made the ratio estimate look 'good'. The flood-gate of unequal probability sampling was then opened and a surprisingly large number of learned papers has been published on the subject. What is even more surprising is that no one seems to worry about the fact that the surveyor can allot only one set of selection probabilities $\pi_1, \pi_2, ..., \pi_N$, but that he has usually to estimate a vast number of different population totals. For each particular population total the surveyor may be able to find an appropriate ratio (or regression) estimate. But how can he possibly make all these different ratio estimates look 'good'?

Of late, a great deal has been written about the Horvitz-Thompson estimate. A little while ago Professor Rao proved an optimum property of the method. But to me the H-T estimate looks particularly curious. Here is a method of estimation that sort of contradicts itself by alloting weights to the selected units that are inversely proportional to their selection probabilities. The smaller the selection probability of a unit, that is, the greater the desire to avoid selecting the unit, the larger the weight that it carries when selected.

The question that Professor Hájek raised in the first part of his comments is exceedingly important and is one that, at one time, had given me a great deal of trouble. As Professor Hájek admitted, the question is hard to formulate and is even harder to answer. In the second part of my essay, I shall discuss the problem in greater detail. To-day, let us try to understand the difficulty in the context of the circus elephants. Suppose the surveyor (the owner) selects three elephants u_1, u_2, and u_3 with probabilities proportional to their past weights (and, say, with replacements) so that the data is $x = [(u_1,y_1), (u_2,y_2), (u_3,y_3)]$. In this case, the selection probability of the labels $u = (u_1,u_2,u_3)$ depends on the past weights of the 50 elephants and, therefore, also *depends* on their present weights—the state of nature $\theta = (Y_1, Y_2,..., Y_{50})$. If the selection probability of u depends on θ, then the very fact of its selection gives the surveyor some information about θ. Should the surveyor ignore this fact and act as if he always wanted to select this set of labels u and analyze the data x on that basis? This is precisely what I am advising the surveyor to do and this is what Professor Hájek thinks to be an error. But let us stop and think for a moment. Does the information that u is selected tell the surveyor anything (about θ) that the surveyor did not know already? When the question is phrased this way, one will be forced to admit that there is no real difference between the above plan and a simple random sampling plan. Indeed, the important point that I am trying to make is this, that even when the sampling plan is sequential, the relevant thing is the data generated by the plan and the likelihood function (which depends only on the data and has nothing whatsoever to do with the plan).

However, contrast the above sampling plan with a plan where the owner asks the elephant trainer to give him the names of three elephants that come first to his mind. If (u_1,u_2,u_3) are the three elephants that are selected by the above plan, then the surveyor does not really know how he got the labels (u_1,u_2,u_3) and so he cannot analyze the data x. Could it be that the three elephants were refusing to eat for some time and that is why they were on the trainer's mind at the time? If the owner must depend on the trainer for the names and present weights of three sample elephants, and if he does not have the sampling frame (so that he cannot select the labels himself), then he may be well advised to instruct the trainer to select the three sample labels at random. Randomness is a devil no doubt, but this is a devil that we understand and have learnt to live with. It is easier to trust a known devil than an unknown saint!

The second point raised by Professor Hájek is easier to deal with. If the surveyor *knows* that the ratios of the present and past weights of the elephants are nearly equal, then why does he not use the ratio estimate itself? I do not see any particular merit in the estimate suggested by Professor Hájek.

Now, let us turn to Professor Godambe's objection to the likelihood principle in the context of survey sampling. It will be easier for us to understand Godambe's point if we examine the following example. In a class there are 100 students. An unknown number τ of these students have visited the musical show *Hair*. Suppose we draw a simple random sample of 20 students and record for each student, not his (or her) name, but only whether he has seen *Hair*. The likelihood is then a neat (hypergeometric) function involving only the parameter of interest τ. Godambe likes this likelihood function. However, if we had also recorded the name of each of the selected students, then the likelihood function would have been a lot messier. It would no longer have been a direct function of τ, but would have been a function of the state of nature $\theta = (Y_1, Y_2, ..., Y_N)$, where Y_j is 1 or 0 according as the student *j* has or has not seen *Hair*. Godambe does not know how to make any sense of this likelihood function. My advice to Professor Godambe will be this: "If the names (labels) are 'not informative', if there is no way that you can relate the labels to the state of nature θ, then do not make trouble for yourself by incorporating the labels in your data". After all, isn't this what we are doing all the time? When we toss a coin several times to determine the extent of its bias, do we record for each toss the exact time of the day or the face that was up when the coin was stationary on the thumb? We throw out such details from our data in the belief that they are not relevant (informative). Statistics is both a science and an art. It is impossible to rationalize everything that we do in statistics. These days we are hearing a lot of a new expression—rationality of type II. It is this second kind of rationality that will guide a surveyor in the matter of selection of his sample and the recording of his data.

The final remark of Professor Godambe seems to suggest that he has not quite understood what I said in the last paragraph of my essay. It is simply this that the constants in the estimating formula of the surveyor need not be (indeed, they should not be) pure numbers like π and e. The estimating formula (estimator) that the surveyor chooses surely depends on the particular inference situation. If the mathematician wishes to find out how the estimator behaves in the altered situation where the population values are measured in a different scale, he should first ascertain from the surveyor whether he (the surveyor) would like to adjust the constants in his formula to fit the new scale. When the surveyor is given this freedom, then it is no longer true that $G = NS/n$ is the only linearly invariant estimator in the class of all estimators that depend only on the label-set and the sample total. The Godambe assertion holds true only in the context of a severely restricted choice.

If I have understood Professor Royall correctly, then he claims that his super-population models for the parameter $\theta = (Y_1, Y_2, ..., Y_N)$ are non-Bayesian in the sense that such models do have objective frequency interpretations. His contention about the tossed coin in the closed palm is somewhat misleading. Let us examine a typical super-population model in which the Y_j's are assumed to be independent random variables with means αA_j and variances βA_j^γ, where A_j is a known auxiliary character of unit *j* and α, β, γ are known (or unknown) constants ($j = 1, 2, ..., N$). To me, such a model looks exactly like a Bayesian formalization of the surveyor's background knowledge or information. Certainly, there is nothing objective about the above model. Indeed, is any probability model objective? When a scientist makes a probability assumption about the observable X, he is supposed to be very objective about it. But as soon as he makes a similar statement about the state of nature θ he is charged with the unmentionable crime of subjectivity. Mr. Chairman, you have always been telling us that the ultimate decision is an 'act of will' on the part of the decision (inference) maker. Isn't it equally true that the choice of the probability model for the observable X is also an act of will on the part of the statistician? Equally subjective is the choice of the 'performance characteristics'. A true scientist has to be subjective. Indeed, he is expected to draw on all his accumulated wisdom in the field of his specialization. My own subjective assessment of the present day controversy on objectivity in science and statistics is this that the whole thing is only a matter of semantics.

If we define mathematics as the art and science of deductive reasoning—an effort at deducing theorems from a set of basic postulates, using only the three laws of logic—then statistics (the art and science of induction) is essentially anti-mathematics. A mathematical theory of statistics is, therefore, a logical impossibility!

THE SPECIFICATION OF PRIOR KNOWLEDGE BY CLASSES OF PRIOR DISTRIBUTIONS IN SURVEY SAMPLING ESTIMATION
V. P. Godambe and M. E. Thompson
University of Waterloo

1

Introduction: The Finite Population Model for Survey Sampling

Consider a finite population whose individuals are labelled by the integers $1, 2, \ldots, N$. Associated with individual i of the population is some real quantity x_i. The statistician is allowed to observe some of these x_i, and is in general interested in making inferences about the vector $\underline{x} = (x_1, \ldots, x_N)$. Part of his task is therefore to devise a way of selecting the elements i of the population for which x_i is to be observed. Let S be the collection of all subsets s of $\{1, 2, \ldots, N\}$. Elements of S will be called *samples*. If $p(s)$ is a probability function on S, the corresponding *sampling procedure*, denoted by

$$(S, p),$$

is to observe $(s, x_i : i \in s)$ with probability $p(s)$. For a sampling procedure (S, p), the probability $P(s, x_i' : i \in s \mid \underline{x})$ of observing a particular *outcome* $(s, x_i' : i \in s)$ is given by

$$
\begin{aligned}
P(s, x_i' : i \in s \mid \underline{x}) &= p(s) \quad \text{if } x_i' = x_i \text{ for all } i \in s, \\
&= 0 \quad \text{otherwise.}
\end{aligned}
\tag{1.1}
$$

Accordingly, we can think of the vector $\underline{x}$ as a *parameter* for the distribution of outcomes. It is convenient to regard the parameter space as being the whole of R_N, and this assumption will be made throughout.

The *population total*

$$\tau = \tau(\underline{x}) = \sum_{i=1}^{N} x_i$$

is a function on the parameter space R_N; and we shall suppose that we are interested in the point estimation of the value of τ rather than the individual unobserved x_i's. Thus we wish to determine a real function e on the set of outcomes $(s, x_i : i \in s)$ which in some sense is appropriate for estimating τ.

Definition 1.1. A real-valued function $e(s, x_i : i \in s)$ on the set of outcomes $(s, x_i : i \in s)$ is called an *estimator*.

Now let ξ be a prior distribution on the Borel subsets of the parameter space R_N. Given an outcome $(s, x_i' : i \in s)$, we have the posterior probability

$$P_\xi(\underset{\sim}{x} \in \underset{\sim}{B} \mid s, x_i' : i \in s) = P_\xi(\underset{\sim}{x} \in \underset{\sim}{B}/x_i = x_i', i \in s) \qquad (1.2)$$

for any Borel subset $\underset{\sim}{B}$ of R_N, the right hand side being the usual conditional probability. If B is a Borel subset of R, then

$$P_\xi(\tau(\underset{\sim}{x}) \in B \mid s, x_i' : i \in s) = P_\xi(\underset{\sim}{x} \in \underset{\sim}{B} \mid s, x_i' : i \in s) \qquad (1.3)$$

where $\underset{\sim}{B} = \{\underset{\sim}{x} \in R_N : \tau(\underset{\sim}{x}) \in B\}$. Let us denote by $\xi(\tau \mid s, x_i' : i \in s)$ the corresponding cumulative function, which gives the posterior distribution of the population total τ.

Note that the posterior probabilities of (1.2) and (1.3) are independent of the sampling procedure. Thus it may be argued that when there is a prior distribution on R_N, inferences about the parameter, given an outcome $(s, x_i' : i \in s)$, should not depend upon the sampling procedure used to obtain the outcome.

Now the statistician's prior knowledge about the population can seldom be expressed as a single prior distribution ξ on the parameter space R_N; it is often more realistic to assume that his prior knowledge is equivalent to some class C of prior distributions ξ on R_N. If this is the case, his posterior knowledge, given an outcome $(s, x_i' : i \in s)$, is expressible as the corresponding class of posterior distributions of the form (1.2). Again, the class of posterior probabilities does not depend on the sampling procedure. In comparing the usefulness of various estimators of τ in such a situation, it is therefore natural to invoke criteria which do not involve the sampling procedure. The primary purpose of this paper is to study estimators based on reductions of the data $(s, x_i : i \in s)$ by certain *sufficiency* and *invariance* considerations, and to try to derive conditions on the statistician's prior knowledge (that is, on the class C) which will suggest particular estimators based on these reductions.

2

Bayesian Sufficiency

Let us suppose that the statistician's prior knowledge is equivalent to a class C of prior distributions ξ on R_N.

Definition 2.1. A function $t(s, x_i : i \in s)$ on the space of outcomes is said to be a *Bayes sufficient statistic* for the population total τ with respect to the class C of prior distributions on R_N if and only if the Bayes posterior distribution $\xi(\tau \mid s, x_i : i \in s)$ depends on $(s, x_i : i \in s)$ only through $t(s, x_i : i \in s)$ for every $\xi \in C$.

A statistic $t(s, x_i : i \in s)$ which is Bayes sufficient for τ with respect to C contains all of the statistician's posterior knowledge about τ. It would seem reasonable therefore to choose an estimator of τ from among functions depending on $(s, x_i : i \in s)$ only through $t(s, x_i : i \in s)$.

For example, let C be any class of probability measures ξ on (the Borel subsets of) R_N under which the co-ordinate random variables are probabilistically independent; equivalently, C is a class of product measures on R_N. Then given $\xi \in C$ and the data $(s, x_i' : i \in s)$, the posterior probability that τ belongs to some set of values A is the same as the posterior probability that $\sum_{i \notin s} x_i$ belongs to $A - \sum_{i \in s} x_i'$ (where $A - \sum_{i \in s} x_i'$ is the set obtained by subtracting $\sum_{i \in s} x_i'$ from each element of A). Hence we have

Theorem 2.1. The statistic $(s, \sum_{i \in s} x_i)$ is a Bayes sufficient statistic for τ with respect to any class C of product priors.

It follows from Theorem 2.1 and the remarks preceding the theorem that an estimator of τ ought to be some function of s and the *sample total* $\sum_{i \in s} x_i$.

Suppose in particular the class C of prior distributions referred to in Theorem 2.1 is such that for each $\xi \in C$, $x_1, \ldots, x_N$ are distributed independently and normally with common mean θ_ξ and common variance σ_ξ^2. We further assume that C is such that all values of θ_ξ, between $-\infty, \infty$, and σ_ξ^2 between $0, \infty$ are possible. Now under this condition most Bayesians would assume a kind of prior distribution on (θ, σ), $d\theta d\sigma/\sigma$ for example (Ericson, 1969), and obtain by integrating on C a single (exchangeable) prior distribution on R_N. This evidently will contradict the assumption of independence of $x_1, \ldots, x_N$ implied in C. Again a fiducialist may obtain the fiducial distribution of (θ, σ) on the basis of the observed sample values $x_i : i \in s$ and use it to predict the x-values associated with the unsampled individuals of the population (Kalbfleisch and Sprott, 1969). But it is well-known that the fiducial distribution just referred to can be computed only by using the knowledge of the parameter-variate relationship, that is, the relationships between θ, σ and x above, (Barnard, 1962). Hence, for the Bayesian, the approach of Bayesian sufficiency would be meaningful only if the class C in Theorem 2.1 is such that he cannot define a prior distribution on C. To a fiducialist it would be meaningful only if the class C of priors does not reveal to him any parameter-variate relationships. In effect the Theorem 2.1 attempts to formalize a rather crude yet very *fundamental statistical intuition* that in some situations for estimating the population total the sample observations $x_i : i \in s$ are relevant only through the sample total $\sum_{i \in s} x_i$. How fundamental this statistical intuition is would be clear by even a brief glance at the history of *arithmetic mean*. For a quick reference for this history we refer to Plackett (1958). To be more specific the Theorem 2.1 tries to formalize in a usable way a very weak sort of prior knowledge on the part of the statistician (which in reality more often than not is the situation) implying in some 'not very well defined sense' that the different values $x_1, \ldots, x_N$ of the population have nothing to do with each other or that possibly they are causally independent.

As another example, let C_1 be any class of exchangeable prior distributions ξ on R_N; that is, each $\xi \in C_1$ satisfies the property that if $A_1, \ldots, A_N$

are real Borel sets,

$$\xi(x_{\sigma(1)} \in A_1, \ldots, x_{\sigma(N)} \in A_N) = \xi(x_1 \in A_1, \ldots, x_N \in A_N) \qquad (2.1)$$

for any permutation σ of the integers $1, 2, \ldots, N$. Then given $\xi \in C_1$ and the data $(s, x_i : i \in s)$, the posterior distribution of τ depends on the data only through the *order statistic* $(X_1, X_2, \ldots, X_n)$, where n is the size of sample s and $X_1, X_2, \ldots, X_n$ are the values of $x_i : i \in s$ in non-decreasing order. Hence in this case $(X_1, \ldots, X_n)$ is a Bayes sufficient statistic according to Definition 2.1.

3

Invariance of Prior Knowledge under Transformations of the Parameter Space

Let s be a fixed sample, and let us consider the class $\underline{T}$ of all transformations T of the parameter space which *preserve* the problem of estimating τ from observations on s. Formally, each $T \in \underline{T}$ is a one-to-one, Borel measurable function of R_N onto R_N. For each $i \in s$, the ith co-ordinate $(T\underline{x})_i$ of $T\underline{x}$ depends only on $x_i : i \in s$; and for each $i \notin s$, $(T\underline{x})_i$ depends only on $x_i : i \notin s$. Finally, for each $T \in \underline{T}$ there is a one-to-one function Γ_T from R onto R such that

$$\sum_{i=1}^{N} (T\underline{x})_i = \Gamma_T(\sum_{i=1}^{N} x_i) . \qquad (3.1)$$

Now for each prior distribution ξ on R_N and $T \in \underline{T}$ there is an *induced* prior distribution $T\xi$ on R_N, namely the prior distribution of $T\underline{x}$. Suppose that the statistician's prior knowledge of the parameter space can be characterized by a class D of prior distributions on R_N. The class of prior distributions of $T\underline{x}$ is then

$$TD = \{T\xi : \xi \in D\} .$$

Let us consider an *estimation rule* which tells us what estimator to use for τ for each of the classes TD of prior distributions on R_N, as T runs through $\underline{T}$. We denote by $e_{TD}(s, x_i : i \in s)$ the estimator to be used when the prior knowledge is the class TD, and by $e_D(s, x_i : i \in s)$ the estimator to be used when the prior knowledge consists of D. It would seem desirable in view of (3.1) that e_{TD} should satisfy

$$e_{TD}(s, (T\underline{x})_i : i \in s) = \Gamma_T(e_D(s, x_i : i \in s)) \qquad (3.2)$$

for every T. For example, suppose that the statistician has available two measuring instruments, differing in scale and origin, for making his observations. It is natural for him to require that the rule which he uses to estimate the population total should yield him the same estimate from sets of measurements which are the same but expressed in different measuring units. Formally speaking, (3.2) ought to be satisfied for all T of the form

246

$$(T\underline{x})_i = cx_i + d \qquad c \neq 0 \tag{3.3}$$

If ξ is a prior distribution on R_N, it is easily seen that if $y_i = (T\underline{x})_i = cx_i + d$ then the posterior distribution

$$T\xi(\sum_{i=1}^{N} y_i | s, y_i : i \in s) = \xi(\sum_{i=1}^{N} cx_i + Nd | s, x_i : i \in s). \tag{3.4}$$

If there is a unique modal value $\hat{\tau}$ of the posterior distribution of $\sum_{i=1}^{N} x_i$ given ξ and $(s, x_i : i \in s)$, then there is a unique modal value of the posterior distribution on the left-hand side of (3.4), and it is equal to $\Gamma_T(\hat{\tau})$, where $\Gamma_T(\hat{\tau}) = c\hat{\tau} + Nd$. Thus, when D consists of a single prior, the process of estimating τ by the modal value of its posterior distribution, which we may call the *modified Bayes estimate* of τ, is invariant in the sense of (3.2) under a linear change of measuring units. On the other hand, the posterior mean value of $\sum_{i=1}^{N} y_i$ (the usual *Bayes estimate* of $\sum_{i=1}^{N} y_i$) is not in general equal to Γ_T evaluated at the posterior mean value of $\tau(\underline{x})$.

Now consider the subclass G of $\underline{T}$ consisting of elements T with the property that $TD = D$. Transformations in G leave the *prior knowledge* invariant. It can be shown that the subclass $\underline{G}$ is in fact a group. The invariance condition (3.2) on the estimation rule becomes

$$e_D(s, (T\underline{x})_i : i \in s) = \Gamma_T(e_D(s, x_i : i \in s)). \tag{3.5}$$

(For convenience we shall drop the subscript D.)

For example, if $\underline{G}$ is the group of all transformations given by

$$(T\underline{x})_i = x_{\sigma(i)} \qquad i = 1, 2, \ldots, N,$$

where σ is a permutation of $1, \ldots, N$ leaving s fixed, then Γ_T in (3.1) is the identity, and the invariance condition (3.5) says in effect that $e(s, x_i : i \in s)$ should be a function of the order statistic $(X_1, \ldots, X_n)$. If D is a class of exchangeable priors, then, the invariance condition yields a second justification (besides Bayes sufficiency) for discarding the labels of the observations when estimating τ.

For another example, suppose that $\underline{G}$ consists of all T of the form

$$(T\underline{x})_i = cx_i + d \qquad c \neq 0 \tag{3.6}$$

Then $\Gamma_T(x) = cx + Nd$. Our invariance condition (3.5) on e becomes

$$e(s, cx_i + d : i \in s) = ce(s, x_i : i \in s) + Nd. \tag{3.7}$$

If from some additional considerations we have decided to base e on s and the sample total $\sum_{i \in s} x_i$ (as we might on the grounds of Bayes sufficiency if the class D, like C in Theorem 2.1, consists of product measures) we can rewrite (3.7) as

$$e(s, \sum_{i \in s} (cx_i + d)) = ce(s, \sum_{i \in s} x_i) + Nd. \tag{3.8}$$

247

Theorem 3.1. The unique solution of (3.8) is given by

$$e(s, \sum_{i \in s} x_i) = \frac{N}{n} (\sum_{i \in s} x_i) \qquad (3.9)$$

where n is the size of sample s.

Proof: Setting $c = 1$ *and* $\sum_{i \in s} x_i = 0$ in (3.8) gives

$$e(s, nd) = e(s, 0) + Nd, \qquad (3.10)$$

and since d is arbitrary,

$$e(s, \sum_{i \in s} c x_i) = e(s, 0) + \frac{N}{n} \sum_{i \in s} c x_i \qquad (3.11)$$

for all $c \neq 0$. But using (3.8) with $d = 0$ and (3.11) with $c = 1$ we obtain

$$e(s, \sum_{i \in s} c x_i) = c e(s, \sum_{i \in s} x_i) = c e(s, 0) + c \frac{N}{n} \sum_{i \in s} x_i. \qquad (3.12)$$

Hence $e(s, 0) = 0$, and (3.9) is immediate.

4

Invariance Criteria Satisfied by the Difference Estimator

Let $\underline{a} = (a_1, \dots, a_N)$ be a fixed (known) vector, and consider the *difference estimator* given by

$$e(s, x_i : i \in s) = \frac{N}{n} \sum_{i \in s} (x_i - a_i) + \sum_{i=1}^{N} a_i. \qquad (4.1)$$

As in the case of (3.9), this estimator is based on the sample total $\sum_{i \in s} x_i$. Following the considerations in Section 3 we may ask whether there is a group $\underline{G}$ of transformations T on R_N, satisfying (3.1), such that the difference estimator is *unique* among estimators based on the sample mean and satisfying (3.2).

The difference estimator is often used in situations where the $\underline{a}$ vector is thought to have come from the same prior distribution as $\underline{x}$, except possibly for a translation by the same amount in each co-ordinate. This suggests requiring invariance of the estimator under the group $\underline{G}$ whose elements are given by

$$(T\underline{x})_i = x_i + d, \qquad (4.2)$$

(and thus $\Gamma_T(x) = x + Nd$). Our estimator ought to satisfy

$$e(s, \sum_{i \in s} (x_i + d)) = e(s, \sum_{i \in s} x_i) + Nd \qquad (4.3)$$

for every real d, which implies the form (as in the proof of (3.8))

$$e(s, \sum_{i \in s} x_i) = \frac{N}{n} (\sum_{i \in s} x_i) + e(s, 0). \qquad (4.4)$$

If we impose the further condition that $e(s, \sum_{i \in s} x_i)$ be *correct* at a, that is, that $e(s, \sum_{i \in s} a_i) = \sum_{i=1}^{N} a_i$, then e is the difference estimator of (4.1).

Alternatively, one's prior knowledge might suggest that, given the a vector, the quantities $x_i - a_i$ are independent and jointly distributed according to some class of distributions. As possible elements of G let us consider one-to-one transformations T of R_N onto R_N of the form $(Tx)_i = f(x_i, a_i)$, where $f(., a_i)$ is Borel measurable. Because of the nature of the prior knowledge, it seems natural to impose condition (3.1) in the following way, that there be a real-valued function ϕ_T such that

$$\sum_{i=1}^{N} f(x_i, a_i) - \sum_{i=1}^{N} a_i = \phi_T \left(\sum_{i=1}^{N} (x_i - a_i) \right). \tag{4.5}$$

for all $x, a \in R_N$.

Theorem 4.1. Functions f and ϕ_T satisfying (4.5) for all a have the forms

$$f(x_i, a_i) = cx_i - (c-1)a_i + d \qquad c \neq 0 \tag{4.6}$$

and

$$\phi_T \left(\sum_{i=1}^{N} (x_i - a_i) \right) = c \sum_{i=1}^{N} x_i - c \sum_{i=1}^{N} a_i + Nd. \tag{4.7}$$

Proof: Fix a, and set $x_2 = \ldots = x_N = 0$ in (4.5).

Then

$$f(x_1, a_1) + \sum_{i=2}^{N} f(0, a_i) = \phi_T \left(x_1 - \sum_{i=1}^{N} a_i \right) + \sum_{i=1}^{N} a_i$$

and since x_1 is arbitrary,

$$f(x_1 + x_2, a_1) + \sum_{i=2}^{N} f(0, a_i) = \phi_T \left(x_1 + x_2 - \sum_{i=1}^{N} a_i \right) + \sum_{i=1}^{N} a_i.$$

Setting $x_3 = \ldots = x_N = 0$ in (4.5) gives

$$f(x_1, a_1) + f(x_2, a_2) + \sum_{i=3}^{N} f(0, a_i) = \phi \left(x_1 + x_2 - \sum_{i=1}^{N} a_i \right) + \sum_{i=1}^{N} a_i.$$

Hence,

$$f(x_1 + x_2, a_1) - f(x_1, a_1) = f(x_2, a_2) - f(0, a_2)$$

$$= f(x_2, a_1) - f(0, a_1).$$

Then $g(x) = f(x, a_1) - f(0, a_1)$ satisfies the Cauchy functional equation

$$g(x_1 + x_2) = g(x_1) + g(x_2),$$

and it follows, since $f(x, a_1)$ is a Borel measurable function of x, that

$$f(x_1, a_1) = c(a_1)x_1 + \sigma(a_1).$$

249

Thus

$$f(x_i, a_i) = c(a_i)x_i + \sigma(a_i), \tag{4.8}$$

and equation (4.5) can now be written

$$\sum_{i=1}^{N} c(a_i)x_i - \sum_{i=1}^{N} a_i + \sum_{i=1}^{N} \sigma(a_i) = \phi_T(\sum_{i=1}^{N} (x_i - a_i)). \tag{4.9}$$

By comparing the form of (4.9) for $(x_i = 1, x_j = 0$ for $j \neq i)$ as i varies from 1 to N, we see that

$$c(a_1) = c(a_2) = \ldots = c(a_N) = \text{some constant } c \neq 0.$$

Finally, setting $x_i = a_i = a$ for all i in (4.9) gives $N(c-1)a + N\sigma(a) = \phi(0)$. If $\phi(0) = Nd$, $\sigma(a) = (1-c)a + d$, and we have (4.6).

Now fix a again and consider the group G of transformations of the form

$$(Tx)_i = cx_i - (c-1)a_i + d \qquad c \neq 0;$$

and suppose that the class C of priors of x is invariant under this group. This is equivalent to supposing that the class of priors of $z = x - a$ induced by the class C is invariant under the group H on the z space whose elements are given by

$$(Hz)_i = cz_i + d.$$

One way of applying the estimator invariance condition (3.2) is to consider $e(s, \sum_{i \in s}^{N} x_i)$ to be the sum of $\sum_{i=1}^{N} a_i$ and an estimator $e'(s, \sum_{i \in s} (x_i - a_i))$ of $\sum_{i=1}^{N} (x_i - a_i)$, and impose the invariance condition on e' corresponding to the group H. As in Section 3 this yields

$$e'(s, \sum_{i \in s} (x_i - a_i)) = \frac{N}{n} \sum_{i \in s} (x_i - a_i),$$

and e is the difference estimator.

5

Invariance Criteria Satisfied by the Ratio Estimator

Again let $a = (a_1, \ldots, a_N)$ be a fixed vector, and consider the ratio estimator

$$e(s, x_i, i \in s) = \sum_{i=1}^{N} a_i \frac{\sum_{i \in s} x_i}{\sum_{i \in s} a_i}. \tag{5.1}$$

The ratio estimator is often used in situations where the a vector is thought to have come from the same prior distribution as x, except possibly for a change of scale. This suggests requiring invariance under the group G whose elements are given by

$$(T\underline{x})_i = cx_i \qquad c \neq 0 \qquad (5.2)$$

so that $\Gamma_T(x) = cx$. If we assume that C, a class of distributions representing the prior knowledge, is invariant under G, then an estimator e (depending on the sample total) of τ ought to satisfy

$$e\left(s, \sum_{i \in s} cx_i\right) = ce\left(s, \sum_{i \in s} x_i\right) \qquad (5.3)$$

for every $c \neq 0$, and hence

$$e\left(s, \sum_{i \in s} x_i\right) = e(s, 1) \sum_{i \in s} x_i. \qquad (5.4)$$

Imposing the further condition that $e\left(s, \sum_{i \in s} x_i\right)$ be correct at $\underline{a}$ yields the ratio estimator (5.1).

Alternatively, one might know a priori that $\underline{a}$ is an observation on a (R_N-valued) random variable, and that conditioned on this random variable the quantities x_i / a_i, $i = 1, 2, \ldots, N$ are independent and distributed according to some class of distributions. Accordingly, as candidates for elements of a suitable group G of transformations on R_N, let us consider one-to-one transformations T of R_N onto R_N of the form

$$(T\underline{x})_i = f(x_i, a_i)$$

where $f(\,.\,, a_i)$ is Borel measurable. It seems natural to impose the condition, analogous to (4.5), that there be a real-valued function ϕ_T such that

$$\frac{\sum\limits_{i=1}^{N} f(x_i, a_i)}{\sum\limits_{i=1}^{N} a_i} = \phi_T\left(\frac{\sum\limits_{i=1}^{N} x_i}{\sum\limits_{i=1}^{N} a_i}\right) \qquad (5.5)$$

for all $\underline{x}, \underline{a} \in R_N$.

Theorem 5.1. Functions f and ϕ_T satisfying (5.5) for all $\underline{a}$ have the forms

$$f(x_i, a_i) = cx_i + ha_i \qquad c \neq 0 \qquad (5.6)$$

and

$$\phi_T\left(\sum_{i=1}^{N} x_i \bigg/ \sum_{i=1}^{N} a_i\right) = c\frac{\sum\limits_{i=1}^{N} x_i}{\sum\limits_{i=1}^{N} a_i} + h. \qquad (5.7)$$

Proof: Using the same argument as in the proof of (4.8), we see that

$$f(x_i, a_i) = cx_i + \sigma(a_i), \qquad c \neq 0. \tag{5.8}$$

Thus (5.5) can be written

$$\sum_{i=1}^{N}(cx_i + \sigma(a_i)) \Big/ \sum_{i=1}^{N} a_i = \phi_T\Big(\sum_{i=1}^{N} x_i \Big/ \sum_{i=1}^{N} a_i\Big). \tag{5.9}$$

Let $x_i = a_i = a \neq 0$ for all i. Then

$$(Nca + N\sigma(a))/Na = \phi_T(1), \text{ and } \sigma(a) = \phi_T(1)a - ca,$$

$$\text{or } \sigma(a) = ha \text{ where } h = \phi_T(1) - c.$$

Now fix $\underline{a}$ again and consider the group $\underline{G}$ of transformations of the form

$$(T\underline{x})_i = cx_i + ha_i \qquad c \neq 0;$$

and suppose that the class C of priors of $\underline{x}$ is invariant under this group. This is equivalent to supposing (assuming $a_i \neq 0$ for all i) that the class of priors of $\underline{z} = (x_1/a_1, \ldots, x_N/a_N)$ induced by the class C is invariant under the group $\underline{H}$ on the $\underline{z}$ space whose elements are given by $(H\underline{z})_i = cz_i + h$.

Again in analogy with the argument of the previous section, we suppose $e(s, x_i : i \in s)$ to be the product of $(\sum_{i=1}^{N} a_i)$ and an estimator $e'(s, \sum_{i \in s} x_i/\sum_{i \in s} a_i)$ of $\sum_{i=1}^{N} x_i/\sum_{i=1}^{N} a_i$. We then impose the invariance condition corresponding to the group $\underline{G}$ on e', and this will take the form

$$e'(s, h + c \sum_{i \in s} x_i / \sum_{i \in s} a_i) = h + ce'(s, \sum_{i \in s} x_i / \sum_{i \in s} a_i). \tag{5.10}$$

From equation (5.10) it is immediate that

$$e'(s, \sum_{i \in s} x_i / \sum_{i \in s} a_i) = \sum_{i \in s} x_i / \sum_{i \in s} a_i,$$

and hence e is the ratio estimator.

6
Unbiassedness and the Role of Sampling Designs

One may note that the foregoing analysis is entirely independent of the sampling procedure (1.1) adopted by the statistician. It has often been said (recent references being Joshi (see Godambe, 1968) and Ericson, 1969) that artificial randomization, that is, a sampling procedure (1.1), is adopted to protect oneself against errors in some basic assumptions. In the same spirit we here say that one may introduce different sampling procedures to make the estimators, obtained in previous sections as appropriate on considerations (such as Bayes sufficiency and invariance) of prior knowledge, unbiassed in the following sense:

Definition 6.1. Given a sampling procedure (1.1), an estimator e (Definition 1.1) is said to be unbiassed for the population total τ if

$$\sum_{s \in S} e(s, x_i : i \in s) p(s) = \tau(\underset{\sim}{x})$$

for all $\underset{\sim}{x} \in R_N$.

It is easy to see that the estimator based on the sample mean, namely (3.9), and the difference estimator (4.1) are unbiassed (Definition 6.1) for the sampling procedure:

$$P(s, x_i' : i \in s \mid \underset{\sim}{x}) = \begin{cases} 1/{}^N C_n \text{ if } x_i' = x_i \text{ for all } i \in s \text{ and if} \\ \qquad s \text{ contains } n \text{ individuals } i. \\ 0 \text{ otherwise.} \end{cases} \qquad (6.1)$$

The above sampling procedure is evidently equivalent to simple random sampling without replacement and with a fixed $(=n)$ number of draws. Again the ratio estimator (5.1) is unbiassed (Definition 6.1) for the sampling procedure:

$$P(s, x_i' : i \in s \mid \underset{\sim}{x}) = \begin{cases} \sum_{i \in s} a_i {}^{N-1}C_{n-1} \sum_{i}^{N} a_i \text{ if } x_i' = x_i \text{ for all } i \in s \text{ and } s \\ \qquad\qquad\qquad \text{ contains } n \text{ individuals } i. \\ 0 \text{ otherwise.} \end{cases} \qquad (6.2)$$

The Definition 6.1 above of unbiassedness in a certain sense implies *even* treatment to all parametric points $\underset{\sim}{x} \in R_N$. In this sense the use of proper sampling procedures provides the statistician some protection in case *uneven* treatment given by him to different parametric points $\underset{\sim}{x} \in R_N$ (through Bayes and invariance considerations) was based on his *faulty* prior knowledge.

7
Reversing the Roles of Unbiassedness and Prior Knowledge

In the foregoing discussion we *first* applied considerations of *a priori* knowledge to obtain in some sense appropriate estimators and *afterwards* in Section 6 resorted to unbiassedness (Definition 6.1) to find appropriate sampling procedures. The following remarks suggest that the results remain unchanged even if the order of application of the considerations of *a priori* knowledge and unbiassedness is reversed.

Let C_2 be a class of prior distributions ξ on R_N, defined as follows:

$$C_2 = \left\{ \xi : \begin{array}{l} \text{(i) } x_1, \ldots, x_N \text{ when distributed according} \\ \quad \text{to } \xi \text{ are probabilistically independent.} \\ \\ \text{(ii) For some } \alpha, \int x_i d\xi = \alpha + a_i, i = 1, \ldots, N; a_i, \\ \qquad i = 1, \ldots, N \text{ being some } \textit{specified} \text{ numbers.} \\ \\ \text{(iii) } \int x_i^2 d\xi < \infty, i = 1, \ldots, N. \end{array} \right\} \qquad (7.1)$$

[*Note:* Except for the trivial condition (iii) the class C_2 above differs from the class C in Definition 2.1 and Theorem 2.1 in so far as it imposes the additional restriction (ii) on the prior ξ. This additional restriction, as Theorem 7.1 (to follow) will indicate, corresponds to the *invariance* arguments in Section 4.] Now for any unbiassed estimator $e(s, x_i : i \in s)$ (Definition 6.1) the variance $V(e, \underline{x})$ is given by

$$V(e, \underline{x}) = \Sigma_s [e(s, x_i : i \in s) - \tau(\underline{x})]^2 p(s), \qquad \underline{x} \in R_N \qquad (7.2)$$

where as before $\tau(\underline{x})$ is the population total and the $p(s)$ are the probabilities given by the sampling procedure (1.1). We state without proof the

Theorem 7.1. If the sampling procedure is given by (6.1) then in the class of all unbiassed estimators (Definition 6.1) for τ the expectation

$$\int V(e, \underline{x}) \, d\xi$$

of the variance (7.2) is minimized for the difference estimator e in (4.1) for every $\xi \in C_2$ in (7.1).

In other words Theorem 7.1 says that *when the sampling procedure is given by (6.1), then in the class of all unbiassed estimators for τ, the difference estimator (4.1) is Bayes (with respect to the squared error as loss) for every prior distribution $\xi \in C_2$ in (7.1).*

It is easy to see that as a special case of Theorem 7.1 we obtain the corresponding theorem for the estimator (3.9) based on the sample mean. The proof of Theorem 7.1 will appear in a subsequent paper. It seems likely that a corresponding result may be true for the ratio estimator (5.1) also, at least in some asymptotic sense.

8
Conclusions

In many situations the statistician's prior knowledge about the survey-population is so crude as to render the usual Bayesian-fiducial-frequency oriented assumptions highly unrealistic. Specifically unrealistic are the following assumptions for such situations:

(a) The survey-population vector $\underline{x}$ has a specific prior distribution or in other words the survey-population has come from a specific super-population (Ericson, 1969).

(b) The survey-population has come from a superpopulation having a known form of distribution involving an unknown parameter (Kalbfleisch and Sprott, 1969).

(c) The survey-population vector $\underline{x} = (x_1, \ldots, x_i, \ldots, x_N)$ itself is such that x_i's (variate values associated with different individuals i of the population) have a parametrized distribution of known form but unknown parameter (Godambe, 1969, p. 250).

In such situations therefore it is impossible to compute on the basis of the sample observations $(s, x_i : i \in s)$ any Bayes posterior distribution or fiducial or confidence intervals for the population total τ.

Yet, in the same situations of rather crude prior knowledge described above the estimator (3.9) based on the sample mean, the difference estimator (4.1) and the ratio estimator (5.1) (and possibly their appropriate extensions to stratified sampling situations and the like) have appealed to the mature statistical intuition of practitioners. These estimators also have shown some desirable frequency properties. The present paper is a formal study of possible motivations of the use of these estimators.

References

1. Barnard, G.A., Jenkins, G.M. and Winsten, C.B., "Likelihood Inference in Time Series," *Journal of the Royal Statistical Society, A,* 125, 321-372, 1962.
2. Ericson, W.A., "Subjective Bayesian Models in Sampling Finite Populations," *J. Roy. Statist. Soc., B.,* 31, 195-233, 1969.
3.*Godambe, V.P., "Bayesian Sufficiency in Survey-Sampling," *Annals of the Institute of Statistical Mathematics,* 20, 363-373, 1968.
4. Godambe, V.P., "A Fiducial Argument with Applications to Survey-Sampling," *J. Roy. Statist. Soc., B,* 31, 246-260, 1969.
5. Kalbfleisch, J.D. and Sprott, D.A., "Applications of Likelihood and Fiducial Probability to Sampling Finite Populations," *New Developments in Survey-Sampling,* John Wiley and Sons, 358-389, 1969.
6. Plackett, R.L., "The Principle of Arithmetic Mean," *Biometrika,* 45, 130-135, 1958.

COMMENTS

D. J. Bartholomew:

The authors' remark on page 254 that some of their estimators, derived on Bayesian arguments, are unbiased in the frequency sense if a particular sampling rule is adopted. This raises the question similar to that discussed in my own paper, of whether there is any meaningful correspondence between the Bayesian family of priors leading to the estimators and the sampling rule which makes those estimators unbiased. For example, could it be said that when a frequentist thinks it right to draw a simple random sample he is saying the same thing as the Bayesian who declares his intention to adopt the author's model? I think it can and I am grateful for a further example to add to my limited collection.

D. R. Cox:

Does the result (1.2) that the posterior distributor is independent of the sampling rule apply if so-called length-biased sampling is used? In this $p(s)$ of (1.1) is a function of the x's. It would be disturbing if the analysis of a set of numbers were identical for both simple random sampling and for length-biased sampling.

*Some of the results in reference (3) above are duplicated in the present paper for the sake of continuity of the discussion.

NOTE ADDED SUBSEQUENTLY: I now see that the results of the paper do *not* apply to length-biased sampling, and hence there is no 'paradox'. Although formally the authors' $p(s)$ could be a function of the x-values of the individuals in s without affecting their $(1.1) \circ (1.2)$, such a dependence is not possible. After renormalization the probabilities would depend also on the x-values of individuals not in s. It might at some point be interesting to extend the analysis to cover length-biased sampling.

J. D. Kalbfleisch:

In Section 8, the authors list as "specifically unrealistic" the assumptions made by Ericson (1969), Kalbfleisch and Sprott (1969) and Godambe (1969). I willingly grant the third of these but I would like to comment on the other two.

The assumption made by Kalbfleisch and Sprott (1969) was that the variate values associated with the units of a finite population were themselves random realizations from one of a family of hypothetical populations indexed by a parameter vector θ. There are several situations where such an assumption would seem realistic; for example, in a finite population of heights it would seem realistic to assume that within a stratum the heights are independent realizations of a process which could be described by a normal distribution. It is important that use be made of such information since otherwise inferences about the finite population will be unnecessarily weak. I can see no reason why such an assumption is more unrealistic here than in experimental design when similar measurements are being made.

Ericson (1969) makes the additional assumption that the parameter vector has a known distribution $p(\theta)$ and applies Bayesian methods to the problem. There is an important difference between this additional assumption and the assumption of a super population which is somewhat cloaked by the tendency of calling both prior distributions in the survey sampling literature. The assumption of a hypothetical distribution with an unknown parameter can, and should be, tested by standard statistical procedures. It should be used only so long as it describes the variation in the data collected. There would seem to be no way of applying similar testing procedures to the distribution $p(\theta)$.

D. V. Lindley:

(This was not given in the discussion but only communicated in writing for publication in the *Proceedings*.)

The role of the likelihood function in sampling from a finite population appears to need clarification, especially in view of the comments, amongst others, by Cox and Zelen. My opinion is that the basic idea is that of informative sampling described by Raiffa and Schlaifer.

In explanation it is convenient to use Rao's notation. Let the finite population be described (i, θ_i), $1 \leq i \leq N$, where the first argument indexes the unit and the second describes the characteristic of that unit. Let X_i be the observation on θ_i made if unit i is selected. (In Godambe and Thompson's notation $X_i = \theta_i$ and the probability distributions which follow are degenerate.) Let $s = (i_1, i_2, \ldots, i_n)$ be the sample selected, containing units i_k, $1 \leq k \leq n$. Then in Cox's example of the fibre sampling the probability of s depends on θ and we should write, not $p(s)$ but $p(s \mid \theta)$. This possibility does not appear to be allowed for in the authors' formulation (nor in that of Basu). Having selected the units, their values are observed and we need $p(X \mid s, \theta)$ where $X = (X_{i_1}, X_{i_2}, \ldots, X_{i_n})$.

Consequently, if $p(\theta)$ is the prior distribution for θ, Bayes theorem shows that the posterior distributions given both the units and their values is

$$p(\theta \mid X, s) \propto p(X \mid s, \theta) p(s \mid \theta) p(\theta).$$

If $p(X \mid s, \theta)$ is the likelihood function we see that the likelihood function contains all the information if, and only if, $p(s \mid \theta)$ does not depend on θ. This special case is called, by Raiffa and Schlaifer non-informative sampling. In general the sampling is informative.

A numerical example may be informative. Let $N = 3$ and each $\theta_i = 1$ or 2. Let $p(\theta)$ assign equal chances to all $8 = 2^3$ possibilities. Let $n = 1$ and $p(s = i \mid \theta) = \theta_i / \sum_j \theta_j$. (The chance of selection is proportional to the true value θ_i, to imitate Cox's example). Suppose unit 1 is selected, and then X_1 observed without error (so that $X_1 = \theta_1$) with the result $X_1 = 1$. The table below lists the prior probabilities, the posterior probabilities given only the knowledge of the unit selected (namely 1), and finally the posterior probabilities given also the observation on the unit. It is seen that there is information provided by s alone and the sampling is informative. In fact $s = 1$ suggests that θ_1 is 2, rather than 1.

θ	$p(\theta)^\S$	$p(\theta \mid s=1)^\S$	$p(\theta \mid s=1, X_1=1)^\S$
(1,1,1)	20	20	52
(1,1,2)	20	15	39
(1,2,1)	20	15	39
(2,1,1)	20	30	0
(1,2,2)	20	12	31
(2,1,2)	20	24	0
(2,2,1)	20	24	0
(2,2,2)	20	20	0

($\S$ all probabilities have been multiplied by 160 for ease in comparison).

J. Neyman:

The following remarks concern three points:

(i) First, I wish to compliment Professor Thompson on her excellently organized and lucid presentation of the paper. Here are my best wishes for continuation in the profession which, I suspect, she joined only recently.

(ii) While agreeing with Professor Cox on many issues, this time I think he did not quite get the sampling schemes envisaged in the papers under discussion. As I understand it, the population sampled contains N members numbered $i = 1, 2, \ldots, N$. Any member is equally available as any other and there are $\binom{N}{n} = M$ available samples symbolized by s_k, with $k = 1, 2, \ldots, M$. The sampler selects M probabilities, say p_k, and proposes to select one sample of n units by a random process governed by these probabilities.

This scheme does not correspond to sampling of fibers, some long and some short. The latter problem is analogous to that of sampling galaxies, some bright and some faint, some nearby and others far away. In our Berkeley terminology, the problem of fibers, as well as the problem of galaxies, involves the so-called selection probabilities.

(iii) My third point is a question which I wish to address to Professors Godambe and Thompson. The question refers to the class of priors they consider. For the sake of clarity I will formulate the question with reference to a concrete example.

Suppose that the population sampled is the set of households in an American city. Suppose that the first r households (out of the total N) are located "on the wrong side of the tracks" and the remainder, beginning with the $r+1^{st}$, on the right side. Finally, suppose that the object of the contemplated survey is to estimate τ the total income of the N households. My question is: Am I right in thinking that the class of priors contemplated does not contain the particular prior that the first r households are "on the wrong side of the tracks"? With this particular prior, the precision of the final estimate of the total income would depend on the sampling strategy (S,p). In olden days I used to work on this problem.

REPLY

We are grateful for all the comments on our paper.

We agree with Professor Bartholomew's interpretation of the relationships between prior distributions and sampling plans.

Professor Cox is right in his feeling that the sampling probabilities are not uninformative in "length-biased" sampling, as Professor Lindley's example shows. It should perhaps be noted that there do exist (sequential) sampling plans in which the probability of selecting sample s depends on x, but only through $(x_i : i \in s)$. For example, if the population is $[1,2,3,4]$, one may choose to observe x_1 and x_2, then x_3 if $x_1 \geq x_2$ and x_4 if $x_1 < x_2$. Then

$$p(1,2,3) = 1 \text{ if } x_1 \geq x_2$$
$$= 0 \text{ if } x_1 < x_2.$$

Such sampling plans were not specifically included in our formulation; however it is easy to see that the posterior distribution of $\underline{x}$ given an observation from such a plan is again independent of the sampling probabilities. The set of individuals s in this case is informative about $\underline{x}$, but only about those co-ordinates x_i for which $i \in s$. In contrast, in Professor Lindley's example the selection of element 1 is informative *not only* about x_1 *but also* about x_2 and x_3. Thus the remarks of Professor Cox are indeed valid for Professor Lindley's example.

Some of the remarks we have made above also apply to the comment (ii) of Professor Neyman. About his comment (iii) it must be admitted that our formulation does *not* deal with stratification. Some observations concerning stratification have appeared in a paper by Godambe (1968). (See References at the end of the paper.)

LINEAR REGRESSION MODELS IN FINITE POPULATION SAMPLING THEORY

Richard M. Royall
The Johns Hopkins University

Summary

An important part of finite population sampling theory is examined under certain linear regression models. The populations considered are of a kind frequently met in practice: associated with the ith unit of the population are two numbers, x_i and y_i, with only x_i known. The problems studied are those of choosing a sample of units, estimating the population total of the y's, and providing a useful measure of the uncertainty in this estimate.

In a previous work [12] both least-squares analysis under a probability model and empirical calculations were shown to suggest that for producing an estimate close to the true population total, standard procedures which entail random sampling plans are often inferior to strategies which call for purposive (non-random) selection of samples. For example, when conventional practice would employ the ratio estimator with a simple random sampling plan, the use of the ratio estimator with a certain purposive sampling plan frequently will produce much better results. The regression models also provide a useful framework for studying certain properties of estimators. The main results in [12], some of which have also been derived by Scott and Smith [13], are sketched here in Section 3.

The problem of constructing, from an observed sample, a measure of the uncertainty in a given estimate of the population total is considered in Section 4. The least-squares approach to this problem produces variance estimates which do not depend on the rule used in deciding which units to include in the sample. These variance estimates can thus be used with both purposive and random sampling plans. From consideration of the ratio estimator, it is argued that even when the sample is selected by a simple random sampling plan, the variance estimate derived from the regression model frequently provides a more appropriate indication of the uncertainty in a given estimate of the population total than is provided by the conventional variance esti-

259

mate. Some empirical support for this conclusion is presented in the Appendix.

The results suggest that finite population sampling theory and practice should put greater emphasis on (*super-population*) probability models, and that there is a need for more careful analysis of when and why artificial randomization should be used in the selection of samples.

1
General Introduction

Perhaps the most striking feature of conventional finite population sampling theory is the fact that the only stochastic component is deliberately introduced and controlled by the sampler. Although *super-population* models have been studied as limiting cases, as the population sizes increase indefinitely, the probability distribution with respect to which such characteristics as bias and standard error of estimates are ordinarily calculated, and with respect to which *large sample* approximations are derived, is that determined by the sampler when he specifies the probabilities of selection for all possible samples. Since no assumptions concerning unknown probability distributions are required, the results obtained appear to be, in some sense, *non-parametric* or *robust*.

Despite its apparent freedom from probabilistic assumptions, which seems at first glance to be an attractive feature, the conventional theory has some serious shortcomings, the most conspicuous of which is its sheer intractability. Two notorious examples of this are:

1. the surprising complications encountered in the study and execution of *probability-proportional-to-size* sampling plans and
2. the awkwardness of almost all probabilistic calculations concerning estimates of ratios.

(Cochran, [1, page 157], wrote that "The distribution of the ratio estimate has proved annoyingly intractable...") More disturbing than its inelegance is the general infertility of the conventional theory. This is not to imply that practitioners do not have quite an assortment of useful tools, but that in the study and application of these tools there is precious little theory to formalize and guide the judgment and intuition of the sampler. Two comments are offered in support of this statement:

1. There is a notable lack of attainable optimality criteria comparable to those which play a useful role in the rest of statistics. Godambe's [4] demonstration, for the case of simple random sampling, of the non-existence of a best (minimum variance) estimator within the class of linear unbiased estimators deserves mention here. In fact, with squared error loss function, it is hard to find any statistic which is not admissible in the class of unbiased estimates of its expected value (Godambe and Joshi [6]).
2. The likelihood function seems to tell the statistician more about the limitations of the conventional model than about *what the data have to say* (Godambe [5], Hartley and Rao [7], Royall [11]).

The fact that conventional sampling theory has some serious shortcomings (in addition to the many finite population versions of more general problems) is widely recognized. Much of the current dissatisfaction with the theoretical foundations of conventional finite population sampling theory has been stimulated by Godambe's 1955 paper [4]. Lahiri [9] has given, from a practitioner's viewpoint, a clear presentation of some of the inadequacies of the conventional theory. Of course there have been many contributions aimed at improving and enriching the conventional theory without tampering with its essential characteristic: the central role of the probability distribution created by the sampler. Recently there have also been contributions, notably by Ericson [3] (Bayesian) and by Kalbfleisch and Sprott [8] (likelihood-fiducial), in which a probabilistic superstructure (super-population model) has been erected atop the conventional model and then systematically exploited. These works differ from the many previous efforts invoking similar (in some instances, identical) super-population models in that their inclusion of such models as integral components of the theory, rather than as optional limiting cases to be studied mainly for their suggestive value, leads to a radical change in emphasis. The probability distributions comprising the super-population models usurp the dominant position of the distribution determined by the sampler. In fact, one of the most important functions of probability sampling, in the contexts of these theories, appears to be that of affording protection against certain types of departure from the super-population model (see [8, Section 8], and [3]).

Although conventional sampling theory has carefully avoided allowing super-population models to play any essential role, Kalbfleisch and Sprott conclude that "It is frequently realistic and *in a sense necessary* to assume a hypothetical population from which the finite population is drawn at random." [Italics are mine.] The present work can be interpreted as an attempt to amplify and support this conclusion. In particular, it is an attempt to show that even if attention is restricted to such notions as linear estimation, bias, variances, and confidence intervals (as opposed to examination of Bayes posterior or fiducial distributions) hypothetical population models are useful and *in a sense necessary* in many situations from which they have been omitted in the conventional theory.

2
Specific Problems

Consider a population of N units labelled $1, 2, \ldots N$. Associated with unit i are two numbers (x_i, y_i) with x_i known and y_i fixed, but unknown. Attention will be confined to the problems of choosing a sample of units, estimating the total $T = \sum_{i=1}^{N} y_i$, and supplying an appropriate indication of the estimate's precision. The sample is selected as follows: a (non-empty) subset s of the labels $1, 2, \ldots, N$ is selected according to some probability function p defined on the collection S of all such sets; the sample consists of those units whose labels are in the selected set s. The set of labels of units not in the sample will be denoted by $\bar{s}$. The sample size is assumed to be fixed, say n; i.e., if $n(s)$ is the number of different labels in s, $p(s) > 0$ only if $n(s) = n$. The function p, which will be called the sampling plan, cannot depend on the unknowns $y_1, y_2, \ldots, y_N$. The estimator can be any real-valued statistic not a function of the unobserved y-values. For a sampling plan p and an estimator $\hat{T}$, the pair $p: \hat{T}$ will be called a strategy.

An estimator $\hat{T}$ is conventionally defined to be unbiased, with respect to a given sampling plan p, if whatever $y_1, y_2, \ldots, y_N$,

$$\sum_{S} p(s) \hat{T} = T.$$

Such an estimator will be called *p-unbiased* here. In conventional theoretical studies the p-bias, $\sum_{S} p(s) (\hat{T} - T)$, and the sampling mean-square error, $\sum_{S} p(s) (\hat{T} - T)^2$, are taken as important performance characteristics of the strategy $p: \hat{T}$.

A class of populations of some practical importance can be characterized roughly as those with scatter diagrams in which $x_1, \ldots, x_N$ are all positive, the N points (x_i, y_i) are concentrated about a straight line passing through the point $(0, 0)$, and the scatter of points about the line increases with increasing x.

The problems of choosing a sampling and estimation strategy and a measure of precision for use in such populations will be studied under the following popular model: The numbers $y_1, \ldots, y_N$ are realized values of independent random variables $Y_1, Y_2, \ldots, Y_N$ where Y_i has mean βx_i and variance $\sigma^2 v(x_i)$. The function v is known, with $v(x) > 0$ for $x > 0$; the constants β and σ^2 are unknown. The joint probability law of $Y_1, Y_2, \ldots, Y_N$ will be denoted by ξ.

The problem area chosen for consideration here is general enough to illustrate what seem to be the essential differences between the present *least-squares predication* approach and the conventional one, is important enough to have received much attention from the conventional point of view, and is realistic enough to permit meaningful empirical comparisons using actual populations. Strategies for point estimation were considered in the previous work [12], the main results of which are sketched in the next section. New work on the problem of providing an indication of the uncertainty in a point estimate is presented in Section 4 and in the appendix.

3
Strategies for Estimation

After the sample is selected and observed, any estimator $\hat{T}$ of T can be expressed as

$$\hat{T} = \sum_{s} y_i + \hat{\beta} \sum_{\bar{s}} x_i.$$

This equation defines $\hat{\beta}$, which will be called the implied estimator for β. The error in the estimate is

$$T - \hat{T} = \sum_{\bar{s}} (y_i - \hat{\beta} x_i);$$

thus the uncertainty in $\hat{T}$, as an estimate of T, is precisely the uncertainty in $\hat{\beta} \sum_{\bar{s}} x_i$ as an estimate of $\sum_{\bar{s}} y_i$. Apparently, for a given sample s, a good estimator for T is one whose implied $\hat{\beta}$ is such that $\hat{\beta} \sum_{\bar{s}} x_i$ is a good *predictor* for the total of the unobserved random variables, $\sum_{\bar{s}} Y_i$. From this point of view it seems natural to call an estimator unbiased, for a given sample s, if $\hat{\beta} \sum_{\bar{s}} x_i$ is an unbiased estimator of the expected value of $\sum_{\bar{s}} Y_i$, that is, if

$$E_\xi (T - \hat{T}) \equiv E_\xi (\sum_{\bar{s}} Y_i - \hat{\beta} \sum_{\bar{s}} x_i) = 0 \qquad (1)$$

where E_ξ denotes expected value with respect to the probability law ξ of $Y_1, Y_2, \ldots, Y_N$. An estimator which satisfies (1) for all s with $n(s) = n$ will be called ξ-*unbiased*. Note that ξ-unbiasedness is defined with reference only to ξ and does not depend on the rule p used in deciding which units should be observed. Clearly $\hat{T}$ is ξ-unbiased if and only if the implied estimator $\hat{\beta}$ is ξ-unbiased for estimating β, i.e. $E_\xi \hat{\beta} = \beta$. An estimator can obviously be ξ-unbiased but not p-unbiased and vice-versa.

In the previous work [12] a sampling and estimation strategy p: $\hat{T}$ was described as *better than* another strategy p': $\hat{T}'$ if

$$MSE\ (p:\hat{T}) < MSE\ (p':\hat{T}')$$

where $MSE\ (p: \hat{T})$ denotes the expected (with respect to ξ) *sampling mean-square error* of the strategy $p:\hat{T}$,

$$MSE\ (p: \hat{T}) = E_\xi \left[\sum_s p(s)(\hat{T} - T)^2 \right]. \qquad (2)$$

In [12] it was shown (Theorem 1) that under the model ξ, if $\hat{\beta}^*$ is the weighted least-squares estimator of β,

$$\hat{\beta}^* = \sum_s \frac{x_i y_i}{v(x_i)} \Big/ \sum_s \frac{x_i^2}{v(x_i)},$$

then for any sampling plan a best linear ξ-unbiased estimator of T is

$$\hat{T}^* = \sum_s y_i + \hat{\beta}^* \sum_{\bar{s}} x_i.$$

Specifically, if $\hat{T}$ is any linear (in the y's) ξ-unbiased estimator and p is any sampling plan then

$$MSE\ (p:\hat{T}^*) \leq MSE\ (p:\hat{T})$$

with strict inequality unless $\hat{T} = \hat{T}^*$ with probability one (ξ) for all s such that $p(s) > 0$. This is a direct consequence of familiar results in linear regression and prediction theory. Thus when $Y_1, Y_2, \ldots, Y_N$ are independent random variables with common mean and variance (the case when $x_i = x_j$ for all $i, j = 1, 2, \ldots, N$), N times the sample mean is a best linear ξ-unbiased estimator of T.

Somewhat more generally, in the case $v(x) = x$ the best linear ξ-unbiased estimator is

$$\hat{T}_1 = \sum_s y_i + \left(\sum_s y_i \Big/ \sum_s x_i\right) \sum_{\bar{s}} x_i$$

$$= \left(\sum_s y_i \Big/ \sum_s x_i\right) \sum_{i=1}^{N} x_i,$$

the familiar *ratio estimator*. However, when the model ξ with $v(x) = x$ seems relevant and a simple random sampling plan is used, conventional sampling theory (a) claims optimality for $\hat{T}_1$ only when N is infinite, (b) suggests that the estimator is good in cases of more modest N, and (c) is generally preoccupied with the p-bias and distributional problems associated with the fact that if a different set of units had been observed, the denominator of $\hat{\beta}$ would probably have been different. (Ref. Cochran [1, Section 6.9])

Procedures in which the probability π_i that unit i is in the sample $(\pi_i = \sum_{\{s: i \in s\}} p(s))$ is proportional to x_i and the Horvitz-Thompson estimator,

$$\hat{T}_{HT} = \frac{1}{n}\left(\sum_s \frac{y_i}{x_i}\right)\sum_{i=1}^{N} x_i,$$

is used have received much attention. Frequently x_i is a measure of the size of unit i (e.g. the units are farms and x_i is the acreage of farm i), so sampling plans with π_i proportional to x_i will be referred to as *probability proportional to size* (pps) sampling plans. Since $\sum_{i=1}^{N} \pi_i$ is the sample size n and no π_i can exceed 1, a $p_{pps}: \hat{T}_{HT}$ strategy exists only if

$$nx_i \leq \sum_{j=1}^{N} x_j \tag{3}$$

for all $i = 1, 2, \ldots, N$. Note that when (3) is satisfied there are generally many sampling plans which select samples of the required size and which have the required inclusion probabilities—p_{pps} refers to any one of these. Since the sample size is fixed, $\hat{T}_{HT}$ is both p_{pps}-unbiased (i.e. p-unbiased with respect to any pps sampling plan) and ξ-unbiased.

It has been shown that $p_{pps}: \hat{T}_{HT}$ procedures are optimal under the model ξ with $v(x) = x^2$ if p-unbiasedness is demanded (Godambe [4], Godambe and Joshi [6]). That is, under ξ with $v(x) = x^2$,

$$MSE(p_{pps}: \hat{T}_{HT}) \leq MSE(p: \hat{T})$$

for any sampling plan p with fixed sample size n and any p-unbiased estimator $\hat{T}$. However, Theorem 1 in [12] shows that the estimator

$$\hat{T}_2 = \sum_s y_i + \frac{1}{n}\left(\sum_s \frac{y_i}{x_i}\right)\sum_{\bar{s}} x_i,$$

which is ξ-unbiased but not p_{pps}-unbiased, satisfies

$$MSE(p_{pps}: \hat{T}_{HT}) \geq MSE(p_{pps}: \hat{T}_2). \tag{4}$$

Furthermore, this last inequality is shown in [12] to hold, not only for pps sampling plans, but for any fixed-size sampling plan and any variance function v such that $v(x)/x^2$ is non-increasing. Inequality (4) comes as no surprise but as a relief to one disturbed by the *optimality*, under the model ξ, of an estimator with

$$\hat{\beta}_{HT} = \sum_s y_i \left(\frac{\sum_{j=1}^{N} x_j}{n x_i} - 1 \right) \bigg/ \sum_{\bar{s}} x_i$$

as its implied estimator for β.

Since the order of the two averaging operations in (2) can be reversed, it is obvious that for fixed-sample-size strategies involving a given estimator $\hat{T}$ there is an optimal sampling plan (not necessarily unique) which is purposive (non-random). If, among samples s of size n, the minimum value of $E_\xi(\hat{T}-T)^2$ is achieved at s^*, then $MSE(p:\hat{T})$ is obviously minimized, for the given $\hat{T}$ and sample size n, by a sampling plan which entails selection of s^* with certainty. If $v(x)$ is non-decreasing, $v(x)/x^2$ is non-increasing, and the optimal estimator $\hat{T}^*$ is to be used, then it can easily be shown that $E_\xi(\hat{T}^*-T)^2$ is minimized by any s which maximizes $\sum_{\bar{s}} x_i$. If p^* denotes a sampling plan under which such an s is selected with certainty, then

$$MSE(p^*:\hat{T}^*) \leq MSE(p:\hat{T}^*)$$

for all p such that $p(s) > 0$ only if $n(s) = n$ [12, Theorem 5].

Thus, although the linear least-squares prediction approach does point to the popular ratio estimator as optimal under a model which appears to represent many actual populations quite adequately, it also points to a quite unconventional sampling plan as optimal for use with this statistic.

Some empirical comparisons, using actual populations for which complete data, (x_i, y_i), $i = 1, 2, \ldots, N$, have been published, support these findings. J. N. K. Rao [10] reported sampling mean-square errors of numerous estimators in sixteen populations under *simple random sampling* (srs) with various sample sizes. The estimators considered by Rao ranged from the classical ratio and regression estimators to various estimators produced by attempts to modify the simple ratio estimator so as to eliminate or reduce its p-bias.

The procedure of using the purposive sampling plan p^* and the ratio estimator was applied to the same populations and for the same sample sizes considered by Rao. For each sample size and each population for which Rao reported results (61 cases) the squared error of the purposive sampling procedure was compared [12, Table 2] to the minimum sampling mean-square error among the ten or so estimators considered by Rao. Of the 61 ratios,

$$\frac{\text{minimum sampling } MSE \text{ among } srs \text{ procedures}}{\text{squared error using } p^*:\hat{T}_1 \text{ procedure}}$$

8 were less than 1 while 52 exceeded 1. (There was one tie.) The minimum ratio was 1/3.7; the maximum was over 1000.

Uncertainty in Estimates

Although the expected sampling mean-square error may be of interest before the sample is selected, under the present model it is not generally appropriate as a measure of the uncertainty in an estimate constructed from an observed sample. The conventional measure of this uncertainty is the sampling mean-square error,

$$\sum_S p(s)\,(\hat{T}-T)^2 = \sum_S p(s)\left(\sum_s (y_i - \beta x_i)\right)^2 \tag{5}$$

where $y_1, y_2, \ldots, y_N$ are held fixed. The present approach suggests that a more appropriate measure of the uncertainty in $\hat{T}$ is the mean-square error, with respect to the distribution of $Y_1, Y_2, \ldots, Y_N$, with the sample s held fixed:

$$E_\xi(\hat{T}-T)^2 = E_\xi\left(\sum_s (Y_i - \beta x_i)\right)^2 \tag{6}$$

In the case of $Y_1, Y_2, \ldots, Y_N$ independent and identically distribution with mean β and variance σ^2 ($x_i = 1, i = 1, 2, \ldots, N; v(1) = 1$) the estimator $\hat{T}^*$ is the simple expansion estimator, $N \sum_s y_i/n$, and (6) becomes

$$\frac{N^2}{n}\left(1 - \frac{n}{N}\right)\sigma^2 \tag{7}$$

when the sample size is n, regardless of the sampling plan. Although all of the $\binom{N}{n}$ samples of size n lead to the same value (7) of $E_\xi(\hat{T}^* - T)^2$, which suggests that among plans for choosing a sample of a given size no one is better than another, there seem to be good reasons for simple random selection. (See e.g. [8, Section 8].) If such a plan is adopted, the expansion estimator is also the conventional choice, and (5) is

$$\frac{N^2}{n}\left(1 - \frac{n}{N}\right)\left(\frac{1}{N-1}\right)\sum_{i=1}^{N}\left(y_i - \frac{T}{N}\right)^2 \tag{8}$$

In practice both of these quantities are unknown and must be estimated. The estimate conventionally used for (8) is

$$\frac{N^2}{n}\left(1 - \frac{n}{N}\right)\left(\frac{1}{n-1}\right)\sum_s\left(y_i - \frac{\sum_s y_j}{n}\right)^2,$$

which is also the ξ-unbiased estimate of (7) based on the residual sum of squares.

More generally, however, for a given sampling plan and estimator the two approaches lead to different measures of uncertainty. The case of the ratio estimator will be considered in some detail. Whatever sampling plan was used in selecting s,

$$E_\xi(\hat{T}_1 - T)^2 = \sigma^2\left[\sum_s v(x_i) + \left(\sum_s x_i\right)^2 \frac{\sum_s v(x_i)}{(\sum_s x_i)^2}\right]. \tag{9}$$

If $v(x)=x$ (this is the condition under which $\hat{T}_1$ is optimal) then (9) reduces to

$$\sigma^2 \sum_{i=1}^{N} x_i \left(\sum_{\bar{s}} x_i \Big/ \sum_{s} x_i \right).$$ (10)

The fact that (10) decreases as $\sum_{\bar{s}} x_i$ increases is consistent with what intuition demands of the behavior of a measure of the uncertainty in $\hat{T}_1$ as an estimate of T in a population with scatter diagram as described in Section 3. For a given n, the greatest uncertainty is assigned by (10) to the sample s'' which minimizes $\sum x_i$, and the least is assigned to the sample s' which maximizes this sum. Since s' involves selection of those units which tend to deviate most from the line of concentration of the N points (x_i, y_i), after the units with labels in s' are observed, the values of those y's which are least predictable are *known*, and the problem remaining is to predict the sum $\sum_{\bar{s}} y_i$ for which prediction is easiest. Also, the sample provides the most stable estimate of the slope of the line. Thus it seems appropriate that the minimum of (10) is attained at s'. Similar considerations suggest that the estimate arising from the sample s'' is, as (10) indicates, the most uncertain of all estimates based on samples of the same size. Of course the preceding comments can be given precise expression in terms of the model ξ with $v(x)=x$. The point here, however, is that the usefulness of the least squares prediction approach is more closely related to how well the model *fits* the scatter diagram than to the degree of difficulty one encounters in imagining an infinite super-population from which the actual finite population was drawn or in conceiving of a stochastic process which generated the values $y_1, y_2, \ldots, y_N$.

A ξ-unbiased estimate of σ^2 is that based on the weighted squared residuals:

$$\hat{\sigma}^2 = \left[\sum_s (y_i - \beta_1 x_i)^2 / x_i \right] \Big/ (n-1)$$

where $\beta_1 = \sum_s y_i \Big/ \sum_s x_i$. If σ^2 in (10) is replaced by $\hat{\sigma}^2$, the result seems to be a useful, if crude, indicator of the after sampling uncertainty in $\hat{T}_1$ as an estimate of T.

Under the conventional theory, with a simple random sampling plan, the ratio estimator is not quite p-unbiased, but the p-bias becomes small as the sample size increases (under appropriate mild conditions) and is neglected when n is *large*. The measure of uncertainty used (at least for large n) is

$$V_0 = \frac{N^2}{n} \left(1 - \frac{n}{N}\right) \left(\frac{1}{N-1}\right) \sum_{i=1}^{N} \left(y_i - Rx_i\right)^2$$ (11)

where $R = \sum_{i=1}^{N} y_i \Big/ \sum_{i=1}^{N} x_i$. Note that all samples s are assigned the *same* measure of uncertainty. According to this criterion none of the possible samples provides a less uncertain estimate of T than does any other. The situation becomes even less satisfactory when the usual estimate

$$V_1 = \frac{N^2}{n} \left(1 - \frac{n}{N}\right) \left(\frac{1}{n-1}\right) \sum_s \left(y_i - \beta_1 x_i\right)^2,$$ (12)

267

is substituted for (11). If the population has a scatter diagram like that described earlier, this statistic will tend to be large when s is such that $\sum_s x_i$ is large, indicating greatest uncertainty when uncertainty is actually least. It will tend to be smallest when the sampler has the great misfortune of choosing the n units having the smallest x's. (If $v(x) = x$, the ξ-expectation of (12) is increased when for any i in s, x_i is replaced by a larger value.)

The conventional confidence interval,

$$\hat{T}_1 \pm z_\alpha \, V_1^{1/2}, \tag{13}$$

is supported by large sample approximations which show that, with simple random sampling, under certain mild conditions on the N points (x_i, y_i), for sufficiently large n, the distribution of $(\hat{T}_1 - T)/V_1^{1/2}$ is approximately normal with mean zero and variance one. Thus under these assumptions on the population values and sample size, the *validity* of (13) is established without reference to a super-population model.

By contrast the interval,

$$\hat{T}_1 \pm z_\alpha \, V_2^{1/2}, \tag{14}$$

where

$$V_2 = \hat{\sigma}^2 \sum_{i=1}^N x_i \left(\sum_{\bar{s}} x_j \Big/ \sum_s x_j \right),$$

is validated by large-sample theory under the model ξ without reference to artificial randomization. Consider the situation in which new units are added to the population and some of these new units are added to the sample. Let s_N denote the sample (actually the set of labels of units in the sample) when population consists of N units, and let $\bar{s}_N$ denote the complement of s_N in the set $\{1, 2, \ldots, N\}$. Formally, the sequence of samples $s_1, s_2, \ldots$ is such that

(a) $s_N \subset s_{N+1}$

(b) $\bar{s}_N \subset \bar{s}_{N+1}$ and

(c) both $n(s_N)$ and $n(\bar{s}_N)$ increase without limit as N increases.

Now under the model ξ the statistic

$$\sum_s \left(y - \beta x_i \right) \Big/ \left(\sum_s v(x_i) \right)^{1/2}$$

has, for sufficiently large N, an approximate normal probability distribution with mean zero and variance σ^2, provided that the sequence of distributions of the random variables Y_i, i in s_N, satisfy the Lindeberg-Feller conditions. In the important special case of $v(x) = x$, if say $E_\xi (Y_i - \beta x_i)^4$ is proportional to x_i^2 and the x's have finite positive upper and lower bounds, then the statistic $\sum_s y_i / \sum_s x_i$ has, for large N, an approximate normal distribution with mean β and variance $\sigma^2 / \sum_s x_i$. If the same argument and conditions are applied to the random variables Y_i, i in $\hat{s}_N$, it is seen that the error

$$\hat{T} - T = \sum_s x_i \left[\left(\sum_s y_i \Big/ \sum_s x_i \right) - \left(\sum_s y_i \Big/ \sum_s x_i \right) \right]$$

has, for large N, an approximate normal probability distribution with mean zero and variance given by (10), and since, under the present model and assumptions, $\hat{\sigma}^2$ is a consistent estimator of σ^2, $(\hat{T} - T)/V_2^{1/2}$ has an approximate normal distribution with mean zero and variance one.

Thus both (13) and (14) are *valid* confidence intervals, but their validities are established with reference to different probability distributions: if the same population is sampled repeatedly, then, under some conditions on n, N, and the actual x's and y's, the relative frequency with which a simple random sampling plan will select an s such that the interval (13) contains T will be approximately $1 - \alpha$; for a given s, under some conditions on n, N, and the probability distribution (ξ) of $Y_1, Y_2, \ldots, Y_N$, the probability with which the interval (14) will contain T is approximately $1 - \alpha$.

Since for a given sample the two intervals can be quite different, it is important to determine which of the two is, under given conditions, the more appropriate for after-sampling analysis of data. Under the model ξ, with $v(x) = x$, (14) seems clearly to be preferable, on the basis of both (i) general arguments regarding the appropriateness of conditional frames of reference for evaluation of observed data (See e.g. Cox [2].) and (ii) analysis of the way in which the statistics V_1 and V_2 tend to vary from sample to sample in the population. Some empirical comparisons are presented in the appendix.

Although important questions regarding the robustness of (14) remain to be answered, it appears that in many problems super-population models are necessary tools for use not only in finding an efficient sampling plan and estimator but also in representing the uncertainty in an observed estimate.

Bibliography

1. Cochran, W. G., *Sampling Techniques,* New York, Wiley, 1963.
2. Cox, D.R., "Some Problems Connected With Statistical Inference," *Annals of Mathematical Statistics,* 29, 357-72, 1958.
3. Ericson, W.A., "Subjective Bayesian Models in Sampling Finite Populations," *Journal of the Royal Statistical Society B,* 31, 195-224, 1969.
4. Godambe, V. P., "A Unified Theory of Sampling for Finite Populations," *Journal of the Royal Statistical Society, B,* 17, 269-224, 1955.
5. Godambe, V.P., "A New Approach to Sampling from Finite Populations I," *Journal of the Royal Statistical Society B,* 28, 310-19, 1966.
6. Godambe, V. P., and Joshi, V. M., "Admissibility and Bayes Estimation in Sampling Finite Populations I," *Annals of Mathematical Statistics,* 36, 1707-22, 1965.
7. Hartley, H. O., and Rao, J. N. K., "A New Estimation Theory for Sample Surveys," *Biometrika,* 55, 547-57, 1968.

8. Kalbfleisch, J.D., and Sprott, D. A., "Applications of Likelihood and Fiducial Probability to Sampling Finite Populations," in *New Developments in Survey Sampling,* N. Johnson and H. Smith, editors, New York, Wiley-Interscience, 1969.

9. Lahiri, D. B., "On the Unique Sample, the Surveyed One," paper presented at the Symposium on Foundations of Survey Sampling in Chapel Hill, N. C., April, 1968.

10. Rao, J. N. K., "Ration and Regression Estimators," in *New Developments in Survey Sampling,* N. Johnson and H. Smith, editors, New York, Wiley-Interscience, 1969.

11. Royall, R. M., "An Old Approach to Finite Population Sampling Theory," *Journal of the American Statistical Association,* 63, 1269-79, 1968.

12. Royall, R. M., "On Finite Population Sampling Theory Under Certain Linear Regression Models," Paper number 459, Department of Biostatistics, The Johns Hopkins University, 1969.

13. Scott, A., and Smith, T.M.F., "Estimation in Multi-Stage Surveys," *Journal of the American Statistical Association,* 64, 830-40, 1969.

APPENDIX

Some Empirical Results

Under the model ξ with $v(x) = x$ the linear least-squares prediction theory described here points to the ratio estimator as optimal. When this model applies and the ratio estimator is used, samples s in which $\sum_s x_i$ is small tend to produce greater errors of estimation than do samples in which $\sum_s x_i$ is large. This fact should be reflected in any function which is to be used to describe the after-sampling uncertainty in a ratio estimate. In particular the theory suggests that:

(a) The optimal sample consists of the n units whose x-values are largest, and the worst possible sample consists of the n units whose x-values are smallest. These extreme samples will be referred to as HI and LO respectively.

(b) The variance V_0 conventionally associated with the ratio estimator under simple random sampling is inappropriate as a measure of after-sampling uncertainty, because it fails to reflect the fact that uncertainty decreases as $\sum_s x_i$ increases. This indicates that even if V_0 were known, confidence intervals based on V_0 would be too narrow when the LO sample is selected and too wide when HI is selected. Although V_0 is only an approximation to the true variance, these comments also apply to the true variance.

(c) The conventional variance estimate V_1 tends to overestimate V_0 in samples containing mostly large x's and to underestimate V_0 in samples containing mostly small x's. Thus in LO samples approximate confidence intervals based on V_1 will tend to be even narrower than those based on V_0, which themselves are too narrow (and conversely for HI samples).

(d) The variance estimate V_2 calculated from the weighted least-squares residuals represents an appropriate measure of after-sampling uncertainty. If both n and N-n are large, nominal $100(1-\alpha)\%$ confidence intervals based on the limiting normal distribution of $Z_1 = \hat{T}_1 - T/V_2^{\frac{1}{2}}$ will contain the actual population total T approximately $100(1 - \alpha)\%$ of the time in LO samples as well as in HI samples.

To what extent these statements apply in practice is questionable on the grounds that no real population of interest is generated strictly in accordance with the model (or, from a subjective point of view, on the grounds that one's knowledge of the population is rarely if ever described precisely by the model). The same sixteen populations studied in [10] and [12] were examined in an attempt to determine the degree to which these four conclusions derived under the model might be consistent with empirical results. On the basis of brief descriptions of the sixteen populations (Table 1) the model (ξ with $v(x) = x$) would appear to be at least plausible. However, inspection of the sixteen scatter diagrams reveals considerable diversity.

For $n = 2, 4, 8, 12,$ and 16 both LO and HI samples were selected from each population. For each sample size the quantity V_0 was calculated. For each sample the actual error in the ratio estimate was determined and the statistics V_1 and V_2 were evaluated. The results for sample sizes 4, 8, and 16 are summarized in Table 2. In this table the three quantities Z_0, Z_1, and Z_2 are the ratios of the actual error $\hat{T}_1 - T$ to the *standard error* $V_0^{\frac{1}{2}}$, to the estimated standard error $V_1^{\frac{1}{2}}$, and to $V_2^{\frac{1}{2}}$ respectively. The results for sample sizes 2 and 12 were consistent with those for the sample sizes reported. For $n = 2$ the results were somewhat more erratic than for $n = 4$.

In some cases, because of ties among the x-values, more than one sample qualified as LO or HI. For example, in population 11 the fifteen smallest x's were all equal; thus there were $\binom{15}{4}$ different LO samples of size 4. In cases in which there were no more than 10 of these samples they were all analyzed. If, as in population 11 for $n = 4$ LO, there were more than 10 possible samples, 10 of these were selected at random. In the cases in which more than one sample was examined, the absolute error and the three ratios $|Z_0|$, $|Z_1|$, and $|Z_2|$ were calculated for each sample and their average values reported in Table 2. These entries are enclosed in parentheses with subscripts used to indicate the number of samples. For example, in population 11 the average value of $|Z_1|$ in the ten samples selected at random from the $\binom{15}{4}$ possible $n = 4$ LO samples was 5.41. Thus the Table 2 entry is $(5.41)_{10}$.

These results seem to be generally consistent with (a)-(d). The HI samples tend to yield more accurate estimates than LO samples. The statistics V_1 and V_2 behave as predicted; in most LO samples $|Z_1| > |Z_0| > |Z_2|$, indicating that $V_1 < V_0 < V_2$, and for HI samples the reverse inequalities tend to hold.

The infinite values of Z_1 and Z_2 in population 13, $n = 4, 8, 16$ LO, and in population 14, $n = 4$ LO indicate that the n sample points (x, y) lie precisely on a straight line passing through the origin; thus for these samples $V_1 = V_2 = 0$.

Table 1. Description of the Populations.   *(From Rao [10] Table 1)*

Population number	Source	x	y	N
1	Sukhatme (1954) p. 165	mean agric. area	mean no. livestock	7
2	Cochran (1963) p. 156	size of city in 1920	size of city in 1930	49
3	Cochran (1963) cities 25-49	size of city in 1920	size of city in 1930	25
4	Cochran (1963) cities 1-24	size of city in 1920	size of city in 1930	24
5	Cochran (1963) p. 183	no. in not inoculated group	no. polio cases — not inoculated group	34
6	Cochran (1963) p. 183	no. in placebo group	no. polio cases — placebo group	34
7	Cochran (1963) p. 204	eye-estimated wt. of peaches	actual wt. of peaches	10
8	Cochran (1963) p. 325	no. rooms in block	no. persons in block	10
9	Sukhatme (1954) p. 183	area under wheat — 1936	area under wheat — 1937	34
10	Sukhatme (1954) p. 183	cultivated area — 1931	area under wheat — 1936	34
11	Sukhatme (1954) p. 279	no. villages in circle	area under wheat in circle	89
12	Sampford (1962) p. 61	crops and grass acreage — 1947	oats acreage — 1957	35
13	Kish (1965) p. 625	no. dwellings	dwellings occupied by renters	270
14	Kish (1965) p. 42	no. dwellings	dwellings occupied by renters	20
15	Yates (1960) p. 163	eye-estimated vol. of timber	measured vol. of timber	25
16	Yates (1960) p. 159	no. persons absent from kraal	total no. persons in kraal	43

References for Table 1

1. Cochran, W. G., *Sampling Techniques,* New York, Wiley, 1963.
2. Kish, L., *Survey Sampling,* New York, Wiley, 1965.
3. Sampford, M. R., *An Introduction to Sampling Theory,* Edinburgh, Oliver and Boyd, 1962.
4. Sukhatme, P. V., *Sampling Theory of Surveys With Applications,* Iowa State College Press, 1954.
5. Yates, F., *Sampling Methods for Censuses and Surveys,* 3rd edition, London, Griffin, 1960.

Table 2. Errors and "Normalized Errors" (Z_0, Z_1, Z_2) of Ratio Estimates in Extreme (LO and HI) Samples.

Pop'n	N	n	Error LO	Error HI	Z_2 LO	Z_2 HI	Z_0 LO	Z_0 HI	Z_1 LO	Z_1 HI
1	7	4	66.8	-18.3	1.22	-1.05	1.46	-0.40	5.25	-0.32
2	49	4	6853	-241	0.65	-1.50	11.72	-0.41	9.48	-0.36
		8	4843	-221	1.12	-1.65	12.27	-0.56	10.13	-0.50
		16	$(2231)_2$	-383	$(1.20)_2$	-3.84	$(8.91)_2$	-1.53	$(9.40)_2$	-1.40
3	25	4	1820	-210	1.57	-1.74	5.38	-0.62	4.54	-0.57
		8	$(1002)_2$	-289	$(1.82)_2$	-4.39	$(4.66)_2$	-1.34	$(5.31)_2$	-1.69
		16	570	-157	2.04	-2.64	5.15	-1.42	5.77	-1.55
4	24	4	3183	107	0.57	1.14	13.89	0.46	9.34	0.37
		8	1010	-87	0.50	-1.10	6.97	-0.60	6.02	-0.42
		16	199	-114	0.28	-3.66	2.75	-1.57	3.18	-1.49
5	34	4	-63.9	-28.5	-1.52	-2.72	-1.67	-0.75	-7.35	-0.63
		8	-26.8	-13.8	-0.79	-1.02	-1.06	-0.55	-2.92	-0.36
		16	6.8	1.5	0.29	0.18	0.46	0.10	0.94	0.07
6	34	4	-8.3	-9.1	-0.18	-0.84	-0.24	-0.26	-0.90	-0.23
		8	10.5	-9.6	0.21	-1.28	0.45	-0.42	0.93	-0.40
		16	$(16.3)_2$	4.6	$(0.73)_2$	0.56	$(1.20)_2$	0.34	$(2.79)_2$	0.24
7	10	4	-18.7	15.1	-2.23	1.67	-1.86	1.51	-2.78	1.36
		8	-2.0	7.0	-0.41	2.21	-0.50	1.70	-0.49	1.86
8	10	4	54.2	7.5	0.85	0.21	1.14	0.16	1.00	0.18
		8	9.1	-38.2	0.38	-2.80	0.47	-1.97	0.45	-2.46
9	34	4	132	-372	0.12	-0.41	0.14	-0.41	0.54	-0.13
		8	319	-238	0.46	-0.61	0.53	-0.39	1.75	-0.20
		16	241	-25	0.83	-0.15	0.68	-0.07	2.62	-0.05
10	34	4	223	291	0.13	0.77	0.15	0.20	0.70	0.31
		8	-538	46	-0.73	0.18	-0.55	0.04	-2.91	0.08
		16	157	117	0.31	0.26	0.27	0.20	0.70	0.15
11	89	4	$(123765)_{10}$	$(13954)_3$	$(1.58)_{10}$	$(0.63)_3$	$(4.87)_{10}$	$(0.55)_3$	$(5.41)_{10}$	$(0.23)_3$
		8	$(105926)_{10}$	-12218	$(2.07)_{10}$	-1.01	$(6.03)_{10}$	-0.70	$(7.18)_{10}$	-0.39
		16	$(91344)_{10}$	$(12729)_{10}$	$(2.69)_{10}$	$(1.60)_{10}$	$(7.75)_{10}$	$(1.08)_{10}$	$(9.22)_{10}$	$(0.71)_{10}$
12	35	4	261	-320	1.34	-2.55	0.99	-1.21	4.42	-1.01
		8	$(90)_2$	-148	$(0.56)_2$	-1.27	$(0.52)_2$	-0.85	$(1.77)_2$	-0.51
		16	96	$(39)_2$	0.72	$(0.68)_2$	0.93	$(0.38)_2$	1.95	$(0.28)_2$
13	270	4	$(4559)_{10}$	831	$(\infty)_{10}$	2.10	$(5.64)_{10}$	1.03	$(\infty)_{10}$	0.43
		8	$(4559)_{10}$	834	$(\infty)_{10}$	2.63	$(8.03)_{10}$	1.47	$(\infty)_{10}$	0.58
		16	$(4559)_{10}$	$(720)_3$	$(\infty)_{10}$	$(3.36)_3$	$(11.54)_{10}$	$(1.82)_3$	$(\infty)_{10}$	$(0.81)_3$
14	20	4	-255.0	25.1	$-\infty$	1.52	-6.75	0.67	$-\infty$	0.33
		8	-160.4	18.8	-3.67	2.44	-6.94	0.82	-28.87	0.56
		16	-34.0	3.0	-1.84	1.51	-3.60	0.31	-5.26	0.28
15	25	4	1558	-960	1.46	-1.46	2.09	-1.29	5.03	-0.83
		8	1620	-392	3.69	-0.92	3.42	-0.83	9.70	-0.54
		16	$(391)_2$	$(360)_5$	$(1.46)_2$	$(2.26)_5$	$(1.62)_2$	$(1.48)_5$	$(2.47)_2$	$(1.32)_5$
16	43	4	-181.0	45.2	-1.73	0.22	-0.90	0.26	-4.56	0.11
		8	118.7	-4.6	0.57	-0.05	0.88	-0.03	1.24	-0.02
		16	0.5	32.8	0.004	0.52	0.006	0.39	0.007	0.29

For $j = 0, 1, 2$ the interval $\hat{T}_1 + kV_j^{\frac{1}{2}}$ contains the actual total T if and only if $|Z_j| \leq k$. For all n and N the Tchebychev inequality guarantees that no more than approximately 11% of the $\binom{N}{n}$ possible samples yield values of $|Z_0|$ greater than 3. (The stated percentage is approximate because the quantity V_0 is an approximation to the mean square error of $\hat{T}_2$ under simple random sampling.) The data suggest that when a simple random sampling plan is used, even this extremely conservative bound does not justify a high degree of confidence that the interval $\hat{T}_1 \pm 3V_0^{\frac{1}{2}}$ contains T when the sample is known to contain a preponderance of units with small x-values. The problem here is, of course, the old one of determining the appropriate probability distribution to use in calculating confidence coefficients. Here the confidence coefficient apparently should be conditioned on the x-values in the sample. Except for some special cases, if the only probability distribution considered is the before-sampling uncertainty concerning which units will be selected for observation, then this approach leads nowhere — the statistician finds that he has conditioned himself right out of the problem. The model considered in this paper seems to provide an appropriate conditional frame of reference for calculation of approximate confidence coefficients, and the results in Table 2 indicate that confidence intervals based in the approximate normality of Z_2 might indeed provide useful, if crude, indicators of the after-sampling uncertainty in $\hat{T}_1$ as an estimate of T.

One explanation of whatever measure of success the interval $\hat{T}_1 \pm kV_1^{\frac{1}{2}}$ may enjoy in practice is suggested by the model. For many populations it can be shown that in samples in which the x-values are well-scattered through the population, the two statistics V_1 and V_2 tend to be approximately equal. Thus in such samples the conventional interval $\hat{T}_1 \pm kV_1^{\frac{1}{2}}$ is approximately the same as $\hat{T}_1 \pm kV_2^{\frac{1}{2}}$.

The *optimality,* in *practical* applications of the ratio estimator, of the sampling plan which selects the *HI* sample with certainty has not been demonstrated, although in the previous empirical study [12] this sampling plan seemed to be clearly superior to simple random sampling in most of the populations examined. The results in Table 2 above suggest that the use of any sampling plan which assigns positive probability to samples s for which $\sum_s x_i$ is extremely low is highly questionable. There are many sampling plans under which performance of the ratio estimator (and of the statistic V_2) is less sensitive to certain departures from the model than it is under the *HI* plan. There are also many estimators which are less sensitive than the ratio statistic to departures from the model. Investigation of these alternative sampling plans and estimators is needed.

Acknowledgment

The valuable assistance of Mr. Jay Herson, who did most of the work summarized in Table 2, is gratefully acknowledged.

274

COMMENTS

D.R. Cox:

There is a parallel with results in experimental design, but there, the extreme designs arising under strong assumptions need to be used very cautiously. It would be interesting to know how much loss of precision there is if some of the sample is used in checking non-linearity. The author's suggestion of concentrating on the individuals with large x values is akin to the common practice of forming a stratum from the large x values and sampling this intensively or completely.

V.P. Godambe:

I do not know why one should restrict to linearity in Gauss-Markoff set-up. Surely linearity is a purely mathematical constraint having no justification or even meaning in terms of statistical concepts.

I.J. Good:

It might be useful to compromise between the uses of systematic and random sampling by a method that is not simply stratified sampling. For example, consider the integration of a complicated but calculable function by sampling (which, as Dr. Basu said, is essentially equivalent to other sampling problems). We could first approximate $\int_0^1 f(x)dx$ by means of a standard integration formula, chosen to be optimal under certain simplified assumptions; and we could in addition use random sampling of x values to detect unexpected departures from the assumptions, such as whether $f(x)$ is smooth enough. (The random sampling could be stratified.)

J.C. Koop:

First I should like to make some remarks on the speaker's two "notorious examples" of the intractability of conventional finite population sampling theory.

In one example, he complains about "the surprising complications encountered in the study and execution of probability-proportional-to-size sampling plans". The execution of such plans, in my view, is not difficult. In the Dominion Bureau of Statistics we have recently completed selection of sets of first-stage units and sets of second-stage units, for the Post Census Survey of Agriculture in 1971 by Lahiri's technique without any difficulty in the randomization procedures involved. The survey is to cover almost the entire country. The probability calculations involved were simple and were done manually. If Dr. Royall means calculations such as those required for the Horvitz-Thompson estimator, namely

$$P_i = \sum_{s \supset i} p(s) \text{ and } P_{ij} = \sum_{s \supset i,j} p(s),$$

surely they can be done by modern electronic computers; the formulae themselves suggest how the machine should go about making the computations.

Regarding his second example on "the awkwardness of almost all probabilistic calculations concerning estimates of ratios" I must say that such calculations can be avoided by changing the strategy of estimation in the following way. Instead of taking a single sample to estimate a ratio, take, for example, five independent replicated samples whose total size is about equal to the size of the intended single sample. If $R_1, \ldots, R_5$ are the ratio estimates, each based on the respective independent replicates (the formula depending on the underlying sampling design), and if R_1 and R_5 are the smallest and largest estimates in the set, then the argument by Savur (*Proc. Indian Academy, Sc., A5*, 1938, 564) applies, that is,

$$P\{R_1 < R_m < R_5\} = 1 - (\tfrac{1}{2})^4 = .9375,$$

where R_m is the median of the distribution of the ratio estimates. If the universe value R under estimation is "close" to R_m this argument is quite satisfactory. Otherwise, the argument in general for k independent estimates can be revised as follows:

$$\begin{aligned} P\{R_{\text{smallest}} < R < R_{\text{largest}}\} &= 1 - [(\tfrac{1}{2} - \delta)^k + (\tfrac{1}{2} + \delta)^k] \\ &= 1 - (\tfrac{1}{2})^{k-1} + 0(\delta^2), \end{aligned} \tag{1}$$

where $\tfrac{1}{2} + \delta$ is the cumulative frequency up to R $(-\tfrac{1}{2} < \delta < \tfrac{1}{2})$.

An approximation to the probability on the right-hand side of (1) will be fairly adequate, even when k is 5 or 6 and δ is small. This will be $1 - (\tfrac{1}{2})^{k-1}$.

Similar probability statements can also be made using other estimators, including those considered in the paper. A characteristic feature of this argument is that (limiting) normality assumptions and variance calculations are not in any way involved.

Finally, exact expressions for the expected value and variance of ratio estimates have been derived without the use of infinite series expansions and will appear shortly in *Metrika*.

J. Neyman:

In attempting to formulate my comments for publication, I must begin by presenting my apologies for what I said at the conference. At the time I did not read Professor Royall's paper. From what I heard, I formed the impression that Professor Royall claimed original authorship of the regression analysis formulae which he wrote on the blackboard. In this connection I mentioned that I learned these and related formulae sometime in the 1920's, I think from a book written by M. Ezekiel. Upon returning to the hotel, I glanced at Professor Royall's paper and found that he is well informed of the least squares literature and hasten to present my deep regrets for what I said.

On my own part, I must confess that of late I drifted away from sampling surveys: my last paper on the subject "Contribution to the Theory Sampling Human Populations" was published in the *J. Am. Stat. Assoc.* sometime in 1938 or so. Thus, I am not familiar with what Professor Royall describes as "standard" or "conventional" procedures. Also, I am in full agreement with him that, when circumstances warrant, the use of regression analysis is strongly indicated.

Now, a few remarks that occurred to me *after* I read the paper.

In approaching the problem of empirical study of a population, any population (whether of humans, or of mice or of galaxies, etc.), it appears desirable to arrange that the outcome of the study does not depend upon anything pre-conceived, which may not be inherent in the population studied. In other words, I find it desirable to use procedures, the validity of which does not depend upon the uncertain properties of the population. In the example treated in the paper, where the population studied is that of farms in a locality, with x_i re-presenting the known acreage of the ith farm, the uncertain property of the population is the linearity of regression of the studied Y (perhaps income) on the known x. In fact, Professor Royall appears ready to act on the further as-sumption that the regression line in question passes through the origin of the co-ordinates. *Ex post facto*, it may well be found that the two hypotheses are (approximately) satisfied. However, to arrange the sampling scheme the valid-ity of which depends on these (or any other) unverified hypotheses, seems dangerous.

My own limited experience in social and economic studies suggests that the regression of income on the size of a farm may well be non-linear and that it might curve up as x is increased (otherwise, there would be no tendency for small farms to disappear in favor of larger ones). At any rate, the possibility of some such thing happening cannot be excluded *a priori*. If this be so, then the design of the survey favored by Professor Royall could lead to an unpleasant bias: The inclusion in the sample of predominantly large farms would result in an overestimate of the total farm income. Furthermore, if the sample does not contain any of the really small farms, it would be impossible to establish the existence of the bias in any *ex post* analysis.

The remedy seems to be in something intermediate between what Professor Royall recommends and what he condemns. This intermediate procedure con-sists in stratifying the sample according to the available values of x: so many, say n_1 farms with acreage between the lowest a_0 and a convenient a_1, so many, say n_2, with acreage between a_1 and a_2, etc. A population of farms worth studying through a sample survey will contain a reasonable number N_i (perhaps 100) of units between a_{i-1} and a_i. Out of these a *random sample* of a predetermined number n_i should be taken for purposes of the survey. By this method, the validity of the survey will not depend upon any preconceived idea regarding the properties of the population studied. The regression analysis can be applied and is likely to be useful.

There are quite a few facets to the theory involved, which I prefer to develop without the assumption that each observable y_i is a particular value of a ran-dom variable connected with the idea of a "superpopulation". In other words, for me y_i is the actual income of the ith farm and all the randomness that I like to consider is the "man made" randomness introduced through random sampl-ing. These various facets of the situation are just a little too long to be described in the present comments and all I must add is that the least squares regression line (linear!) of y's on midpoints of the intervals (a_{i-1}, a_i) deserves particular attention.

Just one point more. Somewhere in his paper Professor Royall insists on the preference for "purposive selection" as contrasted with random sampling. My hope is that what he means is really an appropriate stratification of the sample, somewhat like described above. However, once the sample (and the population!) have been stratified in one way or another, the actual selection of units out of each stratum should be strictly random, perhaps using tables of "random numbers". If the table used is anywhere reasonable (there are quite a few of them now) the statistician will be protected from all sorts of biases that invariably occur when random sampling is dispensed with. At any rate, it was my personal experience covering quite a few cases that frivolity with random sampling invariably led to unnecessary unpleasantnesses. One of them is described in some detail in my first paper on the subject, published in the *J. Roy. Stat. Soc.* of 1934.

Before concluding I wish Professor Royall all success in questioning the "conventional" or traditional methodology and in his efforts to see how it works. This is the spirit of Behaviouristic Theory of Statistics!

REPLY

I thank the speakers for their comments and suggestions.

Many of the comments, both those which appear in print and those voiced at the symposium, express reservations which I share. I see now that these reservations should have received greater emphasis in my paper. The comments to which I refer are those of the form "Yes, but what if such-and-such conditions do not hold?" I enthusiastically agree that the questions of robustness are of utmost importance. They *must* be answered before one takes the important step from determination of theoretically optimal strategies under certain assumptions to recommendation of a particular strategy as being one which represents, for a specific application, a reasonable compromise between optimality under the most plausible assumptions and validity under departures from these assumptions. These questions have not yet received adequate attention. This is especially true of questions of the sort mentioned by Professor Cox concerning the use of part of the sample to check for non-linearity.

For the problem involving farm acreages discussed by Professor Neyman, I did not intend either to *recommend* the extreme purposive sampling plan or to *condemn* simple random sampling — I am sorry if I gave that impression. For the particular response variable (income) chosen by Professor Neyman I surely would not use any strategy which could not be expected to perform satisfactorily if a reasonable degree of curvilinearity is present. Among the factors which must be considered are population size, sample size, and the distribution of acreages in the specific population of interest.

Of course, naive application of super-population models can easily lead to disastrous errors. If the strategy which is optimal under the most likely set of conditions is clearly inappropriate under other conditions which might apply, then a compromise strategy is called for. It is the familiar problem of deciding how much insurance to buy. Here the price of the insurance policy is paid in units of efficiency, and super-population models seem to be useful tools for suggesting sound policies and for analyzing their benefits and costs. Such an analysis is not difficult and can help one to decide, for a particular application, whether the attractive policy favored by Professor Neyman is a better investment than one of the numerous alternatives which are available.

On the other hand, naive application of a formula whose "validity" is "guaranteed" by the statistician's act of randomization can also lead to serious errors. Some examples appear in the appendix. Frivolity with any aspect of statistical activity, from formulating a problem mathematically to drawing a graph, can be expected to lead to unnecessary unpleasantness.

I am somewhat confused by Dr. Koop's remarks, since they seem to me to support, rather than to refute, my statement concerning pps sampling plans and the estimation of ratios.

Professor Godambe's question is an interesting one. I am not aware of a statistical concept which justifies restriction to linearity in the Gauss Markov set-up, but one practical justification which is sometimes mentioned is the fact that when a linear estimate is used, there is a good chance that its distribution (which is unknown) can be approximated well enough for most purposes by the normal probability law and that a reasonably good estimate of its variance can be obtained easily.

APPLICATIONS OF STATISTICAL INFERENCE TO PHYSICS
J. Orear and D. Cassel
Cornell University

In our talk we will briefly summarize the present experimental techniques used in nuclear and high energy physics. On-line computers and high density data storage permit one to record multidimensional events in their original n-dimensional space at a high rate. A typical example is the reaction

$$\pi^- + P \rightarrow W^0 + \Delta^0$$
$$W^0 \rightarrow \pi^+ + \pi^- + \pi^0$$
$$\Delta^0 \rightarrow P + \pi^-.$$

So what is observed is $\pi^- + P \rightarrow (\pi^+ + \pi^- + \pi^0) + (P + \pi^-)$.

The 3 vector momentum and energy of each of the five final particles is recorded (each event is then a point in 20 dimensional space). A very common procedure in the analysis of such data is to calculate the effective mass of a certain combination of final state particles. Then one obtains a distribution of several hundred or several thousand mass values which will have local clusterings as a result of statistical fluctuations and sometimes as a result of a real physical effect. When a new real effect is discovered, it is usually due to a new elementary particle.

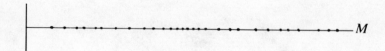

Figure 1.

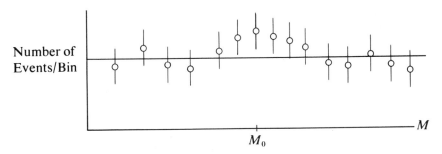

Number of Events/Bin

M_0

M

Figure 2.

Bump hunting is one of the major current activities of experimental physicists. A typical experiment will produce a list of N mass values which could be plotted as points on a mass scale (See Figure 1). Arbitrary bins could be chosen and experimental points could be plotted as shown in Figure 2. Usually each bin will contain over 100 events. Then one is safe in assuming the points are Gaussian distributed and applying Chi-squared tests to the least squares sum. But this practice of bump-hunting raises several questions — most of which we do not know the answers. We appeal to this symposium for help in answering our questions. Some questions:

1

If there is no a priori evidence of a particle at mass M_0, but one finds a large bump at mass M_0, is there any statistical procedure which can be used to judge the significance of that bump? Or can one only use the first experiment to establish the hypothesis of a bump at M_0, and then do a second experiment to test the new hypothesis? If this is all one is permitted to do, it seems somewhat wasteful of information in the first experiment.

2

Shortcomings of Chi-squared tests. We note that in Figure 2 there are 6 successive points lying above the best fit straight line. This correlation of adjacent points is lost in making a Chi-squared test to the no bump hypothesis. Is there any statistical test which will not throw away this kind of information? Figure 2 points will give a reasonably good Chi-squared fit to the straight line, but yet by our eye, the bump looks like a more serious fluctuation than described by the Chi-squared probability (for example, the probability of getting on the same side of the line 6 times in a row is not taken into account). Choosing wider bins makes some use of the adjacent bin correlation, but shape information is then thrown away.

3

Some physics distribution functions in angle, mass, or energy have zeros or near zeros. Then some of the bins will contain small numbers of events which may no longer be approximated as Gaussian distributed. We know how to make the best fit to a set of Poisson distributed points, but we do not know any goodness-of-fit test. Is there any?

4

This raises the more general question of goodness-of-fit tests. In general, the maximum likelihood method can be used to find the best fit to a set of N

experimental events (or N bins of events as discussed above). One will then obtain a numerical value $\mathscr{L}^*$ for the likelihood function. If one knew the true distribution function, one could in principle calculate the distribution function for $\mathscr{L}$. (We do this for a special case below.) One could then state the probability for getting a value $\mathscr{L}^*$ or smaller. However, most physics experiments we know of would require an impossibly large amount of computing time to calculate the distribution function of $\mathscr{L}$. Is there a simpler approach?

5

A statistical problem in physics which may have a simple solution of which we are not aware is that of correlated multiple scattering.

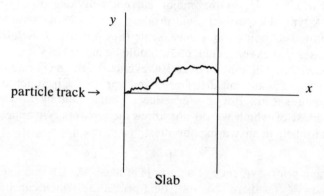

particle track →

Slab

Figure 3.

A particle track which enters a slab of material along the x-axis will do a random walk in y; that is,

$$\bar{y}(x)=0, \ \overline{[y(x)-\bar{y}(x)]^2}= kx.$$

In many experiments a series of measurements y_i at known values of x_i are made. We would like to know the n-dimensional probability density function $d^n p = F(y_1, \ldots, y_n)\, dy_1, \ldots, dy_n$. Clearly the value of y_n is strongly correlated to the value of y_{n-1}, and to a lesser degree y_{n-2}, y_{n-3}, etc. If a physicist knew this distribution function, he could use it to make a best fit to the initial direction of the particle.

6

We have a basic question which gets to the heart of the Bayesian approach. A crude formulation of this question is: in choosing between two hypotheses, does the likelihood ratio give a reliable estimate of how often one would make the wrong choice?

A more accurate formulation of the above question is: Let $R_{AB} = \Pi \dfrac{f_A(x_i)}{f_B(x_i)}$ be the likelihood ratio comparing hypothesis A to hypothesis B. If $R_{AB} \gg 1$, we choose hypothesis A. If $R_{AB} \ll 1$, we choose hypothesis B. (If $R_{AB} \sim 1$, it is clearly not safe to make a choice.) To be specific we put the arbitrary dividing line at $R_{AB} > 10^2$ for choosing A and $R_{AB} < 10^{-2}$ for choosing B. In principle one can then calculate directly the median value of R_{AB} when A is true which we shall call $(R_{AB})_A$. Also one can calculate the probability of getting $R_{AB} > 10^2$ when B is true. Let this probability be P_B. If the likelihood ratio is to serve as a useful guide for betting odds, then $1/P_B \approx (R_{AB})_A$. Can one prove that this is true for all $f_A(x)$ and $f_B(x)$? Or does anyone know of any example where this is not true?

We have explored this relationship for one special case – a case which happens to be famous in physics. It is the determination of the spin of the K-meson. Let hypothesis A be spin 0 ($f_A = 1$) and hypothesis B be spin 1 ($f_B = 2x$) where $x = E/E_{max}$ and E is the energy of the π^- decay product ($K^+ \rightarrow \pi^- + \pi^+ + \pi^+$).

Assume our experiment consists of a random sample of 30 K^+ decays. Then $\ln R_{AB} = \Sigma \ln \frac{1}{2x_i}$. We have made a Monte Carlo calculation of $\ln R_{AB}$ using 5000 trial experiments of 30 events each, first assuming A is true and then assuming B is true. The results are shown in Figure 4.

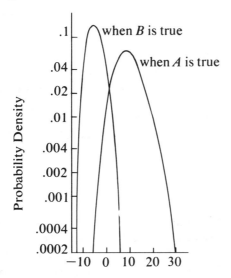

Figure 4. Distribution of $\ln R_{AB}$.

We see from Figure 4 that the median value of R_{AB} when A is true is 7700. We also see from curve B that the chance of making a mistake in choice of hypothesis is $P_B \approx 6 \times 10^{-4}$. This latter determination is not too accurate since only 3 out of the 5000 trials gave R_{AB} greater than 10^2. So $1/P_B \approx 1670$ which is to be compared to $(R_{AB})_A$ which in this case is 7700. These two estimates agree within a factor of 5.

If we do the comparison the other way around the agreement is better. Now assume B is true. The relation we are proposing becomes $P_A \approx (R_{AB})_B$. The median value from curve B is $(R_{AB})_B = \frac{1}{415}$, and curve A gives $P_A \approx \frac{1}{625}$ based on 5 trials out of 5000. Here the two predictions of making a mistake are quite close.

Our question for the experts is: Can this relation be generalized for all probability density functions $f(x)$, or is one forced to Monte Carlo each specific problem? Unfortunately modern computers do not have the capacity to Monte Carlo many specific problems.

COMMENTS

D. J. Bartholomew:

I agree with Dr. Rubin that the Kolmogorov-Smirnoff statistics would provide a better test than χ^2 of the hypothesis of "no bumps" in data such as those illustrated in figures 1 and 2. However, the one-sided version which he proposes would not be particularly sensitive to the presence of a bump. The one-sided test would be more appropriate if searching for a trend.

The χ^2-test used by the authors can be supplemented by a test of significance for the pattern of deviations. A bump will produce a run of positive (or negative) departures from the uniform distribution. David, F. N. "A χ^2 'smooth' test for goodness of fit", (*Biometrika*, 1947, Vol. 34, page 299), proposed such a test which might be useful in this context.

M. S. Bartlett:

Further to the remark by Dr. Good* on the least squares problem, I would suggest it worthwhile to keep the two parts separate in the analysis in order to check on the compatibility and accuracy of the two parts before putting them together.

On the estimation of lifetimes, I might note that my own use of scores based on the log likelihood derivative was first developed in this physical context when I was at Manchester.

D. R. Cox:

The distributions obtained by simulation at the end of the paper refer to sums of independent variables and hence can be approximated by normal distributions whose mean and variance can be calculated; if necessary adjustments based on third and fourth moments can be included. (The theoretical means are 9.2, -5.8 and the variances are 30, 7.5.) The corresponding problem in which there are adjustable parameters in the two models is more complicated.

I. J. Good:

The problem of whether bumps in histograms correspond to physical facts and are not accidental was briefly considered in my book *Probability and the Weighing of Evidence*, Griffin, London, 1950, 85-86, but I should like to make a few further suggestions. (They are based on the mimeographed paper, not the spoken version where a more special and easier question was raised.) They are also relevant to the problems of sampling considered yesterday by several speakers. Basically the problem is one of density estimation *combined* with significance testing.

In the first place, one can try maximum-likelihood fitting of densities proportional to exp $f_{2n}(x)$, where $f_{2n}(x)$ is a polynomial of degree $2n$. Each bump will tend to increase the degree of the polynomial by two. By taking

*Good's comments to follow.

the Neyman-Pearson-Wilks likelihood-ratio, we obtain a significance test for whether the extra degree is justified; but of course it will be necessary to take into account some information or judgment concerning the initial probability that the bump corresponds to a physical reality.

If you feel that your judgment of the initial probability of the bump is too unreliable (too wide, too vague), then you can use the Bayes factor as an indication that further experimenting is worth while, in accordance with Priggish Principle number 20.

One measure of the roughness of a density curve is the expected weight of evidence for distinguishing it from itself when shifted to the right by a short distance ϵ. This is

$$\int f(x) \log \frac{f(x)}{f(x+\epsilon)} \, dx \simeq \epsilon^2 \int \frac{[f'(x)]^2}{f(x)} \, dx,$$

which happens to be proportional to the amount of Fisherian information per observation concerning the location of the curve, if it is subjected to an unknown slide. In order to make this invariant with respect to transformations of the x-axis it seems better to measure the roughness by

$$\int \frac{[h'(x)]^2}{f(x)} \, dx,$$

where $h(x) = f(x)/g(x)$,

$$g(x) = \left\{ -\frac{\partial^2 v(x, y)}{\partial y^2} \bigg|_{y=x} \right\}^{\frac{1}{2}}$$

and $v(x, y)$ is the utility of asserting that the value of a variable is y when it is really x (see Good, "What is the Use of a Distribution?" *Multivariate Analysis II*, 1969, 183-203). We might now choose f so as to maximize the log-likelihood minus a term proportional to the roughness, that is, to maximize

$$\sum_{i=1}^{r} \log f(x_i) - A \int \frac{[h'(x)]^2}{f(x)} \, dx,$$

where $x_1, x_2, \ldots, x_r$ are the observations and $f(x) \geq 0$, $\int f(x) \, dx = 1$.

For quadratic loss functions, $g(x)$ is a constant. It is also a constant when the loss function is *any twice differentiable function* of $(x - y)^2$ such as the *sensible* form

$$\alpha - e^{-\beta(x-y)^2}.$$

When g is constant we wish to maximize

$$\sum_{i=1}^{r} \log f(x_i) - A \int \frac{[f'(x)]^2}{f(x)} \, dx.$$

It is as if we assumed a density in function space proportional to

$$\exp\{-A\mathcal{E}[\frac{\partial}{\partial x}\log f(x)]^2\}.$$

An indication of how to select the constant A is obtained by considering the case where f is assumed to be normal $N(\mu, \sigma^2)$. Then (when g is constant) the estimates of μ and σ^2 turn out to be

$$\hat{\mu} = \bar{x}, \hat{\sigma}^2 = \frac{1}{r}\{2A + \sum_{i=1}^{r}(x_i - \bar{x})^2\}.$$

In particular if $r = 1$ we have $\hat{\sigma}^2 = 2A$ so it is sensible to select A equal to half the initial expected variance. The rule for selecting A would presumably be adequate even if it were not assumed that f is normal. This procedure provides an unconventional and difficult problem in the calculus of variations, but perhaps solutions can be obtained by methods of successive approximation, convenient on electronic computers. A modification of this method works well and will be published elsewhere.

The authors raise the question of how one should choose the size of the *bins* (widths of the intervals of the histograms) when applying the chi-squared test. Of course the same problem arises in connection with the likelihood-ratio test. Here I should like to suggest that several different sizes should be tried and that the tail-area probabilities so obtained might well be combined by the harmonic-mean rule of thumb, (see *J. Am. Stat. Ass.*, 53, 1958, 799-813).

In the case the bins are so small that the asymptotic theory for chi-squared and for the likelihood-ratio test become unreliable, then use might be made of the results in a forthcoming paper in *J. Am. Stat. Ass.* concerning their exact distributions for numbers of categories up to 18 and sizes of samples up to 28. (Down to tail-area probabilities of about 2% Monte Carlo simulation would be adequate.) Outside the range of this work, the mean and variance, and higher moments, of chi-squared can be used to obtain reasonable approximations to its distribution in accordance with some calculations by Haldane (*Biometrika*, 1938).

In their oral exposition the authors raised the question of least squares estimation of x subject to two sets of approximate simultaneous linear equations $Ax \simeq y$, $Bx \simeq z$, the latter being more accurate by, I understand, a known ratio. We can write $Ax = y + \epsilon_1$, $Bx = z + \epsilon_2$, and we can then *abut A* and B into a larger matrix. This reduces the problem to the usual linear model, with the error covariance matrix diagonal and *known* up to proportionality, but not proportional to the identity. [The authors' proposed solution corresponded to the case where the two sets of errors had *equal* variances.]

O. Kempthorne:

It interests me considerably that the problems discussed and the questions raised are of the same *logical* nature as those which have arisen in biology and agriculture over the past seventy years. I say this not to imply in any way that physics is backward, but to give evidence for the view that the basic problems considered in the *discipline* of statistics are universal. It is interesting that for decades or even centuries, a law of large numbers made *small* sample statistical procedures unnecessary in physics. But clearly, a sample is large or small only to the magnitude of error and the magnitude of effects which would have scientific or technological importance. It is interesting that the *bump* problem is the problem of mixtures of distributions on which some very basic work was done by Karl Pearson in 1893, in connection with the description and analysis of human populations. This is, of course, the man who invented the χ^2 test which is now being widely used in physics.

Strong implication has been made by representatives of some "religions" of statistics that the desire of scientists for goodness of fit tests of significance and other tests of significance has been the result of brainwashing by the frequentist advocates. I believe it is a quite untenable hypothesis that physicists have been so brainwashed. If I may be facetious, who but a physicist can brainwash physicists?

On the matter of tests of significance for bumps, it is important to realize that a test of significance is based on an ordering of possible data sets. Many orderings are reasonable. It is moderately clear that the χ^2 test which uses ordering by probability under the model being tested is highly insensitive to bumps. It would be very rash for me (or for anyone else, I think) to give a good off-the-cuff ordering. The problem is stimulating and I would hope that some statisticians would address their attention to it. I would suggest, however, that *spacing* statistics may be useful.

I deem it worthy to note that while our physicist friends are concerned with the likelihood function, they are concerned with power or sensitivity of tests. I suggest the adherents of the likelihood principle note this.

It is important to suggest also that if our physicist friends go to a consultant they should ask if he is a Bayesian, a frequentist, a logical probability man, and that they would be well advised to obtain, in the present state of confusion, a stratified sample of statistical opinion.

R. Prentice:

I would like to pose a question to Professor Cassel. Professor Kempthorne has reminded us fairly often in the past few days that the central problem of inference is estimating the form of a probability distribution (frequency distribution) as opposed to the parameters of such a distribution and yet here you are, experimenters in an active field of enquiry, coming and presenting us with this very explicit frequency distribution depending on five parameters (referring to particular expression on blackboard) and are concerned with estimating one of these in particular. Does this mean that previous experimentation has established this frequency distribution or that there is adequate theoretical justification for it? In any event, I am interested that at this point of time estimation of parameters in this experiment is your primary concern.

REPLY

To Professor Bartlett:

We certainly agree that is desirable to keep the two parts of the problem separate first. In fact, in the experiment mentioned, the data were analyzed separately, and then together.

To Professor Bartholemew:

Some of our comments on our education made in our reply to Dr. Kempthorne can also apply here. It is usual practice, though, to look at a graph of the deviations to try to find by eye a run of positive or negative departures from a uniform distribution. We agree that a more sophisticated test would be useful in this context.

To Professor Good:

The procedures suggested appear to be interesting, and some estimate of their utility in this type of problem could be made. Frequently a resonance will appear with little statistical reliability (from the point of view of our unsophisticated tests) in several experiments, before it is conclusively established in an experiment with a large number of events in the region in question. It would be very interesting to analyze some of the less reliable experiments with these more sophisticated techniques, while knowing the answer by hindsight.

To Professor Kempthorne:

You are not being facetious in asking, "Who but a physicist can 'brainwash' physicists?" As you can see, we and most of our colleagues are not widely read or proficient in the literature of statistics, and we tend to pass information, misinformation, and a certain amount of witchcraft along as though it all had the reliability of Euclidean geometry. The problem, of course, is one of time and commitments. Statistics is one of the dozen or so technical skills that we need for doing our experiments, so we typically do not have the time to spend to learn the subject as we should. An elementary text written by someone who is acquainted with the nature of our problems and our language would be of help. We often find that the biological and social science orientation of the texts which we find, makes it difficult to relate the solutions to our problems.

To Professor Prentice:

In physics we frequently encounter the situation in which the form of the distribution function is *assumed* known from a combination of previous experimental and theoretical results. We are then principally interested in estimation of parameters, the reliability of these estimates, and tests of significance. It is standard practice to calculate χ^2 for the data with the estimated parameters. An unlikely value of χ^2 is likely to send the experimenter or his theoretical friends off to look for a modified distribution function. Frequently this search is aided by the knowledge that certain effects have been neglected, perhaps unwisely, in creating the original model. Some of the most interesting and exciting work is that in unexplored areas, where the form of the distribution is not known. "Bump hunting" is often in this category.

SCIENTIFIC INFERENCES AND DAY-TO-DAY DECISIONS

G. A. Barnard
University of Essex

This symposium is held at a time of relative stasis in the perennial controversy over the foundations of statistical inference. Twenty years ago it looked to many as if statistical decision theory had formulated once and for all the paradigm for our problems and all that remained to do was to develop methods for the solution of the formidable mathematical problems which this formulation opened up. Since then, although some extremely valuable work has been done along these lines — notably in the precise treatment of the experimental design problem by Kiefer — a number of difficulties have arisen which have led some of those who most enthusiastically welcomed Wald's formulation of decision theory as a final version of our problem to welcome, with equal enthusiasm, an entirely different approach in subjective Bayesian terms. Such people have been consistent in their view that the problems of foundations have been solved, though not consistent in the solutions they have adopted. Far be it from me to suggest that no one should change his mind — I was glad to learn, through a third party, that I gained much credit in Fisher's eyes as being, according to him, the only statistician who had ever admitted he had been mistaken. But if we do change our minds, we must admit the possibility that we may change again. It is my main theme in this paper that we should recognise that none of the formulations of our problems now current are without serious difficulties if they are thought of as representing universal truth; at the same time all of them have something to contribute, in their proper sphere.

We have a duty to determine, so far as we can, what is the proper sphere for each formulation, and not to lapse into a lazy and woolly eclecticism. But above all we need to recognise how prone we all are to force all the multifarious types of problem which come our way as statisticians into far too small a number of over-rigid moulds. Indeed, it may be doubted whether we can in any very useful sense speak of a *type* of problem at all — every issue as it comes up tends to call for some special form of treatment. We need to recognise that our categories of *hypothesis testing, point estimation,* and the like, are little more than pedagogic devices which, if taken too seriously, can lead to dangerous over-simplifications.

Yet, of course, to discuss general issues at all we are forced to categorise our problems to some extent. One such category might be called that of the *parametric information problem.* Here we consider an experimental result x, belonging to a class of possible (and in some sense comparable) results S, in relation to a parametrised family of probability functions $f(x, \theta)$, the parameter θ ranging over a set Ω. As an example we may quote one we have discussed before — that of obtaining information about the parity non-conservation parameter in the decay of Λ particles, from measurements x_i ($i=1,2,...,n$) on the cosine of the scattering angle. The frequency function is known from general quantum theoretical considerations together with a careful consideration of the possible experimental biases to be

$$f(x, \theta) = \prod_i \frac{1 + \theta x_i}{2}$$

(we have omitted a slight complicating factor which arises from the impossibility of detecting Λ particle decays for which the scattering angle is near to zero).

In the early days of discussion of results of this kind, interest centred on whether or not parity was conserved in the process — corresponding to the hypothesis $\theta=0$. Thus among 63 tracks recorded in one series of observations, 42 were found for which x_i was positive, against an expected number of 31.5 on the assumption of conservation of parity. The deviation was

$$\frac{10.5}{\frac{1}{2}\sqrt{63}} = 2.6$$

times its standard error and in the light of the general state of knowledge at the time (shortly after the discovery that parity is not conserved in β-decay of cobalt) this was considered good evidence in favour of the idea that parity is not conserved in *weak interactions* — a class of processes to which both β-decay and Λ particle decay belong. Further evidence has since overwhelmingly confirmed this idea, and interest now centres on the (non-zero) value of the parameter θ. Here the problem may fairly be stated, at present, as consisting in the *indicative summarisation* of the observational record. Hopefully, one day, theoretical physicists will have a theory which predicts that the value of θ should be in the neighbourhood of some specific value θ_0; or they may need to choose between, or compare, two theories one of which predicts the value θ_1 while the other predicts the value θ_2; but at present no such theories are available, and so we need a summary of the observations to date, indicating in some way the range of values for θ which the observations suggest as relatively plausible, to help theoretical physicists in their theory building. Two mathematically equivalent ways of providing such an informative summary are, to specify the likelihood ratio-function

$$L(\theta; x) = K(x) f(x, \theta)$$

which may be represented as

$$L(\theta; x) = \prod_i (1 + \theta x_i)$$

or by the score function

$$\mathrm{Sc}\,(\theta, x) = (\partial/\partial\theta) \log L(\theta; x).$$

By the term *ratio-function* we mean an equivalence class of functions, under the equivalence relation $\phi_1 \sim \phi_2$ iff $\phi_1(\theta)/\phi_2(\theta)$ is independent of θ. It is convenient to represent such a class of functions by one of its members, and if the members of the class are bounded functions of θ then the class is often represented by that one of its members whose supremum is 1; in the present case it is more convenient to use the representative which takes the value 1 when $\theta = 0$. Thus a ratio-function may be thought of as a function which contains an arbitrary multiplicative constant. It is one respect in which the score function $\mathrm{Sc}(\theta; x)$ has the advantage over the likelihood, that this arbitary constant disappears.

The summarisation provided by the likelihood or the score function is efficient, in that for two observations x, x' to give raise to the same likelihood it is necessary and sufficient that $f(x, \theta) | f(x', \theta)$ should be independent of θ — that is, that the conditional probability distribution within the set of observations giving the same likelihood should not involve θ. We can express this by saying that the statistic whose value for the observation x is the ratio-function $L(\theta; x)$ (the likelihood statistic), and the equivalent statistic whose value for the observation x is $\mathrm{Sc}(\theta; x)$, are minimal sufficient statistics for θ. We can find a function R such that

$$x = R(L, e)$$

where L is the value of the likelihood statistic and e is a random variable whose distribution is fixed, irrespective of θ. This can be interpreted to mean that the information in x has been split into two parts — L and e — such that the first part L is relevant to θ while the second part e is irrelevant to θ. No information about θ is lost if we ignore the value of e.

In the present case these familiar results can be regarded as almost trivial, because the ratio-function L is clearly (a class of multiples of) a polynomial of degree n in θ with roots $-1/x_i$, so that if we know L we know the observations x_i to within a permutation. The distribution of e in this case is uniform over the set of all permutations of $[1, 2, ..., n]$ (assuming the x_i are all distinct, which will, in the model, be true with probability 1). What is much more important for our present discussion is, that *if we know L to close approximation then we have nearly all the information about θ that is contained in x.* To make such a statement precise we need to specify a metric for likelihoods, and a metric for information. We shall for this purpose restrict generality, in the interests of simplicity of treatment, to cases such as the instance now being considered, where both S and Ω are compact, and $f(x, \theta)$ is continuous in both variables, and is such that, for every θ, every open set in S has positive probability. With these restrictions the set of possible likelihoods becomes the set of equivalence (classes of) functions strictly positive on Ω, and a metric can be defined by fixing arbitrarily on some point θ_0 in Ω and setting

$$\| L \| = \sup_{\theta} | \log L(\theta; x) - \log L(\theta_0; x) |$$

(a definition obviously independent of the function chosen to represent the likelihood ratio-function), with the distance $d(L_1, L_2)$ being defined as

$$d(L_1, L_2) = \| L_1/L_2 \|.$$

It is evident that the topology implied by this metric does not depend on the arbitrary choice of θ_0. And if the norm of the likelihood, given an observation y, is small, this means that y can be imbedded in a sample space T, with a frequency function $g(y,\theta)$, such that the maximum distance, in the usual sense for probability distributions

$$\sup_{A < S} | \Pr(A; \theta) - \Pr(A; \theta') |$$

between the distribution specified by θ and that specified by any other value θ', is small. In other words, y can provide little information about θ.

In these terms, if we have a family of likelihoods with typical member L^* which approximate to the possible likelihoods such as L, in the sense indicated, we can evidently find a function R^* such that

$$x = R^*(L^*,y,e)$$

where L^* is the likelihood aproximating the true likelihood L, y is an observation giving little information about θ, and e is, as before, uniformative about θ. This is the mathematical translation of the statement above, that if we know L to close approximation then we have nearly all the information in x about θ. The requirements as to continuity etc. are evidently met in the specific case we are considering. This justifies the practical procedure of representing $L(\theta;x)$ by a graph. Such a graph will indicate what values of θ are suggested by the data, namely those for which the representing function is large, and it will enable us to compare the relative support provided by the data for one value, θ_1 as against another value θ_2. The relative support is measured by the ratio $L(\theta_1;x)/L(\theta_2;x)$. Finally, when the number of observations is large the quantity (often called Bartlett's pivotal quantity)

$$\xi(\theta) = \mathrm{Sc}(\theta_0; x) / <-\mathrm{Sc}'(\theta_0; x)>_0^{\frac{1}{2}}$$

where the dash denotes the derivative with respect to θ, and $<\ >_0$ denotes mean value when $\theta = \theta_0$, will be normally distributed with zero mean and unit standard deviation if the true value of θ is θ_0, and so this can be used to test whether a specified theoretical value θ_0 is, or is not, compatible with the data.

Thus a record of the likelihood, and/or the mathematically equivalent score function, enable us to answer all questions that are likely to arise in the context we are considering. In particular, it may be noted that statements such as those we could derive by Bayes' theorem from some prior distribution for θ, of the form $\Pr(\theta \in Q) = 1 - \alpha$, are most unlikely ever to be required, since it is almost unknown for a physical theory to imply a statement of this form. Theories typically lead to assertions of the form: $\theta \approx \theta_0$. At the same time, the procedure for combining the information from a set of observations x_1 with that from another set of independent observations x_2 is simplicity itself. The likelihoods are combined by multiplication

$$L(\theta; x_1 \text{ and } x_2) = L(\theta; x_1) L(\theta; x_2)$$

while the score functions are combined by addition

$$\text{Sc}(\theta; x_1 \text{ and } x_2) = \text{Sc}(\theta; x_1) + \text{Sc}(\theta; x_2).$$

It would thus appear that the likelihood provides a fully satisfactory answer to the problem of "indicative summarisation" in the case of a single parameter. The only point of difficulty arises in connection with Bartlett's pivotal, where the denominator $<-\text{Sc}'(\theta_0;x)>_0^{\frac{1}{2}}$ involves taking a mean value for x ranging over the sample space S. This involves the requirement, already mentioned, that the set of results S should be reasonably comparable with S which, in turn, can be taken as meaning that the quantity $-\text{Sc}'(\theta; x)$ should not vary over $S \times \Omega$ in any marked way.

Such a mean value may be widely discrepant from the value observed in any particular case, unless there is some restriction on the range of variation allowed for $\text{Sc}'(\theta; x)$. As an extreme case of this difficulty we may take the problem to which Armitage has drawn attention, arising when we sample $x_1, x_2, \ldots, x_i, \ldots$ from a normal distribution with unit variance and mean θ. When the sample size is N the score function is

$$\text{Sc}(\theta; \underline{x}) = N(\bar{x} - \theta)$$

so that

$$-\text{Sc}'(\theta; \underline{x}) = N$$

and Bartlett's pivotal is

$$\sqrt{N}\,(\bar{x} - \theta).$$

No difficulty arises, of course, if N is fixed; but if N is allowed to vary so that sampling stops just as soon as

$$|\,\sqrt{N}\,(\bar{x} - \theta_0)\,| \geq K$$

for some specified θ_0, then the approximation involved in using Bartlett's pivotal for testing $\theta = \theta_0$ clearly cannot hold. The need to exclude such cases was what underlay the requirement mentioned earlier, that the set S

consisted of observations "in some sense comparable with x". In particular, we cannot always regard S as consisting of all results which might possibly have been obtained by a repetition of the sampling procedure actually used. The reference set S may be to this extent, hypothetical.

It will be, I hope, a feature of the present symposium that we try so far as possible to relate our views on the foundations of statistical inference with the ways in which statistical methods are in fact used by those who are required in practice to make such inferences. It is not usually easy to see just how these inferences are made because so often, especially nowadays, they ultimately involve a vast mass of information which cannot be readily summarised. It may be worth while therefore to draw attention to a reasonably accessible instance in the literature where the inferential processes can be seen at work — and in the mind of someone who, (unlike so many physicists!) had taken the trouble to familiarise himself thoroughly with mathematical statistics. I refer to Eddington's account of the first experimental test of Einstein's prediction about the deflection of light rays in the gravitational field of the sun, in his *Space, Time and Gravitation*, Chapter VII. Einstein had, in his classic paper of 1915, predicted that a light ray grazing the edge of the sun would be deflected through an angle of 1".75, while the corresponding figure from the Newtonian theory was 0".87. An eclipse was due on May 29th, 1919, when the sun was in the midst of "a quite exceptional patch of bright stars...if this problem had been put forward at some other period of history, it might have been necessary to wait some thousands of years for a total eclipse of the sun to happen on the lucky date". In spite of World War I the Astronomer Royal of England drew attention to the opportunity and two expeditions were mounted, one to Sobral in North Brazil and the other to Principe in the Gulf of Guinea. On the day, the weather was unfavourable at Principe, so that although sixteen photographs were obtained, only one finally proved good enough to provide an estimate of the grazing deflection. On the other hand the weather at Sobral was good, but trouble with the optical system meant that many of the images were blurred. In the end the results, with their *probable accidental errors* were

$$\text{Sobral } 1''.98 \pm 0''.12,$$
$$\text{Principe } 1''.61 \pm 0''.30.$$

Eddington says "It is usual to allow a margin of safety of about twice the probable error on either side of the mean. The evidence of the Principe plates is thus just about sufficient to rule out the possibility of the half deflection, and the Sobral plates exclude it with practical certainty".

Eddington's account gives far more detail that I have suggested, and the fact that he had himself been persuaded by Einstein's logic, and was himself a participant in the Principe expedition together with his command of literary style, means that his full account conveys a strong sense of excitement. But for the present purpose it is enough to draw attention to the combination of good luck, in the date of the eclipse, with bad luck, in the weather and other matters, which finally resulted in the sample size being very different from what could have been planned. Yet, we have no hesitation in accepting, as reference set S in this case, a set in which the number of usuable star images is the same as that actually obtained. Any suggestion that the probable errors should be reduced because in a repetition of the experiment it could be expected, on the average, that more images would have been obtained would be greeted with derision.

It would in my view be wrong to suggest that the principles which should govern the choice of reference set S are entirely, and in all cases, clear. The well-worn case of the 2×2 table obtained by making m binomial trials with probability of success θ_1 and n independent binomial trials with probability of success θ_2 illustrates this point. If x_1, x_2 are the observed numbers of successes, and the parameter of interest is $\theta = (\theta_1, \theta_2)$, it is clear that the set S over which $x = (x_1, x_2)$ ranges should be the Cartesian product

$$C = \{0, 1, \ldots, m\} \times \{0, 1, \ldots, n\}.$$

But if we transform the parameters by the logistic transformation and then consider the sum and the difference:

$$\phi_1 = \log\left[\theta_1/(1-\theta_1)\right] + \log\left[\theta_2/(1-\theta_2)\right]$$

$$\phi_2 = \log\left[\theta_1/(1-\theta_1)\right] - \log\left[\theta_2/(1-\theta_2)\right]$$

it may be that ϕ_2 is the only parameter of interest, while ϕ_1 is a nuisance parameter. In this case, B the subset of C in which $x_1 + x_2$ takes a fixed value, equal to that observed, seems more appropriate. In real life it is likely that ϕ_2 will be the parameter of primary interest, but it is unlikely that no interest at all will attach to ϕ_1 and, to this extent, the appropriate reference set is not uniquely defined.

The 2 x 2 table also provides an opportunity to discuss the relaxation of the stringent topological conditions laid down for the sample space S and the parameter space Ω, and the probability functions $f(x, \theta)$. If we allow the parameter space for (θ_1, θ_2) to be the closed unit square, the condition that Ω is compact is met, but for the parameter point $(0,0)$ the only point in S having positive probability is $(0,0)$, so that the condition $f(x, \theta) > 0$ is violated. On the other hand, if we take as parameter space the open unit square, the condition that this should be compact is violated. To resolve the difficulty we may note that for a sufficiently large pair of sample sizes (m,n) if we start with a parameter space consisting of the open square we can, by neglecting events whose total probability is arbitrarily small, restrict our consideration to a closed (and hence compact) subset of the unit square. At the same time our sample space C can be restricted to exclude points in the neighbourhood of the corners $(0,0)$ and (m,n) and as soon as this is done the main differences between inferences using this restricted sample space and the further restricted space B above also tend to disappear. In a similar way, by considering large sample sizes and neglecting sets of arbitrarily small probability, we can bring most cases of parametric inference within the scope of the theory sketch here. Cases where more than one parameter is involved only need the replacement of the definition given above for the score function by the more general definition (as a vector-valued function).

$$\underline{\mathrm{Sc}}\,(\theta; x) = \nabla \log L(\theta; x)$$

(where T denoted transpose), then the generalisation of Bartlett's pivotal becomes the vector

$$s^{-1}(\theta) \, \underline{\mathrm{Sc}}(\theta; x)$$

which has the zero vector for its mean and the unit vector for its variance, when θ is the true value. As before, in sufficiently large samples the distribution is close to normality, provided the sample space is such that $s(\theta)$ does not vary widely. Should there be any doubt about the accuracy of the normal approximation for the distribution of Bartlett's pivotal in a given case with finite sample size, the exact probability level associated with an observation x, in relation to a hypothetical value θ_0, can always be obtained by Monte Carlo or similar methods.

It thus appears that the likelihood function provides an essentially complete answer to *scientific* problems of parametric inference. Why *scientific*? Because in a scientific problem we are entitled to consider what happens when the number of observations becomes large. The likelihood is, of course, always well defined, even in small samples; and its minimal sufficiency property also is exactly valid in small samples. What may give rise to difficulty, in small samples, is the exact interpretation of the likelihood; and it is because, in scientific matters, we can always (indeed we should) await further observations before reaching conclusions, and because any doubts about the interpretation of likelihood can be resolved with sufficiently large samples that we are entitled to claim as we do.

Cases do arise, of course, where the interpretation of likelihood is unambiguous even in small samples. For instance, where the parameter of interest is associated with a group of transformations, invariance considerations may resolve ambiguities which might otherwise be present.

Some of you will at this point, wish to draw attention to Stein's example, on the *estimation* of the mean of a k-variate normal population, $k \geq 3$. What Stein showed was that, in relation to a loss function proportional to the square distance in the parameter space, the group invariant estimate was inadmissible. But in the scientific problem of parametric inference, as I am suggesting we should formulate it, the question of providing a point estimate of the parameter which minimises any particular loss function does not arise. The data will certainly point towards values in the neighbourhood of the usual estimate as being more plausible than values far away from this; care may be needed, when k is large, in making an appropriate use of Bartlett's pivotal. But briefly we can say that Stein's difficulty does not arise here because the problem he was concerned with does not arise.

It is quite otherwise if we turn to day-to-day decisions. Here, in addition to the sample space S and the parameter space Ω we have an action or decision space D and a more or less well-defined loss function $l(\theta, d)$ associated with taking the decision d when θ is the true value of the parameter. We cannot wait for more observations (I am not unaware, of course, that the problem of experimental design can be included with the decision-theoretic formulation of statistical theory; but I ignore it here to keep the discussion free from inessential complications) and must use a decision function of some kind, $\delta(x)$. As a result of the brilliant work of Wald and his associates, we know that the admissible decision functions are those which can be interpreted as having arisen from some guess as to the relative frequency with which different parametric values will turn up in the long run, the decision function being then chosen so as to minimise the expected loss per decision, in the long run. The common sense approach to such a problem will be to classify the given problem with other problems which appear to be related to it in some relevant way, and then to estimate, so far as possible, the distribution of the parameter values within the reference class. This leads naturally to the compound decision problem and the empirical Bayes approach developed by Robbins and critically discussed by Copas.

The instance *par excellence* of the "day-to-day decision" type of problem is that of sampling inspection of batches of goods. As I indicated twenty-five years ago (and as Molina did before that) the natural approach to such problems is via the application of Bayes' theorem, with a parameter distribution which can be plausibly estimated from a knowledge of the production process generating the batches. Fortunately, just as the distributional properties of Bartlett's pivotal do not depend sensitively on the choice of reference set, within limits, so the optimal properties of these Bayes solutions typically do not depend sensitively on the choice of parameter distribution within limits.

It appears to me that an approach to day-to-day decision making along these lines is not in the least inconsistent with an insistence on the likelihood approach to scientific problems. In so far, of course, as such an approach to day-to-day problems involves the data for an individual problem via the likelihood function, the two approaches complement each other.

An example which Lindley has discussed may serve to bring out the contrast between the day-to-day and the scientific approach. He has considered, the problem of estimating the marks which would have been scored by a candidate who has failed (for instance, through illness) to take an examination when others in a class with him have done so. Lindley arrives at an estimate equal to the mean of the marks scored by those who took the examination. Such an estimate is reasonable if it is used in connection with studies on the examination — for instance, concerned with how the examination in question serves to predict future performance. But in relation to the individual himself it seems absurd to give him a high or low mark simply because his fellow students did well or badly. This appearance of absurdity arises I think, because in considering the individual we are aware that as we follow his career we shall inevitably acquire more and more information relevant to the score he might have obtained, so that what we have called the scientific approach rather than the day-to-day approach is more in order. This example illustrates, of course, the fact that many practical situations will lie somewhere intermediate between the extremes of scientific and day-to-day. Where the possibility of gathering further relevant information is an essential feature of the problem, we lean towards the scientific approach; where a decision, more or less final, based only on currently available information, is required, we lean towards the day-to-day approach. Both approaches appeal to a *long run* justification. With day-to-day decisions we are concerned with long run average losses, in a series of similar problems, and these will in fact be minimised in so far as we are able by our guesses (or, in the case of

Robbins' approach, by our strategy) to represent faithfully the actual long run frequency with which the various parameter points in fact arise. With scientific issues, on the other hand, we are concerned with a long run in which more and more information is obtained relevant to the problem we are concerned with. This justifies our appeal, in this case, to asymptotic limiting properties of the score function.

Turning now to the subjective Bayes approach which has been winning so much support from the persuasive advocacy of Professor Savage and others, there is surely a case for attempting, so far as possible, to quantify *all* the considerations which lead us to hold the views we do, tentatively, hold. The passage from Eddington to which I have referred shows him to have been largely persuaded before the eclipse expedition, possibly partly on grounds associated with his own religious experience, of the correctness of Einstein's analysis of gravitational phenomena. The elegance of the mathematical formulation was probably another factor. Of course, the explanation of the advance of perchelion of Mercury was weighty empirical evidence, though the possibility of alternative explanations (such as Dicke's idea of the possible oblateness of the sun) and the rather indirect connection with light meant that it was hard to put a precise figure to the assessment of this evidence, by any objective procedure. The notion of *coherence*, as developed by the subjective Bayesian school, is useful as providing a means of assessing these matters. Though perhaps this is a point at which to sound a protest against this choice of a technical term with emotional overtones. The use of the word *unbiased* for an estimate whose expectation equals the parameter has been criticised similarly. It sounds bad to admit to being biased, but surely it is even worse to be incoherent! Yet we all recognise now that bias in the technical sense, is almost always present in good estimators, and I do not expect it will be long before we come to accept incoherence in the technical sense as a common feature of rational discourse. It may perhaps be worth while to draw attention to a danger of incoherence which lies in a practice common among those who adopt the Bayesian approach. The mathematical convenience of using parameter distributions which are *closed under sampling,* or *conjugate* to the distribution of the observations may have led some to overlook the fact that the same parameter may enter into the distribution of observations of different types. Whatever distribution the parameter has must, for coherence, be the same for all types of observation. But, for example, if in the case of Λ particles we merely classify the x_i as positive or negative, then the parameter θ can be given a conjugate distribution of the Beta form; but, if the x_i are themselves measured, no conjugate distribution exists. As it happens, a uniform distribution for θ over the range $-1 \leqslant \theta \leqslant +1$ seems reasonable in this case (though one would like to hear the views of physicists on this point), so that what we have above taken as the likelihood, given a set of observations x_i, can here be reasonably interpreted (after normalisation) as a posterior probability density. It would perhaps be helpful if we could take the opportunity of this Symposium to try to see what, in the scientific context, the difference of interpretation there is here, between the likelihood and the Bayes approach. As I see it, the posterior distribution suffers from the disadvantage that it does not help in answering the sort of question that scientists need to answer — viz. given a theory for the constants of high energy physics, do the data support the theory or do they tend to contradict it? For it is easy to see that, starting from a uniform prior for θ, we shall always have a positive posterior density for all values of θ; and it is hard to see, in Bayesian terms, why such a density could be said to be any less compatible with one value for θ than with another. If the physical theory produced a prediction for θ of the form θ *lies between a and b*, such a prediction could be assigned a probability in the light of the data; but physical theories rarely are of this form. The way in which the likelihood can deal with this point has already been indicated.

Finally, what of the classical approach of Neyman and Pearson? In one sense, of course, this approach has developed so many ramifications that it can hardly be said to constitute just one approach. For instance, many would, with justice, regard the decision-theoretic formulation, and Robbins's extension of it, for what I have called day-to-day problems as a natural development of the classical approach. I propose, therefore, to comment only on the Neyman-Pearson approach to what I have called scientific issues.

Here the differences between likelihood and the classical approach seem to be mainly those of emphasis. Neyman has, as we all know, put statistics very much in his debt through his insistence on a precise mathematical formulation for every statistical problem. It is all too easy to neglect mathematical rigour and lapse into ineffective vagueness. But one may question whether such insistence can go too far. In a recent discussion on statistical inference one of the speakers remarked that a certain idea had not been fully defined mathematically, *and was therefore useless*. Against this we may put the fact that the concept of *truth* of a mathematical theorem has been shown to be incapable of full mathematical definition; but surely no-one would infer that the concept was useless. In statistical inference we are handling, not only a concept of truth, but also the still more complex notions involved in knowledge of truth, ignorance of truth, and so on; it is not to be expected that any mathematical definitions put forward here could aspire to completeness. We should, of course, try to give these concepts as much in the way of precise definition as we can, just as workers in the foundations of mathematics attempt to do with mathematical truth. But insistence on completeness can stultify much useful work.

To be more specific, I have already indicated elsewhere the close relationship between the Neyman-Pearson notion of *power function* and the notion of *likelihood*. Much of the classical theory of testing hypotheses can be interpreted in terms of finding approximations of the type of L^* above, which come close to preserving all the information in a set of data while having a specific form. For instance if a likelihood function of two parameters is expressible as a product:

$$L^*(\theta_1, \theta_2) = M(\theta_1) N(\theta_2)$$

then inferences are possible about each parameter without regard to the other. If a given L can be approximated by an L^* of this form, we can in this sense make approximate inferences about each parameter separately. There is evidently a connection here with the Neyman approach to the problem of efficient testing in the presence of nuisance parameters.

Perhaps the major difference I have with Professor Neyman is connected with the reference set S. I may be wrong, but he appears to me to insist on regarding S as in some way precisely defined by the conditions under which the experiment was carried out. The example I have quoted from Eddington illustrates the difficulties inherent in this approach. Pratt and Cox have drawn attention to absurdities that seem to flow from its adoption.

I wonder whether the following approach offers hope of resolving this issue? We may say that the mathematical statistician's work begins after a reference set S and a parameter space Ω have been fixed upon. If so, of course, S is by definition to be taken as given once and for all, and we are committed to the classical approach. But in fact, of course, it will be the mathematical statistician who sets up the model, with S, Ω, etc. What considerations will guide him in selecting his S? I have suggested that S must consist of *reasonably comparable results,* and this can be interpreted to mean *results of reasonably similar precision.* I conjecture that it would be enough, to give this concept mathematical expression, to interpret it as meaning that the denominator in Bartlett's pivotal (the information, in Fisher's sense) should be reasonably the same for all points in S. Sometimes (as, for instance, with the location parameter problem) we can find a reference set for which the information (as a function of *both* x and θ) is exactly constant, and this properly marks out such a set as superior to all others. More commonly however, (and the Λ particle problem provides an example) we cannot find such a reference set, and we must make do with an approximation. In this latter case we can and should specify the reference set adopted, and make mathematically precise statements based on it; but we have to admit that the possibility of an alternative, equally mathematically precise, statements are also possible, and to this *limited* extent the inference from the data is not unique. From the scientific point of view this does not matter because, as the volume of data grows, the differences between the possible inferences converge to zero.

In conclusion, may I put up a plea for more scientific doubt, and less dogmatism, in connection with discussion of these issues — even though I well know myself to have expressed matters over-categorically on occasion. I have in mind, more particularly, refereeing. I quote from a recent referee's report (not on a paper of mine): "The examination of likelihood functions does not, by itself, constitute a method of inference. In the few cases where more specific inferences are recommended, there is no attempt to provide a justification". One wonders just what this referee understands by a method of inference, and what he would accept by way of justification. It is hard to see what he could mean by method of inference other than a method of arriving at statements of a form approved by some theory of statistical inference, or a justification in terms of some accepted view. Yet, of course, such an insistence on an interpretation of one theory in terms of another is clearly unreasonable. In refereeing a Bayesian paper, it seems to me, we should criticise from the point of view of the Bayesian approach, however much we may disagree with it, and similarly for other approaches. It is most important that we should not regard these issues as having been fully settled. One may doubt whether they ever will be.

COMMENTS

D. J. Bartholomew:

I was interested in the author's conjecture that the choice of reference set should be such that $< Sc(\theta) > \frac{1}{2}$ is approximately constant. I think this is right and Armitage's example will serve to explain my reason. A person using the rule: continue sampling until $|\sqrt{N}(\bar{x} - \theta_o)| > K$, shows in his choice of this rule, a strong disbelief in the value $\theta = \theta_o$. This could be expressed by saying that he is determined to conclude that $\theta \neq \theta_o$. Any computation of the score function using this reference set is thus building the experimenter's prejudice into the information measure. We therefore need a reference set which has no concealed prejudice on the experimenter's part and it seems likely to me that Barnard's suggestion may be the right one.

However, I would wish to go further and require that the experimenter should adopt the sampling rule dictated by this conjecture. The reference set would then correspond to what would happen in the long run. What we require is a way of deciding when a proposed experiment conceals no experimenter bias. The problem is analogous, and perhaps equivalent, to the Bayesian problem of choosing a noninformative prior.

M. S. Bartlett:

There is much in Professor Barnard's paper that I seem to be in agreement with. In particular, the problem he refers to of estimating the performance of an examinee is somewhat analogous to the estimation problem referred to in the last section of my own paper. I welcome his appeal to identify the situations in which different approaches are most useful, though I do not know how successful his appeal will be!

I was a bit surprised at his references to *Bartlett's pivotal*; I gather this is so named because of my development of this technique in a physical context. I would point out, however, that Fisher first introduced this kind of technique for significance testing; my contribution lay mainly in the use of higher order corrections. With regard to the general use of the likelihood function, I noted in my own paper that I require appropriate sampling properties, which in some problems can break down. Like Professor Lindley, I am not entirely clear how far Professor Barnard himself relies on such sampling properties.

D. R. Cox:

A difficulty with the interesting suggestion of adopting a reference set in which the second derivative of the log likelihood is fixed is that this derivative will in general contain some information about the parameter. While the amount of such information will often be small, there is a need to show this before so conditioning. There is presumably a connection with the conditions under which a locally most powerful test is capable of appreciable improvement in the region of practical importance; see Professor Neyman's paper.

V. P. Godambe:

I think the likelihood principle would be at once appealing even *necessary* under conditions unlike those of Barnard: The parameter space and sample space are essentially undefined or are unknown. Under the condition surely the *only* way to see how an observed sample comparatively confirms (or disconfirms) two possible values of the parameter is through the corresponding likelihood ratio.

The dichotomy of scientific inference and day-to-day decision is mostly artificial. This would be borne out by sample-survey situations. There is an operationally well defined parameter. The statistician wants to infer about the value of the parameter and use his inference for several purposes for which generally no loss function is defined. Yet sample surveys, if not the most widely applied branch of statistics, can be second only in respect of application to regression analysis.

Now I elaborate on the preceding paragraph. In my Ph.D. thesis (1958, London University), I illustrated that by taking into account the actual losses incurred due to the decisions already taken one could successively improve the decision procedure. Thus the overall decision procedure so obtained would incorporate within itself the inferential aspects of every single decision situation involved in it. This again would confirm that dichotomizing the statistical problems as inference or decision is mostly artificial. Almost all the actual problems of statistics would possess both aspects, decision and inference.

I. J. Good:

The question has come up of whether utilities are relevant to scientific inference. Accordingly I should like to recall that Fisher's information matrix emerges from my work on the utility of a distribution in the case that the only use of the parameters is the specified one; that is, when they do not occur as parameters elsewhere and are not given a physical interpretation. This seems to support the view that utilities are useful at least in the *logic* of statistical inference in a scientific context.

Dr. Barnard is not the only statistician who has changed his mind on an important general issue. Three other examples are L.J. Savage and D.V. Lindley who once were frequentists, and J.M. Keynes who conceded that subjective probabilities are primary in his biography of F.P. Ramsey.

Dr. Barnard said that "we can always (indeed we should) await further observations before reaching conclusions". The implication is that we should never reach conclusions. This seems too emphatic although I agree that virtually no empirical proposition is certain.

I also wondered (with Dr. Lindley) whether I was the culprit who said that a certain idea had not been fully defined mathematically and was therefore useless. I might have made such a remark with some exaggeration, about an idea that was *intended* to be mathematically defined. Not knowing the context, I do not know how guilty the culprit was.

O. Kempthorne:

On the whole, I believe Dr. Barnard's assessment is "bang on". I wonder why he attributes pivotal quantities to Dr. Bartlett (though I do not wish to take anything away from Dr. Bartlett). My understanding is that the idea is highly implicit, if not explicit, in essentially all of Fisher's work of the 1920's. The whole process of scoring dates back to this time also, I believe.

I would like to add a comment in writing. Dr. Barnard is regarded as the proponent (or interpreter of Fisher) of the likelihood principle. I am not at all clear on how far Fisher was prepared to go. For years now, we have been hearing of this principle. I believe Dr. Barnard has a real duty to the statistical profession to express *by the written word* just what he has in mind. I, for one, have heard far too much loose, imprecise talk on the matter.

I would have liked to raise orally the question of choice of parametrization in likelihood calculations. It is easy to stumble onto two-parametric situations for which the likelihood contours in a *natural* parametrization are highly non-elliptic. I wonder also whether the natural parameters of the exponential family, as described by Lehmann, for instance, are natural except for a possibly *unnatural* problem.

Also, I am curious about the use of expectation in the scoring idea. The curvature of the actual likelihood function is relevant (of Fisher, 1925). I recall being "blasted" by Dr. Yates when I was a youth for being unsure of what to do. But the question is not, I believe, elementary and there is no obvious and perfect answer.

D. V. Lindley:

The idea put forward in this paper that different statistical techniques should be used depending on the type of problem being studied — Bayesian methods for decision making, hypothesis testing in science, for example — is one that makes very good sense to me. When I investigate these techniques I find that whilst Bayesian methods do not use the reference set (except in designing an experiment when it is well defined) many of the others, such as hypothesis testing, do. I find it hard to see why the reference set is required in one but not the other. Indeed, I have recently become convinced that the most important difference between the two schools is that whereas a Bayesian uses, as a prop for his methods, the prior distribution, the prop for the other approaches is the reference set. Bayesians are often attacked for the arbitrariness of their prior. A reply is that the choice of reference set involves similarly arbitrary selections.

I would therefore like to welcome most enthusiastically Barnard's attempt to describe the reference set by, as I understand it, some condition on the value of $< - \mathrm{Sc}'(x,\theta) >^{1/2}$. If this could be done at the same type of level as the Bayesians explanation of the prior then many objections to the non-Bayesian methods might disappear.

In this connection I would like to put a question: namely, why introduce the expectation operator in discussing the score derivative? If its value at the maximum likelihood estimate had been used instead, no reference set would have been needed. As Hájek was arguing earlier, in order to approximate to the likelihood function, it seems necessary to abandon the likelihood principle.

The phrase used on page 295 in discussing the 2 × 2 contingency table that a certain reference set *seems appropriate* is an unfortunate one. If I have a similar problem how is its reference set to be chosen: by writing to Essex? If I am not the person referred to on page 299 then I would like to support his view that a mathematical concept should be reasonably defined. I am reminded of an occasion when the fiducial distribution of a ratio was being discussed and a young lady who had worked on this had the temerity to ask Fisher a question at a lecture of his. Fisher replied that she should have consulted him first, before starting work. This is surely a negation of the scientific method.

A. Plante:

I find the expression *indicative summarization* appealing in connection with likelihoods mainly because the weak logical status which it implies naturally calls to be complemented by other principles which we might find meaningful.

In reference to Professor Lindley's remarks, I am not sure, in fact, whether the posterior probability statements of a subjective Bayesian should pretend to be more than indicative summarizations either.

Let me explain this point in some details. The relationship $\Pr(E) = \Pr(E')$, where E and E' are random events occuring in different situations, completely specifies the uncertainty concerning E and E' as being equivalent, without reference to anything else. But I doubt we can say the same thing about a posterior probability relationship of the type $\Pr(E_1|x) = \Pr(E_2|y)$, when (x, E_1) and (y, E_2) refer to different situations, where E_1 and E_2 are events in their respective parameter spaces and where x and y are data. It seems to me that if one could give a realistic example where the sampling variance of the statistic $\Pr(E_1|x)$ were uniformly much larger than that of $\Pr(E_2|y)$, then one ought to state this fact together with the statement $\Pr(E_1|x) = \Pr(E_2|y)$ in order to specify the uncertainty involved.

C. R. Rao:

Am I correct in saying that Dr. Barnard would not allow Bayesian techniques in problems of scientific inference? It is not clear from his paper to what extent he would use apriori probabilities in a decision making situation and what the rules for obtaining such probabilities are, if they are not already supplied as datum of a problem.

SIMILARITY AND PROBABILITY
J. A. Hartigan
Yale University

1
Introduction

The word like = similar is close to the word likely = probable, and across the European languages, there is a meaning of probability = similarity to truth, or having the appearance of truth. In this paper, a relationship between similarity and probability is formalized, in order to use similarity judgments to construct probability statements and models, and to interpret probability statements. Similarity concepts appear in early statements of induction, for example Hume (1739) "like objects, placed in like circumstances, will always produce like effects". See Blake (1960), Keyes (1921), Achinstein (1963, 1964), Hesse (1963, 1964) for some discussion of induction and analogy. Similarity plays a critical role in modern definitions of probability, and in statistical practice.

(a) *Tossing a penny*

On tossing a penny the probability of heads is 0.5. One argument for this, is that there is no reason to prefer heads to tails; they differ only in irrelevant markings, and therefore each has equal probability. Or, the event, heads, is similar to the event, tails, so each has equal probability. The similarity judgment is a human one, and a perceptive person might note (it might even be true) that the head side of this coin is more convex than the tail side, and give it lesser probability. The statement that the event heads was not quite similar to the event tails is not enough to imply the probabilities here. Outside evidence is being used: the general behaviour of objects which are slightly convex on one side.

A special type of outside evidence is accommodated in the frequency argument. Toss a penny once and consider whether or not it will give head or tail. This experiment is similar to other coin tossing experiments which

305

(we vaguely remember) gave neither a preponderance of heads or tails. In the frequency definition, the probability of an event following an experiment is the limiting relative frequency of the event in an infinite sequence of similar experiments. There is some embarrassment about the meaning of *similar experiments* since the experiments must be fairly like the original but not identical to it.

A crude application of the frequency definition would then estimate (since only a finite sequence of experiments is available) the probability of heads in the present experiment, by its observed proportion in past experiments, .5. Yet the past experiments will vary in their similarity to the present one. Say

100	tosses of this coin (a penny) today	$p = .55$
1,000	tosses of this coin previously	$p = .51$
10,000	tosses of coins of all sorts	$p = .505$
100,000	tosses of physically symmetrical (two sided objects)	$p = .500$

The present probability will be a weighted combination of these proportions, with weights increasing with the number of experiments and decreasing with the similarity of the experiments. Actually similarity must be judged on the outcomes as well as the conditions of the experiments. For example one of the relevant experiment sequences, that of symmetrical objects, did not involve heads as outcomes at all, yet the outcomes were relevant to the present toss.

(b) *Clusters*

In analysis of large sets of data, a standard operating technique is to split the data into smaller homogeneous sets and analyze within these clusters. (Between them too sometimes). For example in analysis of U.S. election data, it is natural to split the states into clusters consisting of the South, North East, Mid West, West or finer classifications. Thus almost before calculations begin, it is necessary to consider similarity relationships between objects. A rather grandiose general principle, fitting in with the frequency definition of probability, is: reduce error by averaging over homogeneous sets. Similarity considerations, in selecting homogeneous sets, precede the averaging process.

(c) *Analogy*

If two objects are similar in many known properties, they will probably be similar in an unexamined property. (As in the frequency definition, similarity judgments are the stuff of the probability statement.) For example, if two trees are similar in having the same shape and colour of leaves, form of branching, type of bark, rate of growth, the saps will probably be similar in chemical composition. Be it admitted that this expectation is partly based on general knowledge about trees — for example, trees of the same species vary a lot in size, number of leaves and branches so this sort of information doesn't count for much in comparing trees. It should be noted that there are two similarities here (just as there are similarities between events and between experiments in the frequency definition) — similarities between objects, and similarities between properties. If two objects are similar in certain properties, they will probably be similar in an unexamined property which is itself similar to the properties already examined.

(d) *Fitting Models*

Models can be fitted perfectly well, thank you, without any interference by probability theory. Least squares was popular before the Gauss Markov theorem. A model is defined as a set of ideal observations $X(\theta)$ indexed by a parameter θ. A distance function $\rho(x, x')$ is defined on the set of observations. If x is the actual observation then the fitted model $X(\theta)$ minimizes $\rho[x, x(\theta)]$. Thus the fitted model value is most similar to the observation x.

These examples show that similarity considerations are ubiquitous in statistical inference, especially appearing in preprobabilistic formulations. Some of the examples are treated in more detail in the rest of the paper. I am not entirely convinced that the similarity-probability relation I have chosen is the right one, but I feel strongly that some connection should be developed.

2

Definition of Similarity

Let A be a set or sample space; let B be a σ-ring of subsets of A. Typical members of B will be denoted by $E, F, G, \ldots$. Let P be a probability measure on B. The corresponding similarity on B is a real function on $B \times B$ defined by

$$S(E, F) = P(EnF)/P(E)\,P(F).$$

In particular $S(E, E) = 1/P(E)$ so that the similarity function determines P.

If X and Y are two random variables on B, the similarity of $X = x$ to $Y = y$ is

$$S(x, y) = f_{X,Y}(x, y) / f_X(x) f_Y(y)$$

where f_X is the density of X with respect to some underlying measure μ_X, f_Y the density of Y with respect to μ_Y, and $f_{X,Y}$ with respect to $\mu_X \mu_Y$. Equivalently $S(x, y)$ is the density of X, Y with respect to the product of the probability measures of X and Y. This particular case of similarity corresponds with the usual notion of similarity as a point function of two variables, and will be used extensively. An important property of this point function is that it is invariant under $1-1$ transformation of X or Y.

To give an example, let μ be a parameter with distribution $N(\mu_0, \sigma_0^2)$, let X be an observation with distribution given μ, $N(\mu, \sigma^2)$. Then

$$
\begin{aligned}
S(x, u) &= f(x|u) / f(x) \\
&= (1 + \sigma_0^2 / \sigma^2)^{\frac{1}{2}} \exp\{\tfrac{1}{2}[(X - \mu_0)^2 / (\sigma^2 + \sigma_0^2) - (X - u)^2 / \sigma^2]\}.
\end{aligned}
$$

The value of u most similar to x is $u = x$. The value of x most similar to u is $x = u + \sigma^2 (u - \mu_0) / \sigma_0^2$. This optimum value is always further away from μ_0 than u is, because this movement away decreases the prior density of x(centered on μ_0) relatively more than the posterior density of x given u(centered on u).

3
Invariance

In experimental design, non-parametric statistics, fiducial inference, games of chance, the frequency definition of probability, the invariant method of introducing probability judgments requires that a probability distribution be invariant under certain $1-1$ transformations T of the sample space into itself. In approaching probability through similarity, the function S is declared invariant under T if $S(TE, TF) = S(E, F)$ for each $E \in B$, $F \in B$, and this is equivalent to declaring P invariant under T. Since similarity functions are more complicated than probability functions, there is no advantage in going this route in this invariant case. In later cases judgments of equal similarity will be made, which cannot be expressed as invariance of the probability distribution. As an example of invariance, consider the frequency definition of the probability of an event E in an experiment $\mathcal{E}$ as the limiting frequency of occurrence of $\mathcal{E}$ in an infinite sequence of experiments *similar* to $\mathcal{E}$. There are severe difficulties in identifying a similar sequence of experiments. A method of de Finetti (1937) defines similar experiments to be such that the probability of any sequence of events is invariant under permutation of the experiment sequence. De Finetti's purpose was to show that the objectivity of the frequency definition is spurious, and that it requires judgments of equal probability, just like the older definition of probability as the ratio of the number of favourable cases to the total number of cases.

As a second example of invariance, let θ, $0 < \theta < 2\pi$, be an unknown angle, let X, $0 < X < 2\pi$, be an observation on θ. It seems natural to require that the joint distribution of X and θ not depend on the base vector from which the angles are measured. Thus (X, θ) must have the same distribution as $(X + C, \theta + C)$. Invariance requirements then set X and θ uniform and give (x, θ) a density of form $\lambda(x - \theta)$ with respect to lebesgue measure on $(0, 2\pi)$.

4
Prior Distributions

Let $x_1, x_2, \ldots, x_n$ denote a sequence of observations given a parameter θ; these observations are independently and identically distributed with joint density, conditional on θ, $f(x_n|\theta)$. Given the marginal density of θ, the complete joint probability density of θ and x_n is known, (in particular the conditional density of the unknown parameter θ given x_n is known; this is the posterior density of θ given x_n in contrast to the marginal or prior density of θ). A number of techniques have been advanced by Jeffreys (1961) and others for determining the prior density of θ as a function of the given conditional density $f(x_n|\theta)$. Two obvious prescriptions are that the marginal distribution of θ be uniform, or that the marginal distribution of x_n be uniform, but these must be rejected because the first is not invariant under $1-1$ transformations

308

of θ and the second under $1-1$ transformations of $\underline{x}_n$. However $S(\underline{x}_n, \theta)$ is invariant under transformations of $\underline{x}_n$ or θ, so that similarity judgments will not result in contradictions through transformation.

For example a natural requirement is that $S(\underline{x}_n, \underline{x}_n')$, where $\underline{x}_n'$ denotes a repetition of the values $x_1, x_2, \ldots, x_n$ in a further n experiments, is the same for all $\underline{x}_n$. This is a way of saying that any result $\underline{x}_n$ is as good as any other. For example for n observations from $N(\theta, 1)$, this requirement is satisfied if θ is uniformly distributed from $-\infty$ to ∞. (This distribution is improper, a frequent outcome when invariant considerations are pivotal; however, here, the posterior distribution of θ given $\underline{x}$ is proper after one observation). In general, the requirement that $S(\underline{x}_n, \underline{x}_n')$ be constant will be satisfied for no prior distribution, but asymptotically, given θ as the true value,

$$S(\underline{x}_n, \underline{x}_n')/\sqrt{n} \rightarrow J(\theta)/h(\theta)$$

where $h(\theta)$ is the prior density at θ, and $J(\theta)$ is the famous Jeffreys' density

$$J(\theta) = [E\{(\frac{d}{d\theta} \log f)^2 \,|\, \theta\}]^{\frac{1}{2}}$$

for θ real valued, and

$$J(\theta) = |\, E(\frac{\partial}{\partial \theta_i} \log f \frac{\partial}{\partial \theta_j} \log f \,|\, \theta) \,|^{\frac{1}{2}}$$

for θ vector valued.

Thus $S(\underline{x}_n, \underline{x}_n')$ is approximately constant for large n if $h(\theta) = J(\theta)$.

Another way of expressing indifference in similarity terms requires that $\sup_{\underline{x}_n} S(\underline{x}_n, \theta)$ be the same for all θ asymptotically; thus every parameter value is potentially equally well supported by the observations. This requirement also implies Jeffreys density.

The point here is that any judgment or requirement on S will lead to a prior density on θ (not necessarily Jeffreys!) which is invariant under transformation of θ or $\underline{x}$, and so some of the traditional difficulties in setting priors are vanquished.

The set $\{\theta \,|\, S(x, \theta) > C(x)\}$ is a region of θ values which has minimum volume, measured in terms of the prior h, for fixed posterior probability $P(S(x, \theta) > C(x) \,|\, x)$. The same regions are obtained if θ is ordered by the likelihood function. Also the set $\{x \,|\, S(x, \theta) < C(\theta)\}$ is a most powerful critical region for testing θ against the alternative that x comes from the mixture density $\int f(x|\theta) \, h(\theta) \, d\theta$. Asymptotically, for θ fixed, $\log S(\underline{x}_n, \theta) = \log [\sqrt{n} J(\theta) \,|\, h(\theta)] + \frac{1}{2} X_k^2 + O(n^{-\frac{1}{2}})$ where θ has dimensionality k. Thus if $h(\theta) = J(\theta)$, $S(\underline{x}_n, \theta)$ is asymptotically a pivotal function which generates minimum volume confidence regions (with volume measured in a certain way), or most powerful tests against a selected alternative. Here similarity is being used to relate Bayes and non Bayes approaches, something of a digression from the main purpose of demonstrating probability calculations through direct similarity judgments.

5
Fitting Models

In fitting models, a procedure is sometimes used which avoids the explicit assumption of a probability model. For each parameter value θ, an ideal observation $X(\theta)$ is defined. A distance function $\rho(x, x')$ is defined on the space of observations and the model is fitted by finding $\hat{\theta}$ to minimize $\rho(x, x(\theta))$. For example in least squares fitting of linear models the ideal observation corresponding to $\underset{\sim}{\theta}$ is $\underset{\sim}{x}(\theta)$,

$$x_i(\theta) = \sum_{j=1}^{m} a_{ij}\theta_j, j = 1, \ldots, n.$$

Euclidean distance $\rho(x, x') = \sum_{i=1}^{n}(x_i - x_i')^2$ is used on the observation space, and $\hat{\theta}$, the least squares estimate, minimizes $\rho[\underset{\sim}{x}, \underset{\sim}{x}(\theta)]$. It seems plausible to introduce probability by defining $S(x, \theta) = \lambda(\rho[x(\hat{\theta}), x(\theta)])$ where λ is a decreasing function of ρ. It might be thought more natural to use $\lambda(\rho[x, x(\theta)])$, it certainly is simpler, but an observation which is far from any ideal observation will have low similarity to all θ, and average properties of similarity forbid this (specifically $E\{S(x, \theta)|x\} = 1$).

In the least squares model, it will follow (under regularity conditions on λ) that

$$f(\underset{\sim}{\theta}) = 1$$
$$f(\theta|\underset{\sim}{x}) = \lambda(\rho[\underset{\sim}{x}(\theta), \underset{\sim}{x}(\theta)])$$
$$f(\underset{\sim}{x}) = k[\underset{\sim}{x} - \underset{\sim}{x}(\hat{\theta})]$$
$$f(\underset{\sim}{x}|\theta) = \lambda(\rho[\underset{\sim}{x}(\theta), \underset{\sim}{x}(\hat{\theta})]) k[\underset{\sim}{x} - \underset{\sim}{x}(\hat{\theta})]$$

This development is contrary to the usual one, which assumes a simple form for $f(\underset{\sim}{x}|\theta)$, perhaps a uniform prior for $\underset{\sim}{\theta}$, and takes what comes for the posterior density $f(\theta|x)$. Within the least squares context, λ and k may be arbitrarily specified, subject to $f(\underset{\sim}{x}, \underset{\sim}{\theta})$ forming a joint probability density. In all models coincide the least squares estimate of $\underset{\sim}{\theta}$, the maximum likelihood estimate of $\underset{\sim}{\theta}$, the posterior mean and mode, and the value of $\underset{\sim}{\theta}$ most similar to $\underset{\sim}{x}$.

6
Meanings of Probability

In the following, a certain sense of probable, meaning similar to truth (Swedish, sannolik) or having the appearance of truth (German, wahrscheinlich) is traced through the European Languages. In many languages (the English, verisimilar) words with this sense exist, but are rarely used. A number of other languages Crow, Chinese, Japanese, Swahili, Urdu have been explored without finding this sense. Of course, there can be no conclusion that the *real* meaning of probability is similarity to truth. Probable=testable is just as good and also exists in a number of languages.

ENGLISH	LIKELY	LIKE	TRUTH
	verisimilar	similar	verity
DUTCH	waarschÿnligk	gelijk	waar
GERMAN	wahrscheinlich	ahnlich	wahr
NORWEGIAN	sannsynlig	like, lignende	sannhet
SWEDISH	sannolik	lik	sanning
DANISH	sandsynlig	lignende	sandhet
RUSSIAN	pravdapodobny	podobny	pravda
POLISH	prawdapodobny	podobny	prawda
SERBOCROAT	verocatan	jednak, slicann	vernost
FRENCH	probable, vraisemblable	semblable	vrai
ITALIAN	probabile, verosimille	simille	verita
SPANISH	probable, prometedor	semejante	verdad
PORTUGUESE	provavel, verosimul	semelhante	verdade
HUNGARIAN	valoszinuleg	hasonlo	valo
CELTIC	coslach	samhail	fronn
GREEK	$\pi\iota\theta\alpha\gamma o\delta$	$\iota\sigma o\delta$	$\alpha\lambda\eta\epsilon\iota\alpha$

References

1. Achinstein, P., "Variety and Analogy in Confirmation Theory," *Philosophy of Science,* 30, 207-221, 1963.
2. Achinstein, P., "Models, Analogies, and Theories," *The Johns Hopkins University,* 31, 328-349, 1964.
3. Blake, R.M., Ducasse, C.J., and Madden, E. H., *Theories of Scientific Method: the Renaissance Through the Nineteenth Century,* University of Washington Press, 1960.
4. De Finetti, B., "La prevision, ses lois logiques, ses sources subjectives," *Annales de l'Institut Henri Poincaré,* 7, 1937.
5. Hesse, M. B., *Models and Analogies in Science,* London, 1963.
6. Hesse, M.B., "Analogy and Confirmation Theory," *Philosophy of Science,* 31, 319-327, 1964.
7. Jeffreys, H., *Theory of Probability,* Oxford University Press, 1961.
8. Keynes, J. M., *A Treatise on Probability,* London, MacMillan, 1921.

COMMENTS

D.R. Cox:

Does the discussion in the paper of the determination of priors throw any light on the following? There are two (or more) alternative models giving densities $f(y, \alpha), g(y, \beta)$. For a Bayesian comparison we need π_f/π_g the prior odds for the two models and some information about $\pi^{(f)}(\alpha)$, $\pi^{(g)}(\beta)$ the conditional prior densities of the nuisance parameters. Box and Hill gave a solution of a sequential version of the problem and more recently Box and Hensen have given a different solution which seems to give more satisfactory results.

I.J. Good:

I do not see how Dr. Hartigan obtained the Perks-Jeffreys' invariant density in Section 8 and I hope in the final version of his paper he will give more indication of the proof.

I should also like to make a few comments concerning points of terminology.

The expression $P(A.B)/(P(A)P(B))$ is called the *association factor* by J. M. Keynes and others. It is a good name and it is a pity it is not yet universally adopted. Its logarithm is the mutual information between A and B. There has been a little discussion in the literature on the possibility of defining probability in terms of information (See, for example, *Brit. J. Philos. Sc.*, 19, 1968, page 14). I have used the association factor for estimating small probabilities in large pure contingency tables. See, for example, *The Estimation of Probabilities*, M. I. T. Press, 1965.

William Ernest Johnson, the teacher of both Keynes and D. Wrinch, used the expression *permutable* where de Finetti used the word *exchangeable*. Since permutable is at least as good a term and was used by the originator of the idea, I am convinced that it should be adopted, although Johnson did not discover de Finetti's famous theorem. Let's give W. E. Johnson *some* of the credit.

In the last sentence of Section 1, Dr. Hartigan uses the word *odds* where I think he means *probability*. This is no doubt just a slip. The odds ratio is the Bayes or Bayes-Jeffreys-Turing factor.

The ratio of posterior to prior *probabilities* was called the *relative factor* in my 1950 book. (See also my Priggish Principle No. 5)

I mention these points of terminology to indicate my feeling that the choice of terminology should be based partly on good English and mathematical usage and partly on history. Also, when an improved name is invented, some reference to previous names is in order because in this way we might hope to oppose the fractionation of knowledge.

J.C. Koop:

Dr. Hartigan in his paper says "a rather grandiose general principle,... is 'reduce error by averaging over homogeneous sets'". I am not sure whether he also has in mind estimation problems in sample surveys where it is quite usual to put together *similar* or homogeneous units. Dr. Neyman has conclusively demonstrated in his *Lectures and Conferences on Mathematical Statistics* (United States Department of Agriculture, 1938; 2nd edition, 1952) that stratification may fail even for the case of proportional sampling. If we use the variance function of an estimator as a measure of precision (or error), then in general, for finite universes, whether or not one estimator is better than another (given a sampling design) will depend on the sign of the quadratic form in the underlying variates obtained by taking the difference of the variances of the competing estimators. Further if we are considering multivariate characteristics, stratification based on similarity for one or more characteristics may not be successful for controlling error for other characteristics.

D.V. Lindley:

I would like to echo the doubts that have been expressed about the use of prior distributions which depend on the experiment which has been performed. If a Jeffrey's prior, for example, were used then two different physical experiments about the same unknown would have to be entered with different priors, and, furthermore, the final inference from both experiments would depend on the order in which they had been performed.

An even more serious objection, to my mind, is the failure of the Jeffrey's prior to allow for the loss of a degree of freedom for each additional constant that is fitted.

REPLY

I thank the discussants for their comments. A number of persons mentioned the derivation of prior densities, and I will give some more details, although it is only one application of similarity concepts to probability. Given $f(x|\theta)$, the probability density of an observation x given θ, it is necessary to find a suitable probability distribution of θ. Letting $\underline{x}_n = x_1 \ldots x_n$, and letting $\underline{x}'_n$ denote the same observations repeated in a new sequence of n experiments, a similarity judgment is made that $S(\underline{x}, \underline{x}'_n)$ is constant for large n. Let θ be one dimensional with prior density $h(\theta)$, and consider the limiting behaviour of $f(\underline{x}_n)$ given $\theta = \theta_0$.

$$f(\underline{x}_n) = \int f(\underline{x}_n|\theta) \, h(\theta)d\theta$$

$$= \int \Pi f(x_i|\theta) \, h(\theta)d\theta$$

$$= \int \exp \left[\Sigma \log f(x_i|\theta) \right] h(\theta)d\theta$$

Expanding $\log f(x \mid \theta)$ about θ_0 gives

$$f(\underline{x}_n) = \Pi f(x_i \mid \theta_0) h(\theta_0)/J(\theta_0)\sqrt{2\pi}(1 + 0(n^{-1/2}))$$

where

$$J(\theta_0) = \{ E[d/d\theta \log f(x \mid \theta)]^2 \}^{1/2}.$$

Thus

$$S(\underline{x}_n, \underline{x}'_n) = \sqrt{2\pi}J(\theta_0)/h(\theta_0)(1 + 0(n^{-1/2}))$$

if $\theta = \theta_0$; since the similarity judgment is that $S(\underline{x}_n, \underline{x}'_n)$ is constant in the limit, $h(\theta_0) = J(\theta_0)$, Jeffreys' density, as required.

Professor Cox has mentioned a case where it is difficult to apply Jeffreys' density. The reason is that his parameter space is a mixture of two parameter spaces on each of which Jeffreys' density may be computed; but the densities may be improper, and it is not clear how to make up a gobal density by mixing two improper densities. I think this is a real problem. I have no solution.

TWO NONSTANDARD METHODS OF INFERENCE FOR SINGLE PARAMETER DISTRIBUTIONS
James S. Williams
Colorado State University

1
Introduction

I wish to propose two schemes of inference in this paper. The first is similar in some formal aspects to the calculation of logical probabilities using noninformative prior distributions. Ultimately, however, with the second, which I regard as the more promising, I will avoid any recourse to Bayes' formula or postulates about prior information.

The problem in its simplest form involves drawing a random sample from a population in which the measured variate T is continuously distributed over a known range of values t. The distribution function is differentiable and completely specified up to a single parameter θ. This constant can have any value ϑ in a known interval subset of the real line. There is either no sampling distribution of the parameter or it is understood that such a distribution is not to be used in making inferences about θ. The possible inferences concerning values of the constant are to be compared in light of the sample outcome.

Unless it is specified otherwise, only this most elementary problem will be considered. For this there are the following three reasons. The ideas I want to discuss can be brought out fully in the context of estimating a single unknown parameter. The development of these ideas to the point where general solutions can be obtained entails mathematical analyses with considerable difficulties. These have not yet been overcome. Finally, the problem has already been solved in at least three different ways which are well known and to which we need to refer. These are: (a) by use of the standard sampling theory of estimation as proposed by Sir Ronald Fisher (in which the constant is estimated in an efficient manner and the precision of the estimator calculated or estimated), and by J. Neyman (in which confidence intervals with optimal properties are calculated for the constant), (b) by use of a like-

314

lihood theory in which the sample likelihood function is evaluated and compared for selected points in the range of the parameter, and (c) by use of an inverse probability theory in which a prior distribution is selected for the parameter. My ideas have been borrowed in parts from all three of these and were developed by trying to relate one aspect of the third to the other two. The ultimate goal is to measure the effect of a sample outcome on each possible inference.

Very briefly, I understand the continuing interest in the problem of estimating a single parameter to be the following. The outcome of a random sample commonly is used to order the points in the range of the parameter according to their decreasing merits as *approximations* to the value of the unknown constant. It is felt that if any single value is selected as the approximation, if approximations in any pair are compared, or if an interval is chosen for the approximation, one should be able to express quantitatively how strongly he feels or should feel that any one of these is in fact not an approximation but correct. In no way should such a measure be arbitrary. It should depend primarily on the sample outcome and secondarily on information gained independently of the sample or perhaps on beliefs held in advance of any empirical work. These are difficult goals to achieve as can be seen from the objections to each of the solutions described above. Those to the first, center on the lack of unique guidelines for the choice of an estimator or confidence interval and on the absence of results indicating how to measure uncertainty for any chosen set of values in the range of the parameter. Those to the second, mainly involve, I believe, the simplicity of the assumption that a comparison of likelihood values without specific reference to the precision of sampling is adequate to describe one's beliefs about possible approximations. Those to the third concern the concept and choice of a prior distribution function.

I will not discuss the sampling theory and likelihood function approaches further, other than to state my own position as a strong advocate of the former. This comes about from my position as an applied statistician working in the experimental sciences. There I feel the statistician should regard himself as the guardian of objectivity and leave subjective input to the subject matter participants. I am intrigued, however, by the possibility of using a theory of inverse probabilities calculated with noninformative prior distributions. First such a theory greatly simplifies the estimation problem because Bayes formula, in one easy step, incorporates all the informative elements of the sample into the ordering of the points in the range of the parameter and automatically assigns a widely accepted measure of uncertainty to each. Secondly, along with the phrase noninformative prior distribution comes the connotation of a purely objective (scientific) approach to estimation. This the other new or renewed theories of inference descended from Thomas Bayes do not have. However the theory appears to be severely limited by the difficulty and often impossibility of finding noninformative priors and the unsatisfactory interpretations which have been given to these distributions.

The latter of my objections will not stand up if interpretations made with noninformative priors are justified by experience. Despite the fact that I am not convinced at all that these distributions reflect ignorance, or indifference, or a lack of information, they are needed, as Sir Harold Jeffreys has said, to get the theory started and us to the point where calculations can be made.

The problems of defining what a noninformative prior distribution is and describing how to find one are serious. Indeed in the statistical literature of applications of logical probability, one finds the terms logical, flat, vague, indifference, informationless, and noninformative indiscriminantly applied to the same prior distribution function. I have chosen the latter over the others simply because it suggests to me a distribution which fails to identify the parameter, while the others suggest a measurement of belief, knowledge, or preference for the parameter.

In the next section, I take those properties of the established inferential theories that I find useful and attempt to justify two new solutions to the problem of rating inferences. In the following section the first of these is shown to be mathematically equivalent in special cases to the use of posterior density functions derived with noninformative prior densities. In this respect only, it is similar to the inverse probability theories of Sir Harold Jeffreys and Sir Ronald Fisher. Otherwise it is more closely related to likelihood analysis. The second method of inference, which I think is largely new, is discussed in the fourth section where boundary properties of the solution are derived for special cases. The feeling gained from this admittedly inadequate derivation is that the mode of inference is based essentially on comparisons of *pre* and *post* sampling measures of information about the parameter.

The ideas I will discuss are in a formative stage during which it is difficult to present them in precise definitions. Therefore I have used phrases from the vernacular throughout the remaining part of this paper in the hope that these convey the intent behind, if not the exact meaning of, the principal concepts involved. I ask the reader's indulgence for this shortcoming.

2
The Price of Conviction

When one is interested in knowing the true value of a parameter θ, he has open to him inferences of the following types. The value of the parameter θ is ϑ! A satisfactory approximation to the value of the parameter θ is ϑ! The value of the parameter θ is a number in the interval $(\vartheta - \delta_1, \vartheta + \delta_2)$! There seems to be little merit in discussing the first of these if the range of θ is continuous, while the second can be given formal statement by choosing δ_1 and δ_2 properly in the third. Thus only inferences of the form $\theta \epsilon (\vartheta - \delta_1, \vartheta + \delta_2)$ will be considered, and for the remainder of this discussion I will take $\delta_1 = \delta_2 = \delta$ and δ small. These I will refer to as satisfactory approximations for the value of θ and will often denote anyone of them by $\theta \simeq \vartheta$. Since the inferences are not of the form $\theta = \vartheta$, there cannot be just one correct inference.

One who wishes to infer approximate values of θ may adopt either of two points of view. He may be convinced that $\theta \simeq \vartheta$. Then he is faced with the task of proving his point to others and will want to know how difficult that task will be in light of all existing evidence about θ. Or he may consider all possible inferences and regard none as demonstrated to be true. Still he will want to know how difficult it would be to establish each if it were in fact correct.

These two points of view can be united in concept by considering a noninvolved observer (second case) who calculates for each of those who hold equal convictions about different values for θ (first case) his price of conviction. The latter, of course, is to be a numerical measure of the difficulty, in light of existing evidence, of establishing an inference when it is true.

The term conviction is used here as an absolute. Conviction is the upper limit of the grades of belief. Doubt is the lower limit. Thus one cannot be slightly more convinced that $\theta \simeq \vartheta$ than he is that $\theta \simeq \vartheta'$ If he is convinced $\theta \simeq \vartheta$, then he doubts $\theta \simeq \vartheta'$.

In practice one may be convinced, in the light of initial evidence, that $\theta \simeq \vartheta$, his conviction may give way to belief as more evidence is introduced, and finally, if enough evidence is gathered, his belief may return to conviction or terminate in doubt. I contend that this process of human behavior, which has been very successful in scientific inquiry, is largely directed by the evaluations at each step of whether existing evidence points toward ultimate establishment or rejection of an inference.

The problem is that we do not know how best to quantify existing evidence at each step in the scientific process. If variation of θ over the range of possible values is expressed formally in a prior density function, the posterior density function can be analyzed and inferences with high density regarded as likely to be established and those with low density as likely to be discarded. If variation of θ is avoided, even in a formal sense, then the likelihood function based on the possible sampling probabilities of the observed outcome can be analyzed and inferences again evaluated. Or confidence intervals for θ can be examined and points within the intervals regarded as more or less likely to be established and outside the interval as likely to be refuted. However, none of these methods of analysis gives rise to an explicit statement of the price of conviction or how the measure employed in each depends on that price.

Although this price of conviction may not be the only comparative measure of different possible inferences, it is an important one. First, from the viewpoint of establishing an inference, one is interested in the extent to which existing sample evidence is or is not adequate. If precision after sampling is low he will want to sample further until it is high enough to establish or refute the inference. Second, in the comparison of possible inferences, he will want to distinguish from those that still could be established as true, those inferences which are in such discord with the existing evidence that no reasonable amount of future sampling could bring them into accord. Ideally the price of conviction should be graduated on possible approximations to θ so that with experience one could say: these inferences have been ruled out, these are tenable but weakly supported by the existing evidence, and these appear to contain the correct inference!

Now, before we can proceed further to formalize and use the price of conviction, it must be agreed on how one should sample for evidence and exactly what type of evidence is strong enough to establish an inferred approximation.

The latter question will be considered first. My approach is a pragmatic one because I think experience has confirmed the assumption that a hypothesis can be established short of logical certainty. If ϑ is a satisfactory approximation when $|\theta - \vartheta| < \delta$, then there are two essential requirements of sample data of n units that must be met to verify the inference. The first is that relative to the limits of approximation, the estimator θ_n of θ should be close to ϑ, i.e. $|\theta_n - \vartheta| < \epsilon(\delta, n)$ where $\epsilon(\delta, n)$ increases monotonically in n to a limit $\epsilon(\delta)$. The second is that the data should give little credence to any inference that θ is outside the interval $(\vartheta - \delta, \vartheta + \delta)$. (Notice that just as more than one inference is correct, more than one will be verified.) A formal requirement of this type that I find acceptable is that the inference $\theta \simeq \vartheta$ should be justified by a highly favorable likelihood analysis. For example, we could take θ_n as the maximum likelihood estimator, $L_n(\vartheta)$ as the sample likelihood function, choose small ϵ and η, and require

$$\frac{L_n(\vartheta)}{L_n(\theta_n)} > 1 - \epsilon \text{ and } \frac{L_n(\vartheta')}{L_n(\vartheta)} < \eta \text{ for all } \vartheta', |\vartheta' - \vartheta| \geq \delta.$$

Actually the first of the preceding inequalities is not needed for two reasons. One is that if the likelihood function is sufficiently concentrated in the interval $(\vartheta - \delta, \vartheta + \delta)$, the approximation will have been verified even if the function does not spike in the immediate neighbourhood of ϑ. The other is that, under maximum likelihood regularity conditions, the first inequality is equivalent to $\chi^2_{(1)} < 2 \ln(1 - \epsilon)^{-1}$ when $\theta = \vartheta$, i.e. $-2 \ln L_n(\vartheta)/L_n(\theta_n)$ converges to a chi-square random variable rather than to a constant.

The second set of inequalities, which require that the likelihood function be concentrated in the interval of approximation, is the one I will use for establishing inferences.

In answer to the first question of which sampling schemes should be considered, I feel that only sequential sampling, perhaps in batches, is appropriate. Ideally, one would like to continue sampling as long as his belief in an inference is not overthrown, but no longer than he needs to establish his contention.

How then does one measure the price of conviction? Two possibilities come to mind which I will discuss in the following order.

One believes that aside from sampling errors, a precise estimator θ_m when sampled will give an estimate close to ϑ. He then can calculate the probability that $|\theta_m - \vartheta| < \epsilon(\delta, m)$ and relate this to the ultimate fixed sample size needed to establish $|\theta_m - \vartheta| < \epsilon(\delta)$ with high probability, say $1 - \eta$. The calculated probability, which I will call the *chance of success*, is small when $m << n$ due to the imprecision of sampling. However, if it is calculated after each successive sample of size m is drawn, it will change with the accumulated data and will, with certainty, diminish for incorrectly inferred approximations for θ.

More directly related to the inference problem is the expected sample size needed to verify an inference if in fact it is correct. This indeed is the *price of conviction*, and the negative values of successive changes in the price effected by the accumulation of sample evidence are measures of how the data makes easier or more difficult the practical justification of any hypothesis. Such a measure then can be called the *modulus of the sample*. For the noninvolved observer, the distribution of such relative prices and changes over those who hold equal convictions about different values of θ is a formal way of stating the relative difficulties associated with verifying inferred values of θ. When there is a decreasing trend in the modulus for a given ϑ, unshakable conviction will point toward endless sampling, while if the trend is increasing, the prospects of demonstrating $\theta \simeq \vartheta$ will become progressively brighter. One way then, after sample evidence has been gathered, of settling the question of why the inference $\theta \simeq \vartheta$ is better or worse than another inference $\theta \simeq \vartheta'$ would be to compare the expected long run effects of the data on each inference. Although there are fundamental problems yet to be solved before this can be done, the idea I believe can be developed into a useful mode of summarizing how the body of sample data alters the problems of making inferences.

Complete solutions for the chance of success, price of conviction, and sample modulus have not been obtained and are beyond the scope of my present remarks. However in the following two sections I will derive approximate results for special cases which point up some mathematical differences between these two methods of inference.

3

The Chance of Success and Logical Probabilities

The sampling density function of the observable random variable T will be denoted by $f_T(t|\vartheta)$. Assume that maximum likelihood regularity conditions hold for $f_T(t|\vartheta)$. Successive samples of size m are to be drawn and after each, the chance of success on the succeeding samples calculated for each possible value of θ.

Let $\hat{\theta}_m = \theta(T_1, \ldots, T_m)$ be the maximum likelihood estimator of θ and assume m is large enough to use the approximation:

$$\sqrt{m}\,(\hat{\theta}_m - \vartheta) \sim N(0, 1/\sqrt{I(\vartheta)})$$

where $I(\vartheta)$ is Fisher's information function. Let $n(\vartheta)$ be the sample size n for which $P(|\hat{\theta}_n - \vartheta| < \epsilon(\delta)) = 1 - \eta$ when $\theta = \vartheta$ and δ and $\epsilon(\delta)$ are small. Assume that m is very small when compared to $n(\vartheta)$ throughout the range of θ.

It is well known that if η is small, the following solutions are good first approximations.

$$n(\vartheta) = \frac{c^2(\eta)}{\epsilon^2(\delta)I(\vartheta)}, c^2(\eta) < 1,$$

and

$$P(|\hat{\theta}_m - \vartheta| < \epsilon(\delta, m)) = c(\eta)\, \frac{\epsilon(\delta, m)}{\epsilon(\delta)}\, \sqrt{\frac{2}{\pi}}\, \sqrt{\frac{m}{n(\vartheta)}}.$$

The latter, the chance of success prior to any sampling, is proportional to $\sqrt{I(\vartheta)}$ and hence proportional to the noninformative prior density function of θ. It characterizes one aspect of the projected difficulties of inference in the absence of sample data.

After drawing one sample of size m, the chances of success can be recalculated for the draw of a second sample of the same size. The important thing to remember is that if such a sample is drawn, inferences will be judged again in terms of the combined sample outcome. When m is sufficiently large, as has been required,

$$\hat{\theta}_{2m} = \theta(T_1, \ldots, T_m, T_{m+1}, \ldots, T_{2m}) \doteq \frac{1}{2}[\theta(T_1, \ldots, T_m) + \theta(T_{m+1}, \ldots, T_{2m})]$$

$$= \frac{\hat{\theta}_m + \hat{\theta}_{2n-m}}{2}, \text{ say.}$$

Thus for a large part of the range of θ,

$$P(\,|\hat{\theta}_{2m} - \vartheta| < \epsilon(\delta, 2m)\,|\hat{\theta}_m) = P[\sqrt{mI(\vartheta)}\,(\vartheta - \hat{\theta}_m - 2\epsilon(\delta, 2m))$$

$$< \sqrt{mI(\vartheta)}\,(\hat{\theta}_{2m-m} - \vartheta)$$

$$< \sqrt{mI(\vartheta)}\,(\vartheta - \hat{\theta}_m + 2\epsilon(\delta, 2m))\,|\hat{\theta}_m]$$

$$\simeq 4\epsilon(\delta, 2m)\,\sqrt{\frac{mI(\vartheta)}{2\pi}}\,\exp\left[-\frac{mI(\vartheta)}{2}(\hat{\theta}_m - \vartheta)^2\right]$$

$$\simeq \frac{2\epsilon(\delta, 2m)}{\epsilon(\delta)}\,\sqrt{\frac{2}{\pi}}\,c^2(\eta)\,\sqrt{\frac{m}{n(\vartheta)}}\,\frac{L_m(\vartheta)}{L_m(\hat{\theta}_m)}.$$

The relative chances of success after sampling are proportional to $\sqrt{I(\vartheta)}\,L_m(\vartheta)$. This however is the posterior density function of θ derived with the noninformative prior density function.

This solution looks like an inverse probability derivation because δ and hence $\epsilon(\delta, m)$ are small, and for larger values of these parameters, i.e. for common width interval approximations to θ, the relative average chances of success can be calculated by integrating $\sqrt{I(\vartheta)}L_m(\vartheta)$ over an appropriate interval. Since $\sqrt{I(\vartheta)} \propto 1/\sqrt{n(\vartheta)}$, one is tempted then to call the noninformative prior distribution of θ that of the relative ease of proving a point.

Actually analysis of the relative chances of success in this special case turns out mathematically to be a likelihood analysis. The interest in $\sqrt{I(\vartheta)}$ is just for a formal starting point, a statement of relative difficulties unaffected by sample evidence, and not a basis on which to judge the likely truth of inferences. The change effected in this measure, the direct input of the sample evidence, is given by the likelihood function in $\sqrt{I(\vartheta)}L_m(\vartheta)$. Thus when the effect of the data is isolated, it is seen to be expressed wholly through the likelihood function.

The chance of success, in general, has in common with the sample likelihood the disadvantage of being the probability of an event which for every value of θ is most unexpected. It differs from likelihood in being a predictive measure.

4

The Sample Modulus and the Information Function

Two features can be deduced for any solution for the price of conviction. The first is that with the exception of values of θ near the δ neighbourhoods of the sequence of maximum likelihood estimates, the price of conviction will tend to increase with successive sampling. As δ approaches zero then, the price at every point, except the terminal maximum likelihood estimate, will show accumulated increase. This is as it should be because most possible sample outcomes constitute evidence unfavorable to the hypothesis $|\theta - \vartheta| < \delta$. Unless $\hat{\theta}_m$ is in the approximation neighbourhood of ϑ at the end of sampling, the difficulty of establishing $\theta \approx \vartheta$ is greater than it was initially when there was no evidence to weigh against the inference. The overall view of the noninvolved observer then will be that evidence has made inferences about θ values near $\hat{\theta}_m$ easier to establish, progress has in fact been made, and those far from $\hat{\theta}_m$ much more difficult to verify. The second feature is that the price of conviction will be very large if δ and η are small. Then the changes brought about by a few sample observations will be small relative to the initial prices. Strictly for a comparison of the effects of sample evidence on different inferences this is not a disadvantage. For the problem of finding sets of inferences which are untenable because of the greatly increased, projected effort of verifying them, it does pose difficulties.

Some idea of the form of the modulus function can be obtained for density functions which satisfy maximum likelihood regularity conditions by using sample-size bounds on the sequential probability ratio test.

Under initial conditions we want to find the expectation of $N = N(\vartheta)$ where

$$\frac{L_N(\vartheta')}{L_N(\vartheta)} < \eta \text{ for all } \vartheta', |\vartheta - \vartheta'| \geq \delta,$$

first holds and $\theta = \vartheta$. Denote this, the initial price of conviction, by $n(\vartheta)$. Set $\eta = \beta/(1-\alpha)$ and consider the two sequential tests given by

$$\left\{ \begin{aligned} &\text{Accept } \theta = \vartheta \text{ if } \frac{L_N(\vartheta + \Delta)}{L_N(\vartheta)} < \frac{\beta}{1-\alpha} = \eta \\ &\text{Reject } \theta = \vartheta \text{ if } \frac{L_N(\vartheta + \Delta)}{L_N(\vartheta)} > \frac{1-\beta}{\alpha} = \eta + \frac{1-\eta}{\alpha} \\ &\text{Continue sampling otherwise} \end{aligned} \right.$$

where $\Delta = \delta$ or $-\delta$. If η is held constant while $\alpha \to 0$, then $\beta \to \eta, (1-\beta)/\alpha \to \infty$, and each test terminates only when $\theta = \vartheta$ is accepted. Wald's approximation for expected sample size can be applied to each of these tests with the result that

$$\lim_{\alpha \to 0} \xi(N(\vartheta)) = \frac{\ln(1/\eta)}{\xi[\ln f_T(T|\vartheta) - \ln f_T(t|\vartheta + \Delta)]}$$

It is obvious then that

$$n(\vartheta) \geq \max_{\Delta=-\delta, \delta} \frac{\ln(1/\eta)}{\xi[\ln f_T(T|\vartheta) - \ln f_T(T|\vartheta+\Delta)]},$$

and that when δ is very small, say $\delta = d\vartheta$, we can use the bound

$$n(\vartheta) \geq \frac{\ln(1/\eta)}{(d\vartheta)^2 I(\vartheta)}.$$

In order not to stop the development here, I will conjecture from this that for $\delta = d\vartheta$, a satisfactory approximation will be

$$n(\vartheta) = \frac{\ln(1/\eta^*)}{(d\vartheta)^2 I(\vartheta)},$$

for some $\eta^* < \eta$.

Next consider the updated price of conviction after a sample of size m has been drawn. The problem now is to find the conditional expected value of $N(\vartheta)$ for which

$$\frac{L_{N+m}(\vartheta')}{L_{N+m}(\vartheta)} = \frac{L_m(\vartheta')}{L_m(\vartheta)} \frac{L_N(\vartheta')}{L_N(\vartheta)} < \eta \text{ for all } \vartheta', |\vartheta'-\vartheta| > \delta,$$

first holds. Denote this conditional expectation by $n(\vartheta, t_1, \ldots, t_m) = n(\vartheta) + m(\vartheta, t_1, \ldots, t_m)$ and notice that $L_m(\vartheta')/L_m(\vartheta)$ is a constant in the stated inequality requirements.

If $L_m(\vartheta')/L_m(\vartheta) < \eta$, the inference has been verified, $n(\vartheta, t_1, \ldots, t_m)=0$, and $-m(\vartheta, t_1, \ldots, t_m)=n(\vartheta)$. Otherwise the solution can be bounded again by considering the pair of likelihood ratio tests with $L_N(\vartheta+\Delta)/L_N(\vartheta)$ and the bounds

$$\beta^*/(1-\alpha^*) = \eta L_m(\vartheta)/L_m(\vartheta+\Delta)$$

and

$$(1-\beta^*)/\alpha^* = [\eta + (1-\eta)/\alpha]L_m(\vartheta)/L_m(\vartheta+\Delta),$$

i.e.

$$n(\vartheta, t_1, \ldots, t_m) \geq \max_{\Delta=-\delta, \delta} \frac{\ln(1/\eta) + [\ln L_m(\vartheta+\Delta) - \ln L_m(\vartheta)]}{\xi[\ln f_T(T|\vartheta) - \ln f_T(T|\vartheta+\Delta)]}.$$

For $\delta = d\vartheta$, we again will approximate the true solution by a convenient choice of $\eta^* (\eta^* < \eta)$, i.e.

$$n(\vartheta, t_1, \ldots, t_m) = \frac{\ln(1/\eta^*) + d\vartheta \left| \frac{\partial}{\partial\vartheta} \ln L_m(\vartheta) \right|}{(d\vartheta)^2 I(\vartheta)}$$

$$= n(\vartheta) + \frac{\left| \frac{\partial}{\partial\vartheta} \ln L_m(\vartheta) \right|}{(d\vartheta) I(\vartheta)}.$$

322

The modulus function in this solution is

$$-m(\vartheta, t_1, \ldots, t_m) = -\frac{\left|\frac{\partial}{\partial\vartheta}\ln L_m(\vartheta)\right|}{(d\vartheta)\, I(\vartheta)}.$$

If it is convenient to calculate the modulus as percent change in the initial price of conviction we use

$$-\frac{m(\vartheta, t_1, \ldots, t_m)}{n(\vartheta)} = \frac{-d\vartheta}{\ln(1/\eta^*)}\left|\frac{\partial}{\partial\vartheta}\ln L_m(\vartheta)\right|.$$

When the modulus is needed only for comparing values of ϑ, the constant multipliers can be omitted.

The solution for the modulus function is inadequate for all but very small δ because $-m(\vartheta, t_1, \ldots, t_m)$ should be positive for ϑ within the δ neighbourhood of $\hat{\theta}_m$. For the cases where the approximation can be used, the function may be used as a measure of the relative change in information effected by the sample evidence. One can see this by noting that

$$n(\vartheta)\, I(\vartheta) = \frac{\ln(1/\eta^*)}{(d\vartheta)^2}$$

and then by setting

$$n(\vartheta, t_1, \ldots, t_m)\, I(\vartheta, t_1, \ldots, t_m) = \frac{\ln(1/\eta^*)}{(d\vartheta)^2}$$

so that the information value is directly related to the ease of establishing an inference. The solution for the altered information function is

$$I(\vartheta, t_1, \ldots, t_m) = \frac{I(\vartheta)}{1 + \dfrac{d\vartheta}{\ln(1/\eta^*)}\left|\dfrac{\partial}{\partial\vartheta}\ln L_m(\vartheta)\right|}$$

$$\Longleftrightarrow -\frac{m(\vartheta, t_1, \ldots, t_m)}{n(\vartheta)} = \frac{I(\vartheta) - I(\vartheta, t_1, \ldots, t_m)}{I(\vartheta)}.$$

This has a particularly simple form if $f_T(t\,|\,\vartheta)$ is a member of the exponential class of density functions,

$$-\frac{m(\vartheta, t_1, \ldots, t_m)}{n(\vartheta)} = -\frac{md\vartheta}{\ln(1/\eta^*)}\, I(\vartheta)\,|\tau(\hat{\theta}_m) - \tau(\vartheta)|,$$

and shows that the decline in information outside the maximum likelihood neighbourhood is small if m is small relative to $d\vartheta/\ln(1/\eta^*)$ and/or $\theta \simeq \vartheta$ is initially very difficult to establish. For large m the relative loss is great everywhere except in the immediate neighbourhood of $\hat{\theta}_m$ and for inferences with extremely low initial precision. The latter however are easily distinguished from $\theta \simeq \hat{\theta}_m$ by the extreme values of $-m(\vartheta, t_1, \ldots, t_m)$.

323

5
Discussion

The precision measures I have presented have a basic feature in common with inverse probabilities. There is a prior measure which can be combined with sample data to obtain a post sampling measure. The essential difference is that a prior probability is usually thought of as a measure of the truth of an inference while the initial price of conviction is a measure of how difficult it would be to establish an inference if indeed it were true. Thus the inverse probability is a post dictive measure for inference while the price of conviction remains a predictive measure at each step. For the purpose of analyzing inferences, it seems appropriate to separate the initial price of conviction from the post-sampling price and study only the changes effected by the sample. In this respect the measures are similar to relative likelihood which is the sample input function for inverse probability.

Because the initial price of conviction is not a constant unless θ is a location parameter, it is not clear how the modulus function is to be used to rate inferences. For example if $T \sim N(0, \theta)$, our solutions for the modulus functions are proportional to $-|\sum_{j=1}^{m} t_j^2 - m\vartheta|$ and $-|\sum_{j=1}^{m} t_j^2 - m\vartheta|/\vartheta^2$. The first of these is a triangular function of ϑ which indicates that expected sample sizes for confirming inferences on values of θ much larger than the maximum-likelihood estimate have increased greatly over what they originally were. The second however indicates that such increases are negligible on a relative scale, being of the same order as increases in the immediate vicinity of the maximum likelihood estimate. If one originally anticipated the large sampling effort to verify one of these inferences, there has been very little relative change in the expected price even though the expected change due to observing m outcomes has been enormous. On which scale does one compare inferences? Only if we restrict inferences to the scale $\phi(\vartheta) = \int_{-\infty}^{\vartheta} \sqrt{I(u)} \, du$ where $I(\phi)$ is a constant does there seem to be a clearcut answer.

There is no conceptual difficulty to extending functions of the price of conviction to multiparameter problems and less restrictive interval approximations that $\delta = d\vartheta$ although mathematical formulations and solutions will be difficult. There appears to be no problem of uniqueness associated with the price of conviction. For each sequential estimation procedure and width of interval approximation one might use to verify or refute an inference, there is an associated price of conviction. Indeed several such measures could be compared for different estimation methods to determine if one is best for the purpose of discriminating among inferences.

Hopefully, among the possible investigations one can make of functions of the price of conviction, some will result in new procedures for choosing estimators, summarizing data, and quantifying differences in the plausibility of possible inferences.

COMMENTS

G.A. Barnard:

I agree most strongly with Dr. Williams that the statistician should regard him-self as the guardian of objectivity and leave subjective input to the subject mat-ter participants. In connection with the points that Dr. Good and Dr. Rubin have made about consultation with the "client", I think we ought to ask just who the "client" is. Is it, for instance, the individual physicist we are conversing with, or physicists in general? Or is it the scientific community in general? It is rare that the statistician gets consulted by an experimental scientist to find out how the data should influence *him*; he is more often concerned with using sta-tistics in persuading his fellow scientists.

To return to Dr. Williams, I would like to say I feel convinced with him that the possibility of indefinite repetition of experiments is an essential feature of scientific inference, and because his ideas embody this notion I hope he will develop them further; though I must admit that I am not clear on how, for instance, he would deal with nuisance parameters.

D. J. Bartholomew:

I wish to use one of the main ideas of this paper as a stick to beat the Bayesian's. Dr. Williams believes that experimentation should be sequential and I agree with him. This allows us to change our intended sample size in the light of our accumulated knowledge. The following example shows why I think the Bayesian should be prepared to add to his philosophy some means of deciding on an ex-periment, or sampling rule. Suppose that we wish to make an inference about the mean, θ, of a normal distribution with known variance. Let the prior dis-tribution for θ be normal with mean μ and variance σ^2. Now suppose that n observations are obtained and that they are clustered around a point so far from μ that the prior information is clearly wrong. There is nothing in the logic of Bayesian method to cope with this situation. The routine application of Bayes theorem will lead to a posterior density for θ centred somewhere between the prior and the data. If the "weight" in the prior and in the data are of the same order any conclusions based on the posterior density will clearly be misleading.

What kind of additional rule is required to protect the Bayesian from the consequences of his own logic? My own suggestion, which seems to accord with common sense, is that further data should be obtained. Ultimately a point will be reached when the prior information is either swamped or confirmed and the dilemma is resolved. Another possibility is to conclude that the prior informa-tion is wrong (as it obviously is) and to ignore it. However this view seems to strike at the very root of the Bayesian philosophy. Prior information is either *real* information or it is not. If the former, it must be used whatever the out-come of the experiment. If the latter, a basic ingredient of the Bayesian ap-proach has been thrown overboard.

I have one question. Although Dr. Williams has made a point of emphasizing that his priors are non-informative I presume that his ideas would still work if an informative prior were to replace his non-informative prior. Is this correct?

V.P. Godambe:

I find the investigations of the relationships between frequency probability and Bayes probability implied in the present paper and in the previous works of Bartholomew, Jaynes and Pratt very stimulating.

I wonder if these considerations could enable one to determine the prior distribution as follows: Let x be a random variate, with the given frequency function $f(x|\theta)$, θ being the unknown parameter. Now suppose we think that in this situation the appropriate frequency statement is $\Pr(|t - \theta| \leq \epsilon |\theta) \simeq \alpha(\theta) . \epsilon$, t being some statistic and α a known function of θ. Next for a prior ξ on θ let $P_\xi(x)$ be the predictive density of x. Then ξ should be obtained as a solution of the equation $P_\xi(t = \theta) = \alpha(\theta)$.

I.J. Good:

[Comment on page 315] Near the beginning of Dr. William's paper he says "I feel the statistician should regard himself as the guardian of objectivity and leave subjective input to the subject matter participants". I think he ought to have drawn an inference from this position. For, of the three methods of estimation that he lists on pages 314-315, only the likelihood theory is convenient for the incorporation of the subjective input of the customer, and the first method (orthodox interval estimation) is especially inconvenient for this purpose. If I were using the third method in any practical problem, I would cross-examine the customer in order to obtain information concerning his subjective input. It is often convenient however to use orthodox methods in order to save trouble. This is an example of rationality of type II.

There is a disadvantage in leaving all the subjective input to the customer. It is that one of the first responsibilities of a statistician is not to decide dogmatically, in advance, how he is going to analyze data, but to look at the data in an unbiased manner. Let me repeat that: I think that rule 1 for the statistician is *examine the data*. By examining the data the statistician might notice some new hypotheses that are not of undue complexity. They will suggest new experiments, but experiments are usually expensive, so significance tests are needed. Thus the apparently obvious rule "Look at the data" leads at once to the so-called "heresy" of testing hypotheses not thought of in advance.

You certainly would not recommend that an experiment be performed to test every hypothesis however complicated and farfetched it seems to you. If you are at all observant you will notice many absurdly farfetched and complicated features of the data that you would not even *mention* to your customer, whether you apply a formal significance test or not. (I doubt if you could talk fast enough.) This shows that you must have judged their initial (prior) probabilities to be too small, *without* consulting the customer. This shows that *all statisticians are Bayesians*, at least if they use Rule 1. They have been "writing prose all their lives". Sometimes of course you will need to consult the customer in order to arrive at some idea of how farfetched a hypothesis appears, but some customers have very bad judgment about the initial probabilities of complex hypotheses. In such cases I think these initial probabilities should be decided by the statistician and the customer in collaboration by careful questioning. This can be difficult and dangerous but it has to be faced.

326

I should like to ask whether the notion of the price of conviction is related to the device of imaginary results.

R. Prentice:

I would like to comment regarding the concept of non-informative prior as mentioned on page 315 and subsequently. Such a prior seems objective only in that it supposes more information than is available each time (makes the same mistake each time) rather than objective in the sense that it consists only of substantiated information in the experimental situation. Certainly an experimenter who has spent a good deal of work in previous experimentation or examination of how parameter values can be weighted on the grounds of what he knows of the experiment is going to be very disgruntled if the prior distribution thus developed agrees closely with prior deemed non-informative for the response distribution. However, I suspect that this criticism has often been made before and the point I really wish to make is the following: If it is felt worthwhile to consider the problem where there is no pre-experiment knowledge about actual parameter values (either because the objective is in Professor Barnard's terms scientific or pre-experiment knowledge is too diffuse to support a prior) and if we desire to make probability statements about a parameter we should arrive at such statements by means of what is physically present in the experiment. There is a method of inference working in this direction. Fraser's structural model supposes no pre-experiment knowledge about which parameter values (in a specified space) are most reasonable but does suppose some objective physical knowledge as to how the parameter arose (as opposed to was generated). This knowledge (parameter acts on physical error to produce the response) can be used very advantageously to produce a structural distribution for a parameter (and as Fraser mentioned last week the reasons for the successive reductions in the analysis are entirely different from any previous method). That such a derived structured distribution may be the same (or close to) a derived posterior distribution with an "informative prior" is no contradiction. In both cases we have something in addition to the classical model for the response and with respect to the experiment these items have been of similar use in estimating the parameter.

In addition, the fact that we can derive a structural distribution without presupposing a prior distribution tells us that the statement on page 316 attributed to Sir Harold Jeffries requires revision.

REPLY

When I wrote this paper, I realized "guardian of objectivity" was a pretentious phrase which might elicit negative responses from several quarters. It has, both in the open and written discussion, so now my readers must be assured that I still subscribe strongly to what was intentionally penned.

Now in answer to some of the criticism, I allow that a person may be both a statistician and a scientist. Those of us who were at one time trained in Physics or Genetics and retain active interests in these fields will naturally judge and interpret more aspects of our experimental data than the mathematician become only statistician can or should do. We also should be able to construct better models, formulate more meaningful hypotheses, and detect in the data unanticipated irregularities which need to be classified as relevant or irrelevant to the making of subject-matter inferences. However in this dual role, I still want to separate my statistical work, the collection and analysis of the data, from my scientific work of judging what inferences from the data should be made and reported. The former is the basis for the latter, and in my opinion speculations about values of a parameter, which are not formalized in testable hypotheses, may be used to influence one's judgment, but should not affect his statistical calculations except in the following strictly contingent manner. I may ask myself, "If this assumption about certain parameters were true, how would the results of my statistical analysis be altered."

Because I can agree with him almost completely, Dr. Good has convinced me that I do not know what a Bayesian is, a conclusion enforced by much of the discussion at this symposium. Only when he says . . . "you must have judged their initial (prior) probabilities to be too small . . ." do I feel he overstates the operation of my subconscious. I rather think I suggest reasonable explanations based on the methods and concomitant circumstances of sampling that produced my information to judge irrelevant many unusual features of a large set of data. If there is any doubt about the reasonableness of my explanation, I do call the experimenter's attention to the unusual data. This all is Rule 2 common sense which does not require the calculation of conditional probabilities with a modified Bayes' postulate.

The price of conviction is an imaginary result, as Dr. Good suggests, in so far as one is unable to specify how precisely θ must be estimated in order to establish the truth of an inference to everyones satisfaction. This is not as essential, however, for the application of the concept to the comparison of possible inferences. It is not however an example of "the device of an imaginary result" which Dr. Good proposes in his theory of probability as a method of bounding prior probabilities.

I presume Dr. Bartholomew questions the appearance of a noninformative prior in the solution for the chance of success. My point is that this looks like a solution using a prior density function with Bayes' Theorem, but in fact it is not. The resemblance is interesting, but I concluded not significant for the definition, use, and understanding of noninformative priors or of the chance of success. By intention, no use of Bayes arguments was made in developing either the price of conviction or the chance of success.

I owe users of likelihood an apology. Unintentionally I implied they do not strive for objectivity, and Barnard has eloquently refuted this in his reply. Dr. Barnard also raised the question of applying the price of conviction concept when nuisance parameters exist. Undoubtedly this problem will be as difficult to get around as it is in all other inferential set ups except the subjective. In fact it must be the same problem because the price of conviction is calculated in terms of the manner in which one plans to verify an inference.

Is it not correct to say Sir Harold Jeffreys deduced the noninformative prior distribution of the location and scale parameters from reasonable properties of a predictive distribution function for X? At least I interpreted his methods, when they were criticized by Fisher in a heated exchange, as being such. Then in answer to Dr. Godambe's inquiry, I am of the opinion that the use of predictive distributions to define prior distributions is an interesting approach, but no longer a very fruitful one until someone comes out with other clever desiderata, like Jeffreys', for the predictive function.

I want to agree with Dr. Prentice and end to my satisfaction the controversy over Dr. Fraser's work, but I cannot. Probabilities can be calculated from structural distributions if one presupposes a postulate of irrelevance that in the Bayes framework can be accounted for with a noninformative prior. In fiducial probability that postulate must be stated, as I see it, along with the definition of the difference between probabilities given or conditioned on an observed or assumed outcome. The comment attributed to Jeffreys should be taken to heart by the proponents of structural analysis. Please make structural analysis acceptable to most of us by carefully setting out and explaining the key axiom, or call it a rule or convention if you wish, that justifies our using the structural distribution for calculating weights of evidence.

MEASURING INFORMATION AND UNCERTAINTY
Robert J. Buehler
University of Minnesota

Abstract

Several methods are described for assessing subjective judgments about random variables. These involve payoff functions, bets, questioning schemes, and sequences of choices. The same methods can be used with unknown quantities which are not random variables. The method of payoff functions is considered as a way of describing knowledge about unknown parameters.

1
Introduction

Any number of mathematical models can be described involving unknown quantities about which various kinds of information may exist. Thus X may be an unknown value of a random variable whose distribution is known to be described by a density function f. One way of stating our knowledge about X is to state its density function; some other ways will be considered in the following sections. Some other examples of unknowns are: (a) the value of a random variable whose distribution is only partly known; (b) the value of a parameter when data from a corresponding distribution are available; (c) a future value from an unknown distribution when some past values are known.

It is possible to distinguish levels of uncertainty. On the one hand we may not know X, but we may know its density f with certainty. On the other hand, there may also be uncertainty in our knowledge of f—an uncertainty at a different level.

Our approach is subjective and behavioristic. Individuals will be seen to reveal their knowledge or beliefs by choices which are rewarded in various ways. Superior knowledge should be rewarded by a greater expected payoff.

330

In Section 2 we consider several ways of specifying knowledge about a discrete random variable. Section 3 gives extensions to the continuous case. Section 4 considers the application of these techniques to problems of uncertainty about unknown parameters.

2

Some Ways of Describing Knowledge of a Discrete Distribution

Let A_i denote mutually exclusive and exhaustive outcomes and p_i their probabilities, $\Sigma p_i = 1$. Of course the pairs (A_i, p_i) serve to describe completely the distribution in question. However we find it possibly more enlightening to take a rather more subjective or behavioristic approach wherein an individual's beliefs are learned not necessarily from his assertions but preferably from his actions.

2.1 Payoff functions. The subject is informed that he is to give a probability vector $\hat{p} = (\hat{p}_1, \ldots, \hat{p}_n)$, and that after a random outcome A_i is observed, he will be rewarded by an amount $g_i(\hat{p})$. (Negative values would be penalties.) Three payoff functions which have been suggested are

$$\text{logarithmic:} \quad g_i(p) = \log p_i \tag{2.1}$$

$$\text{quadratic:} \quad g_i(p) = 2p_i - \sum_j p_j^2 \tag{2.2}$$

$$\text{spherical:} \quad g_i(p) = p_i / \{\textstyle\sum_j p_j^2\}^{\frac{1}{2}}. \tag{2.3}$$

It is easily verified that all of these, and linear functions of them, *encourage honesty* in the sense that the expectation is maximized when $\hat{p} = p$; that is,

$$\Sigma\, p_i\, g_i(p) \geq \Sigma\, p_i\, g_i(\hat{p}) \tag{2.4}$$

for all probability vectors $\hat{p}$, with equality only when $\hat{p} = p$. The use of payoff functions was first suggested by Good (1952), who favored the logarithmic payoff for information theoretic reasons. McCarthy (1956) characterized payoffs which encourage honesty. De Finetti (1962) suggested the quadratic payoff. Additional references and methodological suggestions can be found in Winkler (1967).

2.2 Bets. A bet can be considered mathematically to be a triple (s_1, s_2, B) where s_1 is the stakes lost by the subject if B fails to occur and s_2 is the stake he wins if B occurs. Ideally we may suppose the subject will accept either side of a fair bet and will accept the favorable side of any non-fair bet — that is one having non-zero expected payoff. A complete list of acceptable bets (as s_1, s_2 range from 0 to 1, say, and as B ranges over all combinations of outcomes A_i) will be more than sufficient to determine the probability vector from which acceptability of bets is determined. In a less idealized case, a consistent subset of acceptable bets might determine a region of values rather than a unique value of the probability vector.

2.3 Questioning schemes. In the context of communication theory, outcome A_i may be thought of as the ith message and p_i its probability. The Shannon information in message A_i is $-\log_2 p_i$ bits, and the average information in a random message is

$$I_S = -\Sigma \, p_i \log_2 p_i \qquad (2.5)$$

bits. The interpretation of I_S is that ideally it is the expected number of binary questions required to determine the message with certainty. Thus with a probability vector $(\tfrac{1}{2}, \tfrac{1}{4}, \tfrac{1}{4})$ we might first ask, "Is the message A_1?" If *yes*, we are finished; if *no* we ask, "Is the message A_2?" The expected number of questions is 1.5, which in this case exactly equals the Shannon information. In general I_S will be a lower bound for the expected number of binary questions; and known results of coding theory give necessary and sufficient conditions the bound can be achieved (see for example Ash (1965), Section 2.5). Similar results are known to hold when binary questions are replaced by m-fold questions for any fixed m; here the coding analogy involves a code in an m-letter alphabet. For any fixed m and any *finite* discrete distribution ($i = 1, \ldots, n$), a constructive procedure is known to give the *Huffman code,* which is optimal in requiring the smallest average number of letters; analogously it produces the questionnaire having the smallest average number of questions.

A natural extension would allow the questioner to change the base m from question to question. The problems here apparently have not arisen in coding theory (the change in size of alphabet is unnatural), but have received some attention in the theory of questionnaires (Picard, 1965). Some extensions of this coding and questionnaire theory are presently under study in which a charge of $\log_2 m$ is made for each m-fold question. It can be shown that I_S then continues to be a lower bound for the expected charge for any questioning scheme (G. Duncan, unpublished).

For present purposes we may suppose that the specification of a questioning scheme reflects beliefs about the probability vector. The subject knowing the vector exactly will, in principle, be able to determine an optimal questioning scheme (whether restricted to binary questions, for example, or in other ways). The performance of a subject's questioning scheme, relative to optimal performance, will be a measure of the accuracy of his knowledge of the true probability vector.

Lindley's (1956) theory of information in experiments compares the Shannon information in the prior and posterior distribution of a parameter. Although the parameters are typically continuous rather than discrete, one of Lindley's results may be paraphrased in the present framework roughly as follows: The true parameter value can be determined with a smaller expected number of binary questions from the posterior distribution than from the prior distribution.

2.4 Evaluation of probabilities by a sequence of choices. To simplify the discussion suppose there are just two possible outcomes: A and the complement of A. To determine a subjective assessment $\hat{p}$ of $p = P(A)$, we first suppose that a fair coin is available which can be used to define outcomes having known probabilities $\frac{1}{2}, \frac{1}{4}, \frac{3}{4}, \frac{1}{8}, \frac{3}{8}, \frac{5}{8}, \ldots$, etc. Let B_r denote an outcome having known probability r. A sequence of choices of two prospects will be offered to the subject. At any step, Prospect A is a payoff of $c(r)$ if A occurs, and Prospect B is a payoff of $c(r)$ if B_r occurs. For the first step we take $r = \frac{1}{2}$. Then choice A presumably indicates $\hat{p} \geq \frac{1}{2}$, while B indicates $\hat{p} \leq \frac{1}{2}$. For the second step we take $r = \frac{3}{4}$ if A was chosen and $r = \frac{1}{4}$ if B was chosen. The second choice will indicate one of the four values $r = j/8$ ($j = 1, 3, 5, 7$) for the third step, and after k choices, the range of $\hat{p}$ is narrowed to 2^{-k}. Assuming that the choices are all consistent with some fixed $\hat{p}$ value, we may calculate at any stage the expected payoff, say $U(\hat{p}, p)$. As in Section 2.1, we can say that the scheme "encourages honesty" if

$$U(\hat{p}, p) \leq U(p, p) \text{ for all } \hat{p}, p. \tag{2.6}$$

Condition (2.6) will not be satisfied automatically, but will depend on the choice of values $c(r)$. This is so because the first choice for example involves not only the first prospects but later ones as well. Thus if $c(\frac{1}{4})$ and $c(\frac{3}{4})$ were equal, the subject would be inclined to choose A at the first step in order to have the chance to choose $B_{\frac{3}{4}}$ rather than $B_{\frac{1}{4}}$ at the second step. It can be shown (A. Hendrickson, unpublished) that the following choice encourages honesty at each step: $c(\frac{1}{2}) = \frac{1}{2}$, $c(\frac{1}{4}) = \frac{1}{4}$, $c(\frac{3}{4}) = (\frac{2}{3})(\frac{1}{4})$, $c(\frac{1}{8}) = \frac{1}{8}$, $c(\frac{3}{8}) = (\frac{2}{3})(\frac{1}{8})$, $c(\frac{5}{8}) = (\frac{2}{3})(\frac{4}{5})(\frac{1}{8})$, $c(\frac{7}{8}) = (\frac{2}{3})(\frac{4}{5})(\frac{6}{7})(\frac{1}{8})$, etc.

For the case of more than two possible outcomes the procedure could be generalized in many ways.

3

Generalizations to Continuous Distributions

3.1 Payoff functions. Knowledge of a continuous random variable X can be described by its density function f. Let f denote the true density and $\hat{f}$ the guessed or estimated density. The payoff to the subject who provides the function $\hat{f}$ can be made to depend on $\hat{f}$ and on an observed value x of X. The continuous analogs of the three payoff functions given in Section 2.1 are:

logarithmic: $\quad g(x, \hat{f}) = \log \hat{f}(x) \tag{3.1}$

quadratic: $\quad g(x, \hat{f}) = 2\hat{f}(x) - \int \hat{f}^2 dx \tag{3.2}$

spherical: $\quad g(x, \hat{f}) = \hat{f}(x) / \{\int \hat{f}^2 dx\}^{\frac{1}{2}}. \tag{3.3}$

It is not difficult to verify that each of these "encourages honesty" by satisfying the inequality

$$Eg(X, f) \geq Eg(X, \hat{f}) \tag{3.4}$$

for all densities $\hat{f}, f$, where expectation is of course taken with respect to the true density f.

We are not aware of any known characterization of honesty-encouraging payoff functions for the continuous case, paralleling the results of McCarthy (1956) for the discrete case.

3.2 Bets. In the continuous case we can again think of a collection of all acceptable bets, based on a presumed distribution, as describing that distribution completely; or we can think of a more restrictive set of (consistent) acceptable bets as providing a partial description of the presumed distribution.

3.3 Questioning schemes. The information theoretic counterpart of a continuous distribution is a conceptual continuous spectrum of messages of which one is transmitted. The information content of such a message must of necessity be considered to be infinite, corresponding to the fact that in general infinitely many binary questions would be needed to determine the message exactly. The Shannon information in a continuous density f nevertheless is customarily defined by the continuous analog of (2.5), that is,

$$I_S = -\int f(x)\{\log f(x)\}\,dx. \tag{3.5}$$

We may refer to Kolmogorov (1956) for further discussion of the relationship of the discrete and continuous definitions and their interpretations. It is pointed out there, for example, that the definition (3.5) is not invariant under transformation; that is, if Y is any transformation of X, the Shannon information of the density of Y will in general be different from the corresponding value for X.

A natural way to relate the continuous and discrete cases is to approximate the continuous distribution by a discrete one. In particular one may subdivide the x-values into equal intervals and consider the collection of probabilities of the intervals. A questioning scheme chosen for this discrete distribution can be thought of as determining an unknown random value x of the continuous variate X to an accuracy of half the interval length. Values of I_S calculated from (2.5) and (3.5) will differ by the logarithm of the interval length plus terms of higher order.

We may again think of knowledge of a distribution of a random variable X as being measured by the performance of a questioning scheme in some specified admissible class (for example, those allowing only binary questions). The expected number of questions to determine an unknown observed value x of X to a given accuracy is the performance criterion.

4

Knowledge About Unknown Parameters

In this section we consider whether the method of payoff functions for describing information about random variables can be applied when an unknown parameter θ takes the place of the random variable. The method of bets has been considered previously (Buehler, 1959; Buehler and Feddersen, 1963). Of course in a Bayesian framework where a prior distribution is assumed, knowledge of θ is described by its posterior distribution, and actions based on this distribution will have optimal properties. If prior knowledge is assumed to be something less than a prior distribution, new problems arise leading to possible generalizations of Bayesian inference.

4.1 Payoff functions. Suppose an observation x is known to have arisen from a known density $g(x; \theta)$, and you are asked to give a density $\hat{f}(\theta|x)$ expressing your knowledge of θ (given x). For definiteness we suppose the logarithmic payoff function (3.1) so that the reward to you will be $\log \hat{f}(\theta_0|x)$ where θ_0 is the true value of θ. The long-run performance (call it W) of any choice f will be given by an expectation over x:

$$W(\theta_0, f) = E \log \hat{f}(\theta_0|x) = \int \{\log \hat{f}(\theta_0|x)\} g(x; \theta_0) \, dx. \qquad (4.1)$$

Any attempt to maximize $W(\theta, \hat{f})$ by choice of $\hat{f}$ meets the difficulty that the choice is not uniform over θ. Given any $\hat{f}$ and its performance function $W(\theta, \hat{f})$, for any value θ' of θ we find an f' with better performance at $\theta = \theta'$ by taking a density which is sharply peaked at $\theta = \theta'$ and does not depend on x at all. While this tends to suggest some averaging or minimax criterion, we will not pursue the matter further at this time.

It is of interest to examine fiducial solutions in the present context. Consider first the case where the likelihood is the location parameter family: $g(x; \theta) = h(x - \theta)$. Here it is known that the fiducial density of θ given x is $h(x - \theta)$, and substituting this for $\hat{f}$ in (4.1) gives

$$W(\theta, \hat{f}) = \int \{\log h(x - \theta)\} h(x - \theta) \, dx = \int \{\log h(t)\} h(t) \, dt, \qquad (4.2)$$

which is incidentally the Shannon information I_S (see (3.5)) associated with h. Here we observe that $W(\theta, \hat{f})$ is actually independent of θ, a fact which of course is closely linked to invariance properties of the solution, and which holds equally for the other payoff functions mentioned in Section 3.

The location-parameter example exhibits another interesting feature. When putting $\hat{f}(\theta|x) = h(x - \theta)$ the subject knows the long-run payoff to him will be the Shannon information in $\hat{f}$, which is the same for all x:

$$W(\theta, \hat{f}) = I_S = \int \{\log \hat{f}(\theta|x)\} \hat{f}(\theta|x) \, d\theta. \qquad (4.3)$$

In this sense he has exact knowledge of his uncertainty about θ.

For n observations $x_1, \ldots, x_n$ from the location-parameter family, the long-run performance W again will be constant over θ if we use for $\hat{f}$ the usual Fisher-Pitman fiducial distribution. The Shannon information I_S of the fiducial distribution will however depend on the value of the ancillary statistic $(x_1 - x_2, x_1 - x_3, \ldots, x_1 - x_n)$. The value of W will be a weighted average of the I_S values with the weights given by the known fixed distribution of the ancillary statistic. Under these circumstances we may say that our uncertainty about θ is known exactly for each given value of the ancillary statistic.

Turning to a scale-parameter family, let x have density $\sigma h(\sigma x)$

$(x > 0, \sigma > 0)$. Taking for $\hat{f}$ the fiducial density

$$\hat{f}(\sigma|x) = xh(\sigma x)$$

we find

$$W(\sigma, \hat{f}) = \int \{\log[xh(\sigma x)]\} \, \sigma h(\sigma x) \, dx,$$

which is not constant over σ. Neither will the Shannon information in $\hat{f}$ be constant over x, so that it is not possible to make the rather strong conclusions available in the location-parameter case. This failure may be viewed as arising from inappropriate group-invariance properties of our chosen payoff function. It is well known that any scale parameter can be logarithmically transformed to give a location parameter; of course if this were done the earlier conclusions would hold. Such a transformation amounts to altering the payoff function so that the subject is not rewarded by $\log \hat{f}(\sigma|x)$ when σ is true, but by $\log \sigma + \log \hat{f}(\sigma|x)$. Invariance properties are thereby restored. Nonlogarithmic transformations and nonlogarithmic payoff functions can be handled similarly. Thus we conclude that in cases where the family of distributions has group structure, payoff functions with suitably matching group structure can be chosen so that performance of the fiducial distribution in asserting knowledge of the parameter has the desirable properties described above for the case of location parameters.

Acknowledgment

This work was supported by NSF Grant GP-9556.

References

1. Ash, Robert, "Information Theory," *Interscience,* New York, 1965.
2. Buehler, Robert J., "Some Validity Criteria for Statistical Inference," *Annals of Mathematical Statistics,* 30, 845-863, 1959.
3. Buehler, R. J. and Feddersen, A. P., "Note on a Conditional Property of Student's *t*," *Annals of Mathematical Statistics,* 34, 1098-1100, 1963.
4. De Finetti, Bruno, "Does It Make Sense to Speak of 'Good Probability Appraisers'?" in Good, 357-364, 1962.
5. Good, I. J., "Rational Decisions," *Journal of the Royal Statistical Society, B,* 14, 107-114, 1952.
6. Good, I. J., Editor, *The Scientist Speculates,* New York, Basic Books, 1962.
7. Kolmogorov, A. N., "On the Shannon Theory of Information Transmission in the Case of Continuous Signals," *IRE Transactions on Information Theory,* 2, 102-108, 1956.
8. Lindley, D. V., "On a Measure of the Information Provided by an Experiment," *Annals of Mathematical Statistics,* 27, 986-1005, 1956.
9. McCarthy, J., "Measures of the Value of Information," *Proceedings of the National Academy of Sciences,* 42, 654-655, 1956.
10. Picard, Claude, *Théorie des Questionnaires,* Paris, Gauthier-Villars, 1965.
11. Winkler, Robert L., "The Quantification of Judgment: Some Methodological Suggestions," *Journal of the American Statistical Association,* 62, 1105-1120, 1967.

ADDENDUM

I will now give two further applications of the method of bets.

Example 1. A finite population contains two unequal integers, positive, negative or zero. Let x denote one value chosen at random. What information do we have about the unknown remaining value y? If $x=75$, in a certain sense we may say $P(y > 75) = P(y < 75) = \frac{1}{2}$. Suppose Peter systematically asserts $P(y > x) = \frac{1}{2}$ on repeated occasions where the population may change on each occasion. Suppose Paul chooses to bet (with equal stakes) on the outcome $y < x$ but only when $x > 50$, otherwise no bet is made. We may distinguish three states of nature: (a) both values ≤ 50; (b) both values > 50; (c) otherwise. In case (a), bets are never made. In case (b), bets are always made and the players break even. In case (c) bets are made half of the time and Paul wins every bet. Because of this advantage to Paul, I say that Peter's assertion $P(y < x) = \frac{1}{2}$ does not have the same validity after x is observed as before.

Example 2. Suppose $x = \theta + e$ where e takes values -1 and $+1$ each with probability $1/2$ and nothing is known about θ except that $\theta = 0, \pm 1, \pm 2, \ldots$. Suppose we observe $x = 3132$. Then Fraser's method of structural inference leads to the assertions

$$P(\theta = 3131) = P(\theta = 3133) = \tfrac{1}{2}.$$

Suppose Peter systematically asserts $P(\theta = x + 1) = \frac{1}{2}$. Paul bets that $\theta = x - 1$, but only when $x > 0$. If $\theta < 0$, bets are never made. If $\theta > 1$, bets are always made, and the expected gain for each player is zero. If $\theta = 0$ or 1, bets are made with probability $1/2$, and Paul wins every bet. Because of the advantage to Paul for certain states of nature, I assert that the structural probabilities do not have the degree of validity which seems to be claimed for them.

COMMENTS

I. J. Good:

Dr. Buehler's third function $g_i(p)$, the one which he described as *spherical*, is the special case $\beta = 2$ of the following formula:

$$g_i(p,\beta) = g_i(p) = \left\{ \frac{p_i}{(\sum_j p_j^\beta)^{1/\beta}} \right\}^{\beta-1} = \lambda_\beta^{-(\beta-1)} (\beta > 1),$$

which could perhaps be called the pseudospherical or quasispherical reward. It too can be seen to have the property of encouraging honesty. Here λ_β is my generalization of Warren Weaver's surprise index ("The Surprise Index for the Multivariate Normal Distribution", *Ann. Math. Statist.*, 27, 1956, pp. 1130-1135; 28, 1957, and "Mathematical Tools" in *Uncertainty and Business Decisions,* Liverpool, second edition, 1957, pp. 20-36). Since λ_β is multiplicative for two independent probability estimation problems it follows that this form of reward is also multiplicative, but multiplicativity is not a desideratum: it is merely interesting. The quadratic reward function (Dr. Buehler's formula 2.2), is neither additive nor multiplicative. It seems reasonable to use an additive reward and one might be tempted to convert the pseudospherical or the spherical reward functions into additive ones by taking logarithms. But this won't do because after taking logarithms the reward would no longer precisely encourage honesty. Note that the form:

$$h_i(p,\beta) = \frac{1}{\beta-1}(\lambda_\beta^{-(\beta-1)} - 1)$$

has the advantage that it tends to log p_i as $\beta \to 1$, and so in effect includes (2.1) as a special case. It encourages honesty but is neither additive nor multiplicative when $\beta > 1$. When $\beta \to 1$ we obtain the logarithmic form which is the only additive one.

In my 1952 paper on rational decisions, in which I proposed the logarithmic reward, I mislabelled the relevant section "Fair fees"; but the description of the problem was correctly stated as "the question of how a firm can encourage its experts to give fair estimates of probabilities", and it was also pointed out that the fee $A \log p + B$ puts it in the interest of the expert to collect additional evidence (if the cost is not too high).

Of course the continuous analogue of the pseudospherical payoff function is

$$h_i(p,\beta) = \frac{1}{\beta-1}\left\{\frac{f(x)}{(\int f^\beta \, dx)^{1/\beta}}\right\}^{\beta-1} - \frac{1}{\beta-1}.$$

Additivity seems fairly desirable because we can regard a pair of estimates of two independent probability vectors as an estimate of the *direct product* probability vector, and additivity is required for the sake of consistency. But if we do not insist on additivity what criterion should be used for selecting β? We should select β so as to maximize the expert's incentive to estimate the probabilities accurately. For the case $n = 2$, I think this principle can be given a precise form. Write $p_1 = x$ and $p_2 = y$ for short. When the reward function is taken as $h_i(p,\beta)$, the expected reward when x is estimated as $x + \delta$ is

$$\frac{1}{\beta-1}\left\{\frac{x(x+\delta)^{\beta-1} + y(y-\delta)^{\beta-1}}{[(x+\delta)^\beta + (y-\delta)^\beta]^{1-1/\beta}} - 1\right\}.$$

We could select β to make this as peaked as possible, that is, as leptokurtic as possible, when regarded as a function of δ. An advantage of using the kurtosis is that it is unchanged when a linear transformation is made, the coefficients being arbitrary functions of β. I wonder whether the kurtosis is maximized by taking $\beta = 1$, that is, by using the logarithmic reward function. Some slight evidence for this is that the coefficient of δ^2 is

$$\frac{(xy)^{\beta-2}}{2(x^\beta + y^\beta)^{2-1/\beta}}$$

and this can be shown to be maximized by taking $\beta = 1$.

Note. By analogy with Rényi's (1961) generalization of my generalized surprise index, we can in effect generalize $h_i(p, \beta)$ to:

$$h_i(p, q, \beta) = \frac{1}{\beta-1} \left\{ \frac{(\hat{p}_i/q_i)^{\beta-1}}{[\Sigma_j \hat{p}_j{}^\beta / q_j{}^{\beta-1}]^{1-1/\beta}} - 1 \right\},$$

where $q_i \geq 0$ $(i = 1, 2, \ldots)$, $\Sigma q_i = 1$. This again encourages honesty. We can think of $(q_1, q_2, \ldots, q_n)$ as a set of initial probabilities for which the consultant deserves no reward. When $\beta \to 1$, $h_i(p, q, \beta) \to \log(\hat{p}_i/q_i)$. When $q_1 = q_2 = \ldots = 1/n$, write $q = q_0$ and we have:

$$h_i(p, q_0, \beta) = \frac{1}{\beta-1} \left\{ \left[\frac{\hat{p}_i}{(n\Sigma_j p_j{}^\beta)^{1/\beta}} \right]^{\beta-1} - 1 \right\},$$

which has nearly the same form as $h_i(p, \beta)$. The continuous form of h_i is

$$h_i(p, q, \beta) = \frac{1}{\beta-1} \left\{ \frac{f(x)/g(x)}{[\int (f^\beta/g^{\beta-1})\, dx]^{1/\beta}} \right\}^{\beta-1} - \frac{1}{\beta-1},$$

which has the advantage of being invariant with respect to a transformation of the x-axis.

I think Dr. Buehler's ingenious refutation of structural inference is incomplete because the probability that $\theta = 0$ or 1 is zero in the light of the Bayes postulate that is implicit in Dr. Fraser's argument. What makes Dr. Fraser's argument seem valid at first sight is that it is reasonable to suppose that $P(n-3133)$ and $P(n-3131)$ are *nearly* equal and so we are not at first shocked that they are assumed to be *exactly* equal. But this is really a very strong and counter-intuitive assumption because the same argument leads to the conclusion that the probabilities of all odd numbers are equal. Also I could have taken $e = \pm 3131$ to make my point.

D. V. Lindley:

The author remarks that there are two kinds of uncertainty (Carnap's probability$_1$ and probability$_2$ perhaps). If so, then one would expect them to be distinct *operationally*: that is, they would obey different axiom systems. Now, so far as I am aware, the axioms of probability are not seriously under dispute (though de Finetti does argue in favour of finite additivity). Does not this therefore strongly suggest that there is only one kind of uncertainty?

May I say that his counter-example to structural inference has alone made this conference worthwhile for me.

H. Rubin:

Prof. Buehler's example can be made even stronger by randomizing his θ. An alternative way to do this, which makes the picture quite clear, is to make the probability of betting $\theta = X - 1$ a monotonically increasing function f of X. The expected gain given θ is then $\frac{1}{2}(f(\theta + 1) - f(\theta - 1))$.

J. S. Williams:

The second betting example is directed as much against fiducial probability, I suppose, as against structural probability. It should be pointed out therefore that the end result of the fiducial argument is not $P(\theta = x + 1 | x = 3132) = \frac{1}{2} \leftrightarrow P(e = -1 | x = 3132) = \frac{1}{2}$, which could be verified or refuted by a frequency experiment, but $P_{x=3132}(e = -1) = \frac{1}{2}$, which is a subjective probability and only trivially subject to frequency verification. The distinction is that $P_B(A)$ is read as the probability of A given B as distinguished from $P(A|B)$ which is the probability of A conditioned on B. In the first case $P_B(A) = P(A)$ if the outcome B can be judged irrelevant in assessing the probability of the event A, and in the second $P(A|B) = P(A)$ if the events A and B are independent. By either of the attempts made by I. Hacking and myself to formulate tests of irrelevance, $P_{x=3132}(e = -1)$ can be $\frac{1}{2}$. The outcome of x is irrelevant if *no prior information on θ* implies *no distribution of θ is ruled out a priori*.

The betting example certainly shows that $P(\theta = x + 1 | x) = \frac{1}{2}$ for all integers is not correct unless there is a flat prior distribution of θ over the integers. If however the latter is ruled out as a possibility, the conditional probabilities cannot all be $\frac{1}{2}$, and the fiducial argument cannot be carried through because x can be shown relevant for making inference on e. The betting example does not show that, without this prior information on θ, there is a better approximation than $\frac{1}{2}$ for one to use as the probability of $\theta = x + 1$ given the observed x of a single trial.

REPLY

Dr. Good's technical comments on payoff functions are welcome. I expect that a number of his points, such as the maximization of incentives, will be valuable for future work.

All four discussants mention my second betting example, and I will take this opportunity for clarification and elaboration. (Incidentally, for a continuous version, see Buehler, 1959, section 4.1.)

In Example 2, Paul has an advantage wherever $\theta = 0$ or 1. Dr. Good's objection that these values may have (prior) probability zero is perhaps satisfactorily answered by Dr. Rubin's modification. By betting $\theta = x - 1$ only when $x > x_0$, where x_0 is random with $P(x_0 = j) = p_j (p_j > 0, j = 0, \pm 1, \ldots, \Sigma p_j = 1)$, the expectation to Paul (for unit stakes) becomes $\frac{1}{2}(p_\theta + p_{\theta - 1})$ when θ is the true state of nature. Dr. Rubin's $f(\theta)$ is easily seen to correspond to $\Sigma_{j=1}^{\infty} p_{\theta - j}$. Dr. Rubin's device spreads out Paul's advantage for $\theta = 0$ or 1 to a smaller, but still positive, advantage for all states of nature.

340

I agree with Dr. Good that putting $e = \pm 3131$ alters the intuitive appeal of Fraser's example.

Dr. Williams correctly states that my example does not show that there is a better approximation than $1/2$ to use for $P(\theta = x+1 \mid x = 3132)$. The example was intended rather to show that the very plausible value $1/2$, which is prescribed by structural theory, does not exhibit all of the properties one might wish a probability value to have. This is of course consistent with the general theme of the paper—an attempt to categorize levels and kinds of uncertainty.

I don't think I understand Dr. Williams' subtle distinction between A given B and A conditioned on B. While this distinction may be quite relevant in its proper framework, I favor viewing the present example in terms of a hypothetical sequence of θ-values, $\theta_1, \theta_2, \ldots$, and the corresponding sequence of x-values, $x_1, x_2, \ldots$. For either the original game or for Rubin's modification there will be a corresponding sequence of payoffs to Paul which we may call $y_1, y_2, \ldots$, and a sequence of average payoffs $\bar{y}_1 = y_1$, $\bar{y}_2 = (y_1 + y_2)/2, \ldots$. We may consider the stochastic behavior of $\bar{y}_1, \bar{y}_2, \ldots$, which will of course depend on the actual sequence $\theta_1, \theta_2, \ldots$. The condition

(A) $$\bar{y}_n \to 0 \text{ almost surely}$$

can be regarded as an indication of one kind of legitimacy of the asserted probabilities. (A) will hold in the original example if and only if $\theta = 0$ and $\theta = 1$ both have limiting relative frequency equal to zero in the sequence $\theta_1, \theta_2, \ldots$. In Rubin's modification: (A) will fail if $\theta_1, \theta_2, \ldots$ are randomly sampled from any prior distribution; (A) will hold for choices like $\theta_n = 10^n$ or even $\theta_n = n$.

Naturally it must be considered a weakness of the example that (A) does hold in some cases. The difficulty is that Paul's expectation is not bounded away from zero for all states of nature. A stronger example not sharing this weakness is known (Buehler and Feddersen, 1963), involving the standard Student t theory for the normal mean. All three theories, fiducial, structural, and confidence interval, lead to the Student intervals; hence the example applies equally to all three.

I believe Dr. Lindley's comments concern the oral presentation more directly than the written version, and it may be helpful to attempt to clarify my position. Rather than defend the assertion, "There are two kinds of uncertainty," I would prefer to go on record with an opinion: "It is likely to be useful to distinguish different kinds of uncertainty in our mathematical theories." Perhaps it is a corollary that, "Non-Bayesian theories can be useful." Not to support the opinion, but only to clarify the meaning, I may cite examples: uncertainty about θ: (i) no prior knowledge, (ii) a prior distribution; uncertainty about event A when B and C have occurred: (i) $P(A \mid BC) = 0.3$; (ii) $P(A \mid B) = 0.2$ and $P(A \mid C) = 0.4$. In a fiducial or structural framework: (i) uncertainty about x given θ; (ii) uncertainty about θ given x.

THE WELL-POSED PROBLEM

E. T. Jaynes
Washington University

Summary

In statistics, the question whether a given problem is well-posed appears more subtle than in other areas of mathematics. Many problems, including some of the most important for physical applications, have long been regarded as underdetermined from the standpoint of a strict frequency definition of probability; yet they may appear well-posed or even overdetermined from the standpoint of subjective probabilities, determined by the principles of maximum entropy and transformation groups. While the distributions found by these methods represent basically a state of knowledge rather than any "objectively real" situation, they nevertheless turn out to have a frequency correspondence weaker than that usually assumed, but still sufficient for many purposes.

In particular, the distribution obtained by invariance under a transformation group is not necessarily the one that would be observed in a real random experiment; but it is often by far the most likely to be observed, in the sense that it requires by far the least *skill* . Consistently to produce any other distribution would require a "microscopic" degree of control over the exact conditions of the experiment. In most physical applications this weaker probability-frequency relation is all that is needed.

These properties are illustrated by analyzing a specific problem, the famous Bertrand paradox. On the viewpoint advocated here, Bertrand's problem turns out to be well-posed after all, and the unique solution has been verified experimentally. We conclude that probability theory has a wider range of useful applications than would be supposed from the standpoint of the usual frequency definitions.

1
Introduction

In a previous article [1], we have discussed two formal principles — maximum entropy and transformation groups — that are available for setting up probability distributions in the absence of frequency data. The resulting distributions may be used as prior distributions in Bayesian inference; or they may be used directly for certain physical predictions. The exact sense in which distributions found by maximum entropy correspond to observable frequencies was given in the previous article; here, we demonstrate a similar correspondence property for distributions obtained from transformation groups, using as our main example the famous paradox of Bertrand.

Bertrand's problem [2] was stated originally in terms of drawing a straight line *at random* intersecting a circle. It will be helpful to think of this in a more concrete way; presumably, we do no violence to the problem (that is, it is still just as *random*) if we suppose that we are tossing straws onto the circle, without specifying how they are tossed. We therefore formulate the problem as follows.

A long straw is tossed at random onto a circle; given that it falls so that it intersects the circle, what is the probability that the chord thus defined is longer than a side of the inscribed equilateral triangle? Since Bertrand proposed it in 1889, this problem has been cited to generations of students to demonstrate that Laplace's *principle of indifference* contains logical inconsistencies. For, there appear to be many ways of defining *equally possible* situations, and they lead to different results. Three of these involve assigning uniform probability density to: (a) the linear distance between centers of chord and circle; (b) angles of intersections of the chord on the circumference; (c) the center of the chord over the interior area of the circle. These assignments lead to the results $p_A = 1/2$, $p_B = 1/3$, $p_C = 1/4$, respectively.

Which solution is correct? Of the ten authors cited, with short quotations, at the end of this article ([2]-[11]), only Borel is willing to express a definite preference, although he does not support it by any proof. Von Mises takes the opposite extreme, declaring that such problems (including the similar Buffon needle problem) do not belong to the field of probability theory at all. The others, including Bertrand, take the intermediate position of saying simply that the problem has no definite solution because it is ill-posed, the phrase, *at random,* being undefined.

In works on probability theory this state of affairs has been interpreted, almost universally, as showing that the principle of indifference must be totally rejected. Usually, there is the further conclusion that the only valid basis for assigning probabilities is frequency in some random experiment. It would appear, then, that the only way of answering Bertrand's question is to perform the experiment.

But, do we really believe that it is beyond our power to predict by *pure thought* the result of such a simple experiment? The point at issue is far more important than merely resolving a geometrical puzzle; for as discussed further in Section 6, applications of probability theory to physical experiments usually lead to problems of just this type; that is, they appear at first to be underdetermined, allowing many different solutions with nothing to choose between them. For example, given the average particle density and total energy of a gas, predict its viscosity. The answer, evidently, depends on the exact spatial and velocity distribution of the molecules (in fact, it depends critically on position-velocity correlations) and nothing in the given data seems to tell us which distribution to assume. Yet physicists *have* made definite choices, guided by the principle of indifference, and they *have* led us to correct and nontrivial predictions of viscosity and many other physical phenomena.

Thus, while in some problems the principle of indifference has led us to paradoxes, in others it has produced some of the most important and successful applications of probability theory. To reject the principle without having anything better to put in its place would lead to consequences so unacceptable that for many years, even those who profess the most faithful adherence to the strict frequency definition of probability have managed to overlook these logical difficulties in order to preserve some very useful solutions.

Evidently, we ought to examine the apparent paradoxes such as Bertrand's more closely; there is an important point to be learned about the application of probability theory to real physical situations.

It is evident that if the circle becomes sufficiently large, and the tosser sufficiently skilled, various results could be obtained at will. However, in the limit where the skill of the tosser must be described by a *region of uncertainty,* large compared to the circle, the distribution of chord lengths must surely go into one unique function, obtainable by pure thought. A viewpoint toward probability theory which cannot show us how to calculate this function from first principles or even denies the possibility of doing this, would imply severe — and, to a physicist, intolerable — restrictions on the range of useful applications of probability theory.

An invariance argument was applied to problems of this type by Poincaré [4], and cited more recently by Kendall and Moran [8]. In this treatment, we consider lines drawn at random in the *x-y* plane. Each line is located by specifying two parameters (u, v), such that the equation of the line is $ux+vy=1$, and one can ask: Which probability density $p(u,v) \, du \, dv$ has the property that it is invariant in *form* under the group of Euclidean transformations (rotations and translations) of the plane? This is a readily solvable problem [8], with the answer $p(u,v) = (u^2+v^2)^{-3/2}$.

Yet this has, evidently, not seemed convincing, for later authors have ignored Poincaré's invariance argument, and adhered to Bertrand's original judgment that the problem has no definite solution. This is understandable, for the statement of the problem does not specify that the distribution of straight lines is to have this invariance property, and we do not see any compelling reason to expect that a rain of straws produced in a real experiment would have it. To assume this would seem to be an intuitive judgment resting on no stronger grounds than the ones which led to the three different solutions above. All of these amount to trying to guess what properties a random rain of straws should have, by specifying the intuitively, equally possible events; and the fact remains that different intuitive judgments lead to different results.

The viewpoint just expressed, which is by far the most common in the literature, clearly represents one valid way of interpreting the problem. If we can find another viewpoint according to which such problems *do* have definite solutions, *and define the conditions under which these solutions are experimentally verifiable,* it would perhaps be overstating the case to say that this new viewpoint is more *correct* in principle than the conventional one but it will surely be more useful in practice.

We now suggest such a viewpoint and we understand from the start that we are not concerned at this stage with *frequencies* of various events. We ask,

344

rather: Which probability distribution describes our *state of knowledge* when the only information available is that given in the above statement of the problem? Such a distribution must conform to the desideratum of consistency formulated previously [1]: in two problems where we have the same state of knowledge, we must assign the same subjective probabilities. The essential point is this: if we start with the presumption that Bertrand's problem has a definite solution *in spite of the many things left unspecified,* then the statement of the problem automatically implies certain invariance properties, which in no way depend on our intuitive judgments. After the subjective solution is found, it may be used as a prior for Bayesian inference whether or not it has any correspondence with frequencies; any frequency connections that may emerge will be regarded as additional bonuses, which justify its use also for direct physical prediction.

Bertrand's problem has an obvious element of rotational symmetry, recognized in all the proposed solutions; however, this symmetry is irrelevant to the distribution of chord lengths. There are two other *symmetries* which are highly relevant. Neither Bertrand's original statement nor our restatement in terms of straws specifies the exact size of the circle, or its exact location. If, therefore, the problem is to have any definite solution at all, it must be *indifferent* to these circumstances; that is, it must be unchanged by a small change in the size or position of the circle. This seemingly trivial statement, as we will see, fully determines the solution.

It would be possible to consider all these invariance requirements simultaneously by defining a four-parameter transformation group, whereupon the complete solution would appear suddenly, as if by magic. However, it will be more instructive to analyze the effects of these invariances separately, and see how each places its own restriction on the form of the solution.

2

Rotational Invariance

Let the circle have radius R. The position of the chord is determined by giving the polar coordinates (r, θ) of its center. We seek to answer a more detailed question than Bertrand's: What probability density $f(r, \theta) \, dA = f(r, \theta) r \, dr \, d\theta$ should we assign over the interior area of the circle? The dependence on θ is actually irrelevant to Bertrand's question, since the distribution of chord lengths depends only on the radial distribution

$$g(r) = \int_0^{2\pi} f(r, \theta) \, d\theta.$$

However, intuition suggests that $f(r, \theta)$ should be independent of θ and the formal transformation group argument deals with the rotational symmetry as follows.

The starting point is the observation that the statement of the problem does not specify whether the observer is facing north or east; therefore, if there is a definite solution, it must not depend on the direction of the observer's line of sight. Suppose, therefore, that two different observers, Mr. X and Mr. Y, are watching this experiment. They view the experiment from different directions, their lines of sight making an angle α. Each uses a coordinate system oriented along his line of sight. Mr. X assigns the probability density $f(r,\theta)$ in his co-ordinate system S; and Mr. Y assigns $g(r,\theta)$ in his system S_α. Evidently, if they are describing the same situation, then it must be true that

$$f(r, \theta) = g(r, \theta - \alpha) \tag{1}$$

which expresses a simple change of variables, transforming a fixed distribution f to a new coordinate system; this relation will hold whether or not the problem has rotational symmetry.

But now we recognize that, because of the rotational symmetry, the problem appears exactly the same to Mr. X in his coordinate system as it does to Mr. Y in his. Since they are in the same state of knowledge, our desideratum of consistency demands that they assign the same probability distribution and so f and g must be the same function:

$$f(r, \theta) = g(r, \theta). \tag{2}$$

These relations must hold for all α in $0 \le \alpha \le 2\pi$, and so the only possibility is $f(r, \theta) = f(r)$.

This formal argument may appear cumbersome when compared to our obvious flash of intuition, and of course it is, when applied to such a trivial problem. However, as Wigner [12] and Weyl [13] have shown in other physical problems, it is this cumbersome argument that generalizes at once to non-trivial cases where our intuition fails us. It always consists of two steps: we first find a transformation equation like (1) which shows how two problems are related to each other, irrespective of symmetry; then a symmetry relation like (2) which states that we have formulated two equivalent *problems*. Combining them leads in most cases to a functional equation which imposes some restriction on the form of the distribution.

3
Scale Invariance

The problem is reduced, by rotational symmetry, to determining a function $f(r)$, normalized according to

$$\int_0^{2\pi} \int_0^R f(r) r \, dr \, d\theta = 1. \tag{3}$$

Again we consider two different problems. Concentric with a circle of radius R, there is a circle of radius aR, $0 < a < 1$. Within the smaller circle there is a probability $h(r) r \, dr \, d\theta$ which answers the question: "Given that a straw intersects the smaller circle, what is the probability that the center of its chord lies in the area $dA = r \, dr \, d\theta$?"

Any straw that intersects the small circle will also define a chord on the large one; so, within the small circle, $f(r)$ must be proportional to $h(r)$. This proportionality is, of course, given by the standard formula for a conditional probability, which in this case takes the form

$$f(r) = 2\pi h(r) \int_0^{aR} f(r) r \, dr, \quad \begin{array}{l} 0 < a \le 1 \\ 0 \le r \le aR. \end{array} \tag{4}$$

346

This transformation equation will hold whether or not the problem has scale invariance.

But we now invoke scale invariance. To two different observers with different size eyeballs, the problems of the large and small circles would appear exactly the same. If there is any unique solution independent of the size of the circle, there must be another relation between $f(r)$ and $h(r)$, which expresses the fact that one problem is merely a scaled-down version of the other. Two elements of area $r \, dr \, d\theta$ and $ar \, d(ar)d\theta$ are related to the large and small circles respectively in the same way; so they must be assigned the same probabilities by the distributions $f(r)$ and $h(r)$ respectively.

$$h(ar) \, ar \, d(ar) \, d\theta = f(r) \, r \, dr \, d\theta$$

or,

$$a^2 h(ar) = f(r), \tag{5}$$

which is the symmetry equation. Combining (4) and (5), we see that invariance under change of scale requires that the probability density satisfy the functional equation

$$a^2 f(ar) = 2\pi f(r) \int_0^{aR} f(u) u \, du \quad \begin{matrix} 0 < a \le 1 \\ 0 \le r \le R \end{matrix} \tag{6}$$

Differentiating with respect to a, setting $a = 1$, and solving the resulting differential equation, we find that the most general solution of (6) satisfying the normalization condition (3) is

$$f(r) = \frac{q r^{q-2}}{2\pi R^q} \tag{7}$$

where q is a constant in $0 < q < \infty$, not further determined by scale invariance.

We note that the proposed solution (b) in the Introduction has now been eliminated for it corresponds to the choice $f(r) \sim (R^2 - r^2)^{-\frac{1}{2}}$, which is not of the form (7). This means that, if the intersections of chords on the circumference were distributed in angle uniformly and independently on one circle, this would not be true for a smaller circle inscribed in it; that is, the probability assignment of (b) could be true for, at most, only one size of circle. However, solutions (a) and (c) are still compatible with scale invariance, corresponding to the choices $q = 1$, $q = 2$ respectively.

4
Translational Invariance

We now investigate the consequences of the fact that a given straw S can in-intersect two circles C, C' of the same radius R, but with a relative displacement b. Referring to Figure 1, the mid-point of the chord with respect to circle C is the point P, with coordinates (r, θ); while the same straw defines a mid-point of the chord with respect to C' at the point P' whose coordinates are (r', θ').

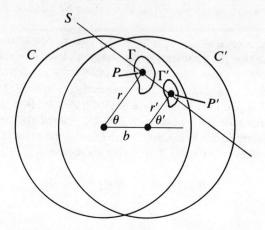

Figure 1. A straw S intersects two slightly displaced circles, C and C'.

From Figure 1, the coordinate transformation $(r, \theta) \rightarrow (r', \theta')$ is given by

$$r' = |r - b \cos\theta| \qquad (8)$$

$$\theta' = \begin{Bmatrix} \theta, r > b \cos\theta \\ \theta + \pi, r < b \cos\theta \end{Bmatrix}. \qquad (9)$$

As P varies over the region Γ, P' varies over Γ', and vice versa; thus, the straws define a $1:1$ mapping of Γ onto Γ'.

Now, we note the translational symmetry. Since the statement of the problem gave no information about the location of the circle, the problems of C and C' appear exactly the same to two slightly displaced observers O and O'. Our desideratum of consistency then demands that they assign probability densities in C and C' respectively which have the same form (7) with the same value of q.

It is further necessary that these two observers assign equal probabilities to the regions Γ and Γ' respectively for (a) they are probabilities of the same event, and (b) the probability that a straw which intersects one circle will also intersect the other, thus setting up this correspondence, is also the same in the two problems. Let us see whether these two requirements are compatible.

The probability that a chord intersection C will have its mid-point in Γ' is

$$\int_\Gamma f(r) r \, dr \, d\theta = \frac{q}{2\pi R^q} \int_\Gamma r^{q-1} dr \, d\theta. \qquad (10)$$

The probability that a chord intersecting C' will have its mid-point in Γ' is

$$\frac{q}{2\pi R^q} \int_{\Gamma'} (r')^{q-1} dr' \, d\theta' = \frac{q}{2\pi R^q} \int_\Gamma |r - b \cos\theta|^{q-1} dr \, d\theta \qquad (11)$$

where we have transformed the integral back to the variables (r, θ) by use of (8), (9), noting that the Jacobian is unity. Evidently, (10) and (11) will be equal for arbitrary Γ if and only if $q = 1$; and so our distribution $f(r)$ is now uniquely determined.

The proposed solution (c) in the Introduction is thus eliminated for lack of translational invariance; a rain of straws which had the property assumed with respect to one circle, could not have the same property with respect to a slightly displaced one.

5

Final Results

We have found that invariance requirements determine the probability density

$$f(r, \theta) = \frac{1}{2\pi R r} \qquad \begin{matrix} 0 \le r \le R \\ 0 \le \theta \le 2\pi \end{matrix} \qquad (12)$$

corresponding to solution (a) in the Introduction. It is interesting that this has a singularity at the center, the need for which can be understood as follows. The condition that the midpoint (r, θ) falls within a small region $\triangle$ imposes restrictions on the possible directions of the chord, but as $\triangle$ moves inward, as soon as it includes the center of the circle, all angles are suddenly allowed. Thus there is an infinitely rapid change in the *manifold of possibilities*.

Further analysis (almost obvious from contemplation of Figure 1) shows that the requirement of translational invariance is so stringent that it already determines the result (12) uniquely; thus the proposed solution (b) is incompatible with either scale or translational invariance, and in order to find (12) it was not really necessary to consider scale invariance. However, the solution (12) would, in any event, have to be tested for scale invariance, and if it failed to pass that test we would conclude that the problem as stated has *no* solution; that is, although at first glance it appears under-determined, it would have to be regarded, from the standpoint of transformation groups, as overdetermined. As luck would have it, these requirements *are* compatible, and so the problem has one unique solution.

The distribution of chord lengths follows at once from (12). A chord whose midpoint is at (r, θ) has a length $L = 2(R^2 - r^2)^{\frac{1}{2}}$. In terms of the reduced chord lengths, $x = L/2R$, we obtain the universal distribution law

$$p(x)\,dx = \frac{x\,dx}{(1 - x^2)^{\frac{1}{2}}}, \qquad 0 \le x < 1 \qquad (13)$$

in agreement with Borel's conjecture [3].

6

Frequency Correspondence

From the manner of its derivation, the distribution (13) would appear to have only a subjective meaning. While it describes the only possible state of knowledge corresponding to a unique solution in view of the many things left unspecified in the statement of Bertrand's problem, we have as yet given no reason to suppose that it has any relation to frequencies observed in the actual experiment. In general, of course, no such claim can be made; the mere fact that my state of knowledge gives me no reason to prefer one event over another is not enough to make them occur equally often! Indeed, it is clear that no pure thought argument, whether based on transformation groups or any other principle, can predict with certainty what must happen in a real experiment; we can easily imagine a very precise machine which tosses straws in such a way as to produce any distribution of chord lengths we please on a given circle.

Nevertheless, we are entitled to claim a definite frequency correspondence for the result (13), for there is one *objective fact* which *has* been proved by the above derivation: any rain of straws which does *not* produce a frequency distribution agreeing with (13), will necessarily produce different distributions on different circles.

But this is all we need in order to predict with confidence that the distribution (13) *will* be observed in any experiment where the *region of uncertainty* is large compared to the circle; for, if we lack the skill to toss straws so that, with certainty, they intersect a given circle, then surely we lack *a fortiori*, the skill consistently to produce different distributions on different circles *within* this region of uncertainty!

It is for this reason that distributions predicted by the method of transformation groups turn out to have a frequency correspondence after all. Strictly speaking, this result holds only in the limiting case of *zero skill*, but as a moment's thought will show, the skill required to produce any appreciable deviation from (13) is so great that in practice it would be difficult to achieve even with a machine.

Of course, the above arguments have demonstrated this frequency correspondence in only one case. In the following section we adduce arguments indicating that it is a general property of the transformation group method.

These conclusions seem to be in direct contradiction to those of von Mises [10], who denied that such problems belong to the field of probability theory at all. It appears to us that, if we were to adopt von Mises's philosophy of probability theory strictly and consistently, the range of legitimate physical applications of probability theory would be reduced almost to the vanishing point. Since we have made a definite, unequivocal prediction, this issue has now been removed from the realm of philosophy into that of verifiable fact. The predictive power of the transformation group method can be put to the test quite easily in this and other problems by performing the experiments.

The Bertrand experiment has, in fact, been performed by the writer and Mr. Charles E. Tyler, tossing broom straws from a standing position onto a five-inch diameter circle drawn on the floor. Grouping the range of chord lengths into 10 categories, 128 successful tosses confirmed (13) with an embarrassingly low value of Chi-squared. However, experimental results will, no doubt, be more convincing if reported by others.

7

Discussion

Bertrand's paradox has a greater importance than appears at first glance, because it is a simple crystallization of a deeper paradox which has permeated much of probability theory from its beginning. In *real* physical applications, when we try to formulate the problem of interest in probability terms, we find almost always that a statement emerges which, like Bertrand's, appears too vague to determine any definite solution, because apparently essential things are left unspecified.

To elaborate the example noted in the Introduction: given a gas of N molecules in a volume V, with known inter-molecular forces, total energy E, predict its molecular velocity distribution, pressure, distribution of pressure fluctuations, viscosity, thermal conductivity, and diffusion constant. Here again, the viewpoint expressed by most writers on probability theory would lead one to conclude that the problem has no definite solution because it is ill-posed; the things specified are grossly inadequate to determine any unique probability distribution over microstates. If we reject the principle of indifference, and insist that the only valid basis for assigning probabilities is frequency in some random experiment, it would again appear that the only way of determining these quantities is to perform the experiments.

It is, however, a matter of record that over a century ago, without benefit of any frequency data on positions and velocities of molecules, James Clerk Maxwell was able to predict all these quantities correctly by a *pure thought* probability analysis which amounted to recognizing the *equally possible* cases. In the case of viscosity, the predicted dependence on density appeared at first to contradict common sense, casting doubt on Maxwell's analysis, but when the experiments were formed, they confirmed Maxwell's predictions, leading to the first great triumph of kinetic theory. These are solid, positive accomplishments; they cannot be made to appear otherwise merely by deploring his use of the principle of indifference.

Likewise, we calculate the probability of obtaining various hands at poker; we are so confident of the results that we are willing to risk money on bets which the calculations indicate are favorable to us. But underlying these calculations is the intuitive judgment that all distributions of cards are equally likely and with a different judgment our calculations would give different results. Once again, we are predicting definite, verifiable facts by pure thought arguments based ultimately on recognizing the equally possible cases; yet, present statistical doctrine, both orthodox and personalistic, denies that this is a valid basis for assigning probabilities!

The dilemma is thus apparent. On the one hand, one cannot deny the force of arguments which, by pointing to such things as Bertrand's paradox, demonstrate the ambiguities and dangers in the principle of indifference, but on the other hand, it is equally undeniable that use of this principle has, over and over again, led to correct, nontrivial, and useful predictions. Thus it appears that, while we cannot wholly accept the principle of indifference, we cannot wholly reject it either; to do so would be to cast out some of the most important and successful applications of probability theory.

The transformation group method grew out of the writer's conviction, based on pondering this situation, that the principle of indifference has been unjustly maligned in the past; what it has needed was not blanket condemnation, but recognition of the proper way to apply it. We agree with most other writers on probability theory that it is dangerous to apply this principle at the level of indifference between *events*, because our intuition is a very unreliable guide in such matters, as Bertrand's paradox illustrates.

However, the principle of indifference may, in our view, be applied legitimately at the more abstract level of indifference between *problems* because it is a matter that is definitely determined by the statement of a problem, independently of our intuition. Every circumstance left unspecified in the statement of a problem defines an invariance property which the solution must have if there is to be any definite solution at all. The transformation group, which expresses these invariances mathematically, imposes definite restrictions on the form of the solution, and in many cases fully determines it.

Of course, not all invariances are useful. For example, the statement of Bertrand's problem does not specify the time of day at which the straws are tossed, the color of the circle, the luminosity of Betelgeuse, or the number of oysters in Chesapeake Bay; thus we infer, correctly, that if the problem as stated is to have a unique solution, it must not depend on these circumstances. But this would not help us unless we had previously thought that these things might be germane.

Study of a number of cases makes it appear that the aforementioned dilemma can now be resolved as follows. We suggest that the cases in which the principle of indifference has been applied successfully in the past are just the ones in which the solution can be *reverbalized* so that the actual calculations used are seen as an application of indifference between problems, rather than events.

For example, in the case of poker hands, the statement of the problem does not specify the order of cards in the deck before shuffling. Therefore, if the problem is to have any definite solution, it must not depend on this circumstance; that is, it must be invariant under the group of 52! permutations of cards, each of which transforms the problem into an equivalent one. Whether we verbalize the solution by asserting that all distributions of cards in the final hands are *equally likely* or by saying that the solution shall have this invariance property, we shall evidently do just the same calculation and obtain the same final results.

There remains, however, a difference in the logical situation. After having applied the transformation group argument in this way, we are not entitled to assert that the predicted distribution of poker hands *must* be observed in practice. The only thing that can be proved by transformation groups is that, if this distribution is *not* forthcoming, then the probability of obtaining a given hand will necessarily be different for different initial orders of the cards; or as we would state it colloquially: The cards are not being *properly* shuffled. This is, of course, just the conclusion we do draw in practice, whatever our philosophy about the *meaning of probability*.

Once again, it is clear that the invariant solution is overwhelmingly the most likely one to be produced by a person of ordinary skill; to shuffle cards in such a way that one particular aspect of the initial order is retained consistently in the final order requires a *microscopic* degree of control over the exact details of shuffling. (In this case, however, the possession of such skill is generally regarded as dishonest, rather than impossible.)

We have not found any general proof that the method of transformation groups will always lead to solutions with this frequency correspondence property. However, analysis of some dozen problems like the above has failed to produce a counter-example, and its general validity is rendered plausible as follows. In the first place, we recognize that every circumstance, which our common sense tells us may exert some influence on the result of an experiment, ought to be given explicitly in the statement of a problem. If we fail to do that, then of course we have no right to expect agreement between prediction and observation; this is not a failure of probability theory, but rather a failure to state the full problem. If the statement of a problem *does* properly include all such information, then it would appear that any circumstances which are still left unspecified must correspond to some lack of control over the conditions of the experiment. But invariance under the corresponding transformation group is just the formal expression of this lack of control, or lack of skill.

One has the feeling that this situation can be formalized more completely. Perhaps one can define some *space* corresponding to all possible degrees of skill and define a measure in this space, which proves to be concentrated overwhelmingly on those regions leading to the invariant solution. Up to the present, however, we have not seen how to carry out such a program; perhaps others will.

8

Conjectures

There remains the interesting, and still unanswered, question of how to define precisely the class of problems which can be solved by the method illustrated here. There are many problems in which we do not see how to apply it unambiguously; von Mises's water-and-wine problem is a good example. Here we are told that a mixture of water and wine contains at least half wine, and are asked: What is the probability that it contains at least three-quarters wine? On the usual viewpoint this problem is underdetermined; nothing tells us which quantity should be regarded as uniformly distributed. However, from the standpoint of the invariance group, it may be more useful to regard such problems as *overdetermined;* so many things are left unspecified that the invariance group is too large, and no solution can conform to it.

It thus appears that the *higher level problem* of how to formulate statistical problems in such a way that they are neither underdetermined nor overdetermined, may itself be capable of mathematical analysis. In the writer's opinion, it is one of the major weaknesses of present statistical practice that we do not seem to know how to formulate statistical problems in this way or even how to judge whether a given problem is well-posed. Again, the Bertrand paradox is a good illustration of this difficulty, for it was long thought that not enough was specified to determine any unique solution; but from the viewpoint which recognizes the full invariance group implied by the above statement of the problem, it now appears that it was well-posed after all.

In many cases, evidently, the difficulty has been simply that we have not been reading out all that is implied by the statement of a problem; the things left unspecified must be taken into account just as carefully as the ones that are specified. Presumably, a person would not seriously propose a problem unless he supposed that it had a definite solution. Therefore, as a matter of courtesy and in keeping with a worthy principle of law, we might take the view that *a problem shall be presumed to have a definite solution until the contrary has been proved.* If we accept this as a reasonable attitude, then we must recognize that we are not in a position to judge whether a problem is well-posed until we have carried out a transformation group analysis of all the invariances implied by its statement.

The question, whether a problem is well-posed, is thus more subtle in probability theory than in other branches of mathematics and any results which could be obtained by study of the *higher level problem* might be of immediate use in applied statistics.

Notes

1. Jaynes, E.T., "Prior Probabilities," *IEEE Transactions on Systems Science and Cybernetics,* Vol. SSC-4, 3, 227-241, September, 1968.

2. Bertrand, J., *Calcul des Probabilités,* 2nd Edition, Paris, Gauthier-Villars, 4-5, 1907. "Aucune de trois n'est fausse, aucune n'est exacte, la question est mal posée."

3. Borel, E., *Eléments de la Théorie des Probabilités,* Paris, Hermann et Fils, 110-113, 1909. "...il est aisé de voir que la plupart des procédes naturels que l'on peut imaginer conduisent a la première."

4. Poincaré, H., *Calcul des Probabilités,* Paris, 118-130, 1912 "...nous avons définie la probabilité de deux manières différentes."

5. Uspensky, J.V., *Introduction to Mathematical Probability,* New York, McGraw-Hill Book Co., 251, 1937. "...we are really dealing with two different problems."

6. Northrup, E. P., *Riddles in Mathematics,* New York, D. van Nostrand Co., 181-183, 1944. "One guess is as good as another."

7. Gnedenko, B. V., *The Theory of Probability,* New York, Chelsea Pub. Co., 40-41, 1962. The three results "would be appropriate" in three different experiments.

8. Kendall, M. G. and Moran, P. A. P., *Geometrical Probability,* New York, Hafner Pub. Co., 10, 1963. "All three solutions are correct, but they really refer to different problems."

9. Weaver, W., *Lady Luck: the Theory of Probability,* Garden City, N.Y., Doubleday Anchor Books, 356-357, 1963. "...you have to watch your step."

10. von Mises, R., *Mathematical Theory of Probability and Statistics* edited by H. Geiringer, New York, Academic Press, 160-166, 1964. "Which one of these or many other assumptions should be made is a question of fact and depends on how the needles are thrown. It is not a problem of probability calculus to decide which distribution prevails—." von Mises, in the preface to *Probability, Statistics and Truth* (New York, Macmillan Co., 1957) also charges that, "Neither Laplace nor any of his followers, including Poincaré, ever reveals how, starting with *a priori* premises concerning equally possible cases, the sudden transition to the description of real statistical events is to be made." It appears to us that this had already been accomplished in large part by Jacob Bernoulli (1703) in his demonstration of the weak law of large numbers, the first theorem establishing a connection between probability and frequency. Reference [1] and the present article may be regarded as further contributions toward answering von Mises' objections.

11. Mosteller, F., *Fifty Challenging Problems in Probability,* Reading, Mass., Addison-Wesley, Pub. Co., 40, 1965. "Until the expression 'at random' is made more specific, the question does not have a definite answer... We cannot guarantee that any of these results would agree with those obtained from some physical process..."

12. Wigner, E. P., *Gruppentheorie und ihre Anwendung auf die Quantenmechanik der Atomspektren,* Braunschweig, Fr. Vieweg, 1931.

13. Weyl, H., *The Classical Groups,* Princeton, Princeton University Press, 1946.

COMMENTS

I.J. Good:

Concerning page 344, it is interesting to apply the invariance argument to a problem concerning continual creation. It provides a physical example of an overdetermined problem. I think the argument was first published by T. Gold or G.J. Whitrow. Consider what distribution of velocities for a newly created particle is invariant under all Lorentz transformations. The answer is that the velocity must be that of light, so that apparently only photons and neutrinos can be newly created. Moreover the distribution of the direction cannot be uniform in all directions in all frames of reference. It is therefore impossible to reconcile the theory of continual creation with the theory of relativity *unless the distribution of the direction of newly created particles depends on the distribution of matter and radiation* as seen from the place where the new particle appears.

I think there must be a flaw in Jaynes' argument on page 350. He argues that if we lack the skill to make our straws intersect a given circle, then we must also lack the skill to produce different distributions for different circles. But what happens if we replace circles by ellipses here? I think this leads to trouble, and that the reason the argument appears correct for circles is that he has *implicitly assumed the uniform distribution of direction for the final position of each straw.* I think what makes this assumption reasonable is that *the straws are in the wind,* by which I mean that the straw is blown around before it arrives on the floor. The theory that would determine the final direction of the straw would be in essence that of a roulette wheel, which I believe was first given by Poincaré in his book on probability. The point here is that provided that the physical distribution of the total angle rotated by the straw is unimodal and wide in comparison with 180 degrees, then the distribution of the final direction will be approximately uniform. This follows by an application of Poisson's summation formula. Likewise, whatever assumption we make about the distribution of the center of gravity, provided that this distribution is unimodal, and wide with respect to the size of the circle, then the center of gravity of the straw will have a uniform distribution in the neighborhood of the circle. (The unimodality is presumably not necessary provided that the distances between successive maxima of the density function are large compared with the circle.) These two distributions, the distribution of the direction of the straw and the distribution of the center of gravity, clearly define the final distribution uniquely.

It seems to me that the basic idea behind the invariance method applied to groups is implicit in the theory of the roulette wheel. The symmetry of the problem determines the logical probability distribution; but it does not determine the *physical* distribution, and hence the long-run frequencies, without an argument similar to Poincaré's.

In the kinetic theory of gases it is usually assumed that the velocity of any particle has a distribution whose direction is uniform. This could be analyzed in a manner similar to that of Poincaré's analysis of the roulette wheel. Underlying the distribution of the direction are other distributions concerned with the collisions between molecules. If we assume that the molecules are spherical, that the collisions are elastic and that the distribution of the position of a molecule which is hit by another is wide in comparison with the diameters of molecules, but that the number of collisions made by one molecule is large in the time concerned, then presumably we can prove that the final distribution of the direction of motion of molecules is nearly uniform. I think this must also follow from the theory of the roulette wheel applied to the polar coordinate angles θ and ϕ. If the molecules are not assumed to be spherical we should use similar arguments to show that the directions of their principal axes are uniformly distributed. This would make the above argument more rigorous.

I think it would be easy to construct models in which the uniform distribution of direction of motion was incorrect. For example, consider a very thin capillary tube, so thin that its diameter is much smaller than the mean free path of the molecules of the gas inside the tube. Then we would presumably find that the distribution of direction was by no means uniform, and that most of the molecules would move approximately parallel to the direction of the tube either in one direction or the other. So I suppose the Maxwell-Boltzman distribution does not apply to very thin capillary tubes. [In the discussion some-one asserted that, allowing for the elastic collisions on the inside of the tube, there would *not* be any tendency for the molecules to run parallel to the tube. He might be right.]

COUNTER-EXAMPLES AND LIKELIHOOD

André Plante
Université du Québec à Montréal

André Plante
Université du Québec à Montréal

1

Introduction

As we all know, *Statistics* has become divided among various approaches which we could call Neyman-Pearson-Wald theory, Bayesian theory, Fisherian theory, and recently, structural probability theory. This plurality of theories, and the fact that controversies between statisticians are not unknown, force the younger generations of statisticians to raise questions concerning the basic principles of statistics, for they can no longer be accepted on trust alone. We are forced to look for a method of comparison of statistical theories and a critique of the foundations of our science.

The basic principles of statistics share with those of the physical sciences a dual character; that is, they are both mathematical axioms from which a theory is deduced, and postulates which assert judgments about reality or the statistician's relationship to reality. In the physical sciences a researcher has an inductive attitude; he is alert to the realization that a new fact may contradict existing theory. Thus, in the physical sciences there is a dynamic relationship between theory and experience or experimentation. The relationship of theory to counter-example in statistics is somewhat similar, or should be. Thus if a counter-example contradicts an existing theory, that theory must be revised in order to accommodate it. This is achieved through examining its basic postulates or principles.

We seem to find six main types of principles in statistical literature (most encountered in several versions). These are Bayesian principles, sufficiency principles, likelihood principles, efficiency principles (minimization, in some sense, of an average loss when a random experiment is repeated), conditionality principles, and symmetry principles (for example, the use of the symmetry of a situation to induce a structural probability distribution). Each type of principle, considered alone, has intuitive appeal but in particular

357

instances they may conflict. Consequently we realize that just because our statistical intuition pushes us to accept a theory based upon one set of general principles or postulates in preference to another, this is no assurance of that theory's validity, for statistical intuition seems to falter at the level of basic postulates. Fortunately, our intuition is more under control when we face particular cases, and this is how we can compare different sets of postulates. For example, Pearson's 1956 principle stating how to determine the reference set by considering a series of repetitions of the experiment which generated the data, while intuitively quite appealing, is clearly seen to be invalid in general in the light of Cox's 1958 counter-example.

We feel it is necessary to stress the *positive* function of counter-examples and to take them as guidelines in our attempts to reformulate existing theories; counter-examples are not merely weapons with which one may attack, but also tools with which one may build.

A counter-example is an instance when one's belief in a specific statistical principle A is in the process of being undermined by the necessity of facing one's simultaneous belief in an alternate principle B in a statistical situation where the two principles are in conflict. One concludes that principle A must be restricted in some manner and principle B must be adopted in some form. The aim of this method of reasoning is to make more precise the scope of validity of different principles and results in a process of re-structuring statistical theories through a trial and error process. One must face, however, that the theory with which one ends may depend, in large extent, on one's point of departure. In an unpublished paper, *The Individual Decision Problem*, I have used counter-example reasoning, and here continue this method to examine the principles of statistical inference. In the present paper, I am examining the concept of likelihood as a measure of rational belief.

I would like to acknowledge my gratitude to George Barnard for having discussed with me in detail several topics related to inference; and while I have borrowed many of his ideas, I would not like to imply that he is in agreement with the interpretation of likelihood or significance tests given here.

2

Collective Decisions, Individual Decisions and Inference

Since Cox's counter-example has shown the necessity of admitting some conditionality principle in any theory of inference, it becomes necessary to distinguish among decision situations a certain type of problem in which a conditionality principle seems desirable as well. I have called these individual decision problems because they are decision problems where the experiment from which the data are obtained is performed only once.

For example, consider a statistician who has to report to government authorities the results of a sample survey connected with the choice of location of a new airport for super-jets. Although the loss function is not easily quantifiable owing to the political aspect of the problem, it is clear, however, that our statistician is taking part in a decision process. It is also clear that he should, if one ancillary statistic is present, apply a conditionality principle in formulating his advice to the authorities.

Sometimes subtle differences distinguish individual decision situations from collective decision situations, that is, from situations where a whole series of repetitions of an experiment is envisaged. The following example is a borderline case, and serves well to illustrate this distinction. Suppose a certain blood test is applied to every patient being admitted to a hospital with the preventative aim of detecting the presence of a grave disease. Suppose also that the response to the test is normally distributed with a variance known from previous experience for a given population of individuals. Furthermore, suppose that when the response is positively significant at the 5% level the patient is intensively examined. If the patients of various blood groups all react with the same variance there is no difficulty in interpreting the results; but if different blood groups have responses which exhibit different variances, then there emerges a mixture structure comparable to that which is found in Cox's counter-example. A physician who considers it his duty to devote his care to each of his patients individually should consider that he has to take, not a collective decision, but a series of separate individual decisions. Consequently, he should interpret each result conditionally, given the blood group of the patient being tested. It is important, however, to notice that, if we consider that we have a series of individual decisions, we would use a procedure which, on the overall, does not detect as many diseased patients as could be detected if we use a procedure with an optimum detection power with respect to the population as a whole. This second procedure, however, denies that an individual has the right to be judged with respect to his own response group.

Individual decision problems are distinct from inference problems in that our first concern is not to express uncertainty but to behave rationally in circumstances where all values are expressible, at least theoretically, in terms of numerical losses. In the present paper we are concerned only with inference, not individual decisions.

3

Likelihood and Standardized Likelihood Functions

The main limitations of the concept of likelihood in situations with a finite dimensional parameter space are: first, likelihood provides us with a relative measure of rational belief for comparing any two parameter values when what we want is an absolute measure for assessing each particular parameter value, given the model; secondly, in itself, likelihood provides no means of eliminating nuisance parameters. These limitations force us to derive from the original concept of likelihood new concepts more appropriate to our goal. One such concept I would call standardized likelihood.

Let $f(\chi; \theta)$ be the density of the observation X with respect to some $\sigma-$ finite measure when the parameter θ takes its value in a finite dimensional space, and let χ be the observed data. In order to obtain a measure of rational belief in a particular value of the parameter, one of the first things that comes to mind is to consider the standardized likelihood function

$$\mathscr{L}(\theta; \chi) = f(\chi; \theta)/\sup_\theta f(\chi; \theta).$$

Indeed, the measure so obtained is relative to the probability model used, and we must assume that a test of goodness of fit is to be applied in order to avoid using too inadequate a model. Given the model, let us therefore tentatively assert the postulate that the numerical value of $\mathscr{L}(\theta; \chi)$ is a measure of rational belief. That is, if another observation χ^* and a corresponding parameter θ^* yield a standardized likelihood $\mathscr{L}^*(\theta^*; \chi^*)$ in a completely different situation, then the equality $\mathscr{L}^*(\theta^*; \chi^*) = \mathscr{L}(\theta; \chi)$ implies that the force of our belief, in a likelihood sense, is the same in both situations.

Although there exists no counter-example against the use of likelihood in general, there are objections to the use of a measure of belief based on standardized likelihood. Let us examine two well known counter-examples to standardized likelihood, in order to make more precise the kind of rational belief, if any, which we may associate with it.

The first counter-example I wish to consider is based on a situation reported by A. Birnbaum (1964). Suppose the observation has a univariate normal distribution, and let the parameter space be $\{(\mu, \sigma): -\infty < \mu < \infty, 0 \le \sigma < \infty\}$ where μ and σ have their usual meaning and where a normal distribution with variance zero is defined to be a distribution concentrated on only one point. Then, for any data χ, the likelihood function is infinite at the point $(\mu, \sigma) = (\chi, 0)$. Our best means of determining the standardized likelihood is to use an argument involving a limiting process when σ tends towards zero, which gives as standardized likelihood function

$$\mathscr{L}(\mu, \sigma; \chi) = \begin{cases} 1 \text{ if } (\mu, \sigma) = (\chi, 0) \\ 0 \text{ otherwise.} \end{cases}$$

However, if we consider a series of observations, all obtained by setting $\mu = 0$, $\sigma = 1$, then a standardized likelihood approach seems to lead us to conclude wrongly with probability one.

A second counter-example to standardized likelihood has been given by Armitage (1961) and refers to its use in a sequential context. In his example Armitage shows that, if for a predetermined value θ_0 of the parameter the stopping rule is *stop when* $\mathscr{L}(\theta_0; \chi) \le \alpha$ *for the first time* where α is an arbitrarily small positive number, then with probability one, the standardized likelihood assigned by the experiment to the value θ_0 is smaller than or equal to α, even when $\theta = \theta_0$.

The above counter-examples, in my opinion, constitute a proof that: first, we must accept some efficiency principle in inference; secondly, we must impose clear limitations to the concept of standardized likelihood. As a consequence of the first counter-example we must at least admit (as Barnard, 1967 has pointed out) that densities are only means of approximating real situations which are always discrete. To cope with Armitage's counter-example, I suggest that the kind of rational belief measured by standardized likelihood be interpreted as *suspicion*.

4

Likelihood as a Measure of Suspicion

Suspicion is an attitude of mind which recognizes that there might be a difference between the way things appear and the way they really are, in short, which expresses a type of uncertainty. A statistician wants to make some judgment of a hypothesis; thus he must *suspect* if his hypothesis is the *true* one. He may be very or slightly suspicious of it, and this variance is the degree to which he believes it is likely or unlikely to be the *true* hypothesis. I assume that standardized likelihood gives us an unbiased index of suspicion, that is, a way of expressing this range of assessment of incertitude but I also assume that we should not act as if a hypothesis were true upon considering only the numerical value of this index without paying due consideration to its random fluctuation, for common sense tells us that it is never wise to act on suspicion alone, and from Armitage's counter-example we see that it is not even wise to stop sampling on suspicion alone.

Economists, stock brokers, even physicians, have problems similar to the statistician's for at times they too must act after considering appropriate indices which they believe summarize quantitatively some aspect of a whole system, yet which exhibit random fluctuations. For them too, the problem of knowing whether an index actually represents what it is thought to represent is distinct from the problem of acting on its basis.

We may distinguish between a suspicion index and an action index. To illustrate this distinction let us recall how suspicion plays a role in the classical detective novel. Given a definite number of suspects, a detective should never confuse his suspicion index with his action index when trying to prove a suspect guilty, (or rather when trying to disprove the null hypothesis that a suspect is innocent), for he does not follow the same set of rules in presenting evidence in order to charge a suspect as he does when expressing his suspicion to himself or to his colleague. For example, at the proof-making stage the detective must attach as little weight as possible to the presence or absence of other suspects, while his (and our) suspicion towards each suspect in turn can be strongly affected by the presence of a more suspicious character.

I suggest that we adopt in inference a suspicion index and also an action index which could take into account the fluctuations of the suspicion index. Such an action index could be

$$Pr\left\{\mathscr{L}(\theta_0; X) \le \mathscr{L}(\theta_0; \chi) \,|\, S; \theta_0\right\} \tag{1}$$

where the null hypothesis is $\theta = \theta_0$, where S is the reference set used, and where χ is the observed data. For a fixed distribution, for given data, and for any sample size, the suspicion index for the true distribution will generally be smaller when the distribution is hidden in a large parameter space because of the way the dimension of the parameter space affects the size of the denominator of Wilks' ratio; whereas the distribution of the action index will be almost independent of the size of the parameter space. However, it might sometimes be impossible to determine uniquely the reference set S, and this may introduce some degree of indeterminacy in the action index. When the action index is smaller than or equal to a pre-assigned fraction α we will consider it a warning not to behave as if the corresponding null hypothesis were true, and we will say that the data are significant at level α.

Expression 1 represents the action index of a cautious person who prefers giving up some efficiency rather than adopting an action index which does not reflect his suspicion in a perfectly coherent manner. In other words, we have made a justifiable compromise between an efficiency principle and a likelihood principle, since Cox's counter-example has shown that, in inference, maximizing the power can be misleading. There is no need to think in terms of alternatives to the null hypothesis, and the associated test is a generalization of tests of goodness of fit which introduces a new distance between observations and hypothesis.

As a measure of sensitivity we could replace the concept of power with a concept of information. The log-likelihood function itself, when we consider it in its curvature aspect, provides us with a point-wise measure of information as Barnard (1951) has pointed out. When we want an overall measure of information, we can easily consider an averaged log-likelihood function, and when we want to know how much information is contained in a given statistic we can look at the partial likelihood function obtained by supposing that the value of that statistic constitutes the only data available.

The main advantage of using the log-likelihood function as a measure of sensitivity or information is that when we have the bad luck of drawing an untrustworthy sample, (that is a sample with a wide open log-likelihood function), our measure of information will also give us some indication how to select the reference set within which we can compute the significance level attained by the data; that is, set S of expression 1. The power function fails to give us such indication which is why Cox's counter-example was possible.

Let us now discuss an apparent implication of Armitage's counter-example to the effect that a standardized likelihood analysis cannot proceed sequentially. I cannot agree with this interpretation, for Armitage's counter-example has the peculiarity that we are asked to build an experiment with the intention of determining the truth value (the standardized likelihood) of a proposition $\theta = \theta_0$ which has its truth value built into the experiment through the stopping rule used. In other words, we have a likelihood analogy to the paradoxical sentence, *I am lying*. We should evidently avoid such a structure in a likelihood approach the same way we avoid it in ordinary life. Instead, I believe that when we proceed to a sequential standardized likelihood analysis, the stopping rule should depend only on an ancillary statistic, or perhaps, if a small amount of distortion is acceptable, on a statistic which has a distribution depending only slightly on the parameter. We could, for example, often continue sampling until the curvature at the top of the log-likelihood function is large enough; at termination we would be assured of a fixed amount of information in our sample, but we would not influence the answer in advance. The corresponding action index should be computed conditionnally, given the ancillary (or quasi-ancillary) statistic used.

5
Elimination of Nuisance Parameters

When the parameter has the structure $\theta = (\mu, \sigma)$ where σ is a nuisance parameter, and μ a parameter of interest, the concept of standardized likelihood does not allow us to attach a value to a set of the form $\{(\mu, \sigma): \mu = \mu_1\}$ for a fixed μ_1. However, having set a correspondence between the mathematical notion of standardized likelihood and suspicion as an attitude of mind, we can refer to our common notion of suspicion to suggest a way of making a summary of standardized likelihoods. To return to our analogy with the detective novel, the suspicion we might feel that some person in a group is guilty cannot be smaller, I believe, than our suspicion of each individual in the group. Furthermore, if we exclude the possibility of collusion, because it has no statistical counterpart, I see no reason why we should suspect the group more than we suspect each person within it. Therefore, I suggest as a summary of the standardized likelihood function, a tentative definition

$$\mathscr{L}(\mu; \chi) = \sup_{\sigma} \, \mathscr{L}(\mu, \sigma; \chi) \tag{2}$$

which is Wilks' ratio (or the Neyman-Pearson 1928 likelihood ratio).

Geometrically, equation 2 means that we simply look at the profile of the standardized likelihood function when we place ourselves so as to see no perspective within the σ-space. Therefore, we can say that, through 2, we still look at the standardized likelihood function itself.

By extension, we may say that the curvature aspect of the graph of $\log[\mathscr{L}(\mu; \chi)]$ is a summary of the information with respect to μ contained in χ. But to visualize the loss of information corresponding to our ignorance of σ, we have to come back to the whole log-likelihood function. The function $\mathscr{L}(\mu; \chi)$ is equivalent to a pseudo-standardized likelihood function $\mathscr{L}^*$ such that $\mathscr{L}^*(\mu, \sigma; \chi) = \mathscr{L}(\mu; \chi)$. Suppose for simplicity that $\mathscr{L}(\mu, \sigma; \chi)$ actually reaches its maximum at a point (μ_0, σ_0). Since we have $\log[\mathscr{L}^*(\mu, \sigma; \chi)] \geq \log[\mathscr{L}(\mu, \sigma; \chi)]$ while $\log[\mathscr{L}^*(\mu_0, \sigma_0; \chi)] = \log[\mathscr{L}(\mu_0, \sigma_0; \chi)]$, we can see that $\log[\mathscr{L}(\mu, \sigma; \chi)]$ has more information content than $\log[\mathscr{L}^*(\mu, \sigma; \chi)]$ and that the loss of information corresponds to our lack of knowledge of σ.

Equation 2 can be the basis of a calculus of likelihoods, and the way in which it summarizes likelihoods is to be contrasted with the additive property of probabilities. Furthermore, if we define the conditional standardized likelihood, given σ, by means of the equation

$$\mathscr{L}(\mu | \sigma; \chi) = \mathscr{L}(\mu, \sigma; \chi) / \sup_{\mu} \, \mathscr{L}(\mu, \sigma; \chi)$$

for a fixed σ, we obtain

$$\mathscr{L}(\mu, \sigma; \chi) = \mathscr{L}(\sigma; \chi) \, \mathscr{L}(\mu | \sigma; \chi)$$

as was pointed out by Box and Cox (1964). If we also define μ and σ as independent, whenever $\mathscr{L}(\mu | \sigma; \chi)$ does not depend on σ we see that μ and σ are independent, if and only if $\mathscr{L}(\mu, \sigma; \chi) = \mathscr{L}(\mu; \chi) \, \mathscr{L}(\sigma; \chi)$; or again, if and only if the probability function $f(\chi; \mu, \sigma)$ is of the form $f(\chi; \mu, \sigma) = g(\chi; \mu) \, h(\chi; \sigma)$. Further similarities with the calculus of probability have been pointed out by Barnard (1949).

Thus by postulating that the profiles of the standardized likelihoods are unbiased indexes of suspicion instead of admitting the existence of prior distributions, we obtain a calculus of likelihoods which can be as useful as Bayesian analysis in dealing with inference problems. Until I am given a clear-cut counter-example against such a calculus of likelihoods, showing the necessity of the Bayesian assumption, I consider that the Bayesian theory of inference introduces an unnecessarily strong assumption.

In a testing situation with nuisance parameter σ, we could say that the data are significant at level α with respect to the null hypothesis that μ has a given value μ_0 whenever

$$Pr\{\mathcal{L}(\mu_0; X) \leq \mathcal{L}(\mu_0; \chi) \mid S; \mu_0, \sigma\} \leq \alpha \qquad (3)$$

for every σ where S is the reference set chosen. If the inequality is satisfied only for some σ's, then the test remains undetermined.

To use test 3, is to use an action index which reflects the way we summarize our suspicion in presence of a nuisance parameter, and it results in giving still less importance to efficiency. Thus to adopt 3 is to behave cautiously rather than efficiently. In order to illustrate this difference let us consider an example which Lehmann (1959, p. 253), ascribes to Stein. If we use Lehmann's notation, and if the point P_i is observed, then the standardized likelihood function for hypotheses H and K and for the nuisance

parameters $p_1, \ldots, p_n$ $(\sum_{i=1}^{n} p_i = 1, p_1, \ldots, p_n \geq 0)$ is given by

$$\mathcal{L}(H, p_1, \ldots, p_n; P_i) = \alpha$$
$$\mathcal{L}(K, p_1, \ldots, p_n; P_i) = p_i$$

where $0 < \alpha \leq \frac{1}{2}$. Therefore, we should have more suspicion towards the larger values of p_i and the relevant profile of the standardized likelihood function is

$$\mathcal{L}(H; P_i) = \alpha$$
$$\mathcal{L}(K; P_i) = 1.$$

Test 3 then implies that at level α it is dangerous to behave as if H were true but to conclude that we must behave as if K were true results in a very inefficient procedure, as Lehmann points out. However, to go on behaving as if H were true, as the most powerful invariant test indicates, is unsatisfactory for one who suspects strongly that $p_i = 1$.

To take an extreme case, would we still feel justified to use an invariance principle if we would know for sure that one of the p's is one, and the others are zero? In this case we might find it preferable to distinguish n alternatives $K_j(j = 1, \ldots, n)$ where K_j is K with $p_j = 1$. We then notice that the union of the rejection regions of the M P tests of H against K_j at level α/n is precisely the rejection region of test 3. But, curiously enough, those rejection regions are all included in the acceptance region of the M P invariant test of level α.

We cannot claim that the M P invariant test is in agreement with our practical intuition either, for if Stein's model were taken to describe a physical phenomenon, I do not think a researcher, having observed P_i, would conclude as the M P invariant test does, that hypothesis H was properly tested when there is so little information concerning the p's. Instead, if he really believes that all the parameter values of his model were plausible *a priori*, he might either invest some research effort in gathering more information concerning the nuisance parameters or else suspend his judgement and do nothing. Thus we see that in reality there is more than one alternative to behaving as if the null hypothesis were true, and in order to neglect none of them, a cautious person is ready to be inefficient with respect to the particular alternative which consists in behaving as if K were true.

6
Conclusion

I do not wish to suggest that the theory of likelihood adopted here is the only one which could be devised when one starts from the concept of likelihood and uses counter-examples as guidelines. However, I believe that, because of its simplicity and its close agreement with traditional statistical practice, through the use of expression 3, we should adopt it as long as we have no clear counter-example to it.

Acknowledgments

I would like to thank the Canada Council, the University of Essex and the Université de Montréal for their assistance to this research.

References

1. Armitage, P., Contribution to the discussion to C.A.B. Smith, "Consistency in Statistical Inference and Decision," *Journal of the Royal Statistical Society, B,* 23, 1-37, 1961.
2. Barnard, G.A., "Statistical Inference," *J.R. Statist. Soc., B.* 11, 115-139, 1949.
3. Barnard, G.A., "The Theory of Information," *J.R. Statist. Soc., B,* 13, 46-64, 1951.
4. Barnard, G.A., "The Use of the Likelihood Function in Statistical Practice," *Proceedings of the Fifth Berkeley Symposium,* University of California Press, 1, 27-40, 1967.
5. Birnbaum, A., "The Anomalous Concept of Statistical Evidence: Axioms, Interpretations and Elementary Exposition," paper presented at the Joint European Conference of Statistical Societies, Berne, 1964.
6. Box, G.E.P. and Cox, D.R., "An Analysis of Transformations," *J. R. Statist. Soc., B,* 26, 211-252, 1964.
7. Cox, D.R., "Some Problems Connected with Statistical Inference," *Annals of Mathematical Statistics,* 29, 357-372, 1958.
8. Kalbfleisch, J.D. and Sprott, D.A. "Application of Likelihood Methods to Models Involving Large Numbers of Parameters," *J.R. Statist. Soc., A,* (to appear), 1970.
9. Lehmann, E.L., *Testing Statistical Hypotheses,* New York, Wiley, 1959.
10. Pearson, E.G., "Statistical Concepts in Their Relation to Reality," *J. R. Statist. Soc., B,* 12, 204-207, 1956.
11. Plante, A., "The Individual Decision Problem," unpublished.

APPENDIX

A few days before reading this paper I learned from Professor Basu a counter-example which definitely dismisses the *null hypothesis* that the profile of a likelihood function can be a substitute for the likelihood function itself in the presence of a nuisance parameter. The counter-example, which Professor Basu attributed to Professor Neyman, is as follows.

Let $x_{ij}(i = 1, \ldots, n, j = 1, 2)$ be independent $N(\mu_i, \sigma^2)$ variates. If we eliminate the μ's by looking at the profile of the standardized likelihood function, we find

$$\mathscr{L}(\sigma^2) = e^n \left[\frac{S^2}{4n\sigma^2} \exp\left\{ -\frac{S^2}{4n\sigma^2} \right\} \right]^n$$

where $S^2 = \Sigma(x_{i1} - x_{i2})^2$. However, for large n, $\mathscr{L}(\sigma^2)$ tends to concentrate on the point $1/2\,\sigma^2$.

As a consequence of this counter-example we must admit some principle of unbiasedness to modify the method of likelihood profile for eliminating nuisance parameters. I would like to suggest one such modification.

Before the observations were taken, we were aware of our intention to eliminate the parameters μ_i by setting the conditions $\hat{\mu}_i(x_{i1}, x_{i2}) = \mu_i$ where $\hat{\mu}_i$ is the maximum likelihood estimate of μ_i. Therefore, there seems to be some grounds to argue that we should consider the likelihood for σ^2 obtained from the conditional distribution of the x_{ij} given $\hat{\mu}_i(x_{i1}, x_{i2}) = \mu_i$. This will automatically eliminate the μ_i from the resulting likelihood function. This approach gives

$$\mathscr{L}^*(\sigma^2) = e^{n/2} \left[\frac{S^2}{2n\sigma^2} \exp\left\{ -\frac{S^2}{2n\sigma^2} \right\} \right]^{n/2}$$

which I will call the modified likelihood profile. For large n, $\mathscr{L}^*(\sigma^2)$ tends to concentrate on the point σ^2.

As a further example, suppose $x_1, \ldots, x_n$ are independent $N(\mu, \sigma^2)$ variates. Then, conditioning on $\bar{x} = \mu$, the modified likelihood profile for σ^2 is

$$e^{\frac{n-1}{2}} \left(\frac{S^2}{\sigma^2} \right)^{\frac{n-1}{2}} \exp\left\{ -\frac{(n-1)S^2}{2\sigma^2} \right\} \quad (n > 1)$$

which has its maximum at

$$\hat{\sigma}^2 = \Sigma(x_i - \bar{x})^2 / (n - 1) = S^2.$$

Note the automatic adjustment for the degrees of freedom. Similarly, conditioning on $\frac{1}{n}\Sigma(x_i - \mu)^2 = \sigma^2$, the corresponding modified likelihood profile for μ is

$$[1 + n(\bar{x} - \mu)^2 / \Sigma(x_i - \bar{x})^2]^{-\frac{n-1}{2}} \quad (n > 1).$$

For $n = 1$, both of these modified likelihood profiles are flat; this seems satisfactory since the conditioning removes all the information from the sample.

The method of modified likelihood profile suggested here seems to lead to

366

the same results as the methods advocated by John Kalbfleisch and Sprott (1970). We can therefore hope that the counter-example method of reasoning can help resolving further difficulties arising in the foundations of statistical inference.

COMMENTS

G. A. Barnard:

I just want to say I agree with Dr. MacKay's* comment, though the point as made by Professor Plante is also sound. In the expression

$$\Pr [x = \mu] = 1 \tag{1}$$

we are tempted to think of x and μ as being restricted to a lattice of values. But in fact x and μ here are supposed to be real. If one had the hypothesis $\mu = \pi$, in (1) above, and observed

$$x = 3.1415926538 \ldots ad\ infinitum$$

then we would have infinitely strong evidence in its favour. On the other hand, of course, such an observation could never really be made.

More generally, the fact that the likelihood increases as σ decreases corresponds to the sensible idea that if an observation is in accordance with two hypotheses it is the more precise hypothesis which receives the stronger support.

D. J. Bartholomew:

Dr. Plante uses Cox's counter-example to argue that Pearson's recipe for choosing the frequentist reference set is unsatisfactory. I think that the real fault lies not in the choice of reference set but in the standard formulation of the Neyman-Pearson theory of hypothesis testing. In that theory it is postulated that the chances of error should be made as small as possible and that the type I error is the more important. This is achieved by fixing the type I error at an acceptably small level and then maximizing the power. An alternative formulation which seems to me more natural and which by-passes most of the difficulties is to minimize

$$\alpha + \omega \beta$$

where α and β are the two error probabilities and ω ($0 < \omega < 1$) expresses the relative importance of the two kinds of error. I explored the consequences of this point of view in my 1967 paper.

The author's way of dealing with Armitage's counter-example is very much the same as my own suggestion in the discussion following Barnard's paper. Within the framework of my own paper this would be expressed as follows: Select a stopping rule such that if the likelihood is combined with an informa-

*Mackay's comments to follow.

tive prior by Bayes' theorem then the resulting statistical procedures agree with frequentist methods.

M. S. Bartlett:

I do not see that Dr. Plante modifies the likelihood function except in ways already recognized. Thus, for large samples and a few nuisance parameters, we can condition asymptotically by the maximum likelihood estimates. In some cases where the number of nuisance parameters increases this method breaks down; but in some cases we can condition exactly. Conditioning on the means of different normal samples leads, for example, to my unbiased modification of the Neyman-Pearson homogeneity of variances test. For unknown variances the sufficient statistics are more complicated as they depend on any unknown means. Dr. Plante refers to the R. S. S. paper by Professors Kalbfleisch and Sprott, to which I sent in a written contribution drawing attention to previous work on this type of problem (including work by James), and noting that the solution should depend on the purpose for which the likelihood function is being constructed.

I. J. Good:

I don't know who first suggested in explicit generality the use of likelihood to be treated as a *statistic* to be used for significance testing by looking at its tail-area probability. One such reference is my paper on the "Surprise Index," *Ann. Math. Statist.*, 27, 1956, pp. 1130-1135; corrections, 28, 1957; but I suppose there are much earlier references, and I would be grateful if somebody would supply one. However my paper is especially pertinent because it suggested $- \log P(E|H) + \mathcal{E}_E[\log P(E|H)| H]$ as a measure of surprise, thus relating likelihood to surprise. Hypotheses could perhaps be chosen by a principle of least surprise, but we would have to add in $-\log P(H)$ to allow for the initial probabilities.

I think *suspicion* and *surprise* are closely associated concepts. As I said in the reference: the "...function of surprise is to jar us into reconsidering the validity of some hypothesis that had previously been accepted", and the same applies to suspicion.

If the tail-area probability of *any* reasonable statistic is small, then this might be grounds for surprise. But if we try several dependent different statistics, all obtained from the same experimental result, there is much to be said for the use of the harmonic mean of the tail-area probabilities (see "Significance Tests in Parallel and in Series," *J. Am. Stat. Ass.*, 53, 1958, 799-813).

Surprise indexes can be introduced based on the evidence provided by *any* statistic, even apparently stupid ones. This corresponds to the fact that you can be surprised by any aspect of an event, including aspects that you judge to have no scientific interest such as the number of H's in names of the people selected in a sample survey. *Coincidences* are often of this kind.

O. Kempthorne:

(1) When one is a member of a jury, how does one conclude that a person is guilty? By a direct probability? I believe NO. One decides that the only tenable

model is that the individual is guilty. This is *not* a direct *probability*. (2) What is the relation of the material of the appendix to the ideas of Maurice Bartlett of 1937 and (I believe) of Quenouille in one of his books? (3) With reference to Dr. Bartholomew's comments, I believe, the concept of errors of the two types is very misleading if interpreted strictly except in a very narrow range of circumstances. I am totally allergic to the use of something like $\alpha + \omega \beta$. (4) What is the likelihood function? How does one write it down properly? The use of continuous likelihood seems very questionable. Does one observe X?

J. MacKay:

Dr. Plante's first counterexample can be resolved, it seems to me, if we examine it carefully. The destiny function is given by

$$\frac{1}{\sqrt{2\pi}\,\sigma} \exp\left\{-\frac{1}{2\sigma^2}(x-\mu)^2\right\} dx \qquad \sigma > 0 \qquad \text{(i)}$$

$$\Pr\{x=\mu\} = 1 \qquad\qquad\qquad \sigma = 0 \qquad \text{(ii)}$$

Given an observation x the likelihood function is

$$K(x)\frac{1}{\sigma}\exp\left\{-\frac{1}{2\sigma^2}(x-\mu)^2\right\} \qquad \sigma > 0$$

$$K(x), x = \mu \qquad\qquad\qquad \sigma = 0$$

$$0, x \neq \mu$$

where $K(x)$ is an arbitrary positive constant. To form the relative likelihood function, we divide by the supremum of the likelihood over all values of μ, σ which in this case is infinite. Therefore

$$L(\mu, \sigma|x) = 0, \ \forall \ \mu, \sigma,$$

which is as expected. It seems important to point out that this example does not seem to lend itself to likelihood analysis in any case since the density in (i) and that in (ii) are not comparable by division as required because the background measures are different.

H. Rubin:

Modified Stein example. The random variable and the parameter both belong to the space of all pairs (i, j), where i is an integer between 1 and n, and j is 1 or 2. Let

$$P(X_1=i, X_2=j \mid \omega_1=i', \omega_2=j') = \begin{cases} \alpha & i=i', j\neq j' \\ \dfrac{1-\alpha}{n} & j=j' \\ 0 & i\neq i', j\neq j' \end{cases}$$

Now if $n = 10^{20}$, $\alpha = 10^{-10}$, the maximum likelihood approach would be right with probability 10^{-10} whereas the converse would be right with probability .9999999999.

REPLY

The need to consider an $\alpha + \omega\beta$-type criterion arises I believe, from the realization that in a dichotomy problem the fixed α- level of Neyman-Pearson theory does not have the same meaning when the two hypotheses are "far" apart as it does when they are "near" each other. But in order to recognize such a situation in Cox's counter-example, as Professor Bartholomew does, one must focus on conditional reference sets which presupposes that one has already admitted to some extent the relevance of a conditionality principle. One can be further reassured of the validity of conditional analysis by considering other examples; for instance, that given by Allen Birnbaum (1962) where one chooses at random one of the two measuring instruments having different precisions. Also, one could, by analogy, consider ancillary information games played between Nature and the Statistician. The optimal strategies for such games are conditional, given the value taken by an ancillary statistic, as I point out in my comment on Professor Rubin's paper presented in this symposium.

Professor Good's "surprise index" and what I have called a "suspicion index" are indeed closely associated, although in other respects different. I am trying to avoid reference to a sample space, while he allows an expectation to be taken. Moreover, our intentions appear to differ. While he is seeking a mathematical definition for the ordinary language concept of surprise, I am looking for a common attitude of mind to associate with the mathematical concept of standardized likelihood. I would like to emphasize that my aim is to stress the *positive* function of counter-examples and show how they can lend support to various statistical theories. I have staked no claim on the theories themselves. Also, examples given to illustrate specific points (the series of medical tests) are, of course, fictitious and do not attempt to cover all the complexity of real situations.

I am tempted to agree with Dr. MacKay's remark that the likelihood approach is inappropriate when different background measures are present; but to be fully convinced, I would require a new counter-example. However, on the basis of Professor Barnard's vivid example, I am very much inclined to find, like Professor Kempthorne, that the use of likelihoods based on continuous distributions is very questionable. If we accept this view, the background measure could only be discrete.

The conditional approach suggested in the appendix is different from Professor Bartlett's μ- criterion for I do not require, as he does, that the conditioning variable, that is the maximum likelihood estimator, be a sufficient statistic. When that estimator is not sufficient, my approach also attempts to justify the process of replacing the remaining nuisance parameter in the conditional distribution by its maximum likelihood estimate. But perhaps the examples given have masked this distinction.

When the likelihood function is *strongly discontinuous* at its maximum, the modified likelihood profile method fails, as illustrated by the counter-example reported by Professor Rubin. If we want to eliminate the parameter i', the method consisting of conditioning on the minimal sufficient statistic (that is the conditional likelihood function statistic for i' given j') also fails.

The positive aspect of Professor Rubin's modification of Stein's counter-example is that it is *intuitively evident* that we must base the summarized likelihood function for j' on the marginal distribution of j. Upon consideration, it seems to me that such evidence comes from the fact that the quantity $(j - j')$ mod (2) is pivotal and that the conditional distribution of i given $j - j' = k$ mod (2) does not depend on j'. We could interpret this as meaning that the pivotal quantity $(j - j')$ mod (2) contains all the information concerning j'. A comparable inference concerning i' is possible on the basis of the pivotal quantity $(i - i')$ mod (n).

More generally, we are led to the principle that in order to eliminate σ, when the parameter has the structure $\theta = (\mu, \sigma)$, we should look for a pivotal quantity $Q(x, \mu)$ such that the conditional distribution of the observations x, given $Q(x, \mu) = k$ does not depend on μ for any k in the range of $Q(x, \mu)$. The likelihood for μ should then be based on the distribution of Q.

With n observations $N(\mu, \sigma^2)$, this approach leads to a likelihood for μ based on the t distribution with n-1 degrees of freedom. Also we are confirmed in our belief in the superiority of this pivotal method over the modified likelihood profile approach (which for this case gives the same result as Bartlett's μ-criterion) because the likelihood based on the distribution of the pivotal quantity t contains more information concerning μ at every point of the sample space.

Thus it seems that the various counter-examples so far considered indicate that we should pay more attention to pivotal quantities, and, perhaps, to the structural aspects of the model when we wish to summarize the likelihood function.

OCCAM'S RAZOR NEEDS NEW BLADES*
Herman Rubin
Purdue University

The principle was announced by William of Occam in the middle ages to be used in theology that "one should not multiply causes without reason". This principle has been adopted by natural philosophers and made a fundamental principle of scientific inference. It is not clear exactly what this statement means in scientific problems. However, certain procedures have been adopted on a possible interpretation of this statement, and in this paper we intend to examine these procedures and to show that they are not valid applications of Occam's Razor. There still remains a problem as to exactly what Occam's Razor means for scientific purposes. We hope to throw some light on this problem and that our observations will lead to a more accurate formulation of the problem of scientific inference.

A blade which has been extensively used in scientific pursuits is to assign a significance level and to test a null hypothesis, usually against a parametric alternative. In some cases the parametric alternative is nothing more than a change in the number of powers of a variable or variables which it has already been decided to include as causes. This latter is not a matter of consideration of exclusion of causes, but exclusion of the complexity of causes, which seems somewhat related, though different from the original principle. However, even when the matter of consideration of which variables to include is encountered, one normally considers including these variables because one has good reason already to believe those variables are causes in the scientific problem under consideration. There are a few problems, like that of the existence of extra-sensory perception or the constancy of the velocity of light in vacuum, where one seriously considers the null hypotheses. However, in other situations, like whether teaching machines have an effect on the per-

*This research was partly supported by the Office of Naval Research Contract N00014-67-A-0226-0008. Reproduction in whole or in part is permitted for any purpose of the United States Government.

formance of students, or whether cloud seeding has an effect on the total amount of rainfall, the problem is not so much the *existence* of the cause, but of the size of the effect, and whether there is any practical importance in including the cause.

However, let us now consider the case when one is really interested in testing whether or not the cause occurs. Two things should be kept in mind. First, we are frequently deciding not whether there is some cause to be included, but whether a particular cause, which we have some reason to believe, should be included. Second, we should keep in mind that the test whether the cause should be included is affected by the correctness of our theory. Nevertheless, in this case, if the sample size is fixed and the sample is not too small, the standard statistical tests are appropriate tests to use. In using these tests there is the problem of deciding what the significance level should be. It has been observed by many authors that the significance level should change with sample size and, in fact, should generally decrease as the sample size increases. As a corollary of the decrease with increasing sample size there should be, of course, an increase with decreasing sample size, so that for very small sample sizes it may be that one should not even consider accepting the null hypotheses no matter what the data is! Several authors have obtained methods of evaluating approximate significance levels, based on the user's assessment of risks. In no case can the appropriate significance level be determined in an *objective* manner.

The main problem in scientific inference is that of deciding when to *accept* or to *announce* a theory. By acceptance of a theory, I mean the taking of a position that, at the present time, it is desirable to proceed as if the theory were true. In many branches of astronomy, the Newtonian theory of gravitation is accepted. There is even a secondary type of acceptance, namely the taking of a position that the action of the *main causes* is described by the theory, and that is desirable to try to further understand the theory. An example of this is Boyle's Law, or Kepler's Laws, which are somewhat crude approximations to the presently accepted theories, but the study of which leads to much of the present development.

The announcement of a theory is the taking of a position that, at the present time, it is desirable to proceed as if the theory might be true. This is the situation, for example, as regards the various approaches to general relativity and cosmology, and in a great many situations in the behavioral sciences.

In both of these situations, the action to be taken cannot even be in principle forced by the data only. A theory which, on certain data, is accepted today, may, on the same data, be considered tomorrow as merely an approximation.

One may ask why it is necessary to accept or announce theories in which one could not believe. This is so as to enable the making of predictions, and the attainment of understanding. Without any theory, nothing could be done — even the cataloging of data requires the acceptance of a theory. The simpler a theory, the more likely it is to lead to understanding.

For this use of Occam's Razor, the appropriate statistical blades (which are necessary except in the simple situations which have prevailed in the physical sciences and occasionally in the biological sciences) have not yet been forged. I believe the forging of these blades will involve the cooperation of theoretical scientists and theoretical statisticians who are as far as possible unprejudiced by their exposure to classical procedures.

There is yet another case in which Occam's Razor is mistakenly used; however, here an appropriate carving knife is available. This is the problem of one-sided testing. Here for moderately large samples, it is only necessary to ask whether the expected gain of introducing the new procedure will outweigh its expected costs, including the necessarily *a priori* assessment of as yet unobserved side effects. The assessment of the prior distribution of nature is relatively unimportant, as is typically the case in estimation.

In summary, the use of α- level significance tests with fixed α as a tool for inference does not seem to have any justification as an application of Occam's Razor. The author has been unable to find any validity for this use except in the certain one-sided cases where the appropriate significance level is usually approximately one-half.

COMMENTS

D.J. Bartholomew:

I have spent a considerable amount of time analysing data from manpower systems. In this analysis, one is often interested in the distribution of the length of time that an employee stays in his job. It was noted a long time ago that such distributions were approximately lognormal in form. This fact is of great practical value because it introduces great simplicity into the subsequent analysis. However, if one applies a goodness of fit test to a particular distribution it is almost always significant in the usual sense. Nevertheless I continue to use the simple law because it seems to represent a basic underlying regularity discernible in a wide variety of situations. The "significant" departures can usually be attributed to transitory features peculiar to the particular system under study. In other words, the lognormal form appears to have something of the character of a natural law. The fact that it may be blurred or distorted in individual cases does not shake my faith in its reality and importance.

M.S. Bartlett:

I am not sure to what extent I agree with Dr. Rubin. When he says that it is important to know the right theory to test, I could not agree more. To me, however, there is a danger of confusion when we discuss this in the general scientific area rather than in a statistical situation in my sense, where I think statisticians have something to say and where of course it is most important to know what the right statistical model to test is. I do not think that the "classical' statistician would be very helpful in the testing of scientific laws, where the statistical element is absent or unimportant, say the theory of relativity, but I do not think that the Bayesian would be either.

D. R. Cox:

The criticism, quite commonly made, that significance tests would in the past have "rejected" very valuable scientific theories seems to me misplaced.

The tests would have indicated unexplained discrepancies between experiment and theory, and this is their usefulness; on any approach to the analysis of data, whether the theory should be *used* involves other considerations.

The discussion in my 1958 paper of the example mentioned by Professor Rubin stressed that from some points of view (Bayesian analysis, decision theory, likelihood plotting) there is no difficulty, but that in other approaches care in choosing the reference set is desirable.

I. J. Good:

I am in accord with much of the basic approach of Dr. Rubin. If statistics seems too constrained, a good field for looking for new ideas is the philosophy of science.

Harold Jeffreys, in his *Scientific Inference*, discussed Kepler's laws and pointed out that Kepler could have obtained a better fit to Tycho Brahe's data by using the epicycle theory with enough parameters. But Kepler's theory was so much simpler that it was preferable as a basis for prediction.

It is interesting that Dr. Rubin and I independently felt the need to sharpen Ockham's razor. Perhaps I need some excuse for raising the question again. I discovered recently that my brother lives within a mile of the village of Ockham in Surrey where William of Ockham was probably born. This gives me a special right to discuss the matter. Actually Ockham's razor was due to John Scotus. The rule I have proposed, with some reasonable justification, is to choose the hypothesis that maximizes the sum of the weight of evidence and a *fraction*, say a half, of the initial log-odds of the hypothesis.

We must usually compromise between explaining the facts and avoiding complexity of explanation. At the same time I should also like to point out an exception. I think that when we are speculating about very complicated adaptive systems, such as the human brain and social systems, we should especially beware of oversimplification—I call such oversimplification "Ockham's lobotomy" (in my "Speculations Concerning the First Ultra-Intelligent Machine," *Advances in Computers,* 6, 1965, 31-88).

O. Kempthorne:

I react only to the last paragraph of Dr. Rubin's written material. The idea expressed here is really rather old. It is discussed in my *Design of Experiments* (Chapter 12, 225-226) in connection with sensitivity of an experiment. I claim no originality thereby. It is desirable, I believe, to express the view that the attempts by some to force the whole of statistical evaluation usefully and totally in decision theory is fallacious. This theory talks as though one can anticipate contingencies like the concentration of poisons by the biological tree, and one can attach costs to the loss of a species, and so on and so on. This arrogance, coupled with stupidity, has plagued statistics and science for years, and is no longer tolerable.

H. E. Kyburg:

The example of Kepler's treatment of Tycho Brahe's data is an interesting one. It is indeed clear that Kepler did not follow the procedures prescribed in the statistical handbooks. But what principles of data analysis *did* he employ? We know the answer. Kepler was an avid astrologer and numerologist, and he was convinced by his reading of ancient law that there was an important connection between the existence of exactly five planets and exactly five regular polyhedra. I forget what the precise connection was that he uncovered, but it was this commitment both to satisfy the requirement of being connected with the regular polyhedra, *and* to do no violence to Brahe's observations that led him to the elliptical orbits and the 2/3 law.

There is a clear message here. It is not that we should consult the stars before constructing scientific hypothesis. It is simply that it doesn't matter a damn bit where a hypothesis comes from. Any source, kooky or otherwise, is all right. Even statistical reduction may be such a source. Whatever the source, however, the hypothesis must fit the data. Kepler was for a long time dissatisfied with his theory because of a very tiny discrepancy between his theory and Tycho's numbers. Finally it developed that Tycho had made a numerical miscalculation in arriving at the number; the calculation was done over; the new number fit the theory; and Kepler was willing to publish.

When a hypothesis fits the data is a knotty question. Far from being as discouraged as everyone else seems to be here, I have the feeling, partly as a result of this meeting, that we are approaching, gradually, an answer to that question. But we can't answer it if we persist in confusing it with the quite different question of how we arrive at hypotheses.

A. Plante:

I have a comment concerning the use of expected gains or losses in scientific situations and, more generally, in individual decision situations. Expected losses appear to be a natural generalization of Neyman-Pearson frequency interpretation. In such a framework, the statistician is asked to think of a series of idealized repetitions of the experiment having generated the data "x". Let us call $S = (\xi_1, \xi_2, \ldots)$ the sequence of results of such repetitions. Also, one could think of the corresponding sequence of losses $S' = (l(\theta, \delta(\xi_1)), l(\theta, \delta(\xi_2)), \ldots)$ which would result from adopting a decision rule δ. The statistician is then asked to act as if S' were real and as if the real, but unknown, loss $l(\theta, \delta(x))$ were a member of S'. However, in presence of an ancillary statistic, we might be able to think of a subsequence of S' which would be more meaningful than S' in the light of the data.

I suspect that to insist on considering S' itself is somewhat similar to assuming that we can force Nature to play the classical zero-sum-two-person game when, in fact, Nature might be playing another type of game like for example, the following.

Suppose the data has the structure $x = (a, y)$, where a is an ancillary statistic. Consider a game played in the following order. (1) The value of the ancillary statistic a is obtained and revealed to the players. (2) Nature chooses θ. (3) The Statistician states his strategy for guessing θ. (4) The value of y is obtained according to its conditional distribution given a and for the θ chosen by Nature.

In such a game the optimal strategies are conditional, given a; and it would be easy to give a numerical example where the statistician would be badly defeated, should he persist to believe he is playing an ordinary zero-sum-two-person game.

K. Price:

The epicyclic theory of planetary motion current at Kepler's time—that the planets move in circles whose centres move on further circles whose centres ...—could have been extended to fit the data as closely as Kepler's model; the value of Kepler's theory was that it required many fewer parameters.

This example is not irrelevant to statistics: it could well be remembered by those who fit models with polynomial expansions without considering other functions such as the exponential.

An example to support the remark that two or more oversimplified models can equally well fit the same data source is that of the wave and particle theories of light. Each is highly useful—neither is commonly believed to be 'true'.

REPLY

To Professor Kempthorne:

I, for one, have never claimed that I can do a good job of anticipating contingencies, nor have I ever claimed that the statistician has any right to interject his opinions — it is the client's evaluation (which the statistician may, by judicious questioning and discussion, help to ascertain) which governs. When a decision is made to use a new feed for laboratory mice, or a new variety of grain, or which insecticide to use, or how to treat cancer, or how to teach statistics, or whether to plan an interstellar expedition, or anything else, these considerations arise.

I maintain it is stupid to say that because these considerations are difficult that they should be swept under the rug of significance levels or any other so-called "objective" criteria. Of course mistakes will be made; the question is frequently between mistakes of omission and commission, and the essence of the decision-theoretic approach is that the overall balance should be considered, and not merely one aspect of it. It is no longer tolerable to ignore side effects; almost by definition, they must be given *a priori* evaluation.

To Professor Plante:

Dr. Plante is apparently unaware that the consistency principle is a restatement of the assumption that Nature is not interested in playing games with the statistician, or more properly, with the client — that the consequences of our poor inference are not the motivating factor for Nature. Of course, if it were assumed Nature were inimical, Bayesian behavior would be irrational.

To Professor Bartlett:

As I have already observed, a reasonable Bayesian approach to testing has not yet been developed. I know of no cases in which the statistical element is that unimportant; errors of measurement are surely present. It seems obvious that any testing procedure which fails in the "clearcut cases" of physics and chemistry cannot be an appropriate one to use anywhere.

LIKELIHOOD METHODS OF PREDICTION

J. D. Kalbfleisch
University College, London

J. D. Kalbfleisch

University College, London

1

Introduction

The use of the likelihood function in problems of estimation has been stressed for some time now. In his writings, Fisher repeatedly emphasized the importance of examining the entire likelihood function and many examples of likelihood analyses have appeared in the literature. To cite but a few, Anscombe (1961), Barnard, Jenkins and Winsten (1962), Feigl and Zelen (1965), and Sprott and Kalbfleisch (1969) have considered such examples. The data are assumed to have arisen from one of a family of hypothetical populations indexed by a parameter vector; these papers are usually concerned with methods of assessing the information contained in the data about this parameter vector. This is a problem of some importance since, if the model is correct, it is the procedure whereby the actual random process giving rise to the data can be ascertained. Once this process is known, exact statements of probability about the outcome of any future trial can be made.

Often, however, one wishes at some point in the investigation to assess the information contained in the data so far collected about the outcome of a pending trial. This is the case, for instance, in medical investigations where the assessment of the prognosis of a particular patient is often of interest. This problem has received little attention from the likelihood point of view; the purpose of this paper is to examine likelihood methods of prediction.

Kalbfleisch and Sprott (1970a) considered this problem within the framework of three structures which gave rise to three different measures of uncertainty. These methods are discussed and slightly generalized in Section 2. In Section 2.1, the method of *fiducial prediction* outlined by Fisher (1935, 1956) is considered briefly. Sections 2.2 and 2.3 contain discussions of likelihood methods of estimating a future observation. The second of these is a generalization of Fisher's method of *likelihood prediction* in the 2 × 2 contingency table (c.f.

378

Fisher, 1956) and this is shown to be an example of second order likelihood as outlined by Kalbfleisch and Sprott (1970b). The likelihood methods are applied to a problem involving survival data in Section 3. Section 4 considers some interconnections and asymptotic properties of these methods.

The notation $f(x; y|z)$ will be used to indicate that z has the logical status of a random variable and y, the logical status of a fixed parameter. This will therefore indicate the conditional distribution of x given z depending on the parameter y. This notation is particularly useful when fiducial distributions are being considered. Suppose that T is sufficient for the parameter θ given the ancillaries A and that its conditional distribution is

$$f(T; \theta \mid A).$$

Assuming that the necessary structure is present, the fiducial distribution of θ is

$$f(\theta; T \mid A).$$

This notation gives a formal way of indicating the change in the logical status of T and of θ.

A number of terms used in this paper were defined by Kalbfleisch and Sprott (1970b). Of particular relevance are the definitions of the relative likelihood function, the maximum relative likelihood function and second order likelihood. I repeat here the definition of an integrated likelihood function:

Definition: Suppose that the distribution of the observations $x = (x_1, \ldots, x_n)$ depends on α and β and is given by $f(x; \beta \mid \alpha)$ where a distribution $p(\alpha; \beta)$ for α is known. The *integrated likelihood function of β* is a function of β proportional to

$$I(\beta) = g(x; \beta) = \int_{\alpha} f(x; \beta \mid \alpha) p(\alpha; \beta) \, d\alpha. \tag{1}$$

When standardized with respect to its maximum, this gives the integrated relative likelihood function of β.

In this paper, the fiducial distribution based on a set of data is used as a distribution *a priori* in Bayes Theorem. This is following the example of Fisher (1956). There is admittedly some question as to the validity of such a procedure as has been illustrated by Lindley (1958). An explanation of these difficulties has been given by Sprott (1960), however, and it is quite conceivable that exact criteria for the validity of this procedure can be ascertained. In any case, the procedure appears to give sensible results and is assumed valid in subsequent sections.

2
The Estimation of Future Observations

Suppose that $x_1, \ldots, x_n$ are independent observations from a distribution $f(x; \theta)$ where θ is an unknown parameter vector. Suppose further that $g(y_1, \ldots, y_m)$ is a function of m additional observations, $(y_1, \ldots, y_m)$, that are about to be taken from the same population. For example, g might be the arithmetic or geometric mean, the minimum or the maximum of the y's. This section is concerned with methods of estimating g. There are involved two sources of uncertainty: the uncertainty about the true nature of the hypothetical population defined by θ and the uncertainty about g for any fixed θ due to the randomness of the process. Any statement about g will have to account for both of these.

Three distinct structures arise and these have been considered by Kalbfleisch and Sprott (1970a). The discussion here is intended to be supplemental to the discussion in that paper; it is, therefore, considerably abbreviated and examples are only indicated. More detailed examples of the last two procedures are considered in Section 3.

2.1 Fiducial Prediction

Fisher (1935, 1956) has considered this approach. Suppose that based on the sample, $(x_1, \ldots, x_n)$, there exists a set of statistics T which (when conditioned on ancillaries A if necessary) is sufficient for θ and possesses sufficient structure to yield the fiducial distribution of θ,

$$f(\theta ; T \mid A). \tag{2}$$

In this distribution, θ has the logical status of a random variable and T, of a fixed parameter. Conditional on a given θ, the distribution of g is

$$f(g \mid \theta). \tag{3}$$

The marginal distribution of g (θ unknown) is therefore

$$f(g ; T \mid A) = \int_{\theta} f(g \mid \theta) f(\theta ; T \mid A) \, d\theta. \tag{4}$$

The distribution (4) accounts for both sources of uncertainty: the uncertainty about θ (summarized by (2)) and the uncertainty due to the randomness of the process (summarized by (3)).

Fisher (1956) has given examples of this procedure in the normal and exponential distributions. Examples which better illustrate the scope of statements possible were given by Kalbfleisch and Sprott (1970a).

2.2 Integrated Likelihood of g

When the necessary structure to construct the fiducial distribution of θ on the basis of the observed sample is lacking, the method of fiducial prediction fails. A related method which is sometimes available is discussed below. The basic idea is to use an hypothesized g to *predict* $x_1, \ldots, x_n$ by the method described in Section 2.1. When $x_1, \ldots, x_n$ is observed, the resulting function can be interpreted as an integrated likelihood function of g. This is really an inversion of the argument in Section 2.1.

Suppose that a value of g has been observed and that no other information about θ is available. If the distribution $f(g ; \theta) \, dg$ has the necessary structure the fiducial distribution of θ depending on the observed g can be obtained as $f(\theta ; g) \, d\theta$. For given θ, the distribution of $x_1, \ldots, x_n$ is $f(x_1 \ldots, x_n \mid \theta)$ and the argument of Section 2.1 can be applied. The marginal distribution of $x_1, \ldots, x_n$ depending on the observed value of g (θ unknown) is

$$f(x_1, \ldots, x_n ; g) = \int_\theta f(x_1, \ldots, x_n \mid \theta) f(\theta ; g) \, d\theta. \tag{5}$$

The distribution (5) is suitable for predicting $x_1, \ldots, x_n$ on the basis of an observed value of g. In particular, for any hypothesized value of g this yields probabilities for the $x_1, \ldots, x_n$ if that value of g were true (that is, observed) and θ were unknown. After $x_1, \ldots, x_n$ are observed (5) can be interpreted as a likelihood function of g (considered as a fixed parameter). Note that (5) is an integrated likelihood function of g as defined by (1).

The following example was considered by Kalbfleisch and Sprott (1970a).

Example 2.1. Suppose that n items are observed over a fixed period, T, and k of these fail at times $x_1, \ldots, x_k$ while the other $n - k$ are still operating at the end of the test period. The likelihood function of θ is proportional to

$$\theta^{-k} \exp \{-(\Sigma x_i + (n - k)T)/\theta\} \tag{6}$$

which does not yield a fiducial distribution for θ. Of interest is the time to failure g of a unit consisting of m items (from the same population) connected in parallel. If y_i is the failure time of the ith item in the unit, then $g = \max(y_1, \ldots, y_m)$ and its distribution is

$$f(g ; \theta) = m e^{-g/\theta} (1 - e^{-g/\theta})^{m-1} / \theta. \tag{7}$$

Applying the above method, (7) yields a fiducial distribution of θ for a hypothesized g and combining this with (6) gives the integrated likelihood of g as proportional to

$$I(g) = \sum_{r=0}^{m-1} \binom{m-1}{r} (-1)^r (\Sigma x_i + (n - k)T + g + rg)^{-k-1}. \tag{8}$$

2.3 Second Order Likelihood

When neither g nor the sample yields a fiducial distribution for θ, the above methods are not applicable. In this section a third method which applies to this case is considered.

Suppose that $x = (x_1, \ldots, x_n)$ is a sample with distribution $f(x ; \theta_x)$ where the underlying value of θ is θ_x. Suppose also, that $g = g(y_1, \ldots, y_m)$ is a function of m observations $y = (y_1, \ldots y_m)$ with distribution $f(g ; \theta_y)$ where the underlying value of θ is θ_y. If x and g are statistically independent, plausible values of the pair (θ_x, θ_y) are those for which the relative likelihood

$$R(\theta_x, \theta_y ; x, g) = R_1(\theta_x ; x) R_2(\theta_y ; g) \tag{9}$$

is large. (R_1 and R_2 are the relative likelihoods for θ_x and θ_y arising from x and g respectively).

The expression (9) can be reparameterized in terms of $\delta = \theta_x - \theta_y$ and θ_y to give

$$R(\delta, \theta_y ; x, g) = R_1(\delta + \theta_y ; x) R_2(\theta_y ; g)$$

so that the maximum relative likelihood function of δ is

$$R_M(\delta; x, g) = \sup_{\theta_y} R(\delta, \theta_y; x, g).$$

In particular, the maximum relative likelihood of $\delta = 0$ is

$$R_M(\delta = 0; x, g) = \sup_{\theta} R_1(\theta; x) R_2(\theta; g) \qquad (10)$$

which gives a measure of the plausibility that $\theta_x = \theta_y$. In fact, the observed value of (10) is the ratio of the maximum value that the probability of the observed (x,g) can attain if the hypothesis $\theta_x = \theta_y$ is true to the maximum probability that (x,g) can attain without this restriction.

If, however, the y_i's are future observations to be drawn from the same population as the x_i's, the hypothesis $\theta_x = \theta_y$ (or $\delta = 0$) is known to be true. Thus (10) measures the plausibility of an event that is known to be true. Considered as a function of g, (10) rates values of g according to how likely (in the sense of maximum relative likelihood) they make an hypothesis that is known to be true. Therefore, the expression (10) is a second order likelihood of g.

Second order likelihood was introduced by Kalbfleisch and Sprott (1970b). Whereas the usual likelihood function rates values of the parameter according to how probable they make an hypothesis that is known to be true, a second order likelihood rates values of the parameter according to how likely (in some well-defined sense) they make an hypothesis that is known to be true. In this case, maximum relative likelihood is used while in the above paper marginal relative likelihood was used. In any case, the logic is the same in both situations.

Fisher, (1956), gave the first example of this procedure in relation to the 2×2 contingency table. If x heads are observed in n binomial trials, the second order likelihood of y heads in m future tosses is

$$\frac{(x + y)^{x+y}(m + n - x - y)^{m+n-x-y} m^m n^n}{x^x y^y (n - x)^{n-x}(m - y)^{m-y}(m + n)^{m+n}}$$

which may be obtained by (10). This rates a value of y according to how likely it would be that that values (if it subsequently occurred) and the observed x both arose from the same underlying Binomial parameter. Fisher merely quotes this likelihood and fails to note its logical distinction from the usual likelihood function.

Additional examples of a somewhat more complex nature were considered by Kalbfleisch and Sprott (1970a). In Section 3, a further example of second order likelihood is given. In this case the method is used to give a plausibility rating to compound hypotheses.

In the next section, these methods are exemplified in relation to an example in which the evaluation of the prognosis of a particular individual is of interest. Such examples are often encountered in medical research. Similar situations arise in business, for example, where the behaviour of the particular machine to be purchased is of more interest than the population from which it arises.

382

Aside from these rather obvious uses of predictive methods, there are several less apparent applications. The standard treatment of discrimination (that is, the classification of a single individual) effectively assumes the population parameters are known to be equal to their estimated values. The use of predictive methods here also accounts for the uncertainty about these parameters. This approach has been considered by Geisser (1964, 1966) using Bayesian predictive distributions. Related to this, the appropriateness of a particular decision often depends on the outcome of a future trial. In this case, the uncertainty about this particular trial, rather than the uncertainty about the population from which it arises, is the appropriate guide to the decision. One other use which can be made of these methods are examined by Kalbfleisch and Sprott (1970a). When the variate values associated with the units of a finite population can be considered as having arisen from one of a family of hypothetical populations, the problem of making an inference about the finite population on the basis of an observed sample is equivalent to the problem of prediction.

3
Survival Time with a Covariate

Feigl and Zelen (1965) and Glaser (1967) have suggested that an appropriate model for human survival data, when concomitant information is present in the form of a covariate, is an exponential distribution with the covariate incorporated. The most appealing of such distributions is the one used by Feigl and Zelen in which the expected survival time, $E(t|x)$ is assumed a linear function of the covariate x. This gives the density function

$$f(t; \alpha_1 \beta \mid x) = \frac{1}{\alpha + \beta x} \exp \left\{ \frac{t}{\alpha + \beta x} \right\} \qquad t > 0. \qquad (11)$$

In the following discussion, however, the logarithm of the expected survival time is assumed linear in x so that

$$f(t; \alpha, \beta \mid x) = \frac{1}{\alpha e^{\beta x}} \exp \left\{ \frac{-t}{\alpha e^{\beta x}} \right\} \qquad t > 0. \qquad (12)$$

This distribution has the advantage of being algebraically and computationally simpler than (11) while illustrating the salient points as well. All of the methods used below could equally be applied to (11).

Suppose that the observations are the pairs (t_i, x_i), $i = 1, \ldots, n$ and that the t_i's given the x_i's are independently distributed with conditional density (12). The joint likelihood function of α and β is therefore proportional to

$$L(\alpha, \beta) = \alpha^{-n} e^{-n\beta \bar{x}} \exp \left\{ -\Sigma t_i \frac{e^{-\beta x_i}}{\alpha} \right\}, \qquad (13)$$

where x is the mean of the x_i's.

Suppose that an individual with covariate value x has just been diagnosed and that his survival time t is of interest. In this Section, procedures for evaluating the information contained in the data about t are considered using the methods of Sections 2.2 and 2.3. These results are illustrated in a numerical example in Section 3.4. It should be noted that in this case, the data do not yield a fiducial distribution for α and β so that the method of Section 2.1 is inapplicable.

3.1 Estimation of Expected Survival Time
The expected survival time of the individual with covariate value x is (from (12))

$$\alpha_x = E(t\ x) = \alpha e^{\beta x}.$$

One way of gaining some information about the survival time t is to estimate this quantity. The likelihood (13) can be reparameterized in terms of α_x and β to give

$$L(\alpha_x, \beta) = \alpha_x^{-n}\, e^{-n\beta(\bar{x}\, -\, x)} \exp\{-\Sigma\, t_i e^{\,-\beta(x_i-x)}\,/\alpha_x\}. \tag{14}$$

The maximum likelihood estimates are

$$\hat{\alpha}_x = \tfrac{1}{n}\Sigma t_i e^{-\hat{\beta}(x_i\, -\, x)} \text{ and } \beta$$

where $\hat{\beta}$ is the solution to the equation,

$$\Sigma t_i(x_i - x)e^{-\hat{\beta}(x_i\, -\, x)} = (\bar{x} - x)\Sigma t_i e^{-\hat{\beta}(x_i\, -\, x)}.$$

The estimate $\hat{\beta}$ can be obtained numerically using a simple iterative technique. Pairs of values (α_x, β) which make the observations relatively probable can be obtained by examining the relative likelihood function

$$R(\alpha_x, \beta) = L(\alpha_x, \beta)\,/\,L(\hat{\alpha}_x, \hat{\beta}). \tag{15}$$

We can summarize $R(\alpha_x, \beta)$ quite simply by plotting contours of constant relative likelihood; those pairs (α_x, β) which make the observations relatively more probable can be read off the graph.

One simple way to summarize the contours in terms of α_x is to plot the maximum relative likelihood function of α_x as defined by Kalbfleisch and Sprott (1970b). This gives

$$R_M(\alpha_x) = \sup_{\beta} R(\alpha_x, \beta) = R(\alpha_x, \tilde{\beta}(\alpha_x)) \tag{16}$$

where $\tilde{\beta}(\alpha_x)$ can be obtained by numerically solving the equation

$$\Sigma t_i(x_i - x)\ e^{-\tilde{\beta}(x_i\, -\, x)} = n\alpha_x(\bar{x} - x)$$

for each hypothesized α_x. The value $R_M(\alpha_x)$ is the maximum value that the relative likelihood can attain when that α_x is put in conjunction with some value for β. In effect, this gives the silhouette of the likelihood contours. This is, of course, not the only way to eliminate β so that we can look at α_x by itself. Another possibility would be to find average likelihoods of α_x by averaging over β with respect to different weight functions. However, neither a marginal nor a conditional likelihood for α_x is available.

Whatever method is used to estimate α_x, this will provide only half an answer to the problem of evaluating the individual's prognosis since it accounts for only one source of uncertainty. In Section 4.2, the problem of estimating the specific lifetime of this individual is considered.

3.2. Estimation of the Individual Survival Time t

In this section the method of Section 2.2 is applied to obtain an integrated likelihood function for t and β.

The survival time t has density function

$$f(t; \alpha_x) = \exp(-t/\alpha_x)/\alpha_x \qquad t > 0$$

which depends only on the parameter α_x. If t were known and no other information about α_x were available, this could be inverted to obtain the fiducial distribution of α_x for the fixed t as

$$f(\alpha_x; t) = \exp\{-t/\alpha_x\}/\alpha_x^2.$$

Applying the method of Section 2.2 for the observed t and for any fixed β, the distribution of $t_1, \ldots, t_n$ given $x_1, \ldots, x_n$ in the absence of knowledge of α_x is

$$f(x_1, \ldots, x_n; \beta, t \mid x_1, \ldots, x_n) = \int_{\alpha_x} f(t_1, \ldots, t_n; \beta \mid x_1, \ldots, x_n, \alpha_x) f(\alpha_x; t) \, d\alpha_x$$

which is proportional to

$$I(t, \beta) = t \exp\{-n\beta(\bar{x} - x)\}/\{t + \Sigma t_i e^{-\beta(x_i - x)}\}^{n+1}. \qquad (17)$$

Having observed $t_1, \ldots, t_n$, this is the integrated likelihood function of t and β. The maximum likelihood estimate of t is $\hat{\alpha}_x$ and the estimate $\hat{\beta}$ of β is not changed. The integrated relative likelihood of (t, β) is therefore

$$IR(t, \beta) = I(t, \beta)/I(\hat{t}, \hat{\beta})$$

and this can be summarized by plotting the contours of constant relative likelihood. The information on t can be summarized by the same methods as were suggested in Section 3.1 for α_x. In particular, we could plot the function

$$IR_M(t) = \sup_{\beta} IR(t, \beta) = IR(t, \tilde{\beta}(t)) \qquad (18)$$

which is analogous to (16). Here again $\tilde{\beta}(t)$ is easily obtained using an iterative technique for each hypothesized t.

It is of some interest to note that the same result (18) is obtained if the method of second order likelihood discussed in Section 2.3 is used. This is often the case when the two methods are applicable at the same time.

3.3 Likelihood of Survival

Second order likelihood (c.f. Section 2.3) provides another method of summarizing the information about t. It may be of interest to obtain a measure of the plausibility that the survival time of an individual with covariate value x exceeds a given time τ. In this case, the method of Section 2.2 does not apply and a weaker measure of uncertainty than likelihood is required.

The future observation is the occurrence or non-occurrence of survival to time τ and

$$\Pr\{\text{survival to time } \tau\} = \Pr\{t > \tau; \alpha_x\} = \exp(-\tau/\alpha_x).$$

If in fact the event $(t > \tau)$ is observed, the above function is the likelihood function for α_x arising from that observed event. It is maximized for $\alpha_x = \infty$ so that the relative likelihood function of α_x is

$$R_2(\alpha_x; t > \tau) = \exp(-\tau/\alpha_x).$$

Now applying the method of section 2.3, the second order likelihood of the event $t > \tau$ is (from (10)).

$$\sup_{\alpha_x, \beta} R_1(\alpha_x, \beta) R_2(\alpha_x; t > \tau) \qquad (19)$$

where $R_1(\alpha_x, \beta)$ is the relative likelihood function arising from the data (15). The expression (19) can be evaluated numerically using a simple iterative technique.

The expression (19) gives the second order likelihood of the event $(t > \tau)$ and may be calculated for each value of $\tau > 0$. The events $t > \tau$ are then rated on a scale between 0 and 1 according to how likely these events make an hypothesis that is known to be true. A clear picture of which of the events $t > \tau$ are plausible can be obtained by plotting (19).

It is of some interest to note that here the second order likelihood of a composite hypothesis has been found. It is possible to do this since t has a probability distribution which may be integrated before it is combined with the observed likelihood to obtain the second order likelihood.

3.4 A Numerical Example

An interesting example of the type discussed above was given by Feigl and Zelen (1965). The dependent variable was the survival time in weeks (from the time of diagnosis) of patients who died from acute myelogenous leukemia and the independent variable (x_i) was the logarithm of the white blood count (WBC) at the time of diagnosis. Only the data for the AG positive group are considered here and these data are summarized in Table 1.

Table 1. Survival times (t) in weeks and white blood counts (e^x) for AG positive group $(n=17)$.

e^x	t	e^x	t	e^x	t
2,300	65	10,000	121	35,000	22
7,500	156	17,000	4	100,000	1
4,300	100	5,400	39	100,000	1
2,600	134	7,000	143	52,000	5
6,000	16	9,400	56	100,000	65
10,500	108	32,000	26		

Assuming the model (12) for these data, the results of Sections 3.1, 3.2 and 3.3 can be applied. In Figure 1, the maximum relative likelihood functions of $\alpha_x = E(t|x)$, (16), and of t, (19), are plotted for a WBC of 10,000. In Figure 2, the corresponding graphs are given for a WBC of 50,000. The data are much more precise about α_x than about the actual life time t. In fact, when the WBC is 10,000, values of t in the range (2.15,305) have maximum relative likelihood of at least 10%; the corresponding range for α_x is (35,101). Not only does the former give a much more realistic picture of the uncertainty involved in estimating the particular survival time under question, but such likelihoods could be of assistance in planning future experiments. They give a much more realistic picture of the variation the experimenter might expect to find.

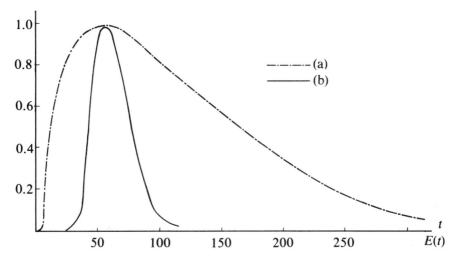

Figure 1. (a) Maximum relative likelihood of t (18).
(b) Maximum relative likelihood of $E(t|x) = \alpha_x$ (16) for a WBC of 10,000.

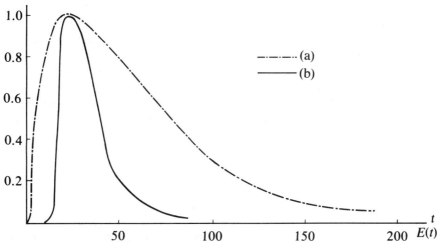

Figure 2. (a) Maximum relative likelihood of t(18).
(b) Maximum relative likelihood of $E(t|x) = \alpha_x$ (16) for a WBC of 50,000.

In Figure 3, the second order likelihood of survival to time $\tau(19)$ is plotted for these data and various WBC's. The second order likelihood of the event that an individual with WBC of 100,000 will survive the 53.5th week is 10%. That is, if it were observed that τ exceeded 53.5, the maximum relative likelihood of the true hypothesis of common values α_x and θ for the sample $t_1, \ldots, t_n$ and this event would be only 10%. The corresponding 10% points for individuals with WBC's 50,000, 35,000, 20,000, 10,000, 5,000, and 2,500 are 70, 81, 100, 142, 203 and 302 weeks respectively.

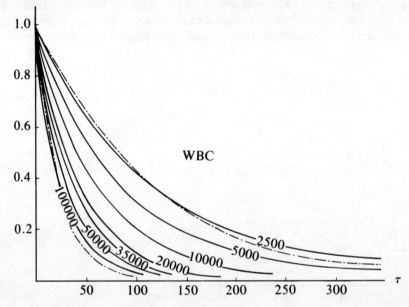

Figure 3. Second order likelihood of survival to time $\tau(19)$ for various WBC's. The broken lines give the maximum likelihood estimates of the probability of survival to time τ for WBC's of 2,500 and 100,000.

In their paper, Feigl and Zelen give a summary of the information about t which is similar to this. They note that the probability that t exceeds τ is $e^{-\tau/\alpha_x}$ and estimate this by $e^{-\tau/\alpha_x}$. Statements about t from this approach are in terms of estimated probabilities with the uncertainty of the estimate not specified. In Figure 3, the estimated probability curves are plotted for WBC's of 2,500 and 100,000. These agree very closely with the curves from the second order likelihood although the differences become more noticeable as the size of x increases. One might expect the curves to agree fairly well since the uncertainty about α_x and β is much smaller than the uncertainty about t for given α_x in this example.

4

Some Remarks on These Methods

The notation used here is that of Section 2. Suppose in the one parameter case that $x_1, \ldots, x_n$ yield a statistic T which is sufficient for the parameter θ and that the fiducial distribution of θ arising from this T exists and is $f(\theta; T)$. If g also yields a fiducial distribution for θ as $f(\theta; g)$ then both the fiducial distribution of g for fixed T (Section 2.1),

$$f(g; T) = \int_\theta f(g \mid \theta) f(\theta; T) d\theta \qquad (20)$$

and the integrated likelihood of g (section 2.2),

$$f(T; g) = \int_\theta f(T \mid \theta) f(\theta; g) d\theta \qquad (21)$$

are available. It would seem reasonable to expect that (20) can be obtained directly from (21) using the fiducial argument.

If $F(g; \theta)$ and $F(T; \theta)$ are the cumulative distribution functions of g and T respectively, then

$$f(\theta; g) = \left| \frac{\delta}{\delta\theta} F(g; \theta) \right| \text{ and } f(\theta; T) = \left| \frac{\delta}{\delta\theta} F(T; \theta) \right|.$$

Substitution for $f(\theta; g)$ in (21) gives the cumulative distribution function of T for fixed g as

$$H(T; g) = \int_{-\infty}^{T} \left\{ \int_\theta f(T \mid \theta) \left| \frac{\delta}{\delta\theta} F(g; \theta) \right| d\theta \right\} dT$$

$$= \pm F(T; \theta) F(g; \theta) |_\theta \pm \int_\theta F(g; \theta) \left\{ \frac{\delta}{\delta\theta} F(T; \theta) \right\} d\theta.$$

Now the first term is a constant since $F(T; \theta)$ and $F(g; \theta)$ are both monotone maps of θ onto the interval $\langle 0, 1 \rangle$ (from the existence of the fiducial distributions $f(\theta; T)$ and $f(\theta; g)$). Also the cumulative distribution $H(T; g)$ is monotone in g so that the fiducial distribution of g arising from (21) is

$$h(g; T) = \left| \frac{\delta}{\delta g} \int_\theta F(g; \theta) \left\{ \frac{\delta}{\delta\theta} F(T; \theta) \right\} d\theta \right|$$

$$= f(g; T)$$

as in (20). Thus the same distribution of g is obtained by using either method.

Suppose that the function $g(y_1, \ldots, y_m)$ approaches a function $h(\theta)$ of the parameter vector θ as the number m of future observations increases. It would then seem reasonable to expect that statements of probability or likelihood about g would approach the classical statements of probability or likelihood about $h(\theta)$. It can be shown quite simply that the fiducial predictive distribution of g (Section 2.1) will approach the fiducial distribution of $h(\theta)$ as m tends to infinity. In the one parameter case, the integrated likelihood function of g and the second order likelihood function of g will approach the likelihood function of θ provided that $h(\theta)$ is a one to one function of θ. The extension to more than one parameter for the integrated likelihood and the second order likelihood requires that there be a number of functions $g_i(y_1, \ldots, y_m)$, ($i = 1, \ldots, s$) which approach functions $h_i(\theta)$, ($i = 1, \ldots, s$) of the vector θ almost surely as m tends to infinity.

Conditions under which the integrated likelihood and the second order likelihood are the same are of some interest. In most simple situations these two likelihoods are identical, but they of course have different interpretations.

Acknowledgment

I would like to thank Professor D. A. Sprott for reading a previous draft of this paper and for making helpful comments and criticism.

References

1. Anscombe, F.J., "Estimating a Mixed Exponential Response Law," *Journal of the American Statistical Association*, 56, 493-502, 1961.
2. Barnard, G.A., Jenkins, G.M. and Winsten, C.B., "Likelihood Inference and Time Series," *Journal of the Royal Statistical Society*, A, 125, 321-372, 1962.
3. Feigl, P. and Zelen, M., "Estimation of Exponential Survival Probabilities with Concomitant Information," *Biometrics*, 21, 826-838, 1965.
4. Fisher, R.A., "The Fiducial Argument in Statistical Inference," *Annals of Eugenics*, London, 6, 391-398, 1935.
5. Fisher, R.A., *Statistical Methods and Scientific Inference*, London, Oliver and Boyd, 1956.
6. Geisser, S., "Posterior Odds for Mulitvariate Normal Classifications," *J. Roy. Statist. Soc.*, B, 26, 69-76, 1964.
7. Geisser, S., "Predictive Discrimination," *Proceedings of the International Symposium on Multivariate Analysis*, New York, Academic Press, 149-163, 1966.
8. Glaser, M., "Exponential Survival with Covariance," *J. Am. Stat. Assoc.*, 62, 561-568, 1967.
9. Kalbfleisch, J.D. and Sprott, D.A., "Applications of Likelihood and Fiducial Probability to Sampling Finite Populations," *New Developments in Survey Sampling*, New York, Wiley, 1970a.
10. Kalbfleisch, J.D. and Sprott, D.A., "Application of Likelihood Methods to Models Involving Large Numbers of Parameters," to appear, *Journal of the Royal Statistical Society*, A, 1970b.
11. Lindley, D.V., "Fiducial Distributions and Bayes' Theorem," *J. Roy. Statist. Soc.*, B, 20, 102-107, 1958.
12. Sprott, D.A., "Necessary Restrictions for Distributions *a Posteriori*," *J. Roy. Statist. Soc., B*, 22, 312-318, 1960.
13. Sprott, D.A. and Kalbfleisch, J.D., "Examples of Likelihood and Comparisons with Point Estimates and Large Sample Approximations," *J. Am. Stat. Assoc.*, 64, 468-484; 1969.

COMMENTS

M. S. Bartlett:

It might be noted that because of the appropriate sufficiency properties holding the author's fiducial prediction quoted from Fisher is also valid in confidence interval terms. The discrete case was discussed by me in a paper "Subsampling for Attributes" published in the *R.S.S.*, and here confidence intervals are also possible although Fisher would not allow the use of the term *fiducial*. I recall that the continuous normal case arose when I was at University College in 1934, in terms of inference about a future sample (or a sample of which the known sample is a part).

R. J. Buehler:

In Section 4, Dr. Kalbfleisch shows that the fiducial distribution of g obtained by two methods is the same. R. B. Hora and I obtained a similar result which appears in the 1967 *Annals of Mathematical Statistics*. I do not think that our result implies his, nor that his implies ours, but nevertheless there is a considerable common area.

S. Geisser:

I would like to suggest another possible approach to prediction when the usual fiducial argument is not available. One may compute a pseudo-fiducial density based on some statistic $T(x_1, \ldots, x_n)$, an insufficient statistic that depends only on θ, whose density $f(T|\theta)$ is inverted to yield $\phi_p(\theta|T)$. The computation of the pseudo-predictive density

$$f(g|T) = \int f(g|\theta)\, \phi_p(\theta|T)\, d\theta$$

may be of some limited predictive value for g.

D. V. Lindley:

I should like to discuss the first of the three methods proposed in this paper (though similar criticisms probably apply to the others). Let $(x_1, x_2, \ldots, x_m)$ be a random sample from some distribution and let x_{m+1} be a single further observation that it is required to predict. The method may then be applied to obtain the predictive density $p(x_{m+1}|x_1, x_2, \ldots, x_m)$. Call this method 1. Now let us use $(x_1, x_2, \ldots, x_{m-1})$ to predict (x_m, x_{m+1}). The method will then provide $p(x_m, x_{m+1}|x_1, x_2, \ldots, x_{m-1})$ and by the usual probability calculations we can derive from this $p(x_{m+1}|x_1, x_2, \ldots, x_m)$. Call this method 2. Now ask the question, will the results of the two methods agree? The answer is, typically no! I have not performed the calculations but I have high personal probability that they will differ for the case where the random sample is from the density

$$\frac{\theta^2}{\theta+1}(x+1)e^{-x\theta}, \qquad x > 0, \theta > 0.$$

J. S. Williams:

I wish to reply to the comments by Dr. Lindley. The notation used in his example obscures the fact that *fiducial probabilities* are not *conditional probabilities*. If one interprets them as such, I believe he always will be able to produce contradictory results if he searches long enough.

I took great pains in a paper published in *Sankhyá* to distinguish as different logical types the probability of event F *conditioned* on the occurrence of the event E, $P(F|E)$, and the probability of event F *given* the occurrence of event E, $P_E(F)$. There it was pointed out that the latter, of which fiducial probabilities are examples, when averaged with conditional probabilities will not, in general, result in valid probability calculations of either type. This was in anticipation that examples such as we have seen to-day would be put forward.

Finally, it should be noted that there is no distribution of θ for the density of x in to-day's example if my prescriptions for the fiducial argument are followed. The development of these, by the way, was influenced by Dr. Lindley's early writings on the consistency of the fiducial distribution with a Bayes' posterior distribution.

REPLY

As Professor Bartlett points out, the fiducial prediction is also valid in confidence theory terms. The examples considered in the paper do not submit easily to confidence interval solutions, however, and it is not clear to me how censored data from an exponential distribution, for instance, can be used to give exact confidence intervals for a future observation from the same distribution.

Professor Geisser's suggestion of using insufficient statistics is an interesting one and worthy of further consideration. What appears to be needed is some way to evaluate the relative loss of information sustained when a reduction to an insufficient statistic is made.

The two methods proposed by Professor Lindley will not, in general, agree. I would suspect that they would differ in exactly those cases where the fiducial distribution is not a Bayes' posterior. The apparent paradox, however, is explained by Dr. Williams. Throughout the paper I have taken care to distinguish between the concepts of "given the occurrence of event E" and "conditional on the occurrence of event E". In method 2 that Professor Lindley proposes, the quantity x_m is given a different logical status from $x_1, \ldots, x_{m-1}$ and it is this that leads to the anomaly. I think it worth re-emphasizing that fiducial probabilities are not conditional probabilities. Those who criticize the fiducial argument should recognize in their arguments this distinction and the two logical types discussed by Dr. Williams.

THE USE OF THE STRUCTURAL MODEL IN COMBINING DATA BEARING ON THE SAME CHARACTERISTICS

J. B. Whitney*
University of Waterloo

Summary

In analyzing some situations, it may be desirable to condense the information given by the data, in such a way as to facilitate combination with other data which relates to the same characteristics. Generally, in parametric models indexed by a parameter, the likelihood function is an appropriate method of combining independent sets of data so as to make statements of inference about the common characteristic. In the special but important cases where the structural model applies, Fraser has shown that independent structural models with the same characteristic from a group of transformations, also allow for such combinations; the statements of inference about the characteristic are in the form of a probability statement.

The combining of data can be extended to the cases where only a component of the characteristic has the same value in each independent model. An analysis of these cases shows that in all but a very special case, the statements of inference about the common component of the characteristic of each structural model cannot be given in the strong form of a structural distribution, but must be stated in terms of a *Marginal Likelihood Function*.

Before combining data, it should be determined whether or not the sets of data are compatible with the hypothesis that the components under consideration have the same value. In some of the above cases, the structural equation allows for the derivation of distributions, which are useful in assessing whether or not the sets of data are compatible.

1
The Model

This section develops the model that will be considered for combining data,

*With a grant from the National Research Council of Canada.

that bears on the same common component of the characteristic in each independent structural model.

A Structural Model

The point of view taken in structural inference is that a response variable y, obtained from a process operating under stable conditions, is derived from an unknown transformation θ operating on realized but unknown value of a random variable e. The random variable e describes the unidentified sources of variation of the process, the internal error of the system. It is assumed that e has a known probability distribution on a space E and that θ belongs to a known set of transformations G. The realized error e and the response variable y are related by the equation

$$y = \theta(e). \tag{1}$$

Since θ is unknown the value of e is unknown, but, by considering the pre-image of y under all the transformations of G, (1) gives the information that the values of e which could have given rise to the observation y must belong to the subset of E,

$$G^{-1}(y) = \{\theta^{-1}(y) : \theta \in G\}.$$

Now consider a random variable which obeys a known probability law. Suppose that certain characteristics of that variable become known. It then seems natural to condition the original probability law if possible in accordance with the added information. If the information restricts the variable to a set whose probability is not zero, then the elementary technique of conditioning can be used. However if the information restricts the variable to a set that has probability zero, the conditional distribution is not defined unless the information is in the form of an event. That is:
  — a partition on the sample space must be specified,
  — the information must state that the variable took its value in one of the members of this partition,
  — no other information must be given.
In the above case the only information given by (1), about e, is that it belongs to the set $G^{-1}(y)$. In many cases the probability distribution of the random variable e has zero probability on the set $G^{-1}(y)$. Whether or not $G^{-1}(y)$ defines an event depends upon the collective action of the transformations in G and not upon the information that y is observed. It is easily shown that if the transformations G form a group, then the collective action of the transformations in G acting upon E define an event on the error space E.

For the basic model in the paper it will be assumed that the error variable $\underline{e} = (e_1 -- e_n)$ takes its values in Euclidean space R^n, and has a known continuous probability distribution with probability element given by $f(\underline{e})\,d\underline{e}$ ($d\underline{e}$ Euclidean differential in R^n). The transformation θ operating on $\underline{e}$ will be a member of a group G. G is assumed to have the following properties:
  (i) G is a subgroup of the positive affine group, which acts upon the n-dimensional Euclidean space R^n,
  (ii) G is an open subset of $R^L, L \le n$.

Properties (i) and (ii) imply that G is a locally compact topological group, endowed with usual topology inherited from R^L.

The response variable $\underset{\sim}{y} = (y_1 -- y_n)$ will also take its values in R^n and its relation to $\underset{\sim}{e}$ is given by the structural equation

$$\underset{\sim}{y} = \theta \underset{\sim}{e} = (\theta e_1, \theta e_2, \ldots, \theta e_n),$$

where θ acts co-ordinate wise on each component of $\underset{\sim}{e}$ or $\underset{\sim}{y}$.

The above can be summarized by the structural model, (Fraser, [2])

$$f(\underset{\sim}{e}) d\underset{\sim}{e}$$

$$\underset{\sim}{y} = \theta \underset{\sim}{e}.$$

The point $\underset{\sim}{y}$ is carried by the elements of G into the orbit $G\underset{\sim}{y} = \{\theta\underset{\sim}{y} : \theta \in G\}$. The sample space of $\underset{\sim}{e}$ is partitioned into G-orbits and the observation $\underset{\sim}{y}$ indicates the particular orbit to which the observation belongs. It has been discussed on several occasions by Fraser [1] [2], that given that $\underset{\sim}{e}$ belongs to a particular member of this partition, conditional probability statements can be made about the realized value of $\underset{\sim}{e}$; since properties (i) and (ii) of G imply a one-one correspondence between elements of G and elements of an orbit, the conditional probabilities about $\underset{\sim}{e}$ can be restated in terms of probabilities about θ; the structural distribution for θ.

A Model for Combining Data

Suppose that the response variable of two independent systems is $\underset{\sim}{y_i}$, $\underset{\sim}{y_i} \in R^{n_i}$, $i = 1, 2$; and that each system is represented by the structural model

$$f_i(\underset{\sim}{e}_i) d\underset{\sim}{e}_i$$

$$\underset{\sim}{y_i} = \theta_i \underset{\sim}{e}_i,$$

where $\theta_i \in G_i$, G_i a group of transformations with properties (i) and (ii).

These two models can be combined to form a composite model

$$f_1(\underset{\sim}{e}_1) f_2(\underset{\sim}{e}_2) d\underset{\sim}{e}_1 d\underset{\sim}{e}_2$$

$$(\underset{\sim}{y_1}, \underset{\sim}{y_2})' = (\theta_1 \underset{\sim}{e}_1, \theta_2 \underset{\sim}{e}_2)' = (\theta_1, \theta_2)(\underset{\sim}{e}_1, \underset{\sim}{e}_2)'$$

where $(\underset{\sim}{y_1}, \underset{\sim}{y_2})'$ and $(\underset{\sim}{e}_1, \underset{\sim}{e}_2)'$ are points in the cross product space $R^{n_1} \times R^{n_2}$, (θ_1, θ_2) belongs to the direct product (G_1, G_2).

Each G_i is a topological group. If H_i and K_i are two subgroups of G_i, and every element θ in G_i can be expressed in a unique manner in the form $\theta = \eta\kappa$, where η is in H_i, κ is in K_i and if the one-to-one mapping of $H_i \times K_i$ onto G, given by $(\eta, \kappa) \rightarrow \eta\kappa$ is a homeomorphism, then G_i is said to be the topological semi-direct produce to H_i and K_i. (Nachbin, [4], page 96). In such cases G_i can be decomposed into the two subgroups and formally expressed as $G_i = H_i K_i$. Clearly the definition of G_i being the topological semi-direct of subgroups can be extended to three subgroups H_i, K_i, M_i, and in this case G_i could formally be expressed as $G_i = H_i K_i M_i$. In the above models suppose that each G_i can be expressed as the semi-direct product of three subgroups such that $G_i = H_i K M_i$, where H_i and M_i are both allowed to be the identity subgroup of G_i, and that each element (characteristic) θ_i has a common component from the subgroup of transformations, K. With these assumptions the composite model becomes

$$f_1(\varrho_1) f_2(\varrho_2) d\varrho_1 d\varrho_2 \tag{2}$$

$$(y_1, y_2)' = (\eta_1 \kappa \mu_1, \eta_2 \kappa \mu_2)(\varrho_1, \varrho_2)'$$

where $\eta_i \in H_i$ $\kappa \in K$, $\mu_i \in M_i$.

When structural models apply it is suggested that (2) is the appropriate model to consider if one wishes to make inferences about a characteristic κ which is common to two independent structural models.

The transformations of the type $(\eta_1 \kappa \mu_1, \eta_2 \kappa \mu_2)$ will form a subset $\gamma \subset (G_1, G_2)$. If the set of inverses of all elements of γ is denoted by γ^{-1}, then from the equation of (2) it is seen that the observation $(y_1, y_2)'$ provides the information that $(\varrho_1, \varrho_2)'$ belongs to the set

$$\gamma^{-1}(y_1, y_2)' = \{\mu_1^{-1} \kappa^{-1} \eta_1^{-1} y_1, \mu_2^{-1} \kappa^{-1} \eta_2^{-1} y_2 \mid (\mu_1^{-1} \kappa^{-1} \eta_1^{-1}, \mu_2^{-1} \kappa^{-1} \eta_2^{-1}) \in \gamma^{-1}\}.$$

It has been shown by Kalotay [3], that the subset of transformations γ will define an event if and only if γ is a group or a coset of a group. In general γ will not form a group or a coset of a group and hence will not define a partition on the error space. In these cases the information that $(\varrho_1, \varrho_2)' \in \gamma^{-1}(y_1, y_2)'$ is not in the form of event and the conditional distribution of $(\varrho_1, \varrho_2)'$ given $\gamma^{-1}(y_1, y_2)'$ is not well defined. Therefore in these cases no structural distribution can be defined on the elements of γ and inferences about κ alone cannot be obtained in the form of a probability statement, obtained by integrating out the nuisance transformations.

2
Inferences About κ

Suppose that γ is not a group or a coset of a group. By assuming that $\kappa = \kappa_0$ additional information about the realized error can be obtained as the assumption restricts the set transformation $\gamma = (H_1 \kappa M_1, H_2 \kappa M_2)$ to the set $\gamma' = (H_1 \kappa_0 M_1, H_2 \kappa_0 M_2)$. In some important cases this additional information is such that the total information from the model about $(\varrho_1, \varrho_2)'$ is in the form of an event. This is so if the direct product components of the restricted set of transformations γ' are cosets of their respective groups G_i. If this is so the realized value $(\varrho_1, \varrho_2)'$ of the error variable belongs to a member of a partition of orbits and the conditional distribution given the orbit can be obtained for the possible values of the realized error variable $(\varrho_1, \varrho_2)'$. The corresponding marginal distribution of the orbits is invariant, with respect to transformations η_i and μ_i and depends at most upon the known error distribution and the value of the transformation κ_0. An unknown element of γ' relates $(\varrho_1, \varrho_2)'$ to $(y_1, y_2)'$. Because of this it can be argued that the conditional distribution of $(\varrho_1, \varrho_2)'$ given the orbit $\gamma'^{-1}(y_1, y_2)'$ gives no information concerning the value of κ_0. Inferences concerning κ_0 can be made from the likelihood function of the marginal distribution, the marginal likelihood function for κ. The marginal likelihood function has summarized all the available information that the model (2) has to offer about κ_0.

Conditions For the Direct Product Components of γ' To Be Cosets

Since there is no relation between the unknown transformations of each direct product component the conditions under which $H_i \kappa_0 M_i$ is a coset do not need to be the same for both values of i. It is easily shown that γ' has direct product components which are cosets if and only if $H_i M_i$ is a group and either $H_i K$ is a group in which H_i is a normal subgroup or $K M_i$ is a group in which M_i is a normal subgroup, for each group of transformations G_i.

The Marginal Likelihood Function for κ

In order to obtain the marginal likelihood function for κ, when γ' is a coset it is necessary to obtain an explicit expression for the conditional distribution. For this purpose it is convenient to transform the data of (2) by the transformation $(\kappa_0^{-1}, \kappa_0^{-1})$ to give

$$(\kappa_0^{-1}, \kappa_0^{-1})\,(y_1, y_2)' = (\kappa_0^{-1} y_1, \kappa_0^{-1} y_2)' = (y_1^{\kappa_0}, y_2^{\kappa_0})'.$$

Then the equation of (2) becomes

$$(y_1^{\kappa_0}, y_2^{\kappa_0})' = (\kappa_0^{-1}, \kappa_0^{-1})\,(\eta_1 \kappa_0 \mu_1, \eta_2 \kappa_0 \mu_2)\,(\ell_1, \ell_2)'$$

$$= (\kappa_0^{-1} \eta_1 \kappa_0 \mu_1, \kappa_0^{-1} \eta_2 \kappa_0 \mu_2)\,(\ell_1, \ell_2)'.$$

The set of elements of the type $(\kappa_0^{-1} \eta_1 \kappa_0 \mu_1, \kappa_0^{-1} \eta_2 \kappa_0 \mu_2)$ which may be denoted by

$$\gamma'_{\kappa_0} = (\kappa_0^{-1} H_1 \kappa_0 M_1, \kappa_0^{-1} H_2 \kappa_0 M_2),$$

will form a group. The partition defined on the error space by this group will be the same as that defined by the corresponding coset. But now (2) is in the form of a structural model, that is, the set of transformations form a group. Thus the conditional and marginal distributions for the model of interest can be obtained as the conditional and marginal distributions of this derived structural model. This is done below.

The position of a point y on its orbit can be described by a transformation variable.

Definition. A variable $[y]$ taking values in a group G is a continuous transformation variable (equivariant map) if $[y]$ is a continuous function on R^n, and

$$[gy] = g\,[y]$$

for all g in G and all y in R^n.

A transformation variable gives a reference point on each orbit: $D(y) = [y]^{-1}y$. The reference points index the orbits and the transformation variable gives position on an orbit: $y = [y] D(y)$. Two transformation variables differ on any orbit by right multiplication by a group element.

In the case under consideration, γ'_{κ_0} is a group and therefore each direct product component is a group. Hence each component model can be written as

$$f_i(\varrho_i) d\varrho_i$$

$$y_i^{\kappa_0} = \kappa_0^{-1} \eta_i \kappa_0 \mu_i \varrho_i, \tag{3}$$

where

$$\kappa_0^{-1} \eta_i \kappa_0 \mu_i \in \kappa_0^{-1} H \kappa_0 M_i = Q_{\kappa_0}, \quad i = 1, 2.$$

Accordingly each model can be analyzed separately, and the marginal likelihood for κ_0 found in each case. The marginal likelihood function for κ_0 from the composite model is obtained by multiplication of the two marginal likelihood functions.

For ease of notation the subscripts i and 0 will be dropped in the discussion below.

Consider the observation space and the error variable space as the same. Let $[\cdot]_\kappa$ be the transformation variable that maps R^n onto Q_κ. Then the associated reference point on an orbit will be given by $[y^\kappa]_\kappa y^{\kappa-1} = [\varrho]_\kappa^{-1} \varrho$ or $D_\kappa(y^\kappa) = D_\kappa(\varrho)$. The associated structural model may be written as

$$f(\varrho) d\varrho$$

$$[y^\kappa]_\kappa = q_\kappa [\varrho]_\kappa, \quad D_\kappa(y^\kappa) = D_\kappa(\varrho)$$

where $q_\kappa \in Q_\kappa$.

The conditional distribution of the realized error variable ϱ on the orbit $Q_\kappa \varrho$, indexed by $D_\kappa(\varrho)$, is easily derived by the use of invariant differentials. On the space R^n the transformation $y^\kappa \rightarrow \tilde{q}_\kappa y^\kappa$ has Jacobian $J_n(\tilde{q}_\kappa)$. Accordingly an invariant differential on R^n is

$$dm(y^\kappa) = \frac{dy^\kappa}{J_n([y^\kappa]_\kappa)}.$$

On the group Q_κ the left transformation $q_\kappa \rightarrow \tilde{q}_\kappa q_\kappa$ has Jacobian $J_L(\tilde{q}_\kappa)$, where L is the dimension of the group Q_κ.

Accordingly the left invariant differential is

$$du(q_\kappa) = \frac{dq_\kappa}{J_L(q_\kappa)}.$$

The probability along a neighbourhood of orbits is

$$f(\varrho) d\varrho = \bar{f}(\varrho) dm(\varrho) = a(D_\kappa(\varrho)) \bar{f}([\varrho]_\kappa, D_\kappa(\varrho)) du([\varrho]_\kappa).$$

The conditional probability element for $[\varrho]_\kappa$ given $D_\kappa(\varrho) = D_\kappa(\underline{y}^\kappa) = D_\kappa$ is obtained by normalization

$$A(D_\kappa)f([\varrho]_\kappa, D_\kappa)\frac{d[\varrho]_\kappa}{J_L([\varrho]_\kappa)}$$

where

$$A^{-1}(D_\kappa) = \int_{Q_\kappa} \tilde{f}([\varrho]_\kappa, D_\kappa)\frac{d[\varrho]_\kappa}{J_L([\varrho]_\kappa)} .$$

The marginal probability element for $D_\kappa(\varrho)$ is given by

$$\frac{f(\varrho)d\varrho}{A(D_\kappa)\tilde{f}([\varrho]_\kappa, D_\kappa)J_L([\varrho]_\kappa)}\frac{d[\varrho]_\kappa}{J_L([\varrho]_\kappa)} = A^{-1}(D_\kappa)\frac{J_L([\varrho]_\kappa)}{J_n([\varrho]_\kappa)}\frac{d\varrho}{d[\varrho]_\kappa} .$$

The marginal probability element at the point $\underline{y}^\kappa$ on the orbit D_κ rather than at the point ϱ on the orbit D is

$$A^{-1}(D_\kappa)\frac{J_L([\underline{y}^\kappa]_\kappa)}{J_n([\underline{y}^\kappa]_\kappa)}\frac{d\underline{y}^\kappa}{d[\underline{y}^\kappa]_\kappa} .$$

Since, in general, the differential element $\dfrac{d\underline{y}^\kappa}{d[\underline{y}^\kappa]_\kappa}$ on the space of the variable D_κ depends upon κ the differential element should be expressed in terms of the variables on the original response space. The differential $d\underline{y}^\kappa$ can be expressed in terms of the differential of the response variable dy;

$$(dy_1^\kappa, dy_2^\kappa, \ldots, dy_n^\kappa) = (dy_1, dy_2, \ldots, dy_n)J(\underline{y}^\kappa)$$

where

$$J(\underline{y}^\kappa) = \begin{pmatrix} \dfrac{\partial y_1^\kappa}{\partial y_1} & & 0 \\ & \cdot & \\ & & \cdot \\ & & \cdot \\ 0 & & \dfrac{\partial y_n^\kappa}{\partial y_n} \end{pmatrix} .$$

Then

$$d\underline{y}^\kappa = |J(\underline{y}^\kappa)|\, d\underline{y}.$$

The differential vector on the group is related to the differential vector on the orbit D_κ by

$$(dy_1^\kappa, \ldots, dy_n^\kappa) = (d[\underline{y}^\kappa]_\kappa)\beta_\kappa$$

where

$$\beta_\kappa = \left(\frac{\partial[\underline{y}^\kappa]_\kappa D_\kappa}{\partial[\underline{y}^\kappa]_\kappa}\right) .$$

The composite transformation is

$$(dy_1, \ldots, dy_n) = (d[\underline{y}^\kappa]_\kappa)\beta_\kappa J^{-1}(\underline{y}^\kappa),$$

which relates the differential vector on the group to differential vector on the inverse image of the orbit D_κ. It is given by Fraser [1], Sprott and Kalbfleisch [5], that relation between the volume elements is given by

$$dl^{-1}([\underline{y}^\kappa]_\kappa D_\kappa) = \sqrt{\beta_\kappa J^{-2}(\underline{y}^\kappa)\beta'_\kappa} \cdot d[\underline{y}^\kappa]_\kappa;$$

$dl^{-1}([\underline{y}^\kappa]_\kappa D_\kappa)$ is differential volume along the inverse image of the orbit D_κ taken at the point $\underline{y}$. The probability for D_κ expressed in terms of cross sectional Euclidean volume at the observed point $\underline{y}$ is

$$A^{-1}(D_\kappa)\frac{J_L([\underline{y}^\kappa]_\kappa)}{J_n([\underline{y}^\kappa]_\kappa)} \cdot |J(\underline{y}^\kappa)| \sqrt{\beta_\kappa J^{-2}(\underline{y}^\kappa)\beta'_\kappa} \cdot \frac{d\underline{y}}{dl^{-1}([\underline{y}^\kappa]_\kappa D_\kappa)}$$

and the marginal likelihood function is proportional to

$$A^{-1}(D_\kappa)\frac{J_L([\underline{y}^\kappa]_\kappa)}{J_n([\underline{y}^\kappa]_\kappa)} \cdot |J(\underline{y}^\kappa)| \sqrt{\beta_\kappa J^{-2}(\underline{y}^\kappa)\beta'_\kappa} . \tag{4}$$

From this the marginal likelihood function for the composite model is easily obtained.

An Example. Consider the case where for each component model the group G of transformations is the set of location-scale transformations. A typical element of G will be denoted by $[\mu, \sigma], -\infty < \mu < \infty$ and $\sigma > 0$. The operation on an element of R^n is defined by

$$[\mu, \sigma]\underline{y} = ([\mu, \sigma]\underline{y}_1, \ldots, [\mu, \sigma]\underline{y}_n) = (\mu + \sigma y_1, \ldots, \mu + \sigma y_n),$$

and the group operation is defined by

$$[\mu_1, \sigma_1][\mu_2, \sigma_2] = [\mu_1 + \sigma_1\mu_2, \sigma_1\sigma_2].$$

The particular cases where $\mu = 0$ and $\sigma = 1$ give the groups of the location and scale transformations respectively.

Suppose that the location transformation is the same for each structural model. Then (2) becomes

$$f_1(\underline{c}_1)f_2(\underline{c}_2)d\underline{c}_1 d\underline{c}_2$$

$$(\underline{y}, \underline{y}_2)' = ([\mu, \sigma_1], [\mu, \sigma_2])(\underline{c}_1, \underline{c}_2)' = ([\mu, 1][0, \sigma_1], [\mu, 1][0, \sigma_2])(\underline{c}_1, \underline{c}_2)'.$$

Suppose $u = u_0$; the transformed data gives the equation

$$([-\mu_0, 1]\underline{y}_1, [-\mu_0, 1]\underline{y}_2) = ([0, \sigma_1], [0, \sigma_2])(\underline{c}_1, \underline{c}_2).$$

If the subscripts 0, 1, 2 are dropped (3) can be written as

$$f(\underline{\varrho})\,d\underline{\varrho}$$

$$[-\mu, 1]\underline{y} = [0, \sigma]\underline{\varrho}$$

or

$$\underline{y}^{\mu} = [0, \sigma]\underline{\varrho}$$

where $[0, \sigma]$ is an element of the scale group. Since the choice of transformation variable is arbitrary, a convenient transformation variable is chosen. Define the transformation variable for the scale group as $[0, Sy^{\mu}]$, where $Sy^{\mu} = \sqrt{\Sigma (y_i^{\mu} - \bar{y}^{\mu})^2}$, $\bar{y}^{\mu} = \dfrac{\Sigma y_i^{\mu}}{n}$. It can easily be shown that Sy^{μ} does not depend upon μ and therefore the superscript μ can be dropped. The reference point associated with $[0, Sy]$ is given by

$$D(\underline{y}^{\mu}) = [0, Sy]^{-1}\underline{y}^{\mu} = \frac{y_1 - \mu}{Sy} \cdots \frac{y_n - \mu}{Sy}.$$

Then according to (4) the marginal likelihood function for μ is proportional to

$$A_1^{-1}(D_{1,\mu})A_2^{-1}(D_{2,\mu}).$$

3
Some Comparisons

Under conditions similar to those above, other methods of combining data can be suggested. Some of these methods are compared with the method of marginal likelihood. For the purposes of this comparison, suppose a location scale probability model on R is given in the traditional form of the probability density function

$$\frac{1}{\sigma} f\left(\frac{x - \mu}{\sigma}\right). \tag{5}$$

Suppose also that data is collected from two such models and that the location parameter μ in each of the two models is assumed to be the same. A possible method is to combine the two models into a composite model, obtain the likelihood function for this composite model and maximize out the nuisance parameters. However this method assumes that the maximum values are the true values for the nuisance parameters. In many cases this method gives an over precise likelihood function. It can easily be shown that in general such a procedure produces the marginal likelihood function modulated by a factor which, in most cases, depends upon the parameter of interest.

It may be suggested that a prior distribution be considered for the parameters, then integrate out the nuisance parameters from the resulting posterior distribution and hence obtain a marginal posterior distribution for the parameter of interest. Suppose this method is applied to the case where both sets of data come from (5), and again μ is assumed to be common to both models. A possible prior distribution for the parameters would be the uniform prior $du \frac{d\sigma_1}{\sigma_1} \frac{d\sigma_2}{\sigma_2}$. A distribution which is identical to the marginal posterior distribution which would be obtained for μ with this prior distribution, can also be obtained through structural models, but only by defining on the error space a partition which depends upon the observations obtained.

A third method would be to average the likelihood function with respect to some weight function of the nuisance parameters. It turns out that the marginal likelihood function for (5) for the common parameter is identical with the result obtained by averaging the likelihood function with respect to the right invariant measure of the group of which the nuisance parameters are elements. In this case $\frac{d\sigma_1}{\sigma_1} \frac{d\sigma_2}{\sigma_2}$.

In general for the cases presented in this paper, and where γ is a coset, the frame work of the structural model leads naturally to a family of distributions indexed by the common characteristic and invariant with respect to the remaining transformations. Statements of inference about the common characteristic can then be based on this family. The discussion above has restricted those statements of inference to the likelihood functions associated with this family.

4

Compatability of Data with the Hypothesis of a Common Characteristic

Suppose first that the two independent structural models are assumed to have the same unknown transformation θ, θ an element of a common group G. As mentioned previously the statements of inference about the transformation can be made in the form of a probability statement. However before combining the data a test should be made to assess the compatability of the data say y_1 and y_2, with the hypothesis that the structural models have the common transformation. An approach to this problem is given below.

The structural equations of the two models in terms of the transformation variables are

$$[y_1] = \theta \, [\varrho_1] \quad D(y_1) = D(\varrho_1)$$

$$[y_2] = \theta \, [\varrho_2] \quad D(y_2) = D(\varrho_2).$$

If the hypothesis is true then it implies that

$$[y_1] \, [\varrho_1]^{-1} = [y_2] \, [\varrho_2]^{-1}$$

or

$$[y_2]^{-1} \, [y_1] = [\varrho_2]^{-1} \, [\varrho_1].$$

Now the distributions of $[\varrho_1]$ given $D(y_1)$ and $[\varrho_2]$ given $D(y_2)$ are known and therefore because of independence of the two models, the joint distribution of $([\varrho_1], [\varrho_2])$ given $(D(y_1), D(y_2))$ is also known. Hence the distribution of $W = [\varrho_2]^{-1} \, [\varrho_1]$ given $D(y_1), D(y_2)$ is known. The mapping

$$([\varrho_1], [\varrho_2]) \rightarrow [\varrho_2]^{-1} \, [\varrho_1] = W$$

defines a partition Ψ on the joint sample space of $([\varrho_1], [\varrho_2])$. If the hypothesis is true, then the various values that can be taken by $[y_2]^{-1} [y_1]$ index the members of the partition Ψ. If the hypothesis is true it is known which value of W has occurred namely the value $[y_1]^{-1} [y_1]$. One is then in the position to determine a suitable discrepancy measure, and to calculate the probability of obtaining an observation of W more discrepant than that observed. On this basis the compatability of the data with the hypothesis can be assessed.

Under certain conditions this approach can also be used to assess the appropriateness of forming the composite model (2). If γ is a group then the case is the same as above. If γ is not a group but for each i, H_iM_i is a group and H_iK is a group with H_i a normal subgroup of H_iK, then for each component model of (2) it is possible to construct a *marginal structural model*. The unknown transformation of the structural equation in each marginal structural model is κ, $\kappa \in K$.

Suppose that the latter condition holds, then the component structural model can be written as

$$f(\varrho)d\varrho$$

$$y = \eta\kappa\mu\varrho ,$$

and because of the assumption of normality the structural equation can be written as

$$y = \kappa\hat{\eta}\mu\varrho$$

where $\kappa \in K$ and $\hat{\eta}\mu \in HM$, HM a group. In terms of transformation variables the structural equation becomes

$$[y] = \kappa\hat{\eta}\mu [\varrho] \quad D(y) = D(\varrho)$$

$$[\varrho], [y] \in G, \kappa \in K, \hat{\eta}\mu \in HM = S, G = KS.$$

The transformation variable is an element of G and it can therefore be uniquely factored into components from K and S;

$$[y] = {}_\kappa[y] [y]_s.$$

New transformation variables can be obtained from old transformation variables by multiplication on the right. A change of transformation variable induces a change of reference point. Obtain a new transformation variable by multiplying on the right by $[y]_s^{-1}$. The structural equation becomes

$$[y] [y]_s^{-1} = {}_\kappa[y] = \kappa\hat{\eta}\mu [\varrho] [y]_s^{-1} = \kappa\hat{\eta}\mu [\varrho]_y$$

with reference point

$$[y]_s D(y) = [y]_s D(\varrho) .$$

The group G can also be written as SK, and $[\varrho]_y$ can be decomposed correspondingly to give

$$\kappa[\underline{y}] = \kappa\hat{\eta}\mu_S[\varrho]_y[\varrho]_{y,\kappa}$$

or

$$\mu^{-1}\hat{\eta}^{-1}\kappa^{-1}\kappa[\underline{y}] = {}_s[\varrho]_y[\varrho]_{y,\kappa}\,.$$

The uniqueness property of the factorization allows the relations

$$\kappa^{-1}\kappa[\underline{y}] = [\varrho]_{y,\kappa}$$

$$\mu^{-1}\hat{\eta}^{-1} = {}_s[\varrho]_y\,.$$

The distribution basic to making inferences about κ alone, is the marginal distribution of $[\varrho]_{y,\kappa}$ given $[\underline{y}]_s D(\varrho)$. Thus as in the first case considered, a distribution that can be used for assessing the hypothesis of a common characteristic is the distribution of

$$[\varrho_2]_{y_2}^{-1},_\kappa [\varrho_1]_{y_1},_\kappa$$

given

$$[\underline{y}_1]_{s_1}\, D_1(\underline{y}_1) \quad [\underline{y}_2]_{s_2}\, D_2(\underline{y}_2)\,.$$

An Example. Consider again the composite model where each component group is the set of location-scale transformations, and the location transformation μ is hypothesised to be the same for both components. A suitable transformation variable for the location-scale group is given by $[\bar{y}, Sy]$. The structural equation for each component may be written as

$$[\bar{y}, Sy] = [\mu, \sigma]\, [\bar{\varrho}, S\varrho]\quad D(\varrho) = D(y)\,.$$

To change transformation variables for the purposes at hand multiply on the right by $[0, Sy]^{-1}$:

$$[\bar{y}, 1] = [\mu, \sigma]\, [\bar{\varrho}, S\varrho]\, [0, 1/Sy]\quad [0, Sy]\, D(\varrho) = [0, Sy]\, D(y)$$

or

$$[0, 1/\sigma]\, [-\mu, 1]\, [\bar{y}, 1] = [\bar{\varrho}\; S\varrho/Sy]\,.$$

The error transformation variable can be factored into a scale-location decomposition and the equation can be written as:

$$[0, 1/\sigma]\, [-\mu, 1]\, [\bar{y}, 1] = [0, S\varrho/Sy]\, [S_yt, 1]$$

where $t = \bar{\varrho}/S\varrho$.

The uniqueness property of the decomposition allows the separation

$$[-\mu, 1]\, [\bar{y}, 1] = [S_yt, 1]$$
$$[0, 1/\sigma] = [0, S\varrho/Sy]\,.$$

Thus the marginal distribution can be found for $S\underset{\sim}{y}t$ given $[0, S\underset{\sim}{y}]$ $D(\underset{\sim}{y})$ in each component model. Then from the above discussion it is seen that the proposed distribution for assessing the hypothesis of a common transformation μ is given by the distribution of $S\underset{\sim}{y_1}t_1 - S\underset{\sim}{y_2}t_2$ given $[0, S\underset{\sim}{y_1}]$ $D_1(\underset{\sim}{y_1})$, $[0, S\underset{\sim}{y_2}]$ $D_2(\underset{\sim}{y_2})$.

References

1. Fraser, D.A.S., "Data Transformations and the Linear Model," *Annals of Mathematical Statistics*, 38, 1456-65, 1967.
2. Fraser, D.A.S., *The Structure of Inference*, New York, John Wiley, 1968.
3. Kalotay, A., "Transformation Models and Statistical Inference," Ph.D. Thesis, University of Toronto, 1968.
4. Nachbin, L., *The Haar Integral*, New York, Von Nostrand, 1965.
5. Kalbfleisch, J.D. and Sprott, D.A., "Applications of Likelihood Methods to Models Involving Large Numbers of Parameters," to appear, *Journal of the Royal Statistical Society, A*, 1970.

COMMENTS

G.A. Barnard:

It appears to me that the structural model of inference constitutes, in a way, a return to attitudes common in the eighteenth and nineteenth centuries, when the results of measurements were thought of as generated by a true value, combined with an error, the latter being, in principle, determinable. It is also a natural model in process control or signal detection theory, where the "noise" process affecting the process, or the signals, has physical reality. Thus Fraser's and Whitney's efforts to explore how wide a range of problems can be brought within the scope of such a model are to be welcomed. Personally, however, I fear that the domain of applicability will prove to be rather narrow.

In connection with the point of criticism raised by Dr. Buehler, I would point out that Bayesian methods are open to objection, on betting grounds, just as are structural or fiducial methods. The fact seems to be that we do not know how to combine arguments, such as the fiducial argument, which make *essential* reference to what is known and what is unknown, with betting considerations. A simple way of seeing part of the difficulty is to note that, to decide the bet, more must be known than is given.

D. Basu:

Professor Kempthorne asked a little while ago about how Professor Whitney would deal with the simplest case of a normal variable X with unknown mean μ. Let me briefly describe to you what actually transpired between Sir Ronald A. Fisher and me when I had asked him the same question in the context of his fiducial distribution theory. This happened in the year 1957 when Professor Fisher was visiting us at the Indian Statistical Institute, Calcutta. My question was the following:

"If X be $N(\mu, 1)$ and it is known that $0 \leq \mu \leq 1$, then what is the fiducial distribution of μ when we have a single observation x on X?"

Professor Fisher said that he will obtain the fiducial distribution of μ from the distribution $N(x, 1)$ by massing all the probability of $N(x, 1)$ to the left of zero at zero and all the probability to the right of unity at unity. That this is logically untenable may be seen as follows:

Let us contemplate the situation where the variance of X is very large, say 10^{10}. In this case, before we observe X, we are practically certain that the observed value x will be either less than 0 or greater than 1. Thus, even before we observe X, we are nearly sure that we are going to allot at least one-half of the total fiducial probability to one or other of the two extremities of the interval $[0, 1]$. If we accept Fisher's fiducial logic, then we are in effect saying that the mere information that μ lies in the interval $[0, 1)$, prepares us mentally to lean heavily to one or other of the two end points of the interval $[0, 1]$. Of course, this is patently absurd. It is clear that, in this case, Fisher's intuition had led him too far away from the reality of the situation. When I pointed out this fact to him, Professor Fisher got furious and said, "Basu, either you believe in what I say or you don't. But do not try to make fun of my theory."

Today, I felt compelled to relate an episode that I never before had said in public. I only hope that my eager young colleagues will now and then contemplate in all humility this fact that even the unparalleled statistical intuition of the great Fisher did not stop him from taking a logically untenable position. I do believe that Fisher's theory of fiducial distribution was conceived in error and was sustained by the tremendous ego of the man. I only hope that Professor Whitney does not fall into such a trap.

D. R. Cox:

I agree with Professor Kempthorne* that very careful analysis of this type of argument is desirable for very simple special cases. It is valuable also to look at cases where different approaches give somewhat different answers. One such problem is the estimation of a transformation in analysis of variance to which at least three solutions have been proposed, maximum likelihood (essentially asymptotic), "Bayesian" (using priors that depend on the observations, at least in the way that G.E.P. Box and I formulated the analysis) and a solution in Fraser's book; I understand that other solutions have been suggested. Dr. D.F. Andrews has produced extreme examples where Fraser's solution is clearly preferable on general grounds. An M.Sc. student of his at Imperial College has done some limited simulations suggesting that in at least some realistic cases the performance of Fraser's solution is not so satisfactory as the others, but further exploration is desirable.

O. Kempthorne:

I have a strong feeling of having been buried in mathematics. I do not mind if I have some feeling that the mathematics leads somewhere. I like "hard" mathematics and not "soft" mathematics. So I ask what is the story with $x = \mu + e$?

I wish also to express my view that most logic which depends totally on

*Professor Kempthorne's comments to follow.

exact observation of a continuous random variable is not relevant to the interpretation of observational data.

With the "classical" problem,

$$y_1 = x_1\beta + e_1, \ e_1 \sim IN(o, \sigma_1{}^2)$$

$$y_2 = x_2\beta + e_2, \ e_2 \sim IN(o, \sigma_2{}^2)$$

there is a considerable variety of possible procedures; for example, approximate point estimation, approximate Neyman-Pearson tests, the fiducial generalization of the Behrens-Fisher problem, the use of historical prior for σ_1^2/σ_2^2, the use of a neo-Bayesian argument etc. I see some logical basis to all these but not to the structural inference ideas.

R.L. Prentice:

To keep open the question of data transformations in analysis of variance, as mentioned by Professor Cox, I would like to draw your attention to some simulations performed by a Ph.D. student, Winston Klass, at the University of Toronto, in the past year in which Fraser's marginal likelihood (filament) provided more favourable estimates than the others mentioned by Professor Cox.

A second comment regarding the application of Fraser's structural inference is that if one knows that a random variable has distribution $N(\mu,\sigma^2)$, this alone is not enough to permit a structural analysis.

REPLY

Professor Kempthorne has asked, "What is the story with $x = \mu + e$?", in regards to structural inference. He also quotes the problem

$$y_1 = x_1\beta + e_1 \qquad\qquad e_1 \sim I, N(0, \sigma_1{}^2)$$

$$y_2 = x_2\beta + e_2 \qquad\qquad e_2 \sim I, N(0, \sigma_2{}^2).$$

Both examples as given, present the variable component (error) of the observation, explicitly. Structural inference is the probability analysis of such models.

The classical approach (the methods suggested by Kempthorne fall under this labelling) examines distributions on the observation space; the space of x in the first case and the space of (y_1,y_2) in the second case. The classical approach gives a probability analysis of the models $x \sim N(\mu, \sigma_0{}^2)$ and $(y_1 \ y_2)$ $\sim I, N(x_1\beta, x_2\beta, \sigma_1{}^2, \sigma_2{}^2)$. The "classical" approach does not analyze the models presented by Professor Kempthorne.

The approach of structural analysis is to examine the distribution on the error space; the space of e in the first case and the space of (e_1,e_2) in the second case. In applications, where models such as those presented by Kempthorne are considered appropriate, it is the error space and its distribution which provide the description of the variation in the system. It is the basic probability space of the system and structural inference is an analysis of this space as dictated by probability theory. Therefore in answer to Professor

Kempthorne's question, structural inference is the probability analysis of models of the form that he has presented.

To apply the methods of structural inference the question arises whether or not the error, the source of variation of the system, has reality in a particular context; (or as stated by Prentice [2], whether or not, the error quantities can be described without reference to the observation or parameter spaces). If it has reality, the methods of structural inference apply. If it has plausible reality then a direct analysis on the basic probability space seems preferable to the classical model with its various reduction principles. Professor Barnard suggests that the range of applicability may prove to be rather narrow. Extensions of the structural model are given by Fraser in his book and at this meeting. The extensions combine, where possible, the classical and structural model and increases the range of applicability of the structural inference. Recent applications may be found in Fraser and Prentice [1] and Prentice [2].

The subject of my talk is not connected with the fiducial probability mentioned by Professor Basu. However, some of the motivation for structural inference did come from the writings of Fisher. Sir Ronald Fisher has contributed a large number of useful results in theoretical and applied statistics. I believe it is obvious that Fisher was more interested in basic ideas than in fine details of formulation. In many instances Fisher took time only to partially formulate his ideas. Because of this, he left open a large number of directions to pursue: we owe him much.

Finally, analysis by structural inference has various theoretical results. The analysis produces a conditional distribution which describes a concealed error variable from a known distribution. The model places possible values of the error in one-one correspondence with possible parameter values; it provides a relabelling of the error variable. And so *one* of the various theoretical results of structural inference is that the conditional distribution can be presented in terms of the relabelled error values i.e. the structural distribution. Other theoretical results concerning structural inference can be found in the papers given by Fraser and myself at the meeting.

I wish to express my thanks to Professor D.A.S. Fraser. I found his suggestions of considerable help in preparing this reply.

References

1. Fraser, D. A. S., and Prentice, R. L., "Randomized Models and the Dilution and Bio-Assay Problems," submitted to *Annals of Mathematical Statistics*, 1970.
2. Prentice, R. L., "Potency Assessment and Comparison in Quantal Response Assay: A Structural Analysis," (unpublished) 1970.

ON HAAR PRIORS*
C. Villegas
University of Rochester

1
Introduction

In recent years there has been an increasing interest in the determination of prior distributions which are based only on the mathematical model considered and do not imply any additional knowledge whatsoever about the parameters under consideration. Prior distributions which represent ignorance are of interest, not only because they lead to inferences exclusively based on the experimental data and the mathematical model, but also because the corresponding posterior distributions may also be used by persons who have only a vague prior knowledge about the parameters. In addition, such prior and posterior distributions may also be of interest to persons who have substantial prior knowledge but want to evaluate more carefully their own personal distributions. Criteria for the determination of prior distributions representing ignorance have been given recently by J. Hartigan (1964), M. R. Novick and W. J. Hall (1965) and E. T. Jaynes (1968). Lindley (1958) has given conditions under which a fiducial distribution can be derived from a prior distribution. For a long time H. Jeffreys (1961) has been a leading advocate of the use of such prior distributions.

As has been emphasized by Fraser [4], in many important cases the mathematical model has an underlying algebraic structure which is important for statistical inference. More precisely, the parameter may be considered to be, in such cases, an unknown element of a locally compact group of transformations of the sample space. If this group has, up to an arbitrary scale factor, only one invariant Haar measure, then it would be natural to choose this measure as a prior measure expressing ignorance, because, up to an

*Supported in part by the Systems Analysis Program, Bureau of Naval Personnel contract number N00022-69-C-0085.

arbitrary scale factor, it would be the only measure naturally associated with the group. One difficulty which is usually encountered when we want to use a Haar measure as a prior measure is that in many cases the measure of the parameter space is infinite. Stone [15] suggests that, in such a case, the Haar measure should be considered only as a quasi prior measure, i.e., as a limit of proper prior distributions. A really satisfactory solution of this difficulty requires a reformulation of the axiomatic foundation of probability theory, for which the interested reader is referred to a paper by Villegas (1967). Another difficulty is that, in many cases, there is not one but two Haar measures, of which one is a left invariant Haar measure and the other is a right invariant Haar measure. Fraser [3] has pointed out that one of these measures leads to posterior distributions which can be derived from a fiducial argument, and Hartigan [8] has shown that the other Haar measure can be derived from an invariance argument. As far as the author knows, the paper published by Barnard in 1952 was the first one in which Haar measures have been used in statistical inference.

In the Euclidean model considered in this paper there is only one invariant Haar measure and it will be shown that it leads to a prior distribution of the covariance matrix which has been considered before by Jeffreys [11], Cornfield and Geisser [6] and Villegas [18] and has been used by Tiao and Zellner [16] and Geisser [5] to develop a Bayesian multivariate theory. In a future paper [19] it will be shown that this Haar prior plays a fundamental role in the development of a Bayesian theory for linear relations.

2

The Euclidean Model

Let the sample space $\mathscr{S}$ be a p-dimensional Euclidean space, and let $\mathscr{G}$ be the group of all nonsingular linear transformations $A: \mathscr{S} \to \mathscr{S}$. Suppose that we observe n random vectors $y_i \in \mathscr{S}$ such that

$$Ay_i = u_i \qquad (i = 1, \ldots, n), \tag{2.1}$$

where $A \in \mathscr{G}$ may be called the *Euclidean precision parameter,* and the u_i are n independent, identically distributed random vectors, whose common distribution is known.

The group $\mathscr{G}$ of all nonsingular linear transformations is a unimodular group, i.e., it has, up to an arbitrary scale factor, only one invariant Haar measure. It would be natural then, to take this Haar measure as a prior distribution because it is the only measure naturally associated with the parameter space. This prior measure can also be derived from the following invariance argument (compare with Hartigan [8] and Jaynes [10]):
Consider the change of parameter

$$B = AT, \tag{2.2}$$

where $T \in \mathscr{G}$ is fixed, and the change of variables

$$y_i = Tz_i. \tag{2.3}$$

Our model is, in the new variables,

$$Bz_i = u_i. \tag{2.4}$$

Since the two models (2.1) and (2.4) have the same structure, it follows that, if no additional knowledge is available, then the prior measure μ_A for the parameter A and the prior measure μ_B for the parameter B should be the same measure μ. But from (2.2) it follows then that this prior measure μ must be a right invariant measure.

Some authors, like Fraser [3] and Hartigan [8], consider as parameter not A but A^{-1} and therefore, for them the invariant prior is the left invariant Haar measure.

A simple fiducial argument can also be given in support of this prior measure. In effect, let $e_1, \ldots, e_p$ be a basis in $\mathscr{S}$ and define the linear transformations $Y, U: \mathscr{S} \to \mathscr{S}$ by

$$Ye_j = y_j, \ Ue_j = u_j, \ (j = 1, \ldots, p).$$

Then we have the model

$$AY = U. \tag{2.5}$$

We assume that the realized Y is a nonsingular linear transformation, and that, almost surely, the random linear transformation Y is also nonsingular and therefore belongs to $\mathscr{G}$. Hence, excluding a null event, we can consider (2.5) as a model, in which the parameter A, the observed random variable Y and the random variable U are all elements of the same group $\mathscr{G}$. Then we can argue that, on purely intuitive grounds, the posterior distribution of U should be the same as its sampling distribution, and therefore the fiducial distribution of A is given by

$$A = UY^{-1} \tag{2.6}$$

where now Y is the realized transformation. It can be easily shown that this fiducial distribution can be obtained as a posterior distribution if the prior is the left invariant Haar measure (which in our case coincides with the right invariant Haar measure).

Let $\mathscr{G}_0$ be the group of all orthogonal transformations $O: \mathscr{S} \to \mathscr{S}$, and let $\mathscr{G}_\nabla$ be the group of all positive upper triangular transformations with respect to an orthonormal basis in $\mathscr{S}$, i.e., the group of all transformations $\nabla: \mathscr{S} \to \mathscr{S}$ whose matrices, with respect to that basis, are upper triangular matrices with positive diagonal elements. Clearly $\mathscr{G}_0$ and $\mathscr{G}_\nabla$ are subgroups of $\mathscr{G}$. It is well known that, for any $A \in \mathscr{G}$, there is a unique factorization

$$A = O\nabla,$$

where

$$O \in \mathscr{G}_0 \text{ and } \nabla \in \mathscr{G}_\nabla.$$

Proposition 2.1. If the prior measure of $A = O\nabla$ is an invariant measure μ on the group $\mathscr{G}$, then the prior measure of O is an invariant Haar measure on the group $\mathscr{G}_0$ and the prior measure of ∇ is a right invariant measure on the group $\mathscr{G}_\nabla$.

Proof: Since $\mathscr{G}_0$ is compact it is unimodular (see [13], Proposition 13, p. 81). Let ϕ be an invariant Haar measure on $\mathscr{G}_0$ and let τ^+ be a right invariant measure on $\mathscr{G}_\nabla$. If $J(O, \nabla)$ is the derivative of μ with respect to the product measure $\phi \times \tau^+$, we have, for any function f defined on $\mathscr{G}$, and any transformations $O_1 \in \mathscr{G}_0$, $\nabla_1 \in \mathscr{G}_\nabla$,

$$\int \phi(dO) \int f(O\nabla) J(O, \nabla) \tau^+(d\nabla) = \int f(A) \mu(dA)$$

$$= \int f(O_1^{-1} A \nabla_1^{-1}) \mu(dA)$$

$$= \int \phi(dO) \int f(O_1^{-1} O \nabla \nabla_1^{-1}) J(O, \nabla) \tau^+(d\nabla)$$

$$= \int \phi(dO) \int f(O \nabla \nabla_1^{-1}) J(O_1 O, \nabla) \tau^+(d\nabla)$$

$$= \int \phi(dO) \int f(O\nabla) J(O_1 O, \nabla \nabla_1) \tau^+(d\nabla).$$

By a comparison of the first and last terms in this chain of equalities, we have

$$J(O, \nabla) = J(O_1 O, \nabla \nabla_1).$$

Since O_1 and ∇_1 are arbitrary, it follows that $J(O, \nabla)$ is a constant c, and therefore $\mu = c\phi \times \tau^+$. Since $\mathscr{G}_0$ is a compact group, $\phi(\mathscr{G}_0)$ is finite and therefore, up to a scale factor c, τ^+ is the marginal measure on the group $\mathscr{G}_\nabla$ (compare this proof with the proof of Proposition 19, p. 85 of [13] and with the proof of Proposition III, p. 31 of [7]).

Proposition 2.2. If τ^- is a left invariant and τ^+ is a right invariant Haar measure on $\mathscr{G}_\nabla$, then

$$\tau^-(d\nabla) = c^- \prod_{j=1}^{p} \nabla_{jj}^{-(p-j+1)} d[\nabla], \tag{2.7}$$

$$\tau^+(d\nabla) = c^+ \prod_{j=1}^{p} \nabla_{jj}^{-j} d[\nabla], \tag{2.8}$$

where c^-, c^+ are arbitrary constants, $[\nabla] = \{\nabla_{jj'}\}$ is the matrix of ∇, and

$$d[\nabla] = \prod_{j \leq j'} d\nabla_{jj'}.$$

Proof: See [9], p. 209.

Note that, since the factors $\nabla_{jj}^{-(p-j+1)}$ and ∇_{jj}^{-j} have improper integrals, the marginal measures of the $\nabla_{jj'}$ are not defined.

In the important case in which the u_i have standard Gaussian distributions, the distribution of y_i is entirely determined by the covariance transformation

$$\Sigma^2 = (A'A)^{-1},$$

where $A' : \mathscr{S} \to \mathscr{S}$ is the adjoint of A. Note that, contrary to the usual notation,

412

the covariance transformation is not denoted by Σ but by Σ^2, leaving the symbol Σ to denote the uniquely defined, positive definite square root of Σ^2.

Proposition 2.3. If the prior measure of A is an invariant Haar measure μ on the group $\mathcal{G}$, then the prior measure of Σ^{-2} has differential form equal to

$$|\Sigma|^{p+1}d\Sigma^{-2}$$

over the set of symmetric, positive definite transformations Σ^{-2}, and O otherwise, $d\Sigma^{-2}$ being the Euclidean differential form for symmetric matrices.

Proof: From the definition of Σ^2 it follows that

$$\Sigma^{-2} = \nabla'\nabla.$$

By Theorem 4.1 of [2], the Jacobian of the transformation $[\Sigma^{-2}] = [\nabla]'[\nabla]$ is

$$\frac{d[\Sigma^{-2}]}{d[\nabla]} = 2^p \prod_{j=1}^{p} \nabla_{jj}^{p-j+1}. \tag{2.9}$$

By Proposition 2.1, the prior measure of ∇ is a right invariant measure. The conclusion follows by (2.8) and (2.9).

References

1. Barnard, G. A., "The Frequency Justification of Certain Sequential Tests," *Biometrika*, 39, 144-150, 1952
2. Deemer, Walter L. and Olkin, Ingram, "The Jacobians of Certain Matrix Transformations Useful in Multivariate Analysis," *Biometrika*, 38, 345-367, 1951.
3. Fraser, D. A. S., "On Fiducial Inference," *Ann. Math. Statist.*, 32, 661-676, 1961.
4. Fraser, D. A. S., *The Structure of Inference*, New York, Wiley, 1968.
5. Geisser, Seymour, "Bayesian Estimation in Multivariate Analysis," *Ann. Math. Statist.*, 36, 150-159, 1965.
6. Geisser, Seymour and Cornfield, Jerome, "Posterior Distributions for Multivariate Normal Parameters," *Journal of the Royal Statistical Society, B*, 25, 368-376, 1963.
7. Gelfand, I. M. and Neumark, M. A., *Unitare Darstellungen der Klassischen Gruppen*, Berlin, Akademie-Verlag, 1957.
8. Hartigan, J., "Invariant Prior Distributions," *Ann. Math. Statist.*, 35, 836-845, 1964.
9. Hewitt, Edwin and Ross, Kenneth A., *Abstract Harmonic Analysis*, Berlin, Springer, 1963.
10. Jaynes, Edwin T., "Prior Probabilities," *IEEE Transactions on Systems Science and Cybernetics*, 4, 227-241, 1968.
11. Jeffreys, Harold, *Theory of Probability*, (third edition), Oxford, Clarendon Press, 1961.
12. Lindley, D. V., "Fiducial Distribution and Bayes' Theorem," *J. Roy. Statist. Soc., B*, 20, 102-107, 1958.
13. Nachbin, Leopoldo, *The Haar Integral*, Princeton, N.J., Van Nostrand, 1965.
14. Novick, Melvin R. and Hall, W. J., "A Bayesian Indifference Procedure," *Journal of the American Statistical Association*, 60, 1104-1117, 1965.

15. Stone, M., "Right Haar Measure for Convergence in Probability to Quasi Posterior Distributions," *Ann. Math. Statist.,* 36, 440-453, 1965.
16. Tiao, George C. and Zellner, Arnold, "On the Bayesian Estimation of Multivariate Regression," *J. Roy. Statist. Soc., B,* 26, 277-285, 1964.
17. Villegas, C., "On Qualitative Probability," *Am. Math. Monthly,* 74, 661-669, 1967.
18. Villegas, C., "On the *A Priori* Distribution of the Covariance Matrix," *Ann. Math. Statist.,* 40, 1098-1099, 1969.
19. Villegas, C., "Bayesian Statistical Inference in Linear Relations," (mimeographed), College of Business Administration, University of Rochester, 1969.

COMMENTS

G.A. Barnard:

Since the history of invariance has been mentioned may I say that the notion has surely been part of the common stock of scientific ideas ever since Einstein's paper of 1905, if not earlier. I recall as an undergraduate hearing Dr. Harold Jeffreys justify the $d\sigma/\sigma$ prior for a scale parameter by essentially invariance arguments in 1934 or 1935; and Pitman's work was done a little later. During World War II many of us discussed, for example, the "justification" of the *t* statistic as the (essentially) unique function of the sufficient statistics and the location parameter which is invariant under change of scale and origin; and when I made use of such notions in my 1949 paper in *J. Roy. Stat. Soc., B,* the point did not seem to call for special remark.

John Tukey and I had a discussion, in the presence of Frank Anscombe and Henry Daniels, on the connection between invariant priors or weight functions and the fiducial argument, some time between 1948 and 1950, but we could not then see how the fiducial distribution of ρ could be obtained. We were also bothered by the questions of groups which are not unimodular. I discussed these points with Fisher himself. He drew my attention to a passage in his 1934 paper on "Two New Properties of Mathematical Likelihood" which was subsequently indexed (by Tukey) under "Group Invariance" in his "Contributions to Mathematical Statistics". It became clear to me that although he had used these ideas, Fisher was not at home with modern group terminology.

It was also around this time that I had some discussions with Sir Harold Jeffreys in which he told me he had heard of the work of Peisakoff in California, and thought it might be highly relevant to his invariance theory. In this he, too, had difficulty with non-commutative groups.

Thus, while I believe Dr. Villegas is right in saying that I was first to use the term "Haar measure" in a statistical paper, I would disclaim any originality in connection with group invariance and statistics in general.

Dr. Rubin's reference to inadmissibility of invariant procedures is worth pondering upon because, as I have said, this seems to me to point up the difference between the scientific and the decision-theoretic approach. It may not be paradoxical, in decision theory, to move one's estimate in an *arbitrarily* chosen direction (since the origin is arbitrary); but scientifically it seems clearly absurd.

I would like to thank Dr. Villegas for a paper which exhibits very clearly a situation which has hitherto been confused and obscure.

I.J. Good

I should like to make a few historical comments.

Harold Jeffreys introduced the idea of infinite probabilities in his book of 1938. (Added in writing: But he did not produce a proper axiomatization.)

Jeffreys and Haldane introduced the idea of invariance of a distribution of a positive random variable, under transformations by powering; and arrived at the density proportional to $1/x$. I think this was one of the first explicit invariance arguments in statistics. In 1946 and 1947 Jeffreys and Perks introduced invariance theories for estimation problems.

In 1952, Barnard suggested the use of Haar measure as an initial probability density, and I think A.G. Laurent made a similar suggestion in a paper "*A priori* Probability" at a weekend conference of the Royal Statistical Society on Sunday, September 23, 1951. Unfortunately I have lost my copy of his paper, which as far as I know was never published.

I recalled in my paper in this symposium that Jeffrey's invariant density could be regarded as a minimax choice, that is, a least-utility choice of initial probability distribution, when the utility is taken as weight of evidence. It would be interesting to work out another invariant density by taking the utility as the expression

$$\frac{1}{\alpha} \log \left\{ \int \left[\frac{dT\,(\mathbf{z}\,|\,\mathbf{x})}{dT\,(\mathbf{z}\,|\,\mathbf{y})} \right]^{\alpha} dT(\mathbf{z}\,|\,\mathbf{x}) \right\} \ (\alpha > 0)$$

in place of

$$\int \log \frac{dT\,(\mathbf{z}\,|\,\mathbf{x})}{dT\,(\mathbf{z}\,|\,\mathbf{y})} \, dT(\mathbf{z}\,|\,\mathbf{x})$$

which is the limit of the previous expression as α tends to 0. The first of these expressions is also additive for a pair of experiments that have nothing to do with each other and as such is a reasonable quasi-utility. Dr. Villegas uses an approach that leads to a unique invariant prior. Presumably we mean different things by the expression "invariant".

J.A. Hartigan:

Mr. Villegas states on page 410 "There is not one but two Haar measures, left and right". Actually there is a family of relatively invariant prior measures, each of which is such that the corresponding posterior distributions are invariant under the same transformations which leave the sampling distribution model invariant.

Secondly the right invariant Haar measure exists only if there is a certain 1-1 relationship between invariant transformations and parameters. Fraser used this right Haar prior in earlier studies of fiducial distributions, which later evolved into the structural method. I am mystified by the reference on page 412 to this fiducial prior as left invariant.

Let me illustrate with one observation from a location parameter model: (1) $x+c$ has the same distribution given $\theta-c$ as x given θ; (2) require that $\theta-c$ has the same distribution given $x+c$ as θ given x; (3) then $\theta-c$ has the same marginal distribution as θ (identifying improper distributions different only in scale); (4) $h(\theta) = e^{k\theta}$ satisfies this requirement — relatively invariant densities are of this form.

I am disagreeing with Mr. Villegas in step (3). He states on page 410 that the two measures over θ and $\theta\text{-}c$ should be identical. I permit them to differ by a scale factor, since this is all that is necessary to ensure invariance in posterior distributions.

D.V. Lindley:

May I simply remark that the concept of invariance in statistics leaves me cold. If someone tells me that object 1 is measured in millimetres and object 2 in miles, then I have fair prior probability that the second object is larger than the first.

[There Good remarked that the same criticism did not apply to orientation.]

Good is correct, but there the group is compact, proper prior distributions will result and the situation looks good.

REPLY

In the first place I want to give Dr. Barnard my sincere thanks for his warm and very interesting comments. In connection with the historical remarks made by him and by Dr. Good, I would like to add that, in a personal communication, Herman Chernoff has let me know that around 1954, Girshick, Rubin and Blackwell, independently of Barnard, developed the use of invariant Haar measures for finding optimal invariant (minimax) procedures, but they did not publish these results.

As was pointed out by J. A. Hartigan, there is also a weaker invariance argument which requires only the invariance of posterior distributions. Of course, the invariance argument described in this paper not only insures the invariance of posterior distributions, but also the uniqueness of the prior. J. A. Hartigan points out that the fiducial prior, which is referred to in this paper as the left invariant Haar measure, is referred to in the literature as the right invariant Haar measure. As is explained in this paper, this difference in terminology arises from the fact that the parameter considered in previous papers is not the parameter A considered in this paper but the inverse A^{-1}. Clearly, if the prior for A is the left invariant Haar measure then the prior for A^{-1} is the right invariant Haar measure.

Finally, in order to give a short answer to Lindley's short comment, I would only guess that his posterior distribution would not differ very much from one obtained from a Haar prior.

A COMPARISON OF FREQUENTIST AND BAYESIAN APPROACHES TO INFERENCE WITH PRIOR KNOWLEDGE

D. J. Bartholomew
University of Kent

D. J. Bartholomew
University of Kent

1
Introduction

Suppose that we have a continuous random variable X with density function $f(x, \theta)$ where θ is a scalar. We are given a sample of observations on X and wish to make inferences about the value of the unknown θ. The Bayesian approach to this problem is to select a prior distribution for θ, to determine the posterior distribution of θ by Bayes's theorem and then to express inferences in terms of probabilities computed from this distribution. The usual frequentist approach is to find a pivotal function of the observations and θ whose distribution in repeated sampling is the same for all θ. Probability statements about the pivotal function can then be converted into statements about whether or not θ lies in a random interval of the real line. Both approaches lead to statements having the same form. For example, an interval estimator for θ will have the form $[\Phi_1(x), \Phi_2(x)]$; in the Bayesian framework we assert that θ lies in this interval with specified probability but the frequentist would treat the end points as random variables and choose them such that the true value of θ is included with the required probability.

Our interest in this paper is in the circumstances under which Bayesian and frequentist methods are formally the same. More precisely, we shall say that the approaches *agree* if

$$\Pr\{\tilde{\theta} \leq \theta_a(\mathbf{x}) \mid \mathbf{x}\} = \Pr\{\theta \leq \tilde{\theta}_a(\mathbf{x}) \mid \theta\} = \alpha \qquad (1)$$

for all α. The tilde is used to denote the random variable in a probability statement; $\theta_a(\mathbf{x})$ is the α-probability point of the posterior distribution of θ; x is the vector of observations. The probabilities in (1) depend on the prior density, $\pi(\theta)$, of θ and on the sample space of x. Our problem is thus to find whether there exists a $\pi(\theta)$ and a sample space such that (1) holds. Some conditions under which agreement occurs are known and are enumerated in Section 2.

However, almost all of the published work on this topic relates to the case when there is no prior information about θ. In fact, it is commonly argued that the great advantage of the frequentist approach is that it requires no arbitrary specification of prior ignorance. This, we shall contend, is a mistaken view because it overlooks the arbitrary choice of sample space used for the computation of frequency probabilities.

There are two obvious advantages to be gained if agreement is attainable. From the practical point of view it means that there need be no difference in the inferential behaviour of statisticians even though they may offer differing explanations of their actions. This makes it possible to teach a body of statistical methods which will find wide acceptance. On the theoretical side, the frequentist is no longer open to the charge, levelled for example by Lindley (1961a), that any inference procedure which is not formally equivalent to a Bayesian method contradicts Savage's axioms of rational behaviour.

The major part of what follows is an exploratory investigation of the extent to which agreement is possible when prior information about θ is incorporated into the inference procedure. In Bayesian theory this is done in a straightforward manner through the prior distribution. There is no general method for taking account of prior information in frequency theory. We shall therefore have to first consider whether and how prior information can be incorporated into frequentist inference procedures. There is one general result, due to Pratt (1963), which provides a partial solution to the problem. He showed that

$$\Pr\{\tilde{\theta} \leq \theta_\alpha(\mathbf{x}) \mid \mathbf{x}\} = \int \pi(\theta) \Pr\{\theta \leq \tilde{\theta}_\alpha(\mathbf{x}) \mid \theta\} d\theta \qquad (2)$$

which implies that agreement is always possible in an average sense. In fact if θ is sampled from a population of θ's with distribution $\pi(\theta)$ then $\theta_\alpha(\mathbf{x})$ will exceed θ on a proportion α of occasions and so Bayesian procedures based on the left hand side of (2) would be acceptable to a frequentist statistician. If, on the other hand, $\pi(\theta)$ is an expression of prior beliefs it is doubtful whether the average on the right hand side of (2) can be given a meaningful frequency interpretation.

In Sections 3 and 4 we shall review methods of incorporating prior information into frequentist and Bayesian inference procedures. The somewhat unsatisfactory nature of these methods leads us to investigate the suggestion of Bartholomew (1965) that the sampling rule is the frequentist equivalent of the prior distribution. In preparation for this we shall summarize in Section 2 the present state of knowledge on agreement in the case of prior ignorance.

2

Prior Ignorance: Summary of Known Results

The principal result on agreement is due to Welch and Peers (1963) who showed that if θ is an unrestricted location parameter then (1) holds if $\pi(\theta)$ is the improper uniform distribution. The case when θ is a scale parameter is included in this result because the transformation $y = \log x$, $\tau = \log \theta$ converts the problem into one involving a location parameter. This generalizes an earlier result of Lindley (1958). Asher, in unpublished work, has shown that Welch and Peers' result can be further generalized to cover some distributions where the range depends on θ.

Welch and Peers (1963) have also shown that approximate agreement can be obtained for a fixed sample size experiment by choosing

$$\pi(\theta) \propto \left| \frac{1}{\sqrt{I(\theta)}} \right| \tag{3}$$

where $I(\theta)$ is the Fisher information function. This is approximate in the sense that

$$\Pr\{\theta \leqslant \tilde{\theta}_\alpha(\mathbf{x}) \mid \theta\} = \alpha + O\left(\frac{1}{n}\right). \tag{4}$$

For any other prior the right hand side of (4) would be $\alpha + O(n^{-\frac{1}{2}})$. The same choice of prior, using other considerations, was arrived at by Jeffreys (1961), Hartigan (1965) and Lindley (1960). Some calculations showing that the approximation can be very good even if $n = 1$ were given by Bartholomew (1965).

It is sometimes possible to obtain partial agreement in the sense that (1) holds for some values of α but not all. Welch and Peers (1963) investigated this point and Bartholomew (1965) showed that $\pi(\theta)$ can sometimes be chosen so that

$$\Pr\{\theta \leqslant \tilde{\theta}_{\alpha_1}(\mathbf{x}) \mid \theta\} - \Pr\{\theta \leqslant \tilde{\theta}_{\alpha_2}(\mathbf{x}) \mid \theta\} = \alpha_1 - \alpha_2 \qquad (\alpha_1 > \alpha_2) \tag{5}$$

even when both probabilities on the left hand side are functions of θ. This means that Bayesian interval estimators can be constructed which are also confidence intervals in the accepted sense. Although the possibility of partial agreement has a practical interest we shall not pursue it further here.

3
Prior Knowledge: A Frequentist Approach

There is no general theory for utilizing prior information in frequentist inference. It is sometimes recommended that prior knowledge should be taken into account subjectively when interpreting a frequentist analysis but it is not clear how or when this is done in practice. A useful starting point for constructing a theory is to ask what effect different kinds of prior knowledge *ought* to have on the performance of the methods. Suppose, for example, that we have reason to believe that θ is near to θ_0. Then if we were testing a hypothesis about θ it would seem sensible to require a test to have higher power near θ_0 than if we had ignored this information. If this means a loss of power in other regions the price is presumably worth paying. If the problem were one of interval estimation we might reasonably ask for a shorter interval near θ_0 with, perhaps, a longer one elsewhere. In both cases we expect to do better when the data support our prior beliefs.

When the prior knowledge imposes a restriction on the range of θ we should expect it to be of most use when the data suggest a value of θ near to the boundary. This should lead to a shorter confidence interval because values of θ outside the region can be ruled out *a priori*.

There have been a number of somewhat unrelated attempts to make use of prior information along the lines suggested above. The best known are in hypothesis testing. If the range of the parameter is restricted the power can be increased in this range by using, for example, a likelihood ratio test based on the reduced parameter space. The simplest example is the one tail test and more elaborate examples with greater gain in power are given in Bartholomew (1961) and elsewhere. When prior knowledge is expressed by means of a weight function, tests based on weighted likelihood ratios can be used.

The major contribution to this problem in the realm of interval estimation is due to Pratt (1963). A solution to the problem when the parameter is known to be non-negative was given by Stein (1962) and commented on by Bartholomew (1962). For brevity we shall confine the ensuing discussion to the case of two-sided interval estimation and consider a family of procedures of which Pratt's and Stein's are special cases.

We require a method of constructing confidence intervals such that their length depends on the prior knowledge in a meaningful way. One approach is as follows. Let s be any statistic based on a random sample of fixed size from $f(x, \theta)$. Then if $H(s, \theta)$ is the distribution function of s, $\hat{y} = H(\hat{s}, \theta)$ is rectangularly distributed between 0 and 1. Hence

$$\Pr\{\alpha_1(\theta) \leq H(\hat{s}, \theta) \leq 1 - \alpha_2(\theta) \mid \theta\} = 1 - \alpha \qquad (5)$$

if $\alpha_1(\theta) + \alpha_2(\theta) = \alpha$. The novelty here lies in allowing α_1 and α_2 to be functions of θ but with constant sum. By inverting the inequalities in (5) we obtain a confidence interval for θ. The length of the interval for a given value of s will obviously depend on $\alpha(\theta_1)$ and $\alpha_2(\theta)$. Our problem is thus to choose $\alpha_1(\theta) = \alpha - \alpha_2(\theta)$ such that the length of the resulting interval varies in the way demanded by our prior knowledge.

In exploring this method we shall suppose that X is normal with known standard deviation. The sufficient statistic for θ is thus $\bar{x}$ and this will be the statistic s. Since n is fixed and σ is known it will simplify the algebra if we put $\sigma/\sqrt{n} = 1$. We shall consider two sorts of prior knowledge about θ; viz, $\theta \geq 0$ and $-\theta_1 \leq \theta \leq \theta_1$. The case where prior knowledge is described by a weight function is covered by Pratt (1963) to whose results we shall refer.

Case (a). The prior knowledge is that $\theta \geq 0$. This problem was discussed by Stein (1962) who proposed taking

$$\left. \begin{array}{l} \alpha_1(\theta) = (\dfrac{\theta}{2c})\alpha, \quad \alpha_2(\theta) = (1 - \dfrac{\theta}{2c})\alpha \quad \text{when } \theta \leq c, \\[2mm] \alpha_1(\theta) = \alpha_2(\theta) = \tfrac{1}{2}\alpha \quad \text{when } \theta > c \end{array} \right\} \qquad (6)$$

for some suitable c. The implications of this choice will be clear from Figure 1. For large values of $\bar{x}$ the interval will be the same as in the unrestricted case but as $\bar{x}$ approaches zero and especially if it is negative the interval is much shorter. To illustrate this remark we give in Table 1 the length of Stein's interval in a particular case.

Table 1. Length of Stein's Interval When $c = 1.645, \alpha = 0.05$.

$\bar{x}$	−2	−1	0	1	2	4	∞
Length	0.45	1.1	1.96	2.96	3.66	3.92	3.92

420

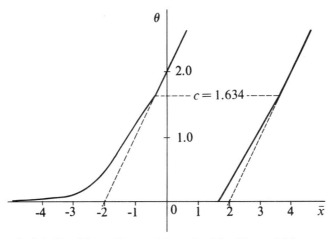

Figure 1. Stein's Confidence Interval for a Positive Normal Mean.
(95% Confidence Coefficient).
For a given $\bar{x}$ the interval is read off vertically against the θ-scale.
The broken line gives the standard interval.

A cruder solution to the problem is to take

$$\left.\begin{array}{ll} \alpha_1(\theta)=0, \quad \alpha_2(\theta)=\alpha & \text{when } \theta \leq c \\[2mm] \alpha_1(\theta)=\alpha_2(\theta)=\tfrac{1}{2}\alpha & \text{when } \theta > c \end{array}\right\} \tag{7}$$

for some suitable c. The advantage of this choice for our purposes is that the calculation of the expected length as a function of θ is much simpler but the general characteristics remain the same. In this case the length is a step function of $\bar{x}$ as follows:

Length

c,	$\bar{x} < 0$
$\bar{x}+c$,	$0 \leq \bar{x} < u_\alpha$
$c+u_\alpha$,	$u_\alpha \leq \bar{x} < c+u_\alpha$
$\bar{x}+u_{\frac{1}{2}\alpha}-c$,	$c+u_\alpha \leq \bar{x} < c+u_{\frac{1}{2}\alpha}$
$2u_{\frac{1}{2}\alpha}$	$c+u_{\frac{1}{2}\alpha} \leq \bar{x}$

where u_α is the upper α-point of the standard normal distribution.

The effect of decreasing c, in both cases, is to give a shorter interval for small $\bar{x}$ at the price of a longer one for larger values of $\bar{x}$. The choice does not appear to be crucial as far as expected length goes as shown in Table 2.

Table 2. The Expected Length of the Interval Given by (7) for $\alpha = 0.05$ with $c = u_{0.025}$ and $c \to 0$.

θ	0	0.5	1	2	3	4	∞
$c = u_{0.025} = 1.96$	2.34	2.60	2.89	3.38	3.64	3.79	3.92
$c \to 0$	1.96	2.43	2.87	3.54	3.84	3.91	3.92

421

The limiting case, $c \to 0$, amounts to using the ordinary confidence interval $\bar{x} \pm 1.96$ but excluding the negative part. The practical objection to this is that the interval is empty if $\bar{x} < 1.96$ which would occur with probability 2.5% if θ were zero. Larger values of c would give intervals of greater average length near $\theta = 0$. If one is ignorant of θ apart from the fact that it is positive there seems to be no uniquely best way of choosing c but a value of 1 or 2 would seem reasonable. Similar remarks apply in the case of Stein's interval.

Case (b). The prior knowledge is that $-\theta_1 \le \theta \le \theta_1$. The natural extension of Stein's interval is to choose $\alpha_1(\theta)$ and $\alpha_2(\theta)$ as follows:

$$\left. \begin{array}{l} \alpha_1(\theta) = \left(\dfrac{\theta_1 + \theta}{2c}\right)\alpha, \quad \alpha_2(\theta) = \left(1 - \dfrac{\theta_1 + \theta}{2c}\right)\alpha \quad \text{when} -\theta_1 \le \theta < -\theta_1 + c, \\[2ex] \alpha_1(\theta) = \alpha_2(\theta) = \tfrac{1}{2}\alpha \quad \text{when} -\theta_1 + c \le \theta < \theta_1 - c, \\[2ex] \alpha_1(\theta) = \left(1 - \dfrac{\theta_1 - \theta}{2c}\right)\alpha, \quad \alpha_2(\theta) = \left(\dfrac{\theta_1 - \theta}{2c}\right)\alpha \quad \text{when} \; \theta_1 - c \le \theta < \theta_1, \end{array} \right\} \quad (8)$$

where $c < \tfrac{1}{2}\theta_1$.

The cruder version becomes

$$\left. \begin{array}{l} \alpha_1(\theta) = 0, \quad \alpha_2(\theta) = \alpha \quad \text{when} -\theta_1 \le \theta < -\theta_1 + c, \\[1ex] \alpha_1(\theta) = \alpha_2(\theta) = \tfrac{1}{2}\alpha \quad \text{when} -\theta_1 + c \le \theta < \theta_1 - c, \\[1ex] \alpha_1(\theta) = \alpha, \quad \alpha_2(\theta) = 0, \quad \text{when} \; \theta_1 - c \le \theta \le \theta_1. \end{array} \right\} \quad (9)$$

The expected lengths of the intervals obtained by this method are given in Table 3 for two values of θ_1.

Table 3. The Expected Length of the Interval Given by (9) for $\alpha = 0.05$, $c = 1.96$ and $c \to 0$ and $\theta_1 = 1.96, 3.92$.

$\theta_1 = 1.96$

θ	0	0.5	1.0	1.5	1.96
$c = 1.96$	2.79	2.71	2.50	2.24	2.00
$c \to 0$	3.12	3.02	2.75	2.36	1.95

$\theta_1 = 3.92$

θ	0	1	2	3	3.92
$c = 1.96$	3.64	3.58	3.30	2.78	2.18
$c \to 0$	3.90	3.84	3.54	3.00	1.87

Here again the smaller value of c gives a shorter interval at the extremes but $c = 1.96$ seems to give a better overall result. As θ_1 decreases the average length of the interval will also decrease and cannot in any case exceed $2\theta_1$. If the prior information had been neglected the length of the interval would have been 3.92 for all $\bar{x}$.

Pratt's work was concerned with the case when the range of θ was unrestricted and the prior knowledge was expressed by a weight function. The approach was via the corresponding hypothesis testing problem using the weighted likelihood method. Full details can be found in the original paper but the essential conclusions can be reached by considering the case when all the weight is placed at $\theta = \tau$. For this case Pratt takes

$$\left.\begin{array}{ll}\alpha_1(\theta)=0, & \alpha_2(\theta)=\alpha \quad \text{when } \theta \leqslant \tau \\[2mm] \alpha_1(\theta)=\alpha, & \alpha_2(\theta)=0 \quad \text{when } \theta > \tau. \end{array}\right\} \tag{10}$$

The length of the corresponding confidence interval when $\tau = 0$ is

Length	
$u_\alpha - \bar{x},$	$\bar{x} \leqslant -u_\alpha$
$2u_\alpha,$	$-u_\alpha \leqslant \bar{x} \leqslant u_\alpha$
$u_\alpha + \bar{x},$	$u_\alpha \leqslant \bar{x}.$

Here we have a shorter interval when $\bar{x}$ is in the expected region but it will be longer, perhaps much longer, if our prior information is wrong. It is an important shortcoming of the method that the interval cannot be shorter than $2u_\alpha$ however much prior knowledge we have. This is a rather modest reduction if we have a high degree of knowledge.

To summarize: the general method we have proposed succeeds to the extent that it can be made to reduce the length of confidence intervals in accordance with our intuitive expectations. However, the intervals do not correspond to posterior probability intervals and so we do not have even the partial agreement which this would imply. In Pratt's problem we have the further disadvantage that perfect prior knowledge represented by putting all the weight at $\theta = \tau$ cannot reduce the length below $2u_\alpha$. This last objection does not, of course, apply if we restrict θ to some interval and let this interval tend to zero as our prior knowledge increases.

4

Prior Knowledge: The Bayesian Approach

The Bayesian approach to inference proceeds by using experimental evidence to change prior beliefs, or knowledge; in consequence such knowledge is an essential ingredient of the analysis. It is also a straightforward matter, in principle, to compute the frequency properties of a Bayesian procedure and so investigate the degree of agreement. We shall do this for the same kinds of prior knowledge as considered in Section 3. Our approach is again exploratory being based on computations for simple examples.

We treat first the case when θ is a location parameter about whose value we are ignorant apart from the fact that it is non-negative. Agreement is obtained if the range of θ is unlimited by taking $\pi(\theta)$ to be the improper uniform distribution. It seems natural to argue, as did Lindley (1961b), that the prior knowledge about the range should not affect the form of $\pi(\theta)$ in the positive half of the real line. The frequentist consequences of this assumption were investigated in Bartholomew (1965) where it was shown that exact agreement in the sense of (1) is never possible — the greatest discrepancy being near $\theta=0$. However, it was also shown that Bayesian intervals could be constructed which were very nearly confidence intervals. The actual calculations related to the logistic distribution but this is so similar to the normal that the conclusion can be taken as applying to both.

The situation may be less satisfactory when θ is known to lie in the interval $(-\theta_1, \theta_1)$. In view of the remarks made above it would seem reasonable for a Bayesian to take $\pi(\theta)$ as uniform over the interval. If this is so then the posterior density function of θ for normal X is

$$p(\theta|\bar{x}) = \frac{\phi(\theta-\bar{x})}{\Phi(\theta_1-\bar{x}) - \Phi(-\theta_1-\bar{x})}, \qquad -\theta_1 \leqslant \theta \leqslant \theta_1 \qquad (11)$$

where $\phi(\cdot)$ is the density function of a standard normal variate and $\Phi(\cdot)$ is the corresponding distribution function. Again we assume $\bar{x}$ to be based on n observations with $\sigma/\sqrt{n}=1$. On Figure 2 we have plotted the shortest 95% posterior interval as a function of $\bar{x}$ for $\theta_1=1$. The frequency probability that this interval will cover the true θ is easily obtained from the diagram by reading off in the horizontal direction the range within which $\bar{x}$ must lie for that value of θ to be included in the Bayesian interval.

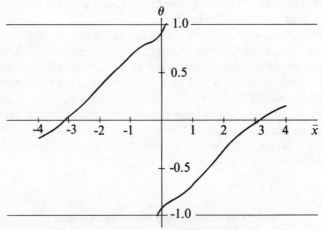

Figure 2. The Bayesian Interval for a Normal Mean with Uniform Prior Over (-1, +1). (95% Confidence Coefficient)
For given $\bar{x}$ the interval is read vertically against the θ-scale beween $\theta = -1$ and $\theta = +1$.

Some values of the coverage probability are given in Table 4.

Table 4. Coverage Probability of the 95% Bayesian Interval for a Normal Mean Known to Lie in $(-\theta_1, \theta_1)$.

θ	−1	−0.5	0	0.5	1
Probability	0.824	0.981	0.999	0.981	0.824

The calculations show that such an interval is quite unsatisfactory from a frequentist point of view.

424

This result does not depend on the choice of $\pi(\theta)$ but is true for *any* continuous prior if θ_1 is sufficiently small. To see this we note that the shortest Bayesian interval must lie inside $(-\theta_1, \theta_1)$ at $\bar{x}=0$. Hence $1-\alpha(\theta_1)$, the frequency coverage probability, must be less than $1-\Phi(-\theta_1)=\Phi(\theta_1)$. Thus provided $\theta_1 < u_\alpha$, $1-\alpha(\theta_1) < 1-\alpha$. Since, by (2), the average coverage probability is $1-\alpha$ it follows that $1-\alpha(\theta_1) > 1-\alpha$ for some other θ_1.

We shall now turn to the case where prior knowledge about θ is expressed by a normal density with mean μ and variance τ^2. Without loss of generality μ may be put equal to zero in which case the posterior distribution of $\tilde{\theta}$ is normal with mean $\xi=\bar{x}\tau^2/(1+\tau^2)$ and variance $\delta^2=\tau^2/(1+\tau^2)$; σ^2/n has again been taken as unity. The value of $\theta_\alpha(x)$ such that

$$\Pr\{\tilde{\theta} \leqslant \theta_\alpha(x) | x\} = \alpha \tag{12}$$

is then $\theta_\alpha(x)=\xi+u_\alpha\delta$. The frequency probability associated with (12) is easily determined since ξ is linear in $\bar{x}$. It is

$$\Pr\{\theta \leqslant \tilde{\theta}_\alpha(x) | \theta\} = 1 - \Phi\{\frac{\theta}{\tau^2} - u_\alpha\delta\}. \tag{13}$$

As we already know, this is only equal to α in the limit as $\tau\to\infty$.

The shortest posterior interval for θ is $\xi \pm u_{\frac{1}{2}\alpha}\delta$ and the associated coverage probability is

$$C_\alpha(\theta) = \Phi\{\frac{\theta}{\tau^2} - u_{\frac{1}{2}\alpha}\delta\} - \Phi\{\frac{\theta}{\tau^2} + u_{\frac{1}{2}\alpha}\delta\}. \tag{14}$$

This probability achieves its maximum at $\theta=0$ and falls away to zero as $\theta\to\pm\infty$. Some numerical values are given in the following table.

Table 5. The Coverage Probability, $C_\alpha(\theta)$, for $\tau=1$, $\alpha=0.2$.

$\pm\theta$	0	0.5	1.0	1.5	2.0	2.5
$C_{0.2}(\theta)$	0.99	0.98	0.94	0.86	0.72	0.53

Our choice of prior density in this case implies that we are fairly confident that θ lies within ±2 of zero yet $\theta=\pm2$ is only included in the interval with probability 0.72 as compared to the nominal level of 0.80. Against this must be set the much higher probability of covering values near the centre of the range which are considered much more likely *a priori*. Nevertheless the essence of the frequentist approach is that it provides a guaranteed coverage probability whatever the true θ. Judged by this criterion the Bayesian method fails.

In spite of this conclusion the Bayesian interval could still be used in a meaningful way by the frequentist. Suppose, for example, that he is sure that θ lies in the interval $(-\theta_1, \theta_1)$ and that he requires his confidence coefficient to be at least $1-\alpha$. He could then construct an interval such that $C(\theta_1)=C(-\theta_1)=1-\alpha$ giving $C(\theta)>1-\alpha$ for $-\theta_1<\theta<\theta_1$. There are two parameters at choice, τ and $u_{\frac{1}{2}\alpha}$ (which determines the confidence coefficient of the Bayesian interval). It is thus possible to satisfy some other requirement simultaneously.

In the following calculations we have chosen to fix $C(0)$ also. Table 6 gives the length of the Bayesian interval satisfying the requirement that $C(\theta_1)$ and $C(0)$ be as indicated in the rows and columns. The second entry in each cell is the confidence coefficient of the standard interval having the same length as the Bayesian interval.

Table 6. Lengths of Bayesian Intervals for the Normal Mean with Specified Values of $C(0)$ and $C(\theta_1)$ (the lower figure in each cell is the confidence coefficient of the standard interval having the same length).

$C(\theta_1)$ / $C(0)$		0.9999	0.9990	0.9900
$\theta_1=2$	0.95	3.96 0.952	3.95 0.952	3.94 0.951
	0.90	3.67 0.933	3.61 0.929	3.51 0.921
$\theta_1=4$	0.95	5.25 0.991	4.94 0.986	4.47 0.974
	0.90	4.98 0.987	4.66 0.980	4.18 0.963

One obvious disadvantage is that in every case the actual coverage probability at the extremes is less than that of a standard interval of the same length. The situation seems more satisfactory for the smaller value of θ_1. Here, for example, if $C(2)=0.95$ one can obtain a guaranteed 0.95 coverage probability over the whole range of interest with much higher values near the middle of the interval. The pay-off from the prior knowledge is reflected not in the length of the interval, which is virtually unchanged, but in the high probabilities of covering values in the region where θ is expected to lie.

The foregoing discussion shows that the frequentist can use Bayesian intervals in a way which makes sense within the terms of his own theory. On the other hand this does not represent the kind of agreement defined by (1). We shall therefore continue the search by relaxing the requirement of a fixed sample size.

5

Agreement Using Sequential Sampling Schemes

The restriction to fixed sample size experiments severely limits the possibility of obtaining agreement. In this section we shall therefore broaden the scope of the enquiry by considering sequential sampling schemes. In Bartholomew (1965) I made the conjecture that it would always be possible to obtain approximate agreement by an appropriate choice of experiment, $\mathscr{E}$, and prior distribution $\pi(\theta)$. We are interested in the conditions under which this conjecture is true and, if so, whether there is any meaningful connection between the appropriate $\mathscr{E}$ and $\pi(\theta)$.

Before producing evidence to support this conjecture let us examine its implications. It could be interpreted to mean that $\mathcal{E}$ and $\pi(\theta)$ were equivalent ways of expressing prior knowledge. Hence a Bayesian having chosen $\pi(\theta)$ ought, if he is to be consistent, to obtain his data by carrying out the corresponding experiment $\mathcal{E}$. By the same token, the choice of a certain $\mathcal{E}$ on the part of the frequentist could be interpreted as an implicit expression of prior beliefs conveyed by the associated $\pi(\theta)$. The acceptability of this view must obviously depend on whether the postulated equivalence of $\pi(\theta)$ and $\mathcal{E}$ is sufficiently clear to be recognized. A full discussion of what this involves would be premature at this stage but one possible requirement having considerable intuitive appeal is the following. If the data confirm our prior expectations we would be satisfied with a smaller sample than if they contradicted them. If the $\mathcal{E}$ satisfying (1) satisfies this requirement we suggest that the interpretation just given has some plausibility.

Some support for the conjecture can be adduced as follows: if we confine ourselves to fixed sample size experiments with independent observations then we know that the result is approximately true (see Section 2). Welch and Peers' (1963) result does not readily generalize to arbitrary sequential experiments. We shall therefore have to be content with showing that some sequential $\mathcal{E}$'s exist satisfying the conjecture.

One such $\mathcal{E}$ was given in Bartholomew (1965). This concerns an exponential life test in which observations are censored at a fixed time. If we agree that prior ignorance of the exponential mean should be represented by the improper prior $\pi(\theta) \propto \theta^{-1}$ then (1) is not true for a fixed sample size experiment. However, a sequential experiment was described in that paper which does satisfy (1). In addition, as I argued there, it seems a more appropriate experiment in that it guarantees a fixed number of failures.

Our next example in support of the conjecture relates to inference about the parameter p in a sequence of Bernoulli trials. A difficulty here is that we are dealing with discrete random variables and exact agreement is not possible. Nevertheless approximate agreement is obtainable and this is sufficient for our purposes. The method is to assume a prior $\pi(p) \propto p^{a-1}(1-p)^{b-1}$ and then to determine a and b so as to give approximate agreement. The first case we shall consider is the fixed sample size experiment; this serves to validate the approximation we shall make but does not add anything to our knowledge. The second case is the sequential experiment in which we continue until a fixed number of successes has occurred. In both cases the posterior density, for r successes out of n trials, is

$$\pi(p|r,n) = \frac{1}{B(r+a, n-r+b)} p^{r+a-1}(1-p)^{n-r+b-1}, \quad (0 \leq p \leq 1). \tag{15}$$

Following Welch and Peers (1963) we shall have agreement if $\int_0^\beta \pi(x|r,n)dx$ is rectangularly distributed over $(0,1)$. Let us denote this integral by $T(r,n|a,b)$ then, by a well-known relation between the incomplete β-function and the binomial series,

$$T(r,n|a,b) = \sum_{i=r+a}^{n+a+b-1} \binom{n+a+b-1}{i} p^i (1-p)^{n+a+b-i-1}. \tag{16}$$

According to David and Johnson (1950), if X is a discrete random variable with distribution function $F(X)$ then $\frac{1}{2}\{F(X)+F(X-1)\}$, and its complement, have a rectangular distribution to a good approximation. If n is fixed in (16) and r is the random variable it follows that

$$S_n = \frac{1}{2}[T(\tilde{r}, n \mid 0, 1) + T(\tilde{r}, n \mid 1, 0)] \tag{17}$$

is approximately rectangular. Since (16) holds only for integer values of a and b we may regard S_{11} as an interpolated approximation to

$$S_n' = T(\tilde{r}, \tilde{n} \mid \tfrac{1}{2}, \tfrac{1}{2})$$

which implies the prior

$$\pi(p) \propto p^{-\frac{1}{2}}(1-p)^{-\frac{1}{2}}. \tag{18}$$

This is the Welch and Peers' result.

Suppose now that we conduct the sequential experiment. The random variable is now n and the approximate rectangular variate is

$$S_r = \frac{1}{2}[T(r, \tilde{n} \mid 0, 0) + T(r, \tilde{n} \mid 0, 1)]. \tag{19}$$

This is approximately equal to

$$S_r' = T(r, \tilde{n} \mid 0, \tfrac{1}{2})$$

giving the prior

$$\pi(p) \propto p^{-1}(1-p)^{-\frac{1}{2}} \tag{20}$$

which is proportional to $\{I(p)\}^{\frac{1}{2}}$ for this experiment. This prior indicates greater ignorance near $p=0$ than does (18) and this seems reasonable because the sequential experiment requires more observations when p is near zero than when p is larger.

Interesting as these results are they all relate to a state of prior ignorance or near ignorance. The difference between (18) and (20) is only as much as would be made by "half an observation". Some work of A. J. Asher's on inference about a normal mean aims to find an $\mathscr{E}$ satisfying (15) when $\pi(\theta)$ is a normal density. Progress so far is encouraging.

Although Welch and Peers' result has not been generalized to cover sequential experiments it is worth noting that Lindley's (1960) agrument based on the concept of information does cover this case. This fact does not bear directly on the frequency interpretation but there may be a deeper connection between the two approaches which carries over to sequential experiments.

The principal conclusions of this investigation may be summarized as follows. First, it is not possible to obtain satisfactory agreement between Bayesian and frequentist inferences with a fixed sample size when prior knowledge is used. Second, agreement can sometimes be achieved by admitting sequential experiments and it has been conjectured that this is always possible – at least approximately. For the frequentist this would mean that when he chooses $\mathscr{E}$ he is giving indirect expression to his prior beliefs and he should be prepared to recognize the fact. The Bayesian, for his part, should be prepared to carry out the experiment determined by his prior. In practice it would be instructive to have a private dialogue between the two points of view in an attempt to harmonize $\mathscr{E}$ and $\pi(\theta)$.

If this point of view were accepted it would have important implications for the analysis of observational as opposed to experimental data. Here the *experiment* has been performed by nature in an unknown way. Hence neither $\mathscr{E}$ nor $\pi(\theta)$ would have a place in the analysis. This would throw us back on the likelihood function as the only basis for inference. In other words, when there is no experiment there can be no meaningful prior and no meaningful sample space. All that remains is the likelihood function.

References

1. Bartholomew, D.J., "A Test of Homogeneity of Means Under Restricted Alternatives," *Journal of the Royal Statistical Society, B,* 23, 239-281, 1961.
2. Bartholomew, D.J., "Contribution to the Discussion of Stein," 1962.
3. Bartholomew, D.J., "A Comparison of Some Bayesian and Frequentist Inferences," *Biometrika,* 52, 19-35, 1965.
4. David, F.N. and Johnson, N.L., "The Probability Integral When the Variable is Discontinuous," *Biometrika,* 37, 42-49, 1950.
5. Jeffreys, H., *Theory of Probability,* 3rd edition, Oxford University Press, 1961.
6. Hartigan, J., "The Asymptotically Unbiassed Prior Distribution," *Annals of Mathematical Statistics,* 36, 1137-1152, 1965.
7. Lindley, D.V., "Fiducial Distribution and Bayes' Theorem," *J. Roy. Statist. Soc., B,* 20; 102-107, 1958.
8. Lindley, D.V., "The Use of Prior Probability Distributions in Statistical Inference and Decisions," *Proceedings of the Fourth Berkeley Symposium on Mathematical Statistics and Probability,* 453-468, 1960.
9. Lindley, D.V., "The Robustness of Interval Estimates," *Bulletin of the International Statist. Inst.,* 38, 209-220, 1961a.
10. Lindley, D.V., "Contribution to the Discussion on Bartholomew," 1961b.
11. Pratt, J.W., "Shorter Confidence Intervals for the Mean of a Normal Distribution with Known Variance," *Ann. Math. Statist.,* 34, 574-586, 1963.
12. Stein, C.M., Confidence Sets for the Mean of a Multi-variate Normal Distribution," *J. Roy. Statist. Soc., B,* 24, 265-296, 1962.
13. Welch, B.L. and Peers, H.W., "On Formulae for Confidence Points Based on Integrals of Weighted Likelihoods," *J. Roy. Statist. Soc., B,* 25, 318-329, 1963.

COMMENTS

G. A. Barnard:

I very much hope Professor Bartholomew will continue with his work, but I would refer to an example I have written about before, concerning Mrs. Haldane and the genetics of cats, which illustrates that our experimental designs are fixed partly by ourselves, partly by Nature, and partly by social factors, including politicians. Thus we can never hope for a completely objectively fixed reference set. Perhaps, therefore, Professor Bartholomew should be satisfied with Bayesian frequentist agreement to order $n^{-\delta}$, $\delta > 0$.

We should note that provided the prior knowledge is derived from observational evidence, classical methods are available for combining it with a given experiment. A prior likelihood may sum up rather vague information. The saying, often heard, that frequentist methods cannot use prior information is not in general true.

R. J. Buehler:

I congratulate Dr. Bartholomew on a very stimulating paper and wish him success in his efforts to harmonize $\mathscr{E}$ and $\pi(\theta)$.

In Section 3 we would probably go beyond the limits envisioned by the author if we were to associate five percent confidence to that portion of a 95 percent interval not contained in a 90 percent interval, but not, of course, 2.5 percent confidence to each segment. By letting the confidence level vary from 0 to 1 we generate a *quasi-fiducial* distribution whose measurable sets are clearly a subclass (depending on x) of the usual Borel sets. Thus the use of the nonstandard two-sided procedure makes many sets nonmeasurable. I do not wish to imply, however, that this carries implications for the main aims of the theory.

The result of Welch and Peers mentioned in Section 2 has been generalized by R. B. Hora and myself (*Ann. Math. Statist.*, 1966). We were able to establish the author's condition (1) in models having a general group invariant structure whenever θ is a parametric function satisfying: $\theta(\omega) = \theta(\omega')$ implies $\theta(g\omega) = \theta(g\omega')$ for all group elements g. Some location-scale examples are: $\theta = \mu + \sigma$; $\theta = \sigma_1/\sigma_2$; $\theta = \mu_1 - \mu_2$, but only when $\sigma_1 = \sigma_2$.

D. R. Cox:

It is arguable that much of the difficulty connected with confidence regions in restricted parameter spaces disappears if one regards them simply as giving parameter values consistent with the data at a given level. Thus it is not surprising if sometimes all parameter values are inconsistent with the data at a particular level, or if, in the problem of the ratio of two means, all values are consistent with the data. There are, of course, difficulties if one thinks of inductive behaviour in the narrow sense; we then have some realized intervals that are either certainly false (or certainly correct).

The correspondence between sampling rule and prior distribution is interesting. I am not clear whether it is being suggested that the physical sampling rule that should be used is related to the prior distribution, or whether it is intended that hypothetical sampling rules should be used. The stopping rule considered by Armitage: stop when $|x_1 + \ldots + x_n| > k\sigma_0 \sqrt{n}$, may be quite reasonable (if not exactly optimum) for some Bayesian decision problems. What then is to be done by someone with a quite different prior distribution wishing to interpret the data so obtained?

V. P. Godambe:

I find Professor Bartholomew's numerical comparisons between Bayes and Frequency probabilities very interesting. When two probability statements,

430

namely, Bayes and Frequency concerning some unknown parameter, agree numerically, for some sampling plan the implication is as follows: The sampling plan provides a frequency validation of the Bayes probability statement even when the presupposed Bayes prior probability distribution (on the parameter space) was unobjective or wrong. In other words, the appropriate sampling plan renders Bayes inference robust. This *robustness* we can study more flexibly by asking the question: Can we mix Bayes and Frequency concepts so that the resulting statistical inference is robust to the variations of Bayes prior distributions and variations of the sampling plans? For interesting illustrations we refer to some recent literature on survey-sampling (see Godambe, 1955, *J. Roy. Statist. Soc.*; or Godambe and Joshi, 1965, *Ann. Math. Statist.*.

I. J. Good:

The notion of *rejecting a prior* on the basis of experimental results is at first a little paradoxical. Like some other paradoxes, it can be resolved by use of rationality of type II. *In principle*, that is under rationality of type I, such rejection cannot occur: you ought to envisage all possible results in advance and adjust your prior accordingly (the device of imaginary results). In practice you cannot do this exhaustively; for even in a sequence of assumed Bernoulli trials you might detect unenvisaged but significant patterns. Thus rationality of type II has to be used and this means that it is possible after all to change a prior in the light of *actual* experimental results.

W.J. Hall:

I wish to register disagreement with some of the previous comments. It seems to me that we are going to have to live with at least two general theories of statistical inference for quite awhile yet and any investigations, such as some of those reported by Dr. Bartholomew, that may provide some kind of interpretation in the framework of one school of what is being done by another, can only help in understanding the various points of view—at least by the vast group of people not fully committed to one particular philosophy.

This is especially true with regard to quantifying prior knowledge. Frequentists surely use prior knowledge too, though they may be reluctant to quantify it.

I will take a minute or two to present a half-baked—no, an ϵ-baked—proposition in this regard. (Professor Good interjected that anything ϵ-baked deserves no more than a minute.) This proposition is directed towards building a bridge between frequentists and Bayesians in quantifying prior information in terms of fictitious prior data. (I don't claim that it's original although the only reference I know is to Novick & Hall, *J. Am. Stat. Ass.*, 1965.)

First, from a frequentist's point of view, let's consider sampling from a normal population. Presumably, the frequentist can summarize his state of knowledge by the triple $(n, \bar{x}, s^2)$, based on the data and with no prior knowledge. If instead he has some prior knowledge, might it not be possible to summarize it by a corresponding triple $(m, \bar{y}, t^2)$, as if he had done a corresponding experiment (with no knowledge prior to that)? His combined (posterior) knowledge would then be represented by the pooled *data* represented by the triple $(M + m, \bar{x} + \bar{y}, \text{ pooled } S_w^2)$. I don't see that this is impossible for a frequentist to accept.

The same formalism may also be acceptable to a Bayesian, though his interpretations and uses of the various triples may be different. If he can resolve what is meant by no prior knowledge, he can construct a prior distribution by combining a non-informative prior with the fictitious prior data (m, $\bar{y}$, t^2) in the usual Bayesian way. Bayes Theorem applied once again will yield the same posterior triple as attributed above to the frequentist. (The Bayesian also assures himself of the convenience of conjugate priors.)

A final note: I am not convinced that Dr. Bartholomew is taking account of the difficulties inherent in lengths of confidence intervals. We want them short only when correct, and otherwise we want them long.

J. A. Hartigan:

The Bayes and frequency intervals agree sometimes with improper priors, but do they ever with proper priors? Dr. Bartholomew has answered this question negatively with location parameters and I have a simple argument to suggest exact agreement is never possible.

We need $P(\theta < \theta_\alpha(x)|\theta) = \alpha$ all θ.

Assume $\theta_\alpha(x)$ has positive density over line, all θ.

Then $P(\theta < \theta_\alpha(x) \mid \theta, \theta_\alpha(x) > 0) = \alpha(\theta) > \alpha$.

So $P(\theta < \theta_\alpha(x) \mid \theta_\alpha(x) > 0) > \alpha$ averaging over θ.

So $P(\theta < \theta_\alpha(x) \mid x) = \alpha$ is impossible.

O. Kempthorne:

I think Dr. Bartholomew is asking questions that are interesting to those of us who are not committed to any ideology. In this respect, I find the writings of Dr. Pratt informative also, though there are some aspects of his writings which on quick (but not sustained) reflection bother me in a non-negligible way. I also wish to express the view that the comments of Dr. Cox are "bang on". (I understand "right on" is the "in" phrase in the U.S.A.)

To add to my oral discussion, I suspect that the upshot of investigations like Dr. Bartholomew's investigations will be that Bayesian ideas result in so-called non-informative priors doing great injustice to the data. Perhaps some of the remarks in my own essay are relevant.

John W. Pratt:

I found this an interesting paper. Some of Dr. Bartholomew's earlier work has a lot of appeal for me, particularly as it points up the way conditional significance levels ought to vary, which implies difficulties for interpreting P values. But here we are being asked to distort our designs to obtain a numerical kind of agreement with those of a different philosophical persuasion. I wouldn't think that this would really appeal to those of any persuasion.

I would like to say a few words about what I thought I was doing in my work on the length of confidence intervals, both in the paper mentioned here and an earlier paper in *J. Am. Stat. Ass.*, since this has come in for attack before and may again. One point of this work was to clarify ideas about the intuitive concept of length of confidence intervals, which had received little attention. It was prompted by Kendall and Stuart's Sections 20.16-19 and, I believe, advanced our understanding of various questions in the area. In particular, it related Neyman's concept of shortest and the intuitive concept of length, which seem entirely unrelated. Another point of the work was to show that frequentists' theoretical justifications can't be taken seriously as they wouldn't really use procedures which minimize the weighted average of the expected length, although these procedures are entirely valid and seem desirable according to their theory. I am not advocating the use of these procedures, but would like frequentist theoreticians to explain why they shouldn't be used.

The shortcoming Dr. Bartholomew mentions in the last two paragraphs of Section 3 is not clear to me. The method gives the lower bound on expected length; no method can do better. Is the shortcoming one of the method or the problem or what?

The case where θ is known to lie in a very short interval is interesting. Here it is clear that the sample has provided essentially no information. If you insist on covering each value of θ some of the time, it seems clear that a ridiculous procedure is almost bound to arise, with confidence intervals which are sometimes empty, and/or cover the whole possible range, or even fail to be intervals.

REPLY

I am most grateful to the contributors for their interesting and useful comments. As far as possible I shall answer the questions raised but in many cases my reply must be an interim one which may need modification after further reflection.

It may help to preface my detailed replies with a general observation about the relationship between the theory and practice of inference. Few of us practice what we preach down to the last detail. We bend the rules and modify our principles a little when common sense seems to demand it. Nevertheless the *pure* theory is important as a point of reference for our intuitions and a general guide for practical conduct. It is therefore inevitable that one sounds more dogmatic than one would wish when expounding ideas within the framework of a particular theory of inference. I therefore welcome comments such as those of Professors Cox and Barnard which draw attention to the limitations of such a position.

Professor Pratt is right in thinking that we are being asked to distort our designs to obtain agreement but what he calls distortion I would call correction. My ultimate intention is to persuade the Bayesian that the experiments required to yield agreement are better than those he might have used without the benefit of my advice! The shortcoming queried by Professor Pratt at the end of Section 3 was intended as one of method. I am saying that if this is the best one can do with a confidence interval then it is not good enough.

I fully accept Professor Cox's point that sensible interpretations can be put upon confidence intervals even if they can be recognized to include none (or all) of the possible parameter values. His second point is an important one. I am suggesting that the sampling rule should be physical rather than hypothetical. I would then say that data can only be meaningfully interpreted by a person whose prior knowledge convinces him that the sampling rule used was the *right* one.

Professor Godambe's idea of using frequency properties to establish the robustness of Bayesian procedures seems, to me, a potentially fruitful one and I look forward to seeing further developments.

I am grateful for Professor Buehler's reference to his own work and for the thought about quasi-fiducial distributions derived from the approach of Section 3. The latter is something on which I am not ready to offer an opinion.

Professor Good's remark relates to a point which was amplified in the verbal presentation of the paper. I drew attention to the fact that a Bayesian confidence interval could have a very small frequency coverage probability if the true value of θ was a long way from our prior expectation. In these circumstances our data will tell us that the prior information was wrong; yet, the logic of the Bayesian method allows us to go on and incorporate this apparently erroneous prior into the inference. Good argues that we should have foreseen this possibility and adjusted the prior accordingly. I am suspicious of this solution because it seems to me to lead us to dilute our prior knowledge to a point where it is hardly distinguishable from prior ignorance. I am quite unable to see what an 'adjusted' prior might look like. His justification for adjusting the prior after the data have been obtained appears to demolish whatever credibility the prior knowledge might have had.

I have often thought that Professor Hall's idea for quantifying prior knowledge in terms of fictitious prior data was attractive. It is a practice which I would heartily commend to convinced Bayesians. From a frequentist point of view I am more doubtful. The cynic would say that dressing up subjective introspection to look like real data is nothing more than sophisticated dishonesty. The purist would object that the proposal does nothing to make the probability statements involved empirically verifiable.

Professor Hartigan's point is interesting and far-reaching in its implications. I take the point to be as follows: if the prior is proper, there is a sub-set of outcomes of repetitions of the experiment which can be recognized to have a probability different to α. The random variable $\tilde{\theta}_\alpha$ (x) is almost certain to lie in this sub-set for sufficiently large θ. Hence for sufficiently large θ the required probability exceeds α and so exact agreement is impossible. I should be interested to see a rigorous proof of this result. For practical purposes, of course, approximative agreement will suffice.

THE ESTIMATION OF
MANY PARAMETERS
D. V. Lindley
University College, London

Summary

This paper is concerned with estimation problems where there are several parameters all of the same type — for example, a set of means. It often happens that the knowledge of the parameters is exchangeable in the sense of de Finetti. This prior knowledge can be used to obtain estimates alternative to the usual ones. The particular problem studied here is the familiar analysis between- and within-groups. New estimates of the group means are obtained, and when the assumption of equality of variances within groups is omitted, new estimates of the individual variances are also derived. The paper begins with some introductory comments on Bayesian ideas.

1
Introduction

The present paper is devoted to the formulation and solution of a technical problem in statistics by means of Bayesian methods. It may be felt that at a conference concerned with principles of inference a more philosophical or fundamental paper might be more appropriate. In this introduction I should like to explain why I have not offered such a paper. There are two reasons. First, it seems to me that the major problem of inference has been solved in its essentials by Bayesian ideas. Second, I feel there is an important role to be played in assessing the merits of different philosophies by using their principles and seeing how they work out in practice. This paper is a working out of Bayesian methods in a familiar situation. In my view they work out rather well — others may think differently. This we might discuss.

If the second reason is easily substantiated, the first needs more consideration. Why is Bayesian inference correct? In the context of a decision problem the principle of consistency or coherence clearly demonstrates that

435

decision-making must be based on a Bayesian analysis, using a (prior) distribution and a utility function, and selecting the optimum decision by maximizing expected utility. There are numerous justifications for this; the one that appeals most to me is that given by Pratt, Raiffa and Schlaifer (1964); perhaps the best-known is that of Savage (1954). No substantial counterargument is known to me. Coherence is such a simple, basic requirement that it is difficult to see how it could be denied as a principle. One reply has been to say that it is better to be incoherent but right sometimes than to be coherently wrong. But if one of the incoherent decisions was right — though I am not sure what right means in this context — coherence would make the others right too. No decision-maker who discovers himself being incoherent would fail to correct himself. Unfortunately most of us never notice our incoherence.

Granted the relevance of the Bayesian method to decision-making, what has this to do with inference? The basic reason why we make inferences is that we want them to help our decision-making. The maker of the inference need have no particular decision in mind — a laboratory scientist need not be concerned with the uses of his knowledge — but for his inference to be of interest it must ultimately assist in decision-making. Now any decision that depends on the data that is being used in making the inference only requires from the data the posterior distribution. Consequently the problem of inference is effectively solved by stating the posterior distribution.

This is the reason why I feel that the basic problem of inference is solved. There are those who advocate the use of methods based only on the likelihood function. This I would not quarrel with, for it is through the likelihood that data enters into Bayes's theorem. But in many situations the use of the likelihood alone seems an unattainable ideal because of the number of nuisance parameters involved. I see no justification for marginal, conditional or other quasi-likelihoods that are not derived from an honest use of integration with respect to a prior distribution.

It is methods that are not based on the likelihood function that are suspect. In particular, unbiased estimates, minimum variance properties, sampling distributions, significance levels, power, all depend on something more — something that is irrelevant in Bayesian inference — namely the sample space. Bayesian methods are often criticized for their dependence on an "arbitrary" prior. It is not usually appreciated that other methods depend on an "arbitrary" sample space. A statistician faced with some data often imbeds it in a family of possible data that is just as much a product of his fantasy as is a prior distribution. A good example occurred recently in a paper of Edwards (1970). The data here are the distributions of human blood groups in the world at the present day. What repetitions of this experiment are envisaged to provide a sample space?

The remainder of this paper is devoted to the study of a situation in which there is substantial prior knowledge, so that its use should significantly improve on the usual estimation procedures. In it I have used the coherence principle, allied to a choice of distributions that is analytically tractable. Thus if a prior leads to an unacceptable posterior then I modify it to cohere with properties that seem desirable in the inference. It may well be that some of the results do not cohere with other considerations that have escaped my attention. If so, I should be glad to have them pointed out, for coherence is basic.

2

Variances Known

The type of problem to be discussed in this paper is one in which there is a substantial amount of data whose probability structure depends on several parameters of the same type. For example, an agricultural trial involving many varieties, the parameters being the varietal means, or an educational test performed on many subjects, with their true scores as the unknowns. In both these situations the parameters are related, in one case by the common circumstances of the trial, in the other by the test used so that a Bayesian solution, which is capable of including such prior feelings of relationship, promises to show improvements over the usual techniques.

More precisely, let data x_{ij} ($i = 1, 2, \ldots m$; $j = 1, 2, \ldots n_i$) be independent, given θ_i ($i = 1, 2, \ldots m$) and σ_W^2, and normally distributed, x_{ij} having mean θ_i and variance σ_W^2. This is the situation usually referred to as analysis of variance between and within groups. In either of the practical circumstances described above, the means θ_i will be thought to be alike and it might therefore be appropriate to assume that, given μ and σ_B^2, the θ_i are independent and normally distributed with common mean μ and variance σ_B^2. (The suffixes on the two variances serve to remind the reader that they are usually called within- and between-variance components respectively.) With the additional structure assumed for the means, the problem has the structure of a Model II, components of variance, situation. There are, however, two main differences between our approach and that usual in Model II analyses. First, we are primarily interested in the means, θ_i, and not in the variance components, σ_W^2 and σ_B^2. Second, our attitude towards the distributional assumptions concerning the means is somewhat different from the usual one, and we pause now to discuss this point.

The point of view adopted here is that the data, $\{x_{ij}\}$, having the structure described above, dependent on parameters, $\{\theta_i\}$ and σ_W^2, imply that it is necessary to express our prior opinions concerning these. In experiments of the two types mentioned above it often seems reasonable to say that there is the same amount of information about each of the means: that is, *a priori*, there is no reason to think that one candidate or variety is superior to the rest. Consequently the prior distribution of the means must have the property that de Finetti (1964) has called exchangeable; that is, the joint distribution of the means remains invariant under a permutation of the suffixes. A famous result of de Finetti, later generalized by Hewitt and Savage (1955), says that exchangeability implies that the θ's have the probability structure of a random sample from a distribution. If we add the additional assumption that this distribution is normal — for the usual reasons associated with a normality assumption in modern statistics — we have the prior distribution described above. It seems to me conceptually important to realize that only the symmetry of our knowledge of the means is being assumed here, and that there is no need for the concept of a hypothetical infinite population from which the means are being sampled. It is the subjective assessment of exchangeability that is fundamental, and not the frequentist sampling result which is derived from it. The idea is, of course, not new, but its implementation in this context may be.

A further point that is relevant in considering the prior distribution of the means is that the point of view taken here does not distinguish sharply between Model I and Model II analyses. I feel that genuine Model I situations, where the assumption of exchangeability (or some modified form of it) does not apply, occur rarely and that most Model I problems arise when the prior knowledge is sparse, that is, when σ_B^2 is large. In any case the distinction between the two models is one between the types of prior that are appropriate.

To complete the specification of the prior distribution it is necessary to discuss μ, σ_W^2 and σ_B^2. In this part of the paper it is supposed that the two variances are known and that the prior knowledge of μ is weak, so that, over the range for which the likelihood is appreciable, the prior density of μ is constant.

This completes the Bayesian specification of the problem and we can proceed to the posterior distributions. Before doing this it is interesting to make a remark concerning the orthodox estimation of the means using a model that only incorporates the normal distribution of the x's given the θ's and σ_W^2. The usual estimate of θ_i is the sample mean, $x_{i.} = \sum_j x_{ij}/n_i$. However Stein (1956, 1962) has shown that such estimates are inadmissible, provided the number of populations, m, is greater than two. Consequently the usual Model I analysis is open to criticism. The estimates proposed in this paper would appear to avoid the difficulty.

Another point worth noting is that if the model is rewritten, $E(x_{ij}) = \theta_i$ with var $(x_{ij}) = \sigma_W^2$, then it fits within the usual least-squares framework and the estimates $x_{i.}$ are the least-squares estimates. There is a Bayesian justification of the least-squares procedure: see, for example, Lindley (1965, §8.3). However it requires both an assumption of normality for the x's and an assumption of little prior knowledge of the means – in the notation of this paper, σ_B^2 is large. It is sometimes held that these two assumptions make the Bayesian approach more restrictive than the classical least-squares one. However, it should be noted that the latter requires two assumptions not needed in the Bayesian analysis, namely, a restriction to unbiased estimates (or some similar restriction to reduce the class of estimates being considered) and a requirement of minimization of variance (or mean-square error). The former restriction is a severe one and hard to justify. It leads to the inadmissible estimates already mentioned. It is possible to formulate a more complete Bayesian approach to least-squares that is not restricted to weak prior knowledge, and the analysis in the first part of this paper is a special case of it with $E(\theta_i) = \mu$ and var $(\theta_i) = \sigma_B^2$. Details have been given elsewhere, Lindley (1969).

3

The assumptions above provide the following density for the data and parameters of the problem, namely

$$p(\mathbf{x}, \theta, \mu) \propto \exp\left[-\sum_{ij}(x_{ij} - \theta_i)^2/2\sigma_W^2 - \sum_i(\theta_i - \mu)^2/2\sigma_B^2\right], \qquad (1)$$

supposing σ_B^2 and σ_W^2 to be known. The integration with respect to μ is easily performed, and the first summation rearranged, with the result that

$$p(\mathbf{x}, \theta) \propto \exp\left[-\sum_i n_i (x_{i.} - \theta_i)^2/2\sigma_W^2 - \sum_i (\theta_i - \theta_.)^2/2\sigma_B^2\right]$$

where $\theta_. = \sum\theta_i/m$. This is proportional to the posterior distribution of the θ's, which is easily seen to be the exponential of a quadratic form in the θ's and therefore multivariate normal. The precise form is

$$p(\theta|\mathbf{x}) \propto \exp\left[-\{\tfrac{1}{2}\{\sum(a_i + b)(\theta_i - t_i)^2 + b\sum_{i \neq j}(\theta_i - t_i)(\theta_j - t_j)\}\right] \qquad (2)$$

where

$$t_i = \frac{x_{i.}n_i\sigma_B^2 + \bar{x}\sigma_W^2}{n_i\sigma_B^2 + \sigma_W^2}, \qquad (3)$$

$$\bar{x} = \sum w_i x_{i.}/\sum w_i, \qquad (4)$$

$$w_i = n_i/(n_i\sigma_B^2 + \sigma_W^2), \qquad (5)$$

$$a_i = (n_i\sigma_B^2 + \sigma_W^2)/\sigma_B^2\sigma_W^2,$$

and

$$b = -1/m\sigma_B^2.$$

In particular the posterior distribution has means given by (3). These are of the form of a weighted average of the mean x_i (the least-squares estimate) and an overall mean $\bar{x}$, (4), the weights, (5), depending in a natural way on the sample sizes and the ratio of the two variance components. The effect of this weighted average is to shift all the estimates for the sample means towards the overall mean. These estimates have a form similar to those proposed by Stein, designed to avoid the inadmissibility already referred to, and generalize an earlier result of mine for the case of equal n_i (see the discussion to Stein, 1962).

From the expressions for a_i and b given above it is possible to verify that the posterior variance of θ_i is

$$\frac{\sigma_B^2\sigma_W^2}{n_i\sigma_B^2 + \sigma_W^2} + \frac{\sigma_W^4}{(n_i\sigma_B^2 + \sigma_W^2)^2\sum w_s} \qquad (6)$$

and it is easy to see that this is less than the sampling variance (equal to the posterior variance under the Bayesian substitute for the usual least-squares assumptions) of the usual estimate $x_{i.}$, namely σ_W^2/n_i. Naturally the incorporation of the prior information has resulted in an improvement in the estimation procedure.

The usual least-squares analysis is unaffected in essentials if the variances are all known up to an unknown constant; in this case, if the ratio σ_W^2/σ_B^2 is known. The same remark holds for the analysis just presented, as is easily appreciated from the fact that the estimates (3) do not depend on knowledge of both variances but only on their ratio. The situation is different if both variances, σ_B^2 and σ_W^2, are unknown. The problem then becomes a degree more difficult.

4

Group Variances Unknown But Equal

Consider now the situation exactly as in the last section but with σ_B^2 and σ_W^2 both unknown. It is then necessary to assign a prior distribution to them. The usual prior for a variance about which little is known has density inversely proportional to the variance. This is perfectly sensible here for σ_W^2 but there are severe difficulties if this is used for σ_B^2. The point has been discussed by Novick (1969). The difficulty can be viewed mathematically by remarking that if a prior proportional to σ_B^{-2} that is, which is improper, not integrating to one – is used, then the posterior remains improper whatever size of sample is taken. Heuristically it can be seen that the between-sample variance provides information directly about $\sigma_B^2 + \sigma_W^2/n_i$ — that is, confounded with σ_W^2 — and not about σ_B^2 itself, so that the extreme form of the prior cannot be overcome by sampling.

Box and Tiao in a series of papers (the 1968 reference is particularly relevant to the present discussion) consider the case where the sample sizes n_i are all equal to n and use a density inversely proportional to $(\sigma_B^2 + \sigma_W^2/n)$, but this has two difficulties despite the analytical convenience that it undoubtedly offers. First, it does not obviously extend to the case of unequal sample sizes; second, and more seriously, it depends on n, the sample size. In most applications I can see no reason for thinking that the experimenter's prior opinion is affected by the amount of data that he has. It may sometimes happen that the prior information influences the choice of sample size but in this situation it is by no means clear that the connection is of the form implicitly implied by this prior distribution.

As an alternative procedure we use the conjugate priors for σ_B^2 and σ_W^2, namely inverse χ^2-distributions. Specifically we suppose $\nu_B\lambda_B/\sigma_B^2$ is χ^2 with ν_B degrees of freedom, and $\nu_W\lambda_W/\sigma_W^2$ is χ^2 with ν_W degrees of freedom, the distributions being supposed independent. There is a discussion of inverse χ^2 in Lindley (1965, §5.3). Essentially the λ's describe the values one expects the variances to have *a priori*, and the degrees of freedom measure the sureness of this prior conviction. ($\nu \to 0$ provides the usual prior proportional to σ^{-2}.) It would be possible to allow ν_W to be zero, but the general value is retained for convenience. As explained above ν_B cannot be allowed to approach zero.

The joint distribution that now replaces (1) is given by

$$p(\mathbf{x},\theta,\mu,\sigma_B^2,\sigma_W^2) \propto \sigma_W^{-(N+\nu_n+2)} \sigma_B^{-(m+\nu_n+2)} \exp\{-\tfrac{1}{2}[\Sigma_{ij}(x_{ij}-\theta_i)^2 + \nu_W\lambda_W]/\sigma_W^2\}$$

$$\times \exp\{-\tfrac{1}{2}[\Sigma(\theta_i-\mu)^2 + \nu_B\lambda_B]/\sigma_B^2\}, \qquad (7)$$

where $N = \Sigma n_i$. Again the integration with respect to μ is straightforward and has the effects of replacing $\Sigma(\theta_i-\mu)^2$ in the second exponential by $\Sigma(\theta_i-\theta)^2$ and multiplying the whole expression by σ_B. The sum in the first integral is conveniently simplified by rewriting it in the form

$$S^2 + \Sigma(x_{i.} - \theta_i)^2 n_i + \nu_W \lambda_W,$$

where

$$S^2 = \sum_{i,j}(x_{ij} - x_{i.})^2 ,$$

the within sum of squares. The integrations with respect to σ_W^2 and σ_B^2 are also straightforward using the result that

$$\int_0^\infty \frac{1}{(\sigma^2)^k} \exp\left(-\frac{A}{\sigma^2}\right) d\sigma^2 \propto A^{-(k-1)}.$$

Consequently the posterior distribution of the θ's, $p(\theta \mid \mathbf{x})$, is proportional to

$$\{S^2 + \nu_W \lambda_W + \Sigma n_i(x_{i.} - \theta_i)^2\}^{-\frac{1}{2}(N+\nu_W)} \times \{\nu_B \lambda_B + \Sigma(\theta_i - \theta.)^2\}^{-\frac{1}{2}(m+\nu_B-1)} \quad (8)$$

This is easily seen to be the product of two multivariate t-distributions. Such distributions have been discussed by Tiao and Zellner (1964) using methods based on expanding in inverse power series of the new degrees of freedom, $N + \nu_W$ and $m + \nu_B$, developed originally by Fisher. Using their methods it is possible to find the means and second-order moments of the θ's. However a simpler, though less accurate, approach is possible.

The mode of the posterior distribution can be found by differentiating (8) — or more easily, its logarithm — and equating the result to zero. This avoids the necessity of evaluating the missing constant of proportionality. The result is that at the mode the θ_i satisfy the equations

$$\frac{(N+\nu_W)n_i(\theta_i - x_{i.})}{S^2 + \nu_W \lambda_W + \Sigma n_s(\theta_s - x_{s.})^2} + \frac{(m+\nu_B-1)(\theta_i - \theta.)}{\nu_B \lambda_B + \Sigma(\theta_s - \theta.)^2} = 0. \quad (9)$$

Write

$$s_W^2 = \{S^2 + \nu_W \lambda_W + \Sigma n_s(\theta_s - x_{s.})^2\}/(N+\nu_W) \quad (10)$$

and

$$s_B^2 = \{\nu_B \lambda_B + \Sigma(\theta_s - \theta.)^2\}/(m+\nu_B-1), \quad (11)$$

then after a little rearrangement (9) may be rewritten

$$\theta_i = \frac{x_{i.}\, n_i\, s_B^2 + \bar{x}'\, s_W^2}{n_i s_B^2 + s_W^2} \quad (12)$$

where

$$\bar{x}' = \Sigma w_i' x_{i.}/\Sigma w_i' \quad (13)$$

and w_i' is defined exactly as (5) but with s_B^2 and s_W^2 replacing σ_B^2 and σ_W^2. The modal values given by (12) have the same form as the modal values — equal to the means — in the case where σ_B^2 and σ_W^2 were known (3) except again that the s-values replace the σ's. Now s_W^2 and s_B^2 are easily seen to be reasonable estimates of σ_W^2 and σ_B^2 respectively, so consequently we have a Bayesian justification for the common classical procedure of proceeding by replacing the true variances by suitable estimates. Of course, equations (9), or (12), cannot be solved directly — in the latter the notation hides the dependences of s_W^2 and s_B^2 on the θ's — but various iterative procedures are possible. These are currently being studied by Novick (personal communication).

It is not easy to obtain the variances of the posterior distribution, and even the second derivatives of the logarithm of the distribution at the mode are not simple expressions. However, it seems eminently reasonable to estimate the variances by taking the formula when σ_W^2 and σ_B^2 are known, (6), and replacing these two values by s_W^2 and s_B^2 (and w_s by w_s').

The form of the posterior distribution, (8), is interesting; the first multivariate - t has its maximum at the usual estimates $\theta_i = x_{i.}$, the second has maxima at all points where the θ_i are equal. Consequently the effect of the second t-distribution is to displace the first from $x_{i.}$ to a position where the θ's are more nearly equal. This is supported by the equation, (12), for the modes where the estimates are shifted towards the overall mean $\bar{x}'$.

Notice, that without the term $\nu_B\lambda_B$ derived from the prior distribution of σ_B^2, the second factor in (12) is singular, tending to infinity as the θ's become equal. This illustrates the discussion given above, where the effect of an improper prior for σ_B^2 (corresponding to $\nu_B = 0$) was explained. With any but very small samples the exact value of ν_B, provided it is small, will not matter. In any case, with very small samples the equation (12) may not have solutions and a more delicate study of the posterior distribution is indicated. The difficulties over the prior for σ_B^2 reflect a genuine inadequacy in the experimental design, the data being incapable of providing direct evidence about σ_B^2 unconfounded with that for σ_W^2. These points have been more fully discussed by Hill (1965) in connection with the estimation of σ_B^2 and σ_W^2.

5
All Variances Unknown

The situation discussed so far in this paper can alternatively be described by saying there are m samples from normal distributions with a common variance, the distinctive feature being the assumed prior knowledge of the exchangeability of the means. The assumption of a common variance, σ_W^2, is usually made for convenience, rather than because it necessarily occurs in practice, the corresponding Fisher-Behrens's type of argument when the assumption is dropped being considerably more difficult. We therefore consider the more general situation of m samples from normal distributions but suppose that, like the means, the variances are also exchangeable; that is, such information as we have about them is symmetric. The variances therefore, form a random sample from some distribution and it only remains to consider the form this takes. The simplest analytically, is the conjugate family, again the inverted χ^2, so this is assumed.

The model may conveniently be described in three stages.

Stage 1. Given $\{\theta_i\}$, $\{\phi_i\}$ the data x_{ij} are independent and normally distributed with $E(x_{ij}) = \theta_i$ and $var(x_{ij}) = \phi_i$ ($i = 1, 2, \ldots m : j = 1, 2, \ldots n_i$).

Stage 2. (Exchangeable). Given μ, τ, ν, σ^2 the $\{\theta_i\}$, $\{\phi_i\}$ are independent. The θ's are normal with mean μ and variance τ. Also $\nu\sigma^2/\phi_i$ is χ^2 with ν degrees of freedom.

442

Stage 3. A priori μ is supposed uniformly distributed and $\nu_0\sigma_0^2/\tau$ distributed as χ^2 with ν_0 degrees of freedom. The prior distribution for ν and σ^2 will be discussed below.

The first stage describes the immediate generation of the data — giving what is usually called the likelihood function. The second stage deals with the exchangeability property. Notice that it has been assumed that the means and variances are independent. There may be situations where this is known not to be true as, for example, when the coefficient of variations $\phi_i^{\frac{1}{2}}/\theta_i$ are stable. It is hoped to study this situation elsewhere. The third stage deals with the prior knowledge of the exchangeable distributions. For the parameters in the normal distribution we have taken the usual conjugate priors with vague prior knowledge on the mean. Vague prior on τ would imply $\nu_0 = 0$, but remembering the difficulties we had above with σ_B^2 (now called τ) it is best to retain $\nu_0 > 0$ for the moment. The prior on ν and σ^2 present novel features to be considered below. Notice another notational change, ϕ_i instead σ_W^2, the common within variance.

In a rigorous probabilistic description of the situation the conditional probability statements are made in the reverse order to that given above, starting with stage 3. Symbolically the joint distribution of all quantities is given by

$$\prod_{i,j} p(\mu, \tau, \nu, \sigma^2) p(\phi_i \mid \nu, \sigma^2) p(\theta_i \mid \mu, \tau) p(x_{ij} \mid \theta_i, \phi_i).$$

Inserting the assumed distributions we obtain a result proportional to

$$\prod_i \frac{1}{\phi_i^{\frac{1}{2}n_i}} \exp\left\{-\tfrac{1}{2}\sum_j (x_{ij} - \theta_i)^2/\phi_i\right\}$$

$$\times \frac{1}{\tau^{\frac{1}{2}m}} \exp\left\{-\tfrac{1}{2}\Sigma(\theta_i - \mu)^2/\tau\right\}$$

$$\times \frac{(\nu\sigma^2)^{\frac{1}{2}\nu m}}{2^{\frac{1}{2}\nu m}\{(\tfrac{1}{2}\nu - 1)!\}^m} \exp\left\{-\tfrac{1}{2}\nu\sigma^2\Sigma\frac{1}{\phi_i}\right\}\frac{1}{\Pi\phi_i^{\frac{1}{2}\nu+1}}$$

$$\times \frac{1}{\tau^{\frac{1}{2}\nu_0+1}} \exp\left\{-\tfrac{1}{2}\nu_0\sigma_0^2/\tau\right\} p(\nu, \sigma^2). \tag{14}$$

We require the joint distribution of the θ's and ϕ's, given the data $\{x_{ij}\}$, and to obtain this (14) has to be integrated with respect to the remaining parameters μ, τ, ν and σ^2. Integrations with respect to μ and τ are straightforward, proceeding as in the earlier case. The effect of these is to replace the second and fourth lines of (14) by

$$\{\Sigma(\theta_i - \theta_\cdot)^2 + \nu_0\sigma_0^2\}^{-\frac{1}{2}(m+\nu_0-1)} p(\nu, \sigma^2), \tag{15}$$

the first and third lines remaining the same.

Consider next the integration with respect to σ^2. Before this can be done we have to specify the form of $p(\sigma^2 \mid \nu)$ and it would undoubtedly be simplest to take this inversely proportional to σ^2 as is usual with a variance. However, for what are essentially the same reasons as those discussed above in connexion with σ_B^2, this gives unsatisfactory answers. We therefore take a conjugate distribution to (14), which, since ϕ_i is inverted χ^2, is direct χ^2. Precisely we suppose

$$p(\sigma^2 \mid \nu) \propto (\sigma^2)^\kappa \exp \{-\tfrac{1}{2}\nu\sigma^2\lambda\}, \tag{16}$$

where λ and κ are two constants describing the prior (as ν_0 and σ_0^2 do for τ). Combining this with the third line of (14) gives

$$\frac{(\tfrac{1}{2}\nu)^{\frac{1}{2}\nu m} p(\nu)}{\{(\tfrac{1}{2}\nu - 1)!\}^m} \exp\left\{-\tfrac{1}{2}\nu\sigma^2\left(\frac{m}{\theta}+\lambda\right)\right\} \frac{(\sigma^2)^{\frac{1}{2}\nu m + \kappa}}{\eta^{(\frac{1}{2}\nu + 1)m}}, \tag{17}$$

where

$$m/\theta = \Sigma\phi_i^{-1},$$

and

$$\eta^m = \Pi\phi_i,$$

so that θ and η are respectively the harmonic and geometric means of the ϕ's. (These two quantities arise because the two means form a set of sufficient statistics for ν and σ^2 in the inverted χ^2-distribution.)

The integration of (17) is easy and the result is, after a little simplification,

$$\frac{p(\nu)}{m^{\frac{1}{2}\nu m + \kappa + 1}} \frac{(\tfrac{1}{2}\nu m + \kappa)!}{\{(\tfrac{1}{2}\nu - 1)!\}^m} \frac{1}{(\tfrac{1}{2}\nu)^{\kappa + 1}} \frac{1}{(\theta^{-1} + \lambda/m)^{\frac{1}{2}\nu m + \kappa + 1}} \frac{1}{\eta^{(\frac{1}{2}\nu + 1)m}}. \tag{18}$$

The final task is the integration of this with respect to ν. (Although ν is naturally thought of as degrees of freedom, mathematically there is no reason why it should not be any positive number.) Exact evaluation does not seem possible, so we first use Stirling's formula to replace the factorials in (18). Although this formula is asymptotic as $\nu \to \infty$ it is remarkably accurate even for ν as low as 2 so that, provided $p(\nu)$ is not large there, the result should be reasonable. The result of the replacement is, after some algebra,

$$\frac{1}{\eta^m} \cdot \frac{1}{(\theta^{-1} + \lambda/m)^{\kappa + 1}} (\tfrac{1}{2}\nu)^{\frac{1}{2}(m-1)} \exp\{-\tfrac{1}{2}\nu m \log \eta(\theta^{-1} + \lambda/m)\} p(\nu).$$

If $p(\nu)$ is not too far from uniform, the integration with respect to ν is simple, being just a gamma integral, and the result is

$$\eta^{-m}(\theta^{-1} + \lambda/m)^{-(\kappa+1)}\{\log \eta(\theta^{-1} + \lambda/m)\}^{-\frac{1}{2}(m+1)}, \tag{19}$$

apart from constants.

Collecting together the results from (14), (15) and (19), the joint posterior density of the θ's and ϕ's is

$$\Pi_i \frac{1}{\phi_i^{\frac{1}{2}n_i + 1}} \exp\{-\tfrac{1}{2}[S_i^2 + n_i(x_{i.} - \theta_i)^2]/\phi_i\}$$

$$\times \{\Sigma(\theta_i - \theta_.)^2 + \nu_0\sigma_0^2\}^{-\frac{1}{2}(m + \nu_0 - 1)}$$

$$\times (\theta^{-1} + \lambda/m)^{-(\kappa+1)}\{\log \eta(\theta^{-1} + \lambda/m)\}^{-\frac{1}{2}(m+1)}, \tag{20}$$

where $S_i^2 = \sum_j (x_{ij} - x_{i.})^2$ is the usual sum of squares within the ith sample.

The conditional density of the means, θ_i, given the variances, ϕ_i, involves only the first two lines of (20). The first provides a normal density with mode at the sample means $x_{i.}$. The second gives a multivariate-t with modes where the means are all equal. Consequently the total density effects a compromise, exactly as in the case of a common within-variance σ_W^2, (8), between the sample means and some common value. The effect of the exchangeability is still to shift the usual estimates towards some overall mean.

The conditional density of the variances, ϕ_i, given the means, θ_i, involves only the first and last lines of (20). The first provides the usual inverted χ^2-distribution with mode involving $\sum_j (x_{ij} - \theta_i)^2$, or S_i^2 if $\theta_i = x_{i.}$, although the divisor is $(n_i + 2)$ instead of the usual n_i due to a contribution from the η^{-m} term in (19). The last line is of a novel form. It involves only the harmonic and geometric means of the variances. Consider it for fixed value of η. Since the harmonic mean, θ, never exceeds the geometric mean, η, it easily follows that the last line has its maximum when $\theta = \eta$, which only occurs when all the ϕ_i are equal. Variation over η is not important since although the maximum occurs as η, the common value of the ϕ_i, tends to zero, the exponential in the first line swamps any tendency to divergence. Consequently the mode of the conditional distribution of the variances will be a compromise between the usual estimates based on S_i^2 and a tendency to make all the variances equal. Hence, exactly as with the means, the effect of the exchangeability on the estimates of variance is to shift them towards a common value.

It is now possible to see the effect of the assumed prior density, (16), for σ^2, given ν, on the final result. The usual prior corresponds to $\kappa = -1$, $\lambda = 0$. The former value causes no trouble in (20), indeed it simplifies it by removing the first term in the final line. The vanishing of λ, however, would cause divergence difficulties when $\eta = \theta$, namely at the mode. Consequently the non-zero value of λ needs to be retained. Notice that in (16) we are dealing with a χ^2-distribution, not an inverted χ^2, and the convergence difficulties arise as σ^2 tends to infinity, not to zero, as it does with the latter. The value of λ will only be important if the mode is near equality of all the variances. Any substantial evidence from the sample of differences will make its (small) value of minor importance. (Similar remarks apply to ν_0 and σ_0^2 in the second line of (20).)

Consider next the mode of (20). We put $\nu_0 = 1$, $\kappa = -1$ for simplicity. Write $T_i^2 = S_i^2 + n_i(x_{i.} - \theta_i)^2$ and $u^2 = \{\sum(\theta_i - \theta_.)^2 + \sigma_0^2\}/m$. Then the equations for the mode are

$$\frac{n_i(\theta_i - x_{i.})}{\phi_i} + \frac{(\theta_i - \theta_.)}{u^2} = 0, \tag{21}$$

and

$$-(\tfrac{1}{2}n_i + 1) - \frac{\tfrac{1}{2}(m+1)/m}{\log \eta/\theta}\left(1 - \frac{\theta}{\phi_i}\right) + \frac{T_i^2}{2\phi_i} = 0, \tag{22}$$

where we have put $\lambda = 0$. As with (9), (21) may be rearranged in the form (compare (12))

$$\theta_{i.} = \frac{x_{i.}n_i u^2 + x^* \phi_i}{n_i u^2 + \phi_i} \tag{23}$$

445

with

$$x^* = \Sigma w_i^* x_{i.} / \Sigma w_i^* \tag{24}$$

and

$$w_i^* = n_i / (n_i u^2 + \phi_i). \tag{25}$$

The form (23) clearly shows the same structure as before, namely a weighted average of $x_{i.}$ and an overall mean x^*, the only substantial difference being that the weights attached to x^* depend on the sample variance, and not, as before, on the common variance σ_W^2.

Equation (22) is similarly rewritten

$$\phi_i = \frac{T_i^2 + \{\theta(m+1)/m\}/\log \eta/\theta}{(n_i+2) + \{(m+1)/m\}/\log \eta/\theta} \tag{26}$$

which is a weighted average of $T_i^2/(n_i+2)$ and θ, the harmonic mean of the ϕ's. The former is a natural estimate of within-variance for the ith sample — apart from the divisor (n_i+2) in lieu of the usual (n_i-1) — and the latter is the overall value towards which each ϕ_i is displaced.

The modal equations in the forms (23) and (26) are easily solved iteratively. With first approximations to the θ's and ϕ's, the weights in the two equations may be calculated. Equations (23) and (26) then provide linear expressions for new approximations and the cycle may be repeated until convergence. With small samples or very similar values of T_i a suitable λ, other than zero, may have to be included.

The posterior variance of θ_i can be found from (6) with the modified weights and estimates replacing the known variances. No reasonably simple expression for the posterior precision of the variances has been found, though in any numerical case the second derivatives of the log-likelihood at the mode could be found and the matrix of these inverted to give some indication of precision. This is unlikely to be satisfactory unless the sample sizes, n_i, are fairly large.

Acknowledgments

I am extremely grateful to Melvin R. Novick who stimulated my interest in this problem, encouraged me to work on it and contributed much in discussion. Part of the work was carried out at Educational Testing Service, Princeton, under Research Grant 1 PO1 HDO1762 awarded ETS by the National Institute of Child Health and Human Development. I am grateful to Paul Jackson for valuable comments on an earlier draft of the paper.

References

1. Box, G.E.P. and Tiao, G.C., "Bayesian Estimation of Means for the Random Effects Model," *Journal of the American Statistical Association,* 63, 174-181, 1968.

2. De Finetti, B., "Foresight: Its Logical Laws, Its Subjective Sources," in *Studies in Subjective Probability,* edited by H.E. Kyburg Jr. and H.E. Smokler, 93-158, New York, Wiley, 1964. (The original French version appeared in *Ann. de l'Inst. Henri Poincaré,* 7, 1937.)

3. Edwards, A.W.F., "Estimation of the Branch Points of a Branching Diffusion Process," *Journal of the Royal Statistical Society, B,* 32, 1970, (to appear)

4. Hewitt, E. and Savage, L.J., "Symmetric Measures on Cartesian Products," *Trans. American Math. Soc.,* 80, 470-501, 1955.

5. Hill, B.M., "Inference About Variance Components in the One-Way Model," *J. Am. Stat. Ass.,* 60, 806-825, 1965.

6. Lindley, D.V., *Introduction to Probability and Statistics from a Bayesian Viewpoint,* Part 2, Cambridge, University Press, 1965.

7. Lindley, D. V., "Bayesian Least Squares," *Bull. Inst. Internat. Statist.* 43, 2, 152-152, 1969.

8. Novick, M.R., "Multiparameter Bayesian Indifference Procedures," *J. Roy. Statist. Soc., B,* 31, 29-64, 1969.

9. Pratt, J.W., Raiffa, H. and Schaifer, R., "The Foundations of Decision Under Uncertainty: An Elementary Exposition," *J. Am. Stat. Ass.,* 59, 353-375, 1964.

10. Savage, L.J., *The Foundations of Statistics,* New York, Wiley, 1954.

11. Stein, C.M., "Inadmissibility of the Usual Estimator for the Mean of a Multivariate Normal Distribution," *Proceedings of the 3rd Berkeley Symposium,* 1, 197-206, 1956.

12. Stein, C.M., "Confidence Sets for the Mean of a Multivariate Normal Distribution," *J. Roy. Statist. Soc., B,* 24, 265-296, 1962.

13. Tiao, G.C. and Zellner, A., "Bayes's Theorem and the Use of Prior Knowledge in Regression Analysis," *Biometrika,* 51, 219-230, 1964.

COMMENTS

M.S. Bartlett:

Professor Sprott* has raised the same general point as I wish to raise. Prof. Lindley says in his paper: "A statistician faced with some data often imbeds it in a family of possible data that is just as much a product of his fantasy as is a prior distribution"! I would say that the statistician's model is different in principle from a prior distribution in that it can be tested. Where it cannot be tested this is to me unsatisfactory. Prior distributions are, as I understand it, in general untestable. What does Professor Lindley mean when he says that 'the proof of the pudding is in the eating'? If he has done the cooking it is not surprising if he finds the pudding palatable, but what is his reply if we say that we do not. If the Bayesian allows some general investigation to check the frequency of errors committed, or even real losses, this might be set up; but if the criterion is inner coherency, then to me this is not acceptable.

I.D.J. Bross:

As several comments have already noted, some of the results derived here were previously obtained in standard genetics problems where "prior populations"

---

*Prof. Sprott's comments to follow

or "superpopulations" have been used for a long time. This coincidence of the two approaches allows us to see the differences more clearly. The main difference is that whereas the prior populations were here on a personal basis — the speaker has repeatedly described them as "naughty" and "nice" — the research workers in genetics have based their choice on the question: Does this distribution provide an adequate representation of what is going on in the actual experimental situation? So it would seem that the essential novelty of the work just presented would be in the claim that we are under a special compulsion to accept this derivation. This claim is based on the Principle of Coherence.

I want to take this opportunity to flatly repudiate the Principle of Coherence which, as I see it, has very little relevance to the statistical inference that is used in the sciences. The main force of this principle comes from a play on words. The linguistic game here is the following. I say: "I reject the Principle of Coherence". Then someone can reply: "Well then you are being incoherent".

The remaining attraction of the Principle of Coherence arises from a failure to make the distinction between coherence in ordinary language and coherence in some specialized sublanguage or jargon. While we do want to be coherent in ordinary language, it is not necessary for us to be coherent in a jargon that we don't want to use anyway—say the jargon of L. J. Savage or Professor Lindley. To make this distinction somewhat clearer let me use Hempel's paradox about the observation of a white shoe supporting the hypothesis that all crows are black. This paradox arises because of the incompatibility of coherence in a logical jargon and coherence in ordinary language.

While an ordinary statement such as "all crows are black" can be translated into the language of symbolic logic, the linguistic operations in the logical jargon are not compatible with those in ordinary language. More specifically the rules for negation are different. In both logical and natural language *crows* represent a class of birds. In logical languages the negation *non-crows* is an equally good class. However in natural language this is not the case. When we speak of objects which are non-crows it is not immediately clear what we are talking about. Indeed, unless we make restrictions such as "black birds which are not crows" the class of non-crows will not make much sense to a native speaker of English.

When we translate a set of sentences into English and operate upon them according to the rules of the logical languages, the statements that are generated will be coherent so far as the logical languages are concerned. However when we translate the statement back into ordinary language we get a statement which is not coherent with the usage of natural language and Hempel's paradox is the result. Exactly the same thing will happen to anyone who insists on being coherent in the language that L.J. Savage uses. So the *Principle of Coherence* produces *incoherence* in the language in which the inferences themselves are expressed.

D. R. Cox:

There clearly exist priors for which, under certain circumstances, rather extreme observations are moved away from the general mean, rather than towards it. Are there non-normal priors of the general type considered in the paper for which observations outside a certain range are left essentially unchanged?

V.P. Godambe:

I find Professor Lindley's justification for his prior distribution of θ_i's in terms of Hewitt and Savage (1955) theorem confusing. The theorem is true only on infinite dimensional space while Lindley appears to use it for finite ($= m$, since θ_i's are m in number) dimensional space.

I am interested in knowing how far Lindley's Bayesian analysis would be robust to the changes in the prior distribution of θ_i's? For instance, can we replace the normal distribution by any bell-shaped distribution?

I.J. Good:

I think this paper by Dr. Lindley is a useful advance in Bayesian methods, dealing as it does with an important classical problem. The paper is an example of Type II rationality in so far as it makes use of such practical devices as *conjugate priors*. Also when Dr. Lindley says that "the proof of the pudding is in the eating" this shows that he is prepared to make use of posterior judgments when considering a Bayesian model. In other words he is making use of the device of imaginary results. It is largely through ignoring this device that some Bayesian analyses in the past have been defective. It will be interesting to see some numerical results, when they are available, as a further check.

[Dr. Cox raised the question of whether there are circumstances where all observations sufficiently far from a middle value (the *origin*) should be taken at their face value, and not squashed towards the origin.]

In reply to Dr. Cox's question, suppose that all the distributions in the model are analytic. Then the distance to be "squashed" will be an analytic function, so if it vanishes in an interval it will vanish everywhere.

[Here Dr. Godambe raised the question of what is a Bayesian.]

I think the essential feature of a Bayesian is his willingness to talk about the probability of a hypothesis. Also, if he thinks that $P(A|B)$ is meaningful then he also thinks that $P(B|A)$ is meaningful.

W.J. Hall:

I wish to make four comments on this paper — on independence in the priors, on nuisance parameters, on dependence in the θ_i's and on invariance.

Professor Lindley's priors are preferable to so many proposed elsewhere in the literature in that they recognize dependence among the parameters in their prior distributions. More about this later. By utilizing exchangeability as a 'first-order' assumption, he has only needed independence assumptions at a 'second-order' level, among the parameters in the mixing distributions and between the two mixing distributions. Some of this second-order independence is not clearly stated, however. Specifically, in Stage III of Section 5, he tacitly assumes independence of the pairs (μ, τ) and (ν, σ^2).

In his oral presentation he referred to μ as a *nuisance parameter*. This says something about his formulation that he hasn't made explicit; namely, that he is interested in the relative sizes of the θ_i's, or rather contrasts in them rather than $\bar{\theta}$ (or its conditional expectation μ). Presumably, some of the σ^2 and degrees of freedom parameters are also nuisance parameters. He seems to want to place relatively flat priors on all of these nuisance parameters. Is this an implicit principle or a coincidence?

Professor Lindley has required his θ_i's to be dependent, and rightly so in his context. I think it is interesting to explore the nature of this dependence. By using the formula for covariance as expected conditional covariance plus covariance of conditional expectation, we find cov $(\theta_i, \theta_j) = \tau^2$ (my notation for the variance of μ), for $i \neq j$, and var $\theta_i = \sigma_B^2 + \tau^2$, so that $\rho = $ corr $(\theta_i, \theta_j) = \tau^2/(\sigma_B^2 + \tau^2)$. Lindley's assumption of a flat prior for μ might be construed as implying that τ^2 is large, and hence ρ is near unity. This is a far cry from assuming independent θ_i's, a not uncommon assumption, and Lindley's is more appealing to me in this context. (This and what follows refers to his "variances known" section.)

Calculation shows that the posterior correlation is $(mn_i\lambda + m - 1)^{-\frac{1}{2}} \times (mn_j\lambda + m - 1)^{-\frac{1}{2}}$ where $\lambda = \sigma_B^2/\sigma_W^2$, implying that with even modest amounts of data the θ_i's become essentially independent *a posteriori*. This too seems reasonable.

Non-Bayesians may react negatively to the possible mysteries of the flat prior for μ. I wonder about the relative merits of a parallel but proper Bayes treatment that I believe will yield the same results. Specifically, suppose we assume the θ_i's are symmetric multinormal with common mean μ, variance $\sigma_B^2 + \tau^2$, and covariance τ^2 — a specific exchangeable prior. My calculations indicate that the same posterior distribution as obtained by Lindley will be obtained when τ^2 is sent off to infinity (μ conveniently disappears).

Finally, I was pleased to find Professor Lindley using an invariance argument (though I suspect he will not like the term), by recognizing symmetries in his prior information and arguing therefore for an exchangeable prior. As noted in my discussion earlier in this conference, I feel it important to recognize invariance, or group structure, in the formulation of a problem and to contemplate (at a minimum) whether one should react to it.

J.D. Kalbfleisch:

I would like to compliment Professor Lindley for a very interesting paper. The results which are obtained do seem to be quite appealing; Professor Lindley has shown a great deal of intuition in applying Bayesian methods to this rather difficult problem. I agree with the suggestion that the various methods be applied to particular problems like this and the results compared. I would, however, like to comment on a few points in the paper which to my mind require some clarification.

In the third paragraph of Section 1, it is assumed that the relevance of Bayesian methods to decision problems is granted and it is argued that this implies their relevance in inference. The argument here depends on the expression *the posterior distribution* which is curious given the subjective Bayesian approach. There is nothing unique about a posterior distribution, it is personalistic, and if one is quoted its appropriateness for decision making depends on who is making the decision. The question then arises as to whose posterior distribution is to be published. Perhaps Professor Lindley means here that the problem of inference is effectively solved by quoting the observed likelihood function (with a reference to Pratt, Raiffa and Schlaifler).

Professor Lindley states that he sees no reason for assuming a prior which depends on n, the number of observations. I agree with this statement, but find it hard to reconcile with the methods used in this paper. There is no doubt that the priors in this paper are functions, not of the number of observations, but of the particular parametric model assumed. These two lead, however, to similar difficulties. If priors are allowed to depend on the mathematical model, different answers will in general be obtained depending on the order in which two experiments are performed.

No doubt this will be answered by stating that the priors are only approximations to the true prior belief and are used for convenience. This granted, however, a prior which depends on n and approximates prior belief has as much justification as the priors used in this paper when they approximate prior belief. It should, perhaps, be made more clear in subjective Bayesian writings that the method used is only approximate to an untenable method that is consistent.

In the subjective Bayesian approach, I see no reason why personal beliefs should be evaluated before rather than after the experiment. Perhaps the statement "Thus if a prior leads to an unacceptable posterior then I modify it to cohere with the properties that seem desirable in the inference" indicates that Professor Lindley agrees. Yet somehow this seems to be assuming what one wishes to prove. This, together with the statement "...the major problem of inference has been solved in its essentials by Bayesian ideas", reminds me of a story which is currently going the rounds in economics. A physicist, a chemist and an economist were marooned on a desert island with no food but coconuts when a can of beans washed up on shore. The chemist and the physicist discussed how they might get the beans out of the can; the chemist said that his technical knowledge would suggest lighting a fire under the can so it would explode; the physicist responded enthusiastically that his training would allow him to predict the trajectories of the beans and figure where they should stand to catch them. At this point the economist interjected "Wait, I have a better idea! Let's assume we have a can opener."

O. Kempthorne:

I shall give an expansion of some very brief points made orally.

The neo-Bayesians write as though the world is dichotomized, with one group using Bayesian ideas all the time, and the other *never* using Bayesian ideas and never examining the possibilities. Dr. Lindley seems to fall in this class. Additionally, there is a vast amount of ignorance of the past. We all have the problem of keeping track of the burst of intellectual activity of the past 100 years. But this does not make any of us less culpable for not being aware of work intrinsically related to what we are doing. I feel that there is a strong element of intellectual arrogance present in all of us, *not excluding myself*. We tend to think that we are "God's gift to the world," and that we are highly original in posing problems and suggesting solutions. I regret that I find elements of this in Dr. Lindley's paper. It is true that he covers himself a little, but, in my opinion, *not enough by far*.

The history of the sort of thing he is doing goes back some 60 or 70 years. The process of "pulling in" means of lower precision is as old as this. It is for this reason that some people, including myself, were not as surprised at the

Stein results as were others. And, please note, that I do not wish to denigrate Stein, whose work is very interesting. I look at the references of Dr. Lindley and I find it appalling in its narrowness—I can find no more appropriate word.

To trace the history of "pulling in" records would be a very arduous task. The argumentation used is Bayesian, and is regarded by some, including myself, as *legitimate* Bayesian, based on a *historical prior*. My own contact with the idea extends over the past twenty-three years. I first encountered it (sad to say!) when I became interested in animal breeding. A short characterization of this field, which is of the greatest importance in human affairs, is that it is based on a model, the Mendelian model, a paper by Fisher (*the Fisher*) of 1918, work of Sewall Wright from about the same period, plus others before, going back indeed to Karl Pearson and G. Udny Yule, G. H. Hardy (who will be remembered a few hundred years from now, I hazard, not only for his mathematics but also because of a letter published in 1908 in *Science* which was then an obscure journal) and Weinberg and several others. Please note that I do *not* include myself. I am a "latter-day" worker, though my *Introduction to Genetic Statistics* is, perhaps, not a bad introduction. Further, the whole process of animal selection is Bayesian, with the difference that the model is justified partly by Mendelism, and the procedures involve estimation of components of variance. In this respect, the procedures have an appearance of being two-stage. The notion of repeatability is described very briefly in Section 13.3 of my *Genetics Statistics* (now in print with the Iowa State University Press), but I claim no originality at all. The paper by Henderson, Kempthorne, Searle and von Krosigk (*Biometrics 15*) is perhaps partially relevant. The book, *Animal Breeding Plans*, by J. L. Lush, predates all this by many years.

In connection with the pooling of means from experiments of unequal precision, a "prior" distribution of error variances was incorporated in the whole process in an I.S.U. Ph. D. thesis by B. A. Rojas of 1958. It is unfortunate, I think, that some of this material was not put into the literature. More recently in connection with the components of variance problem, a thesis was written by S. T. M. Naqvi, in 1969, examining a little the use of several of the so-called *natural* or non-informative priors, as well as a *realistic* prior on the intraclass correlation. As an alternative to these, a little investigation was made of the use of probability goodness of fit test for establishing degrees of consonance of parameter values with the data. Writing up of this has been delayed by illness of the author.

I mention these to indicate that some individuals have not ignored the Bayesian work, even though they have a strong *personal* opinion that much of the recent Bayes work is not relevant to the real problems. In this connection, the use of a prior depending on the size of the experiment was criticized by Naqvi. It is this sort of work that I had in mind when I used the adjective (or expletive) "shoddy" in my own essay.

It is also worthwhile mentioning, I believe, that a former colleague, D. L. Harris, prepared a manuscript entitled, "Estimation of Random Variables" in 1963. The title "bugged" some people, and the manuscript was rejected by *Biometrics* in all its wisdom. The argumentation used was Bayesian, *of course*. Perhaps, a manuscript with such a title will receive a slightly more open reception nowadays. There is some evidence, I believe, of our "open society" being somewhat less open than we would like. Dr. Lindley cannot be criticized, of course for not being aware of this more recent work.

If Dr. Lindley's paper has the effect of making some of the past work respectable in the statistical profession, it will have served a useful purpose. Obviously, also, Dr. Lindley will advance our understanding.

A final point: As a citizen and perhaps intelligent-man-in-the-street, I have wondered seriously about the Bayesian process according to which one's evaluation of an individual, a cow or a bull or whatever, depends on the group in which that individual occurs. There is, I think, a substantial dilemma here. How about justice for the individual?

D. A. Sprott:

The implication that the "arbitrariness" in the choice of sample space is somehow equivalent to the "arbitrariness" in the selection of a prior distribution ought to be queried. This would appear to mean that because two approaches have an arbitrary element, they are necessarily equally arbitrary.

However, even the Bayesians seem to need some aspects of the sample space. For instance a normal likelihood function

$$f(\mu, \sigma) = \frac{1}{\sigma} \exp[-\frac{1}{2\sigma^2} (x_0 - \mu)^2]$$

where x_0 is the observed value of x does not arise out of thin air. Even a Bayesian must presumably know something of the physical process giving rise to the observed x_0 such as, for instance, that values of $x<0$ were not precluded and so alter the likelihood. How, in other words, does a Bayesian arrive at the above likelihood except by deriving that x in a $N(\mu, \sigma^2)$ variate, thus specifying the sample space (even though he may not use its properties)?

Also, what is the relative arbitrariness in specifying a sample space vs specifying a prior distribution? There seems to be infinite latitude in choosing priors; no Bayesian text has laid down rules for this exercise. On the other hand, how many sample spaces will, in any practical sense, give rise to the above normal likelihood? Even if conditioning on ancillaries is taken as arbitrary, how much latitude is there? In spite of Basu's examples, ancillary statistics are not customarily noted for their abundance. I should like therefore to question the view that selection of a sample space is as arbitrary as selection of a prior, and that the Bayesian indulges only in the latter while completely abstaining from the former.

This paper contains an interesting example of where the customary "noninformative" prior $d\mu$ is not so non-informative, and indeed may drastically affect the solution. This seems characteristic of problems with many parameters.

Finally, why not estimate the parameters $\mu, \sigma_B^2, \sigma_W^2$ etc. from the data?

REPLY

In reply to Sprott, may I say that there is no statement in the paper intended to convey the idea that the Bayesian 'arbitrariness' is *equivalent* to the orthodox one. I was merely trying to point out that the usual approach typically contains an arbitrary element in the choice of sample space and that this is not always recognized. In some contexts, the space is unambiguous: in others (an example was given) it is highly imaginative. Similar things are true of a prior distribution. His point about the likelihood function is answered by saying that it can be written down knowing only that $x_1, x_2, \ldots, x_n$ is a random sample from $N(\mu, \sigma^2)$. His final point is answered in the paper: equations (10) and (11) provide such estimates.

Hall is correct and I apologize for having omitted to say that (μ, τ) and (ν, σ^2) are independent. Some of his points about the correlation are answered in Lindley (1965) on page 119 (Example 2). If μ is given a proper prior, for example, normal about μ_0, the final estimate will be a weighted average of $x_i.$, $\bar{x}$ and μ_0. If the proper prior has large variance, the weight attached to μ_0 is small and the estimate given in the paper is a reasonable approximation. The only invariance I plead guilty to using is on a finite space where (I hope) no technical difficulties arise.

Kalbfleisch is also correct: it should be *a* posterior probability, not *the*. I apologize for the stupid error. I share his doubts over conjugate priors. It is always difficult to discuss a Bayesian analysis without reference to a specific application where θ_i is, say, a candidate's ability and not just a symbol, because the prior needs such knowledge. What I try to do is to assess my prior and then ask whether it is reasonably described by a member of the conjugate family. If it is, then the mathematics is going to be easier. (Also such introspection will not suggest any dependence on sample size.) If any crime has been committed then surely it is of the same type as is perpetrated when considering the likelihood function and I don't know what he means by 'untenable'. His anecdote is spoilt by the fact that Ramsay and others have proved that can-openers exist, even on desert islands.

I am sorry that Kempthorne considers me arrogant, narrow and myopic, to mention only three of the derogatory adjectives he has used to describe my work over the period of this conference. It was not my intention to be any of these and I will try to correct these faults in future. His references to Genetics and unpublished Iowa theses are useful and I look forward to reading them. I know that I am very prone to ignore pre-war writings and it is good to be pulled up sharply.

Godambe has spotted a major flaw in the exposition. I am assuming exchangeability for *every* finite *m*. This is not stated in the paper and I am most grateful to him for pointing this out. This question about robustness is a difficult one: in ten years' time we might know the answer. Rubin raised a similar point and I can only say that the analysis for unequal variances ϕ_i does allow for more flexibility than is usually provided. Why does anyone nowadays use a *t*-test in the two sample situations when Fisher-Behrens is available?

Bross and Bartlett both dislike coherence. The choice of an axiom system is to some extent a matter of preference. If Bartlett prefers 'testability' to coherence as his basis then we can, I hope, agree to differ on the friendliest terms. For me, 'testable' probabilities are useless in inference and I would welcome an explanation of how to use them to make practically meaningful statements about, for example, examination candidates. It is *in*coherence that convinces me and I await a monthly cheque from Bartlett and Bross: for incoherent people are, as Savage has remarked, perpetual money-making machines.

Of Cox's point, I can only reply that I don't know. What little information there is is contained in another reply to a discussion (*J. Roy. Statist. Soc., B,* 30, 1968, 65).

I am extremely grateful to all the discussants for the stimulus they have provided in making me think about the issues raised in the paper, and in the whole conference. One can better understand an argument by having to defend it. The conference has reinforced my conviction that the Bayesian argument is basically the correct one. But I am ready to abandon it if anyone can produce convincing counter-examples to demonstrate it is unsound. Where are the equivalents of Basu's and Buehler's elegant, but damaging demonstrations?

THE INFERENTIAL USE OF
PREDICTIVE DISTRIBUTIONS

Seymour Geisser*
State University of New York

Introduction

I understand this to be a conference where it is desirable to present papers not only on new technical developments in inference but also of a semi-philosophical nature in terms of ideas not necessarily fully matured. When one attempts to be a semi-philosopher he generally resorts to soap boxes on street corners, rather than publication. However our conference master is determined that these proceedings be published perhaps on the basis that all the news is fit to print—so be it. Apology aside, what I would like to do is urge you in the handling of data to consider the possibility that you may be overlooking the inferential use of predictive distributions. The Likelihood Brotherhood admonishes us to plot the likelihood, integrate it, average it, interpret it, squeeze it if necessary to extract something informative. My exhortations are roughly in the same vein but in the direction of the predictive distribution. Firstly, I shall point out some interesting current applications and then indicate that a location parameter comparison could be handled by a predictive probability comparison of observables. Then I will propose that you might try doing unto the predictive distribution what you would have done unto the original sampling distribution. That is the gospel—unfortunately not for those of the frequential persuasion but for Bayesians, Fiducialists if there are any left and also for those who adhere to the Higher Structuralism. I suppose the view enunciated here could be termed as predictivism, observablism or even aparametricism—if an *ism* is necessary.

1
Current Applications of Predictive Distributions

For purposes of this paper, we shall designate a predictive distribution as

*This research was supported by NIGMS Grant GM14031.

the distribution of an observable which is completely specified as to form and constants. In other words, one which is immediately exploitable for the calculation of any of its characteristics subject only to the difficulty involved in exact algebraic or numerical evaluation. Hence a distribution function $F_X(x|\theta)$ is predictive when the random variable X is actually or potentially observable and θ is a known set of constants. Distributions are not rendered predictive (pseudo predictive, perhaps) by substituting estimates for the parameters, nor shall a posterior distribution of a parameter set θ, though an integral part of this framework, attain predictive status unless θ is truly an observable entity. Within the usual frequential framework only when the parameter set θ is known do predictive distributions exist that shall be of interest from the point of view taken here. It is clear that predictive distributions in the frequential framework do exist; for example, if X and Y are both $N(\mu, 1)$ and observable then $X - Y$ is observable and its distribution which is $N(0, 2)$ is completely specified but is not particularly informative as to the only unknown ingredient μ of the original $N(\mu, 1)$ distribution. Therefore we shall deal exclusively within the Bayesian framework but remark that the Fiducial or Structural approach, Fraser and Haq (1969), when appropriate will also be relevant.

The simplest example is the predictive distribution of a future observation. Suppose a set of N independent observations summarized by D are available on $F_X(\cdot|\theta)$ where θ is assumed to have the prior distribution $G(\theta)$. Hence from the posterior distribution $P(\theta|D)$ when it exists, we can derive the predictive distribution of a future observation x

$$F(x|D) = E_\theta F(x|\theta) = \int F(x|\theta)dP(\theta|D).$$

This is obviously useful for prediction problems concerning a future observation. There are some less obvious uses for this distribution that are equally interesting and relevant for inference. One such example is given by Guttman (1966) for *Goodness of Fit Tests*. Mutually exclusive and exhaustive categories $A_1, \ldots, A_k$ over the range of X are defined. Each of the available observations, denoted as a set by D, are characterized as to the category to which it belongs. The frequencies $f_1, \ldots, f_k$ are then compared with

$$a_i = N \int_{A_i} dF(x|D)$$

by means of the statistic

$$\sum_{i=1}^{k} \frac{(f_i - a_i)^2}{a_i}.$$

Under the hypothesis that the observations are all from $F_X(\cdot|\theta)$, the statistic has asymptotically a *sampling* distribution which is bounded by χ^2_{k-1} and χ^2_{k-1-r} where r is the dimensionality of θ. Under certain pleasant conditions the asymptotic distribution is χ^2_{k-1-r}. This is a rather interesting revival of a significance testing approach in a Bayesian or predictive context to the elusive *Goodness of Fit* problem.

An innovative application of predictive distributions to sample surveys is given by Kalbfleisch and Sprott (1969) in a fiducial setting. They assume that $X_1, \ldots, X_N$ are independent and identically distributed random variables from $F(\cdot|\theta)$. From the original set they select at random

$$X_{i_1}, X_{i_2}, \ldots, X_{i_n}$$

with values $x_1, \ldots, x_n$. Let $x = \sum_{i=1}^{n} x_i$ and $T = x + X$ where X represents the sum of the unselected values and T is to be estimated. From the Bayesian viewpoint we assume $G(\theta)$ as the prior distribution of θ and calculate the predictive distribution of the unobserved values given the observed values

$$F(x_{n+1}, \ldots, x_N | x_1, \ldots, x_n) = \int \prod_{i=n+1}^{N} F(x_i | \theta) \, dP(\theta | x_1, \ldots, x_n).$$

From the above we calculate $F(T | x_1, \ldots, x_n)$ which provides a distribution for T based on the known values.

Problems of classifying observations using predictive densities also have been given attention in a general framework by the author (1964, 1966). For simplicity I shall give the highlights of that approach confining our attention to two populations Π_1 and Π_2 (later they shall also be referred to as two hypotheses H_1 and H_2). Let Π_i be specified by $f_X(\cdot | \theta, \Pi_i)$ either a probability density or mass function though we shall hence refer to it as a density. It is assumed that data $D = (D_1, D_2)$ are available where here the subscript on D indicates the population from which the data have been sampled. A new observation x is to be classified as belonging to Π_i with known prior probability q_i, where $q_1 + q_2 = 1$. Hence for $q = (q_1, q_2)$

$$\Pr[x \in \Pi_i | q] = q_i f(x | D, \Pi_i) / \sum_{i=1}^{2} q_i f(x | D, \Pi_i)$$

where

$$f(x | D, \Pi_i) = E_a f(x | \theta, \Pi_i)$$

and the expectation is over $P(\theta | D)$, the posterior distribution of the parameter set θ assuming a prior distribution $G(\theta)$. Thus we have the relevant information for making the inference as to which population the newly observed x is more likely to have arisen.

This scheme can be adapted for sequentially or jointly classifying a series of future observation depending on the need. Here we shall only introduce a joint classification scheme as it very naturally leads into some other applications of predictive distributions to the problem of hypothesis testing.

Suppose we wish to classify jointly n independent observations $x_1, \ldots, x_n$. We then calculate the joint predictive density on the hypothesis that $(x_1 \in \Pi_{i_1}, \ldots, x_n \in \Pi_{i_n})$ where i_j is either 1 or 2, $j = 1, \ldots, n$. Hence

$$f(x_1, \ldots, x_n | D, \Pi_{i_1}, \ldots, \Pi_{i_n}) = E_\theta \prod_{j=1}^{n} f(x_j | \theta, \Pi_{i_j})$$

and

$$\Pr\{x_1 \in \Pi_{i_1}, \ldots, x_n \in \Pi_{i_n} | D, q\} \propto f(x_1, \ldots, x_n | D, \Pi_{i_1}, \ldots, \Pi_{i_n}) \prod_{j=1}^{n} q_{i_j}.$$

Again this is the relevant quantity for inferring the joint origin of $x_1, \ldots, x_n$. Now, more specifically, suppose it is known that the set of data $D' = (x_1, \ldots, x_n)$ is entirely from Π_1 or Π_2. Hence we have as a special case of the above

$$\Pr\{D' \in \Pi_i | D, q\} \propto p_i f(D' | D, \Pi_i)$$

with the obvious notational replacement of q_i^n by p_i and $f(x_1, \ldots, x_n | D, \Pi_i, \ldots, \Pi_i)$ by $f(D' | D, \Pi_i)$. Now note that the posterior odds ratio

$$\frac{\Pr\{D' \in \Pi_1 | D, p\}}{\Pr\{D' \in \Pi_2 | D, p\}} = \frac{p_1 f(D' | D, \Pi_1)}{p_2 f(D' | D, \Pi_2)},$$

where $p = (p_1, p_2)$, is relevant for testing the hypothesis that the set of new data belongs entirely to one or the other population assuming only these alternatives. In many cases an actual specification of p is not possible. This has led to the consideration of the ratio of the posterior odds to the prior odds

$$\frac{\Pr\{D' \in \Pi_1 | D, p\}}{\Pr\{D' \in \Pi_2 | D, p\}} \bigg/ \frac{p_1}{p_2} = \frac{f(D' | D, \Pi_1)}{f(D' | D, \Pi_2)}$$

denoted by Cornfield (1969) as the relative betting odds (rbo). This of course is the ratio of predictive *densities*. Cornfield has utilized this ratio as a guide to the sequential evaluation of alternative hypotheses H_1 and H_2 (or Π_1 and Π_2 in our terminology) and applies this to clinical trials.

If observations D whose origin is certain is not a feature of the set-up then the predictive density, provided it exists, is just

$$f(D' | \Pi_i) = E_\theta f(D' | \theta, \Pi_i),$$

the expectation over the prior distribution of θ, and the rbo or predictive density ratio is

$$\frac{f(D' | \Pi_1)}{f(D' | \Pi_2)}.$$

This latter ratio has been extensively exploited by Dickey (1969) in a variety of situations for testing which of two hypotheses is more appropriate to a given set of observed data. He designates these tests as weighted likelihood ratio tests and explores their use.

2
Problems of Comparison

A large proportion of statistical applications are essentially in the realm of comparison – populations, treatments, fertilizers, drugs, etc. In its simplest form we may have a population of experimental units some of which are subjected to a treatment designated by X and some to a treatment designated by Y. After response on the two treatments is in, the experimenter may ask the statistician to determine the *better* treatment and perhaps how much better. Depending on the assumptions and statistician's current philosophical bias or as some would have it, level of naiveté, he may perform a test of significance, test of hypothesis, or plot likelihoods or perform some Bayesian analysis or provide by one means or another an interval for the estimated mean difference. Another way which has not been utilized too often is to consider $\Pr\{Y + a < X | \theta\}$ where a is some prescribed constant and θ the parameter set. If, for example, $Y \sim N(\mu, \sigma^2)$ and $X \sim N(\eta, \sigma^2)$ then for Φ the standard normal distribution function

$$\Pr\{Y + a < X \mid \eta, \mu, \sigma^2\} = \Phi\left(\frac{\eta - a - \mu}{\sigma\sqrt{2}}\right).$$

In general the statistician might want to estimate this quantity $\Pr\{Y + a < X \mid \theta\} = \alpha$ and give an uncertainty interval for it and plot this for an interesting range of a. This certainly can be and has been attacked in a frequential manner. A Bayesian way of handling this would be to find the posterior distribution of α depending on a. The predictive approach stressed here would entail focusing interest not on α but on the predictive (unconditional) $\Pr\{Y + a < X\} = E \Pr\{Y + a < X \mid \theta\} = E\alpha$ the expectation computed over $P(\theta | D)$ or $P(\alpha | D)$ which yields a single number for each a. If there is a single value of a that is most likely to be of interest, no doubt it would be $a = 0$. At any rate the stress is on a comparison of the observables themselves rather than on particular parameters. Conceptually this is a valid and informative comparison even when X and Y do not belong to the same family of distributions whilst parametric comparisons may not be inherently informative in such a situation.

We may also carry this one step further to the comparison of k populations $\Pi_i \, i = 1, \ldots, k$ specified by a distribution function $F(\bullet | \theta_i, \Psi)$ where θ_i is an unknown location scaler and Ψ is an unknown set of nuisance parameters. The problem is often framed as a comparison amongst the θ_i. The simplest case is where normality and homoscedasticity is assumed and the setup is essentially a one-way analysis of variance where an F-test is utilized to test the null hypothesis that the θ_i are all equal. However this will usually be an initial attempt at comparing the θ_i and consequently many multiple comparison procedures have been developed to *complete the analysis*, in particular Tukey (1953), Scheffé (1953) Dunnett (1955). Bechhofer, Sobel and others in a large number of papers too numerous to mention have argued cogently that we should be most interested in ranking the θ_i and entirely skirt the usual analysis of variance procedures. They and others have provided procedures for accomplishing this in certain cases subject to various restrictions within a frequentist framework. It is clear that an analogous program can be carried through in the Bayesian mode by first finding the posterior distribution $P(\theta_1, \ldots, \theta_k, \Psi | D)$ and then the joint marginal distribution $P(\theta_1, \ldots, \theta_k | D)$. Once this has been accomplished then one may, at least conceptually, obtain

$$\alpha(i_1, \ldots, i_k) = \Pr[\theta_{i_1} < \theta_{i_2} < \ldots < \theta_{i_k} | D]$$

for all permutations of $i_1, \ldots, i_k$ amongst the integers $1, 2, \ldots, k$. For the *best* ranking we may choose

$$\max_{i_1, \ldots, i_k} \alpha(i_1, \ldots, i_k) = \alpha$$

or attach weight $w(i_1, \ldots, i_k)$ to each of the $k!$ rankings and in combination with $\alpha(i_1, \ldots, i_k)$ select an optimum ranking. At the same time the predictive approach is available, and seems to me somewhat more informative. Consider future observations X_i from Π_i $i = 1, \ldots, k$ and calculate

$$F(x_1, \ldots, x_k \mid D) = \int \prod_{i=1}^{k} F(x_i \mid \theta_i, \Psi) \, dP(\theta_1, \ldots, \theta_k, \Psi \mid D).$$

We now focus on the ranking of X_i with respect to their joint predictive distribution via

$$\Pr[X_{i_1} < X_{i_2} < \ldots < X_{i_k}] = \gamma(i_1, \ldots, i_n)$$

and claim that this is always a meaningful and valid comparison and quite often more informative to the experimenter since we are dealing with observables or potential observables.

3

A Proposal

Statistical problems as posed by statisticians generally involve making an inference about a characteristic of some distribution function $F_X(x \mid \theta)$ where X and θ may both be multidimensional. Let this characteristic be denoted by $C[F_X(x \mid \theta)]$ or when there is no possible danger of confusion simply $C(x, \theta)$. Most often $C(x, \theta)$ depends only on θ, though there are important cases where x is a vital component – obviously if the characteristic of interest is the distribution function itself or the hazard function $f_X(x \mid \theta)/(1 - F_X(x \mid \theta))$ or as previously the (conditional) posterior probability $\Pr[X \in \Pi_1 \mid \theta]$.

For an assumed distributional form, once the characteristic of interest has been determined, the problem, if there is one, is due to the fact that θ, or a subset of θ, is unknown. However, sample data D are usually at hand to help resolve some of the indeterminacy in our knowledge of the unknown $C(x, \theta)$. In the usual estimation of a characteristic $C(x, \theta)$ there are a variety of principles that could be applied, maximum likelihood, least squares, minimum mean square error to name a few. For a Bayesian, the *modus operandi* is to calculate the posterior distribution $P(C(x, \theta) \mid D)$ and use whatever principle he deems appropriate to grind out a single estimate or an estimating set for $C(x, \theta)$. The predictive approach suggests the following possibility: first calculate the predictive distribution $F(x \mid D)$ and then compute the characteristic with regard to the predictive distribution, say $C(x, D)$. In some of the previous examples it turned out that

$$C(x, D) = E_\theta C(x, \theta)$$

but this will not always be the case, and in fact, $C(x, D)$ may not even exist.

A simple example of this is the following: suppose we have two observations from $N(\mu, \sigma^2)$, and we assume the prior density $g(\mu, \sigma^2) \alpha \frac{1}{\sigma^2}$, then the predictive density of a future observation x is

$$f(x \mid x_1, x_2) \propto [1 + \frac{4(x-\bar{x})^2}{3(x_1-x_2)^2}]^{-1}$$

for $\bar{x} = \frac{1}{2}(x_1 + x_2)$. If our characteristic was defined as the mean of the $N(\mu, \sigma^2)$ distribution which is μ it is clear that the mean of the predictive density does not exist since we are dealing with a Cauchy density. However, and this illustrates another point, μ can also be derived by other operations; that is, if we define our characteristic to be either the mode or median of the $N(\mu, \sigma^2)$ distribution we also obtain μ. These latter characteristics do exist for this predictive density and turn out to be $\bar{x}$, which implies that in many situations there may be room for maneuver. Anomalies of this sort may be ground out *ad nauseam* and they obviously depend on the fact that the predictive distribution will not in general or even rarely in particular belong to the same immediate family as the original assumed distribution that it is *estimating*. Further if the characteristic is some peculiar function, say $h(\theta)$ one may even be hard pressed to determine an operation on $F_X(x \mid \theta)$ that leads to $h(\theta)$, and we may be unable to calculate $C(x, D)$. In many cases there should be no difficulty in thinking up imaginative remedies. However, I think that it ought to be made clear that what is proposed is not an inferential panacea but a reminder that Bayesians ought also to look at and *experiment* with the predictive distribution, and relieve themselves of unnecessary parametric hang-ups. Too often, we have a tendency in statistical practice to reframe an experimenter's vague question about some data in parametric terms. Also, and this is no tendency but a fact which no doubt induces the previous inclination, we are subjected in our literature to solutions of problems which begin—suppose we are interested in the estimation of (testing a hypothesis about) the parametric function blah of θ.

4
Attributes of Predictive Distributions

We have been stressing, aside from the usual and important predictive uses, that if the sample data D_N consists of N independent and identically distributed realizations of the sampling distribution $F(x \mid \theta)$ then the predictive distribution $F(x \mid D_N)$ may serve as an estimate or better yet, as a surrogate for $F(x \mid \theta)$. We shall then be interested in the properties that $F(x \mid D_N)$ should have in order to serve as a suitable substitute for $F(x \mid \theta)$. What would certainly be required is that $F(x \mid D_N)$ converge to $F(x \mid \theta)$ as N increases, and hopefully in a monotonic manner, where, roughly speaking, the larger N the less uncertain is $F(x \mid D_N)$. One way to look at it is to require that the measure of dispersion $\text{Var}(X \mid D_N)$, be such that

$$\text{Var}(X \mid D_{N+1}) \leq \text{Var}(X \mid D_N).$$

It turns out that this may not be realistic in many cases. First, $\text{Var}(X \mid \theta)$ may not exist, and even if it does, $\text{Var}(X \mid D_N)$ need not exist. If we assume the restrictive situation where almost all variances exist, the $\text{Var}(X \mid D_N)$ is a function of D_N and it is clear that to require $\text{Var}(X \mid D_{N+1}) \leq \text{Var}(X \mid D_N)$ is too stringent since the $N+1$ observation can be anything in the range of X. It would then be more sensible only to require that the monotonic dispersion property be only on the average; that is,

$$v_{N+1}(\theta) \leq v_N(\theta)$$

where

$$E_{D_N}\text{Var}(X \mid D_N) = v_N(\theta)$$

and the expectation is over the sampling distributions $F(D_N \mid \theta)$ and $F(D_{N+1} \mid \theta)$.

Perhaps, a more useful yardstick of the uncertainty inherent in a distribution is the entropy or information measure of Shannon (1948) which was adopted by Lindley (1956) for comparing the information in experiments. Hence we could let the uncertainty of $F(x \mid D_N)$ be the entropy

$$h(D_N) = -E_X \log f(x \mid D_N)$$

where for simplicity $f(\bullet \mid D_N)$ is a continuous density or a probability mass function. The smaller $h(D_N)$ the less uncertainty is associated with $F(x \mid D_N)$. Again since $h(D_N)$ will depend on D_N we shall average over the sampling distribution of D_N. Hence the measure of uncertainty is the average entropy

$$h_N(\theta) = E_{D_N}[h(D_N)].$$

Further it would be desirable for the sequence $\{h_N(\theta)\}$ to be non-increasing and converge to

$$h(\theta) = -E_X \log f(x \mid \theta),$$

the entropy of the sampling distribution. In summary before we take seriously a predictive distribution as a surrogate for the sampling distribution it seems sensible that it should satisfy

$$\lim_{N \to \infty} F(x \mid D_N) = F(x \mid \theta),$$

or better yet

$$\lim_{N \to \infty} f(x \mid D_N) = f(x \mid \theta).$$

In addition, for some defined measure of uncertainty say $u(D_N)$ when applied to the predictive distribution, and its average over the sampling distribution of D_N,

$$u_N(\theta) = E_{D_N} u(D_N),$$

we would require that

$$u_N(\theta) \geq u_{N+1}(\theta) \quad \text{for all } \theta, N = 1, 2, \ldots$$

$$\lim_{N \to \infty} u_N(\theta) = u(\theta)$$

where $u(\theta)$ is the uncertainty associated with $F(x \mid \theta)$.

In certain symmetric cases there is a natural measure of the uncertainty of the predictive distribution in terms of its concentration about a central point. Suppose $f(x \mid \theta)$, $f(x \mid D_N)$ are symmetric unimodal densities about $a(\theta)$, $a(D_N)$ respectively for all N for which $f(x \mid D_N)$ exists. Let

$$p(b, \theta) = 1 - \Pr[-b \leq X - a(\theta) \leq b] = \Pr[(X - a(\theta))^2 > b^2]$$

$$p(b, D_N) = 1 - \Pr[-b \leq X - a(D_N) \leq b) = \Pr[(X - a(D_N))^2 > b^2]$$

for any $b > 0$ be the uncertainty measures of $f(x \mid \theta)$ and $f(x \mid D_N)$ respectively. Further let

$$E_{D_N} p(b, D_N) = p_N(b, \theta)$$

so that a desirable property in this case would be

$$p_{N+1}(b, \theta) \leq p_N(b, \theta)$$

$$\lim_{N \to \infty} p_N(b, \theta) = p(b, \theta)$$

for all $b > 0$.

As an example consider the situation of N observations drawn from $N(\mu, \sigma^2)$ where σ^2 is known and $g(\mu)$ is a uniform prior. The predictive density of a future observation is

$$f(x \mid \sigma^2, \bar{x}) = \frac{\sqrt{N}}{\sigma \sqrt{2\pi(N+1)}} e^{-\frac{N(x-\bar{x})^2}{2(N+1)\sigma^2}}$$

where

$$\bar{x} = N^{-1} \sum_{i=1}^{N} x_i.$$

464

It is clear that $f(x \mid \sigma^2, \bar{x})$ tends to the density of a $N(\mu, \sigma^2)$ variate and

$$v_N(\theta) = \sigma^2 (N+1)/N$$

$$h_N(\theta) = \tfrac{1}{2} \log \left(\frac{N+1}{N} \right) \sigma^2 \, 2\pi + \tfrac{1}{2}$$

$$p_N(b, \theta) = 1 - G \left(\frac{b^2 N}{(N+1)\sigma^2} \right)$$

where G represents the distribution function of a χ^2 with 1 degree of freedom. All of these uncertainty measures are decreasing in N and tend to their appropriate limits.

If σ^2 is unknown and we assume that $g(\mu, \sigma^2) \propto \frac{1}{\sigma^2}$ then the predictive density is

$$f(x \mid s^2, \bar{x}) = \left[\frac{N}{(N^2-1)\pi} \right]^{\frac{1}{2}} \frac{\Gamma \left(\dfrac{N}{2} \right)}{s\Gamma \left(\dfrac{N-1}{2} \right)} \left(1 + \frac{N(x-\bar{x})^2}{(N^2-1)s^2} \right)^{-\frac{N}{2}}$$

for $N \geq 2$ and $s^2 = (N-1)^{-1} \sum_{i=1}^{N} (x_i - \bar{x})^2$ and tends to the required normal density as N increases. For $N \geq 4$

$$v_N(\theta) = (N+1)(N-1)N^{-1}(N-3)^{-1}\sigma^2$$

and is a monotonically decreasing function of N tending to σ^2. The average entropy measure of the uncertainty is

$$h_N(\theta) = \tfrac{1}{2} \log \frac{2\pi\sigma^2(N+1)}{N} + \log \frac{\Gamma \left(\dfrac{N-1}{2} \right)}{\Gamma \left(\dfrac{N}{2} \right)} - \frac{N-1}{2} \Psi \left(\frac{N-1}{2} \right) + \frac{N}{2} \Psi \left(\frac{N}{2} \right)$$

where

$$\Psi(x) = \frac{d \log \Gamma(x)}{dx} .$$

Since both $\log \Gamma(x)$ and $x\Psi(x)$ are convex functions for $x > 0$, $h_N(\theta)$ decreases with N and tends to $h(\theta)$. For large N a suitable approximation is

$$h_N(\theta) \sim \tfrac{1}{2} \log \frac{2\pi\sigma^2(N+1)}{(N-1)} + \tfrac{1}{2}$$

since

$$\log \Gamma(x) - x\Psi(x) \sim \tfrac{1}{2} \log 2\pi - x - \tfrac{1}{2} \log x + \tfrac{1}{2} .$$

Because we are dealing here with symmetric densities we can also calculate

$$p_N(b, \theta) = 1 - E_Z\left[F_{1, N-1}\left(\frac{b^2 N}{Z(N+1)\sigma^2}\right)\right]$$

where $F_{1, N-1}$ is the distribution function of an F variate with 1 and $N-1$ degrees of freedom and Z is distributed as a $(N-1)^{-1}\chi^2_{N-1}$ variate. Its explicit evaluation is too complicated to present, but for large N, Z tends to 1, so that

$$p_N(b, \theta) \sim 1 - F_{1, N-1}\left(\frac{b^2 N}{(N+1)\sigma^2}\right)$$

which tends to $p(b, \theta)$.

References

1. Cornfield, J., "The Bayesian Outlook and Its Application," *Biometrics,* 617-657, 1969.
2. Dickey, J.M., "The Weighted Likelihood Ratio, Linear Hypotheses on Normal Location Parameters," *Research Report 30,* SUNY at Buffalo, 1969.
3. Dunnett, C., "A Multiple Comparison Procedure for Comparing Several Treatments with a Control," *Journal of the American Statistical Association,* 50, 1096-1121, 1955.
4. Fraser, D.A.S. and Haq, M.S., "Structural Probability and Prediction for the Multivariate Model," *Journal of the Royal Statistical Society, B,* 31, 317-331, 1969.
5. Geisser, S., "Posterior Odds for Multivariate Normal Classifications," *Journal of the Royal Statistical Society, B,* 26, 69-76, 1964.
6. Geisser, S., "Predictive Discrimination. Multivariate Analysis," *Proceedings of the International Symposium,* New York, Academic Press, 149-163, 1966.
7. Guttman, I., "The Use of the Concept of a Future Observation in Goodness-of-Fit Problems," *Journal of the Royal Statistical Society, B,* 29, 83-100, 1966.
8. Kalbfleisch, J.D. and Sprott, D.A., "Applications of Likelihood and Fiducial Probability to Sampling Finite Populations," *New Developments in Survey Sampling,* Wiley-Interscience, 358-389, 1969.
9. Lindley, D.V., "On a Measure of the Information Provided By an Experiment," *Annals of Mathematical Statistics,* 27, 986-1005, 1956.
10. Scheffé, H., "A Method for Judging all Contrasts in the Analysis of Variance," *Biometrika,* 40, 87-104, 1953.
11. Shannon, C.E., "A Mathematical Theory of Communication," *Bell System Technical Journal,* 27, 379-423, 623-656, 1948.
12. Tukey, J.W., *The Problem of Multiple Comparisons,* unpublished manuscript, 1953.

COMMENTS

V.P. Godambe:

In general, I like Professor Geisser's informal use of Bayes Theorem. Now his approach based on predicative densities seems to have been motivated by a general operational philosophy: the properties of the world around are meaningful only to the extent they affect our *observations*. The history of the development of theories in physics clearly indicates that though in some stages of development operational philosophy can be of great help, a scientist cannot go a long way without assuming that some sort of a real world has some permanent or semi-permanent *properties* which are observable. This for statistics may mean not only *frequencies* (observations) exist but *parameters* (properties of the world) also exist. For instance if I go to a medical doctor for treatment, and if he wants to know my weight he may estimate it on the basis of a few readings on a weighing balance. He would hardly be interested in using these readings to infer how the future readings would vary. For the doctor, parameter called 'weight' exists.

I.J. Good:

de Finetti, in his lecture at the Salzburg conference of 1968 (published in *Synthese*, 1969; and to appear in the proceedings of the conference) argued that probability judgments should be confined to observable events. Priors for parameters could then in principle be inferred. I argued at that conference that this position was a little extreme.

W.J. Hall:

Primarily, I wish to second some of the points made by Professor Geisser and by some of the discussants. I consider them very important and deserving of re-emphasis.

Practical problems I deal with fall into one of two broad categories. In the first, we want to learn something about some specific constants — such as Professor Godambe's weight. In the second, we want to learn something about the operation of a system. An intermediate stage is to learn something about parameters we have put into a mathematical model for the system, but a more fundamental stage is concerned with the future performance of the system. Looking at problems from a predictive point of view helps bridge the gap from the mathematical model back to the real world. The proof of the pudding etc. It is regrettable that our textbooks and our literature are concerned almost entirely with inference about our mathematical models whereas, in my experience, the predictive aspect deserves essentially equal time.

Two more specific comments are the following: It is tempting to look at a predictive density (say a Bayesian one) as an estimate of the model density, but it is only warranted in some kind of conservative sense. The predictive density tends to be broader than the model in that it incorporates lack of complete knowledge about parameters — as noted by Professor Geisser. In some contexts, their estimates *may* be preferable.

Finally, Geisser's example is one with a very substantial invariance structure. If (μ, σ^2) are location and scale parameters, with the assumed prior, then $(x-\bar{x})/s$ has a *t*-distribution in two senses — in the frequentist's sense (for fixed μ and σ^2) and in the predictive Bayes sense (for fixed $\bar{x}$ and s^2). This kind of peculiarity can occur only in the presence of substantial group structure, Haar priors, and the rest. It also makes possible (at least partially) the simple comparison that Geisser has presented.

D.A. Sprott:

Is the method of predictive inference for testing goodness-of-fit on pages 457-458 really valid?

To test the goodness-of-fit of n independent observations $x_1, \ldots x_n$ from a frequency distribution $F(x|\theta)$ by actually testing their goodness-of-fit to a different distribution (which may not even be a frequency distribution, and for which the observations are not independent) seems curious. As a distribution to predict $x_1, \ldots, x_n$, $F(x|D)$ is appropriate. When $x_1 \ldots x_n$ have actually been observed, it would seem they should be used to specify $F(x|\theta)$ more precisely and so test goodness-of-fit to $F(x|\theta)$ and not to $F(x|D)$. $F(x|D)$ would seem appropriate only if after observing x_1, a new value $\theta = \theta_{i+1}$ were selected at random from its prior distribution $G(\theta)$. Then x_{i+1} is selected from $F(x_{i+1}|\theta_{i+1})$. In that case, $F(x|D)$ actually is the objective frequency distribution of the x's.

REPLY

Perhaps in the brief presentation of Guttman's use of predictive distributions, it was not made adequately clear that he was merely utilizing the predictive distribution to generate estimates. In the usual applications

$$p_i = N \int_{A_i} dF(x \mid \theta)$$

is estimated by

$$\hat{p}_i = N \int_{A_i} dF(x \mid \hat{\theta})$$

while Guttman employs

$$a_i = N \int_{A_i} dF(x \mid D)$$

as the estimate of p_i. Further, he finds the asymptotic distribution of the relevant statistic under the assumption that the original observations $x_1, \ldots, x_N$ are a random sample from $F(.\mid \theta)$ and not from $F(\cdot|D)$ as Professor Sprott was apparently misled to believe.

I cannot disagree with Professor Godambe's view that there are occasions (although with far less frequency than current statistical theory and misguided practice would indicate) when parametric estimation may be useful and even informative. On the other hand I would reiterate the point, which is supported to a degree by Professor Hall's comments, that predictive distributions are a relatively untapped reservoir that should be vigorously exploited.

I regret my unfamiliarity with DeFinetti's lecture as mentioned by Professor Good, but I am fully in accord with the notion that we should concentrate much more on the prediction of observable events — that was the major burden of my paper. I find DeFinetti's use, according to Good, of this method to infer priors for parameters academically interesting but somewhat oblique to my point of de-emphasizing those Platonic Ideals, parameters.

Finally, I am somewhat at a loss to interpret the sparsity of critical comment on my paper—rather unusual at this Symposium. The hypotheses of newly developed mellowness or pervasive approval of the espoused views seem less likely than the desire to repair quickly to the ensuing banquet arranged by our host.

PROBABILITY, STATISTICS AND THE KNOWLEDGE BUSINESS*

Oscar Kempthorne
Iowa State University

1

Introduction

The holding of sessions such as the present one is important for the subject of statistics at the present time. Our subject is no different from other academic subjects such as biology, physics, or from our society as a whole. The amount of thought going into any one of these is much greater than a few decades ago. The old beliefs are being examined critically and many of them being discarded. We have in the profession of statistics as in all other aspects of society internal frictions of an order of magnitude quite unknown in the past. Some internal friction is essential. But we seem to be approaching rapidly the situation in which we are losing any communality of outlook and are unable to present any sort of unified picture of our subject. Groups of individuals in statistics and in our professional societies appear to be quite incapable of communicating with each other. A phenomenon was labelled very aptly by Dantzig in the early 50's as statistical priesthood, and he gave what he considered to be two examples. We seem to be rapidly approaching the case in which everyone is a priest with his own private religion.

It is possible, of course, that our subject may split into several unconnected new subjects, but the history of science suggests that this would be contrary to the whole trend. A feature remarkable to me of the new high school physics and chemistry courses is the extent to which ideas have been brought together. It is also clear that the boundary between biology and the physical sciences has become much more vague than it appeared to be 50 years ago.

So I am of the opinion that we must try to pull together. We must try to abstract the underlying basic features of our profession. To do so will require

*Prepared in connection with research supported partly by Aerospace Research Laboratories, Office of Aerospace Research, United States Air Force, Wright-Patterson Air Force Base.

much more tolerance of views different from our own. It will require the development by us of the ability to subject our own views to critical examination and evaluation by others, and, worst of all, to reach the position of admitting that some of our past and present views are narrow, misleading, or even erroneous.

The present situation is critical and becoming gradually more so, because of the breaking up of our subject and the appearance of books at all levels purporting to present statistics which are such that the intersection of pairs of books so titled may be almost zero. This is really quite deplorable, particularly at the initial level whether it be in undergraduate or graduate teaching.

I believe there is no area with comparable disunity, with the exception of philosophy. The analogy is not surprising, perhaps, because the boundary between statistics and philosophy, particularly the philosophy of knowledge, seems very vague.

2
What are the Basic Disagreements?

Is there a subject which can be called statistics? If so, what is it? What is the role of probability in statistics? Going down deeper, what is the meaning of probability? Going down even deeper, what is knowledge and what is the role of probability and statistics in the accumulation of knowledge — what I call with levity, but I believe with a certain aptness, the knowledge business? I must apologise for invading the area of philosophy and particularly the philosophy of knowledge, but I have no alternative.

3
Knowledge versus Decisions

It is critical in my opinion to make a differentiation between the meaning of and the acquisition of knowledge on the one hand and the making of decisions on the other hand. In the making of a decision we have to admit the idea that there is a correct decision and an erroneous decision and that there are costs involved in the acquisition of data and losses involved in the making of incorrect decisions. The simple example of breaking two eggs into a frying pan, or of first breaking each separately into a container illustrates the decision problem. The study of the mechanism of, say, photosynthesis exemplifies the knowledge problem. In this case, there is no finality in the whole study. There is a gradual building up of a model which is consonant with data that have been obtained. Indeed, the history of study of this sort is a sequence of models, none of which is asserted to be the true model, all of which have at one time or another been thought to be consonant with all data that have been obtained, but later shown to be erroneous in part. The nature of this type of research is then that the model is never thought of as being completely determined and a settled question. As we teach in our first courses on statistics, we can reject models but we do not accept models. I use the words *accept* or *reject* in the sense that they are used in acceptance sampling. It is true that we will accept a model for the time being as the best predictor that we have, but we shall find that any predictor we obtain will be found to have deficiencies.

In contrast, there are cases in which one can legitimately talk about the true state of nature and one can then talk of losses resulting from acting according to a state of nature which is not the true one. The simple cases are those of acceptance sampling. If we have a consignment of 1000 electrical switches on which we are to accept or reject delivery, we can agree on a test of whether each switch is acceptable or not. We could subject every one of the switches to test and determine exactly the true quality of the consignment. We can then envisage loss and risk functions and choose a plan of observation and decision with the properties we desire. It is possible, I suppose, to regard the accumulation of knowledge as a sequence of decisions. Certainly it is possible to regard the choice of the experiment one will do as a terminal decision. The drawing of conclusions from a study can be regarded as the choice of a set of conclusions in the space of the sets of conclusions. The publishing of one's conclusions in a professional journal is a terminal decision and so on. But there are aspects in the whole procedure that are not amenable to the type of decision argument that is applicable to acceptance sampling, for instance. The most crucial, I believe, is that in the knowledge business one cannot, in general, specify the space of the sets of conclusions beforehand. The Watson-Crick model of DNA is regarded, very appropriately in the opinion of all humanity who have some background for understanding it, as one of the big jumps in the advance of knowledge. The application of decision theory requires, it seems, that the scientists had a list of the possible models in their mind. Nothing, it appears, is farther from the truth in this case. The model arose by interaction of the scientists with data, and no one had the model in mind *before* the collection of the data. A second aspect is that the model is not a complete model which is the truth for all time. It is a partial model, which, in the judgment of the relevant part of the scientific profession, appears to be consonant with all the data so far obtained. There are still great obscurities in the whole matter which are the subject of very extensive research. The model, it seems then, is a part of the truth in the sense that it seems that any new model which replaces it will have to contain its essential ingredients. I speak here *not* as an expert in molecular biology.

Another example extending over a long duration is the theory of light. I am not expert on this matter either. But the high school student of physics of today is presented with two models for the theory — the corpuscular theory and the wave theory. The corpuscular theory was regarded for decades, or even more, as a true model for light. But then it was found that it could not explain some observable phenomena. Now, I believe, the physicists do not talk about a model as being the *true* one. They say, rather, that a certain model has explanatory and predictive value with regard to some aspects of nature, and another model has similar values with regard to other aspects. Even if there is a model which manages to combine these two models into one, the view of science must be that it is simply the best model that we have at the present time, and that it will surely be superseded by another model.

Even if we move from the idea of *a true model* to *a model which we regard as the best with our present knowledge,* and even if we could delimit our space of possible models, the application of decision theory in any formal way would require some partial specification of losses resulting from using one model rather than another. Whether this can be done, except in very simple cases, is a very moot point. The consequences of using one model rather than another in the accumulation of knowledge are very obscure. Indeed, one can go further, I believe, and state that the consequences of using a poor model may be better than those of using a good model. What, indeed, are the purposes of models? One purpose, to be sure, is to use a model to make a choice of action with regard to a problem of humans. Should we treat *this* cancer patient by drugs, by radiation, or by surgery? Note, incidentally, that in this, as in many problems (or perhaps most), we cannot find out if our choice was the best in our space of possible actions. We can perform a comparative study (perhaps using physical randomization as part of our design) to obtain ideas of whether one action is better than another for a defined and definite group of patients. But another basic purpose of a model, *for the accumulation of knowledge,* is that it enables us to ask questions which we can try to answer. The process of trying to answer the questions will generate new data and new models in our mind, and the process of model building continues with the obtaining of better models to insert in the process. A model is, then, a good one for the purposes of science if it suggests good questions. It must not be so vague that it suggests a huge number of questions for which we cannot invent informative studies.

It is clear that there are losses and risks to the individual worker, whatever his role, but these are so vague that the application of a formal theory incorporating a complete enumeration of the possible states of nature and losses from actions seemed quite impossible. But it is also quite clear that each and every one in the "knowledge" business makes a personal evaluation, more or less vague, of these aspects in deciding what he will do.

The knowledge business has two fundamental aspects, discovery on the one hand, and verification and justification on the other. These are not totally distinguishable, in that processes of discovery have to use some ideas of verification. But there is also a separation because a critical part of verification is the making of predictions, and the checking-out of the predictions on new observations.

4

Applied Statistics

What is statistics, or to ask the *important* question, what is applied statistics? I imagine we cannot even get a modicum of agreement on this. But in order for me to proceed I need a definition. It is: the set of ideas and procedures for the collection, condensation, interpretation of data, and the use of the results to develop useful descriptions of our factual knowledge both for general interest (as in purely descriptive statistics), for the purpose of making predictions and judgments about the future and judgments about unknown events in the past, and for suggesting new avenues of study.

From a certain viewpoint, this definition is so general that it includes almost all intellectual activity. I surmise that the definition I have given would work moderately well as the definition of the nature of history. In this case, the making of judgments about the future is based on the idea that one can learn from history (which may be debated, perhaps, as regards human political affairs). I am strongly of the opinion that the basic processes of historical research are the basic processes of statistics, with the difference that the data of history are usually qualitative and very difficult to incorporate in a formal model. Even the term model occurs in the domino theory as applied to South-east Asia in our recent past. It is the case in historical research that the abstraction of ideas into concepts to which one can apply the processes of mathematics is so difficult and so likely to be incredibly naive, as to make formal rather than verbal theory useless. Here the distinction between *formal* and *verbal* also seems somewhat artificial, the only difference being that in a formal theory the rules of language manipulation and of adjoining symbols to make new sentences are very tight whereas in a purely verbal theory they may be so loose as to give rise to deep disagreements.

The definition I have given is so broad as to be inadequate for present discussion. I make, therefore, a brief listing of aspects of statistics:

(a) the collection of data;
(b) the reduction or condensation of data;
(c) the formation of opinion from data;
(d) the choice of decision on the basis of data; and
(e) probabilistic inference.

The nature of (a) and (b) is clear. In connection with (a), however, there is an obscurity in the minds of some Bayesians, but not others, on the role of *random* sampling. In connection with (b), the development of probability models may be quite critical. I wish, however, to discuss a little, item (c), the formation of opinion. I do not think that opinion is necessarily in the form of a direct probability statement. I may be of the opinion that phenomenon X is due to cause Y, because I have no other explanation that is reasonably consonant with what I can observe. I would certainly judge the guilt of a person on this basis if I were on a jury. The modification of this opinion to a probability that the person is guilty is not necessary to me, I think. The judgment that data are consonant with a particular model is an opinion not in direct probability form. I have to state that I find such opinions informative. I find also the idea of ordering tenability of a set of models by means of, say, a goodness of fit significance test to be useful in my thinking.

Part of the problem leading to the present session is that a number of people take a very limited view of statistics and of inference. Lindley in 1953, for instance, wrote a paper on statistical inference which is to all intents and purposes a paper on decision making. L. J. Savage wrote his book entitled *Foundations of Statistics,* and interpreted statistics as the theory of personalistic decision. Fisher was almost unable to grant any status to the matter of making decisions, or at least dismissed the activity as a rather low one. Most of the writers on inference ignore totally the problem of choice of probability model, so for instance we are supposed to know that $x_1, x_2,...,x_n$, are a random sample from $N(\mu,\sigma^2)$, but we know nothing about μ and σ^2. Really, I say to you, just how artificial can one be?

Real progress in our mutual understanding would be achieved if members of our profession would not take a particularly highly limited view of statistics and confine their teaching *solely* to this view. All of the topics I have mentioned are important parts of statistics.

5

Probability

We have the calculus of probability as an axiomatic mathematical theory in which probability is not defined except as a numerical property of sets in a space of sets. There is, presumably, no argument about the nature of this calculus (though because it is so strongly based on the idea of sets it may be subject to some of the basic unresolved questions of set theory). To apply this theory to the real world we have to make some sort of correspondence and the one made by many is that one envisages the possibility of an infinity of similar trials and a certain proportion of these will have a designated outcome.

This way of proceeding has been distasteful to a number of thinkers, particularly in this century, and we have seen the development of a theory of probability as a theory of reasonable ways to choose between wagers. The probability then is determined by one's point of indifference in a continuous family of wagers. This appears to be the nature of the so-called theory of subjective or personal probability. It seems that this theory leads to a calculus of probabilities identical in form to the axiomatic theory (as given by Kolmogorov, for instance). There has been much controversy about the difference between the approach of tying the axiomatic theory to the real world by means of relative frequency and the so-called subjective approach. I think that the controversy has been totally sterile. Whether one prefers to think in the one way or the other seems to be totally a matter of irrelevant personal taste. I prefer to think in terms of relative frequency in an infinite population of repetitions and can then make a choice from the family of wagers. I can, also, translate someone's choice in terms of wagers into my ideas of relative frequency. What indeed is all the fuss about?

I mention, merely, the idea of logical probabilities of Keynes with the remark that they carry no force for me. I also mention that there appear to be some attempts to relate probability to the structure of language, but I do not comment because I do not understand them (for example, Leblanc). I am disposed a little, however, to the view that this type of theory may be very helpful in improving understanding of the idea of probability. (See, however, some remarks below in the Addendum.)

It seems clear to me that ideas of probability have two distinct uses which need presentation and discussion. These are:

(a) the use of probability ideas as a basis for the condensation of data and to develop models for data,
(b) the use of probability ideas as a basis for prediction, including choice of action or behavior.

These uses are closely connected, but I shall try to convince my audience that there is value in separating them. Because I am concerned with the application of probability ideas to real situations, I shall not discuss the uses of probability in pure mathematical probability theory.

Probability as a Basis for Condensation of Data

It seems completely undebatable that from one viewpoint every happening or fact of the real world is unique, and we can easily describe every happening so that it is unique and not like any other happenings with regard to the attributes we use in describing it. A mode of condensing the whole history into a statement which we, as individuals who have been trained in probability and exposed to gambling devices, find informative, is to use the idea that we can represent our data as the realization of the actions of a designated set of forces and of a sort of Maxwell demon—a probability demon—who determines outcomes by adjoining a random error. This demon is often called *chance*, and it is unfortunate, in my opinion, that the term *due to chance* is used widely by humans, as though it has a logical (or verbal basis) equivalent to "due to a break in the circuit". I note that many people I admire use the statement. What, parenthetically, does this statement mean? It means, I believe, "due to a concatenation of forces which I am unable to resolve in any other way meaningful and useful to me". At least, this is what I mean if I happen to use the phrase.

Let me give a few examples of the use of probability ideas in condensing a segment of history.

Example 1. Consider the records of whether it has rained or not in Ames, Iowa on December 1, and suppose there are 117 records. Certainly one reporting of the data is to give the totality of the record. But this is very tedious. A second is to give merely the numbers of days on which it rained and did not rain. This is a condensation which is meaningful immediately to most of us. A possible third condensation is to record the frequencies of rain and no rain in each decade, and so on. A condensation that is suggested by ideas of probability and gambling devices, is to say that the records of the past 117 years are *like* the possible realizations of a set of 117 independent Bernoulli trials. In written presentation I place the word, "like", in quotation marks. It is a basic word in my probability vocabulary.

Example 2. I observe the heights of 500 fathers and sons. Now there is nothing random about these records, for the fathers and sons are known to me. However, I find it useful to condense my records into the statement that the 500 records are like a random sample from a certain bivariate normal population.

Example 3. I observe N pairs of numbers (y_i, x_i) $i = 1,2,...,N$, in which I have chosen the x_i and observed the y_i. I can present the totality. But a representation that has appeal to me is that the observations are like a realization of

$$\ln y = 0.2 + 0.51 \ln x + \epsilon$$

with the ϵ being like a random sample from a normal distribution with mean zero and variance 0.04.

Example 4. I examine the sex of new born infants for a large variety of cases. I find that I can represent all my data by stating that they are like independent realizations of a Bernoulli trial with P (male) = 0.51.

I find the above way of discussion useful. It explains, very nicely to me, a phenomenon we all see. We see papers in good journals entitled "A reexamination of X's data on twins." The situation is usually that X was led by his data analysis techniques to a certain probability model to represent his data. Another individual has different and perhaps better ideas and gives his representation of the data.

In a real sense, this is a use of probability ideas which has no inferential content. The results of the analysis may be used for prediction purposes but very often not.

In the case of a situation for which one can get more data "like" the present data, one will use the probability model one has obtained as a basis for choosing observations to obtain evidence against it.

7

Probability as Related to an Unknown Outcome

It is here that there has been extensive argument over the centuries. I am sure many of you have had the same experience as I. I have read all the well-known writers (I believe) on the business and find their writings uniformly unsatisfying.

It was natural and still is, perhaps, to hope that there is such a thing as a logic of uncertain inference analogous to the logic of certain inference. (But superficial reading suggests that even the logic of certain inference is obscure.) We see the attempt by Keynes to formulate such a notion. We see the paper of R. A. Fisher, "The Logic of Uncertain Inference".

The problem of uncertain inference has troubled thinkers for hundreds of years. It can be formulated, I imagine, in several ways. I believe I am correct in saying that one of the standard forms is the following. We have observed N objects and found that r of them have an attribute. What is the probability that the next object we observe will have the attribute? Answers to the problem have been given but none has been judged acceptable. Apparently for several decades in the 19th century, the answer that was accepted was the Laplace rule of succession, that the probability is $(r+1)/(N+2)$.

There are at least two possible views about the problem:

(a) that humanity has not been astute enough to determine the answer, or

(b) that the question is not a real question: that it has the form of a question but is in fact meaningless.

I am rather disposed to take the second view. We are all familiar with the Russell paradox of the barber who shaves all those in the village who do not shave themselves. This shows that we can construct a collection of words which has all the linguistic form of a question but is in fact nonsensical. Let us examine the form of the question. The premise seems quite clear: we have observed ten swans and they are all white, for instance. But the question is what we should think about the eleventh swan that we shall observe. We are to express our thinking by means of a probability — indeed, *the* probability. There seem to be two completely obvious problems:

(a) How can we possibly form a view about the next swan that we shall see? At times it is really quite fantastic to me that one should pose the problem.

and (b) What do we mean by probability and what do we mean by *the* probability?

A seemingly more precise version of the question is the following. We are given a history of N trials and r successes, and we are allowed to assume that these are a random sample of N independent Bernoulli trials with parameter p. What is a rational way of expressing uncertainty about p?

If one takes the view that the only rational way of expressing uncertainty is by means of probability, one is in great difficulty. The probability calculus in its most elemental form exhibits with total clarity that the only way to get a probability by a deductive argument is to insert probabilities at the basis. It seems obvious then that if one wishes to end up with a probability, one must begin with a probability.

The work of de Finetti, and others, including the paper of J. Cornfield presented this spring at Iowa City, seem to have the conclusion that if one is to have a way of constructing probabilities over the parameter space for all possible outcomes of trials, which has an elementary property of consistency, the probabilities must be obtained by computing posterior probabilities by the use of Bayes' theorem and some (it appears, any) prior distribution. That is, if one is to obtain a complete battery of consistent probability statements from the possible observations, one must inject probabilities into the system. From a slightly different viewpoint, there is no way of passing from probabilities of data given parameter values to a probability of parameter values given the data without injecting probabilities into the argument. One can only build probabilities from probabilities.

Our society in both its statements of its knowledge, whatever that is, and in its actions depends critically on statements like the following: the weight of object A is 712 grams; the density of lead is 11.4; the velocity of a particular automobile at a designated time and place was over 25 miles per hour, and so on. Can we have similar statements about probability? In the examples I have enumerated and all like it, there is an operational procedure for establishing the result. This operational procedure has objectivity in the sense that the operational procedure could be applied by many investigators and the procedure is accepted by humanity in general. If we accept that we can obtain a probability only by inserting a probability, can we bring logical reasons to bear which suggest that a particular prior is appropriate for each particular probability situation? It seems that Harold Jeffreys thinks so, but it appears that he has been unable to advance even partly compelling reasons for the use of a particular prior. Should we adopt a convention for the Bernoulli case that we will always use the Laplacian prior? Conventions with regard to the interpretation of data are common, but they must have some interpersonal appeal. It is the problem of choice of prior that led Fisher to give at the beginning of his books his arguments for rejecting the use of prior probabilities and I find his statements compelling.

The answer of the Bayesians to the whole dilemma consists, it seems, of two parts. The first is that there is no such thing as *the* probability. In this I think they have done us a great service. The second is that if you want your probability, you must inject your prior and use the Bayes' rule.

I wonder seriously whether it is really necessary to go beyond the premises of the problem. Is it necessary from the viewpoint of knowledge or belief to say anything additional to the statement that our history is in our opinion, that we have observed r successes in N independent Bernoulli trials? I do not thereby imply that there are not decision problems still to be solved by some procedure or other.

From another viewpoint, the premises of the problem are highly questionable. What permits us to assume that the data are a random sample from some distribution? What do we mean by saying that 30 cases we have observed are "like" a random sample from the binomial distribution $B(30, p)$? We are using the word, "random", which we have not defined. We can give it operational meaning by saying that if we use the theory of probability and interpret frequencies in our data as realizations of sets in our probability calculus. But what do we mean by a history being "like" a random sample? It seems to me that we have no alternative here but to derive some measure of distance of the data from the distribution of possible data points. We can make up a measure of distance in many ways but only by ignoring almost all of our historical records. We can express this distance only, it appears, in the probability scale itself, and we are led to significance tests.

The pure problem of inference seems therefore to be totally artificial, and it seems that the great bulk of humanity has so regarded it for centuries without ill effects.

8

Theories of Inference

What are the theories of inference that should be discussed? Are there really any theories of inference? I think that actually there are no theories of inference: there are theories of aspects of inference. What are these? Here is my list:

- (a) the ideas of tests of significance and inversion thereof (going back to Arbuthnot and K. Pearson, and developed most extensively by Fisher, with sporadic invention of new tests);
- (b) the Neyman-Pearson theory of rules of inductive behavior;
- (c) fiducial inference of Fisher;
- (d) likelihood inference;
- (e) Jeffreys' inference;
- (f) Bayesian inference;
- (g) structural inference of Fraser.

My views on the relevances of these to applied statistics are almost obvious from what I have said already.

I have a few very brief remarks about Fraser's structural inference. I can only give my interpretation of what his book says and it is this. Suppose we know that x is normally distributed around μ with unit standard deviation. Then Fraser says it is clear that μ is distributed normally around x with unit standard deviation. In this case, it is agreed that μ is an unknown constant, and presumably when he says that μ is distributed normally around x, he is transferring a statement of the distribution of x about μ to one of μ around x. If the meaning of the statement that μ is normally distributed around x with unit standard deviation is that x is a realization of a random variable, X, that is normally distributed around μ with unit standard deviation, I do not see that any progress has been made. If Fraser prefers to rewrite the one statement in the alternative form, he is certainly free to do so. But I do not see that he has advanced the cause of making rational probability statements—or of making sense of data.

I have no understanding of Fisher's ideas of fiducial inference. I have yet to see any writing on this topic that makes any sense for me.

I have no empathy for the ideas on inference of Harold Jeffreys. I found his book entirely frustrating in that he does not appear to me to impose on himself the discipline he expects from other writers.

I have a few brief remarks about likelihood inference. It is not at all clear what the nature of this is. We are told two things:

(a) the only aspect of the data that matters is the likelihood function; and

(b) look at the likelihood function.

On the first, I have to reject the idea out of hand, totally. The reason is that the likelihood function is a condensation of the data based either on unverified assumption or on data examination. In either case it cannot be regarded as a sufficient condensation of the data. As regards the second aspect — to look at the likelihood function, this is so inadequate a prescription *for me* that I cannot accept it. I have to add that there appear to be many cases and, given one, I presume we can imagine an infinity in which the likelihood function is totally uninformative. This appears to be the case, for instance, in inference from a random sample from an arbitrary finite population and there are many other cases.

But, if we change the question a little and ask if likelihood ideas are relevant or useful to the purposes of applied statistics, the answer has to be quite different. The likelihood function is an exhaustive function given a parametrically specified model. Examination of this function with application of theory of sampling, applied to the maximum are useful in my thinking. I feel also that it is entirely reasonable to accept as a basis for making a probability assessment of a future observation, what is given to me by the model and the maximum likelihood estimate. Such a procedure is somewhat subjective and there is no escape from this. It does seem, furthermore, that this is popularly accepted, though this may be due to the persuasiveness of Fisher's writings rather than goodness of the process *per se*.

I now turn to the ideas of test of significance and inversion thereof — the early ideas of Fisher. It seems to me that whether or not these have a good logical basis and whether they can be given any good axiomatic basis in a theory of learning from observation, they are used often even by those who claim that the ideas have no relevance. Anyone who plots data and then replots in another way does so because the first plot was unsatisfactory. In what sense was it unsatisfactory? The person doing the plotting will have to give his answer, but I surmise that it will be of the form: The plot does not give an appealing simple picture; there is curvature, or there is differential spread, or whatever. How does one decide that there is curvature, that there is differential spread of the points, that a certain point is an outlier and should be treated as a mistake? I know of no way of addressing these questions except by tests of significance. Finally (to abbreviate my picture) the inversion of a significance test with regard to a parameter gives a statement of degree of consonance of parameter values and data, as judged by the particular significance test.

Which significance tests should one use? There is no answer to this except a subjective one. (It is curious that personal views intrude always.) But it is crystal clear that there are not best tests except in very special scalar parameter cases. Certainly in the case of a vector parameter there is no best test. There appears to be no best test which is two-sided except by the introduction of requirements that are artificial, in my opinion.

I now have a few remarks about the Neyman-Pearson rule of inductive behavior outlook. It is the case, perhaps, that the Neyman-Pearson theory of tests of hypotheses arose out of a desire to give a good theory of significance tests. In the process, however, certain aspects were lost. The easy way and perhaps the only way of making a theory of significance tests was to convert the significance test, which provides a measure of strength of evidence against a hypothesis or a model, into an accept-reject rule. The questions of how one is to choose α in a test of hypothesis or how one is to choose a confidence coefficient, and how one is to construct a confidence interval, and how one is to interpret the result, are questions which led to the resurgence of Bayesian ideas, it appears. It is clear that there are deficiencies in confidence intervals which were recognized by Fisher early in the game but have not been recognized by the Neyman-Pearson school. The whole question of what to do about conditioning is obscure. The concept of power is of questionable *general* value. The work of Buehler and others is very informative in this area but has not, it appears, been recognized by the Neyman-Pearson school.

Finally I come to Bayesian inference. In the first place I have to state my opinion that like the other formal theories of inference it ignores the main problem. I have already posed the question: How does one know that the data are a random sample from $N(\mu, \sigma^2)$, for instance? I also pose the question: should one make any evaluation of goodness of fit of model? The Bayesians have not discussed these questions as far as I know, and from this viewpoint alone I have to reject their outlook as carrying total force. Let us pass over these questions, however. It seems clear that there is no satisfying or even partially satisfying process for choosing a prior distribution. So to consider the utility of Bayesian processes I have to take into account the complete arbitrariness and subjectivity of the choice of prior distribution. Note that in the framework of non-Bayesian statistics, there is certainly some subjectivity in choice of model, but there are generally accepted tests of goodness of fit which must be applied before the analysis is accepted by any part of the public.

It is relevant also to note that quoting L. J. Savage (page 59) the theory of personal probability "is a code of consistency for the person applying it, not a system of predictions about the world around him." This is, perhaps, the most informative statement in the whole discussion and makes my position rather clear. I am interested in making predictions about the real world. The processes of science are aimed, I believe, at the making of predictions about the real world. The criteria of value of science are the quality of its predictions. I could continue indefinitely almost talking about this but the point seems obvious.

In discussing Bayesian inference, I am not discussing its use for the case in which the prior distribution is based on a model (as in some genetic problems) or is based on history. I object totally to the intrusion of arbitrary prior distributions and the use of priors which merely serve to make the Bayesian computation simple. This seems to me to be intellectual charlatanism.

If one does the Bayesian computation with some natural prior as an intellectual experiment, and something to be used with other computations from the data, for example, with other priors, and with non-Bayesian computations and one sees how these mesh, there is little basis for argument — except that I would tend strongly to ignore the Bayesian part of such computations that are presented to me.

But to close, it seems clear that if one has to make a real terminal decision one may have to have recourse to a Bayesian argument based on one's personal prior. The models that one examines as candidates for the description of data are given by some very vague Bayesian processes. The selection is based very much on one's personal experience of what was useful with data sets which are "like" the data set actually being examined. There seems no way out of the use of the word "like"—the processes of statistics are the processes of imperfect analogies.

On the use of the Bayesian argument I find the following statement very compelling:

> My third reason is that inverse probability has been only very rarely used in the justification of conclusions from experimental facts, although the theory has been widely taught, and is widespread in the literature of probability. Whatever the reasons are which give experimenters confidence that they can draw valid conclusions from their results, they seem to act just as powerfully whether the experimenter has heard of the theory of inverse probability or not.

This was written by R. A. Fisher in 1935, and 35 years later it carries great force for me.

In the interval between writing the above and the Symposium, I have had additional thoughts which I wish to include in the record. These I label as *Addendum*.

ADDENDUM

To resolve the obscurities about probability and inductive inference is equivalent in my opinion to laying out a philosophy of knowledge. Workers in statistics and particularly those working on theories of statistical inference are on the boundaries of philosophy often, and usually, I fear, without being aware of the fact. To resolve the difficulties, it is critical to obtain some sort of resolution of the nature of language. It is obvious that this is incredibly difficult, perhaps, or even, probably, impossible, because we have to use language itself to explain language. I am firmly convinced that this basic difficulty has not been faced by all the so-called thinkers through the ages, except perhaps by very recent workers. I am highly impressed, perhaps as a result of ignorance, by the example of Wittgenstein, who, it appears, decided that all he had written, and presumably, all the philosophers before him, was wrong, and yet, the old stuff is taught and the errors compounded to the nth degree. I was very struck by a quotation from Bertrand Russell given in *Time* magazine, from his reflections on his 80th birthday:

> I set out with a more or less religious belief in a Platonic eternal world, in which mathematics shone with a beauty like that of the last Cantos of the Paradiso. I came to the conclusion that the eternal world is trivial and that mathematics is only the art of saying the same thing in different words.

This is obviously a devastatingly strong statement. It is a statement that seems obviously wrong from some points of view. But it seems to contain clearly some elements of correctness.

I would like to use a moment to express a thought about the whole business. It is *obvious* to the scientist and it is obvious to the historian of science, that science is not a matter of formulating sentences from a fixed pool of elements. It is *obvious* that science is adding to its vocabulary daily; it is changing the meaning of old words. It finds that the old words were *poor* and it makes modifications and restrictions. This tells me that a theory of knowledge based on what one can do by logical processes with a *given fixed* language cannot get at the real nature.

Who am I to enter these very deep waters? Am I professionally accredited? Do I have a Ph. D. in philosophy? The answers are in the negative. But I make a somewhat reasonable claim which is the following. For 30 years I have been a practicing statistician. What sort of a *beast* is this? I deem it necessary to give a short explanation. For this period, I have spent a considerable amount of time talking to scientists about how they should collect data and how they can try to interpret their results. I have also tried to add a little to the science of genetics. How does a consulting statistician work? Does he say to the scientist: You must do this. You must collect data in this way. You must analyze the data in this way. And so on. You must make a t-test and if t is greater than 2.31, you must reject the null hypothesis? The answers are: Of course not! How idiotic can you be? And yet, the great bulk of workers on foundations of statistical inference seem to think in terms of such answers. Who am I to criticize this work? My answer is that I am a citizen of science and have a right to my views however queer they are. But I do not demand that others agree with me. I merely request the opportunity to place my opinions in the marketplace and then to let them sink or swim.

I believe the great bulk of workers on foundations have essentially no understanding of the processes by which science builds up its collection of empirical facts and models to explain these. I will go further and say that I have yet to read a philosopher who gives me the feeling that he has any understanding of the processes. I will make one exception, Karl Popper, but I have the impression that he is regarded as being *out in left field.*

If you want to get any understanding of what goes on, study the activities of someone like Cuthbert Daniel, or of any of a number of consultants. I am totally scornful of the idea that one can understand by pure thought, whatever that is, and to get some understanding of processes, you should open your mind to real situations, not the stupid hypothetical situations one finds in many papers; let me give some examples:

> to obtain some understanding of the growth of Easter lilies. (Why? Because some people in our society value these and others try to satisfy a reasonable need);

> to develop ideas of the nutrition of hogs;

> to get some understanding of the role of farmyard manure in the growth of plants;

> to try to develop better washing machines, better autos, better razor blades, and whatever.

There is the idea around that it is wise and noble to work on the big problems of humanity such as cancer, or education, but to work on the little problems I mention above is beneath the dignity of man. I believe one must become involved in big problems and small problems. In the last resort, science is an ill-defined collection of human intellectual processes, and to have some understanding of how scientists work, one must work with scientists at all levels. I have yet to meet a trivial problem. I have met scientists who have been brainwashed by statisticians to the view that their problems amount to the calculation of a linear discriminant and the scientists want to know how to do this. But when one really examines the real problem, one finds that the problem is much larger. Take the simple case of comparison of two groups. A simple problem? No! A very difficult one, made simple only by introducing unverified and unverifiable assumptions that lead to a *textbook answer* — often an "answer" that is more misleading than helpful.

I take the viewpoint that here at this symposium there is no point in any of us hiding our real opinions. If the aim is not free unrestricted presentation of one's views, the symposium is a waste of time. Additionally, I personally "may as well be hung for a sheep as a lamb". I have no desire to be nasty and unpleasant with colleagues in the profession. But I believe that statistical processes, data reduction, statistical inference, whatever it is, are critical to humanity. They always were critical, but recognition of this was slow in arriving. As an example which has caused already considerable human anguish and will cause still more for decades to come, I mention what I call the Jensen affair. For those who do not know of the matter, statistical reasoning (so-called) has been used to "show" that the Negroes of the United States are inferior to the Caucasians in mental abilities. The argument used is based on quite unreasonable assumptions. *Lies, damned lies, and statistics.*

The examination of data, reduction of data and discovery of explanatory and predictive models is, I believe, the basis of all knowledge, whether it be in the area of pollen, politics, or poetry. The degrees of quantification and the utility of any quantification vary considerably over this spectrum. Whether one talks about the growth of pollen in botany, or the use of imagery in poetry is not at an elemental level, a real difference in type of explanation. The vocabularies are totally different and so on. Of course, data reduction is based on a model (in poetry as much as biology) and there are many ways of getting models, including deductive processes, but the last resort is the appeal to data whether in the case of pollen or in the case of W. B. Yeats.

To attempt to write a good account of all that is involved is impossible. It is impossible for me because I do not know enough. But more important, it is impossible because each thought one has modifies the import of previous thoughts. The only way to write down a precise statement is to close one's mind to any new ideas, and then one produces models that are so simplified that they are not merely useless, but more important, are dangerously misleading. This has happened so much in statistics in my opinion, that the result is almost tragic.

All that I can do is to give a few short summary statements, based on the highly inadequate language I have, which are meaningful to me. I hope they will have some meaning to some of you who hear or read them. I shall list these views under short titles which hopefully contain a little information.

1. The Failure of Philosophy of Knowledge

I regard this as the crux of all our problems of understanding statistics and statistical inference. I feel that the failure is *abysmal*. I see no aspects of nobility in the failure. I see only a reflection of the arrogance of human minds which thought that they could encompass the whole of an unfolding universe by a few simple words and phrases. I was comforted recently by seeing the comments of Gauss (the Prince of mathematicians, and I imagine, a Prince of science) on the ideas of Kant, given in Bell's book, *Men of Mathematics*.

You see the same sort of thing (mathematical incompetence) in the contemporary philosophers Schelling, Hegel, Nees von Essenbeck, and their followers; don't they make your hair stand on end with their definitions? Read in the history of ancient philosophy what the big men of that day — Plato and others (I except Aristotle) — gave in the way of explanations. But even with Kant himself it is often not much better; in my opinion his distinction between analytic and synthetic propositions is one of those things that either run out in a triviality or are false.

Until I saw this, I had a strong doubt of the correctness of my views. Now I know, at least, that an intellectual giant whose contributions cannot be questioned, had views which give me comfort.

It is quite fantastic to me how individuals can try to develop theories of science and knowledge without doing science, like someone who works on the foundations of mathematics without actually doing any mathematics.

I have been struck in recent years by the role of historians of science. I wish that someone would write a history of statistics. I wish also that someone would write a history of the Bayesian idea. I wish that writers and teachers and lecturers on Bayesian ideas would give some of this history to their students. I wish they would tell their students about the *big* men who rejected the idea totally, *as well* as the *big* men who bought the ideas. The amount of brainwashing of students that is being done appalls me. The contribution of statistics to this is small but significant.

2. The Beauty of Mathematics

It would be easy to go from remarks I make to the idea that I despise mathematics and mathematicians. So before proceeding, let me lay that notion to rest. Insofar as I understand mathematics and this is probably a very debatable matter, I believe mathematics to be the most beautiful creation of the human intellect. I am continually floored by its beauty and power. I do not want to hang mathematicians up by their thumbs. I do not want to do this with anyone.

3. The Problem of Inference

Humanity has heard of this problem in a formal way for a few hundred years. I ask, and I claim no uniqueness for so doing:

Is this a real question?

A real question is a question for which the questioner has some idea of the form of an answer which will be admissible as a possible answer. (I seem to get the impression that this is the real domain of philosophy.) I would like to strip the question down to its elemental form. Here it is:

> I have two small index cards on my person. On each card is written a number. I will tell you what one number is. You are to make an inference about the number on the other card.
>
> The number on the first card is 3.1. What is your inference about the number on the other card?

If there is anyone in this audience who is prepared to make an inference, I would like to hear from him.

I am totally unprepared to answer the question. I do not understand the question. How can I possibly hazard any sort of guess about the unknown fact? I believe this is a false problem. But can I prove this? I have no idea of how one proves that a set of words which has all the linguistic form of a question is a meaningless *collection* of words.

So if I have seen 10 swans and they are all white, I have not the foggiest idea of the color of the 11th swan that I shall see.

Can we modify the question so that it has a partial answer? I believe, yes. If we say that we have looked at a random sample of 10 of 11 swans and we note that these are all white, can we say anything about the 11th swan?

But before we proceed, what do we mean by random sampling? Am I to believe that the 10 swans I see are a random sample of the 11 swans? Do I merely introspect and then say, "My personal probability of being presented with any one of the 11 sets of 10 swans is 1/11?"

When I try to say this, I find I am unable to do so. I don't know what my personal probability is.

486

It appears that some neo-Bayesians are prepared to inject personal probabilities at all stages. They have no problems and they obtain as a result a personal probability.

My only response to this is the standard one associated with computers: *Garbage in — garbage out.*

I know what random sampling is. It is the sort of thing that the Bureau of the Census and all sorts of groups use. The Bayesians may think the process is pure foolishness.

If we use random sampling and if we are merely told the color of each swan as it is drawn, I will use the ordinary elementary processes to obtain some probabilities, but I shall not obtain *the* probability (or even *a* probability) that the 11th swan is white. I go farther. I do not think any numerical probability has utility.

4. What Is Probability?

To the mathematician, this is not a problem (perhaps). The mathematician deals with an entity defined by his axiomatic system, and, for various reasons, calls this *probability.* He might as well call it *mass.* One can take any probability book and replace the word "probability" by the word "mass" throughout without affecting the beauty of the whole development. Or better still, pick a letter from the Japanese alphabet.

So the question is: What is probability in statistics? When Mr. Neo-Bayes says, for instance: "My probability is 0.3, that Nixon will have a second term," what does he mean? If I tell you that the *weight* of my 14 year old son is 170 pounds, you know what I mean but do you know what Mr. Neo-Bayes means? Suppose he says (I do not know that he does), "The figure 0.3 is a measure of my belief, put on a scale from 0 to 1, such that 0 and 1 have such and such meanings." Am I any further forward? I have to say, and I speak with authority about myself, that I am not.

Perhaps Mr. Neo-Bayes will say, "You are being difficult just to annoy. You surely are aware of a two-sided coin." And, of course, I am. And I say, "So? I see two possible results of a flip of the coin." And so we proceed, and we get nowhere for me until the idea of independent tosses and relative frequencies enters. I really don't know how such a conversation would proceed. I believe we should have such a conversation.

My point is simply that the words of Mr. Neo-Bayes have to possess some meanings for me, or it is a complete waste of his time talking to me and of my time, listening to him. Am I alone? That is not for me to answer.

But what does Mr. Neo-Bayes appear to be saying? Something like the following: "I really don't care what my words mean to you. I am concerned only about how I reach decisions when I have to make a choice. I am talking about a personalistic theory of decision *for me.*"

I am not trying to be funny or nasty. That is, indeed, what some members of our profession seem to be saying. I have to express my view that while a personalistic theory of decision is useful to me for my decisions, the essence of science is its depersonalization. To go to the other extreme and to take the position that science is totally objective is equally fallacious. But to assert that all of science is personalistic because it contains some personalistic elements is to remove from science all the interpersonal partial objectivity it possesses, which is its very core.

Where and how does Mr. Neo-Bayes get *his* prior probabilities? I have available to me in writing this, black pens, red pens, yellow pens, pink pens, blue pens. Mr. Neo-Bayes can give you *his* probability on which one of these I am using. He will give you or me his personal probability on any verifiable fact of nature. How? I do not know. Or am I incorrect? Will he sometimes say, "I am sorry, but I do not have any relevant information?" I shall then say, "What do you mean by that? You know what you had for breakfast this morning. Is that piece of information relevant?" A very basic problem in this area at present is that the channels of human intercommunication have broken down. (And this is happening in aspects of society more important than statistics.)

In this connection, it is appropriate, I believe, to note that much of the controversy of the last 40 years has revolved around the purely semantic difficulty that a relative frequency in an aggregate of cases should not be interpreted as a relative frequency in different aggregates of cases, and should not be interpreted necessarily as a *probability,* whatever that is. The relative frequencies obtain in the Neyman-Pearson theory of confidence intervals are completely correct. The wonder of the matter with hindsight, is that anyone should think these relative frequencies would be correct for identifiable sub-aggregates. So, in my opinion, part of the attack on the theory of confidence intervals, which has led to the resurgence of Bayesian ideas is *totally* misdirected. Other parts of the attack, which I discuss below are, in my opinion, very well aimed.

The same type of difficulty occurs, I think, in Fraser's structural inference. Fraser, it appears to me, passes from the fact that the frequency in repetitions of x such that x exceeds μ by 1.96 is 2½%. I am completely at a loss. Dr. Neyman wrote what seems to me to be an excellent critique of this argument which was first proposed by Fisher 30 years ago.

Really, the overall situation seems hopeless. We in the profession of statistics are essentially totally unable to communicate meaningfully with each other. The whole fabric of the society of statistics is falling apart. Let us not, I plead, ignore what is happening. Everyone of us is doing *his own thing.* And why? The answer is obvious. *It is our natural birth right!*

This is the real issue this conference should face. Anything else fades to nothingness in importance. Am I wasting my time expressing these views? I am sorry that I do in fact believe so. I will state my conclusion at the end.

5. The Idea of Tests of Significance

This idea is simple enough. We have data. We have history that we judge to be relevant. On this basis we obtain a model which incorporates the ideas of randomness, the lack of reproducibility of coin flipping, or whatever other analogy appeals to you.

We then map the data on the unit interval (0,1) and the result is a significance level. The mapping can be done in a zillion ways. From examination of possible rules of mapping, we conclude that certain modes of mapping tell us more than others. We use the mapping to give us a sort of standardized distance of the data from the class of data sets that are possible under the model or class of models.

We use the mapping rule to obtain distances of the actual data from different models in a class as indexed by a parameter, and so on and so on.

It is evident that scientists find this procedure informative. One has only to look at an area like psychology to see that some scientists bemoan the fact that they do not have any significance tests for some aspects of their tentative models.

There are vast obscurities in the whole matter, but these are not resolved by converting the procedure into another which has superficial resemblance. Nor are they resolved by pointing to obvious misuses. Nor are they resolved by setting up the straw man, that users of tests of significance regard them as a universal panacea.

6. Accept-Reject Rules

In particular, the nature of tests of significance is not resolved by converting them into accept-reject rules, which is the basis, it appears, of the whole Neyman-Pearson approach which has dominated mathematical statistics for the past 40 years. I exclude all distribution theory *per se,* of course.

This has led to a vast array of very fine mathematical work. But an array of mathematical work which has very limited relevance (note — I do not say "no relevance") to any problem that is not an acceptance sampling problem. This has led in my opinion to the essentially complete avoidance of the problem of learning from data, to the problem of arranging a vast array of empirical facts into an organized body of knowledge — a collection of models — that summarize the facts and have predictive value.

The criticisms are well-known. For example, where does one get α? What can be the role of messing-up the data by introducing the result of a coin-flip? What can be the role of unbiassed tests, when one cannot obtain an unbiassed test even for the simple case of binomial probability of p equals 0.3 without introducing extraneous noise? The situation is, I believe, pathetically stupid. A zoologist comes to me with data, and I, as a statistician, say to him, "Excuse me, please. To give you my suggested answer, I have to see what this coin toss gives. I assure you I have checked out the coin, and it gives heads with frequency of one-half."

There is little point in my giving a long talk on these aspects. They have been flogged, one would think, to death. But have the proponents of these procedures bothered to make any attempt to explain their motivation and to attempt to reconcile their approaches to what seem to be very reasonable questions? I believe not. Instead, the general drift of their developments appears to be to embed their approach in the theory of games, a theory which, it seems to me, is not at all a theory of how to play games, but merely a theory of the "value" of a game. As though, indeed, the problems of humanity are merely a matter of the value of games.

I think the situation is absolutely appalling.

7. Other Approaches

I suppose the remaining candidate, the panacea to some, is the likelihood principle. Perhaps I will have learned something of this here at this symposium. If not, I have nothing to say. Instead of mapping the data into the very dull space consisting of the two elements 0 or 1, or of mapping the data onto the unit interval, we map the data into a function. If the idea of a mapping rule is introduced, I want to know some properties of the mapping rule.

8. Data Mapping Rules

Perhaps this mode of expression aids in some sort of resolution of the enigmas.

To take the case of accept-reject rules, it seems clear that if we assume that the size of test is fixed, there is a very simple requirement that the power be maximized. I see no possibility of improving on this recipe for this problem, with the restrictions that are built into it.

To turn to the case of using accept-reject rules for the evaluation of data, for the purpose of getting suggestions from the data, it seems clear that it is not possible to choose an α beforehand. To do so, for example, at 0.05, leads to all the doubts which most scientists feel. One is led to the untenable position that one's conclusion is of one nature if a statistic t, say, is 2.30 and one of a radically different nature if t equals 2.31. No scientist will *buy* this unless he has been brainwashed and it is unfortunate that one has to accept as fact that many scientists have been brainwashed.

It seems intuitively totally obvious that one does not gain more from data polluted by a random device than from the unpolluted data.

It seems clear to me, if one uses accept-reject rule theory as a basis for data evaluation, rather than for yes-no decisions, that if a set of data leads to rejection with a 5% rule it must lead to rejection with a 10% rule. I would like to see a theory which incorporates this and the non-use of random devices as basic essential ingredients and not as afterthoughts, or as aspects of a process which are dismissed with a wave of the hand or vague apologies.

It seems to me that a mapping rule must be based on a knowledge of the properties of the mapping rule in some set of repetitions that are possible under the models envisaged. It is here that there are great difficulties, which have been discussed by Barnard, Cox and Basu. The only answer that I can see for the big dilemmas that arise in this context, is for the statistician to explain to the scientist the concept of classes of repetitions. It is entirely obvious to me, incidentally, that the concept of a class of repetitions is easy to explain — or, indeed, that there is no need to explain because the idea is inherent in scientific work. But the idea that there are several possible classes of repetitions is more difficult. I think the scientist has to decide what classes of repetition are of interest to him. If I am unable to raise his interest in any class of repetitions, I shall take the view that he has a unique set of data which cannot be related to other *potential* sets of data. I shall then say that I am unable to help him, and that I have the opinion that no statistician can help him.

What ideas can one bring to bear on other mappings of the data? I suppose one can think of mappings of data into almost any class of functions. What classes of functions are candidates? The neo-Bayesian answer is the class of *probability* distributions defined on the parameter space. The utility of this if one has obtained the data by random sampling of a population of populations and then random sampling within the selected population, is obvious and elementary. Here by *random sampling,* I do not mean anything you or Mr. Neo-Bayes wishes to regard as random sampling. I mean the random sampling that the Bureau of the Census does. With random sampling, I get a frequency distribution. If you want to call this a probability distribution, you can. My point is simply that a posterior distribution is in fact a statement of the properties of a defined family of repetitions *in this case.*

But what if you use some prior probability distribution and go through the process, what do you have?

It seems to me that the claims of the neo-Bayesians in this respect are about as shoddy as a supposed intellectual argument can be. They say that it doesn't matter what prior you put in (as long as it is somewhat reasonable?) then the posterior distribution converges to the unit spike corresponding to the population being sampled. This is presented as the grand and great attribute of intellectualized Bayesianism. I have yet to see any force in this argument — the talk about non-informative priors, diffuse priors or whatever.

The answer to me is very simple. If you choose an accept-reject rule to construct a $(1-\alpha)$ confidence interval, then that interval will narrow down to the truth. A reasonable significance test will give a significance level which converges to zero for all but the true value. And so on.

So we have a proposed discriminant between this method and others, which carries no force for me. I believe it carries no force for scientists in general unless they have been brainwashed.

What is the relevance of this type of asymptotic argument anyhow? I wonder. I understand asymptotics as an approximative device for a finite problem. But beyond that, I become very queasy.

So I repeat my question. What are interesting and relevant properties of any rule for mapping data into some space, whether the space of 2 points 0 and 1, the unit interval, or some space of distribution functions, or some space of more general functions. I understand some properties of the mapping rule in Accept-Reject Theory. Some I accept as relevant; others I reject as totally irrelevant. I understand some properties of the mapping rules of tests of significance, and others not. I understand from the literature a little about properties of the mapping rule from the data to the mean or median of a posterior distribution. I understand a little of the mapping rule which takes the data into the likelihood function.

But the adherents of the likelihood *principle* and of arbitrary Bayesian processes, tell me that properties of the mapping rule are irrelevant. You are supposed to have a *gut* (I apologize for this colloquialism) feeling that the mapping rule is good.

What is more, you are to be myopic. You are to consider one mapping rule, and then you are to live and die by it. (*Obviously,* I do not expect anyone to agree with me. Please recall my beginning remarks. I give you my opinions just as they occur to me). I ask: Just how idiotic can one be? Consider the scientist who says beforehand, "I will do just such and such with my data." It is really obvious that all useful theories of the real world have arisen by interaction of humans and data. One must allow the data to make suggestions which had not occurred to one's mind before one obtained the data. It is here also that I become completely ill at ease with the Neyman-Pearson theory of inductive behavior — summarized by the rule: "Choose the most powerful test and then accept or reject." I go further. I say that Dr. Neyman does not do this either. The relationship between his theory and his application to me as a viewer (and please note that I am as complete an authority as there is on this) is very obscure.

9. Relevance to Data Analysis and Interpretation

It is clear to me that statistical methods, as one finds them in the standard books, are useful to scientists. It is clear also that they can be misleading and can be used badly. But the empirical fact I state is clear to anyone who' cares to read science, for example, the magazine *Science*. That these methods may be misleading bothers some, but I believe with total conviction that the only ones who believe that the whole process can be reduced to application of a bunch of rules are those who have no contact with science as it is.

These methods need to be augmented and, indeed, they are being augmented by data analysers. The problems of how to look at data are being worked on by those who have the data. If the profession of statistics and academic departments are not very careful, they will become a mere adornment to universities with no relevance to the huge problems of society which have always existed, but are now appreciated by the bulk of our citizens and not only by an intellectual elite, many of whom *had* values which are no longer acceptable.

To conclude, I think probability inference is a will-of-the-wisp, except in a very limited context of random sampling from given finite populations, and even in that area very few inferences are possible. But I think statistical inference (which I characterize loosely as the collection of processes by which we learn from data) is a valuable battery, which cannot be given the type of *hard* properties that some are looking for.

COMMENTS

V.P. Godambe:

I congratulate Professor Kempthorne for naming 'the finite population sampling' as the genuine problem of statistical inference. Personally I think it is a great pity that this historical conference should have failed completely to recognize this genuine problem of statistical inference.

I am sure Professor Kempthorne does not want us to take his references to the Census Bureau seriously.

I.J. Good:

I hope we all agree with Dr. Kempthorne's comment that the likelihood function is exhaustive (of the evidence from an experiment) given a parametrically specified model. I also agree with his putting in a good word for significance tests, and for saying that personal views always intrude. I now come to my critical comments.

When L.J. Savage said (page 59 of what book?) that personal probability "is a code of consistency for the person applying it, not a system of predictions about the world around him" I think that this remark might have been misleading in regard to his own views; certainly in regard to mine. A person makes various probability judgments mostly about the world but sometimes about mathematics, and the purpose of the theory of personal probability is to *improve the objectivity* of these judgments by checking them for consistency and for drawing deductions from them (I.J. Good, *Probability and the Weighing of Evidence*, 1950). Thus subjective views are usually about the world.

Quantum physicists have often argued that statements about the world can be interpreted only in terms of predictions concerning observations. Some of them even deny the existence of an objective world. This is a more extreme form of subjectivism than is required in subjectivistic statistics, but is closer to the views of de Finetti and Savage than to mine. (See *Synthese*, 20, 1969, 16-24). Nevertheless de Finetti has a point because we cannot prove that physical probabilities or even the world exists. Like robots, *we can only process information and exert some control over our future inputs*, but we cannot explain why this control is effective without forming the concept of an external world together with an internal representation of it. De Finetti proved that if our subjective probabilities are consistent then they will be just *as if* physical probabilities exist and have a subjective prior distribution. I find it easier on the imagination to suppose that the physical probabilities really do exist. In this connection Popper once sent me a preprint of an article on "probability magic" in which he criticized de Finetti, Savage, and myself. When I showed him that his criticisms did not apply to my views he simply deleted my name from the article!

The remark about using initial distributions which merely serve to make the Bayesian computation simple could be misleading. Consider for example the use of the beta distribution, as suggested by G.F. Hardy in 1889, when estimating a binomial physical probability. The use of the distribution is not *merely* to make the Bayesian computation simple. It is *intended* to represent an adequate variety of *unimodal* distributions. This kind of model is similar to the kinds of models introduced by non-Bayesians for other purposes, and the justification is intended to be similar, namely robustness as well as convenience. Moreover if we estimate the physical probability p of a binomial distribution by a linear function of the number of successes, and if the estimate is interpreted as a final subjective probability, then this is *equivalent* to the assumption that the initial distribution of p is of the beta form. So sometimes the *convenient* prior might have indirect non-Bayesian appeal. But I'm sure even Bayesians sometimes sin. Incidentally I think a mixture of beta distributions is usually preferable to a single one, certainly this is true for Dirichlet priors when one is investigating multinomial "primaries". (*J. Roy. Statist. Soc.*, B, 29, 1967, 399-431. On a point of terminology, perhaps what I call distributions of types 1,2,3,... might also be called "primary", "secondary", "tertiary"....)

The quotation from R.A. Fisher raises a paradox. For Fisher in his later years tried to blame Bayes for the fiducial argument; and certainly it was concerned with the probabilities of hypotheses. Apparently Fisher felt the need to talk about probabilities of hypotheses, and this is inconsistent with the implication that there is no need for inverse probability. His remark should not have continued to carry great force for himself.

I have a number of miscellaneous smaller comments. On page 474, Dr. Kempthorne says that he would condemn a man without apparently having any idea of the probability that the man is guilty. I cannot really believe that Dr. Kempthorne is that irresponsible. If he were *forced* to choose between taking a bet of 10 to 1 against or 10 to 1 on presumably he would select the one that was consistent with his judgment of the probability of the prisoner's guilt. Perhaps he meant he would have no *precise* idea of the probability: that would be consistent with partial ordering of probabilities, that is, with upper and lower probabilities.

Regarding page 475, I agree to some extent that it is largely a matter of personal taste whether the frequency definition of probability is used or the subjective approach. But it is not entirely so because there are some propositions for which it is difficult to see any frequency interpretation; for example, the probability that some physical theory is true. I don't know whether anyone finds it helpful to imagine an infinite population of worlds having a variety of sets of physical laws.

Later on the same page, Dr. Kempthorne states that the idea of logical probability carries no force for him but he then goes on to commend to some extent a theory in which probability is related to the structure of language. This approach, used by Carnap, is precisely an attempt to give values to logical probabilities. So I think Dr. Kempthorne has contradicted himself on this point, but it is perhaps only a semantic matter.

He says on page 479 that a Bayesian has to inject his *prior* and use the Bayes rule. But a Bayesian might make use of the device of imaginary results, in which case it is just as true to say that he has injected a *posterior*. (That is not intended in a medical sense.) Also he can use the Bayes/non-Bayes compromise (see Priggish Principle number 13).

On page 479, Dr. Kempthorne says that the pure problem of inference seems to be totally artificial. Here I think the *pure problem of inference* means the estimation of a binomial probability. If so, I think it is basic, and, if it is ever solved to everybody's satisfaction, a theory of non-subjective statistics will have at last begun.

Dr. Kempthorne claims that neo-Bayesians say it does not matter what prior is used, within reason. This is *not* the current position with regard to priors in many dimensions, especially when hypothesis testing is the problem. In such problems I think the device of imaginary results is essential. (See, for example, my paper "A Bayesian Significance Test for Multinomial Distributions", *J. Roy. Stat. Soc., B,* 1967).

By the way, I think I mean the same by *random sampling* as Dr. Kempthorne, although I sometimes call myself a neo-Bayesian.

I agree with Dr. Kempthorne's main conclusion that statistics cannot be embodied in a finite set of rules. This is why I used the classification of axioms, rules, and suggestions (*Probability and the Weighing of Evidence*, 1950). The suggestions are informal and imprecise and unlimited in number.

I also agree that it is useful to improve razors.

H. E. Kyburg:

I am very gratified that Professor Kempthorne has come to see that statistics and epistemology are essentially one subject. Several years ago the question arose at the University of Rochester as to whether or not statistics should be

494

split off from the mathematics department and made into a distinct department. Allen Wallis, our president, argued that statistics should no more be part of the mathematics department than should physics and that if statistics should be included in any other department, rather than being a department in its own right, that department should be philosophy. Whereupon, of course, (rather to my disappointment) all my colleagues in philosophy voted to establish a separate department of statistics.

Now it is very hard for most philosophers to read statistical literature. It's hard for me, and I have the advantage of having been a professional engineer and of having been a professional mathematician before I became a professional philosopher. Professor Kempthorne believes, and he may well be correct, that it is easier for a statistician to become an epistemologist than for an epistemologist to become a statistician. But I don't think it is quite as easy as he thinks it is. Everybody knows how easy it is to lie with statistics; not everybody knows how hard it is to tell the truth in philosophy.

Statistics and philosophy are much alike: From the layman's point of view they consist in belaboring the obvious, or else in pursuing devious lines of reasoning to conclusions which seem nearly as dubious after the argument as before.

Simply as an illustration of this, I should like to remark on Professor Kempthorne's philosophical addendum. On the first page, he quotes Bertrand Russell with apparent approval as saying that "mathematics is only the art of saying the same thing in different words". Professor Kempthorne regards this as devastatingly strong, though I think most philosophers of science not only agree with it, roughly, but find it very useful and illuminating. "Roughly" covers a multitude of both exegetical and philosophical detail not worth going into.

I should now like to contrast this statement, and Professor Kempthorne's attitude toward it with the statement of Gauss which Professor Kempthorne quotes with even greater approval. We all deplore mathematical incompetence — no less in philosophy than in statistics. But included in the quotation is Gauss's opinion that Kant's "distinction between analytic and synthetic propositions is one of those things that either run out in a triviality or are false". Analylicity, in precisely Kant's sense (almost), is what Russell is attributing to mathematical propositions when he says that mathematics is the art of saying the same thing in different words. It is hard to see how one can accept both Gauss's view of the analytic synthetic distinction (now questioned by many philosophers) and Russell's view of pure mathematics, at the same time!

It might be mentioned parenthetically that the classical philosophers Leibniz and Descartes were not altogether without mathematical competence.

On page 478, Professor Kempthorne writes that "Our society in...its statements of its knowledge...depends critically on statements like the following: the weight of object A is 712 grams; the density of lead is 11.4; the velocity of a particular automobile at a designated time and place was over 25 miles per hour, and so on. Can we have similar statements about probability?" This is a puzzling question. Are these statements not probabilistic? When I say that the weight of the object A is 712 grams, am I not making a statement about the mean of a distribution? If not, what is the "operational procedure" for establishing that this is the weight? Isn't that operational procedure just a particular case of a procedure for establishing the mean of a distribution? When you and I agree that the weight of A is 712 grams, it seems to me that we are regarding our measurements of A as producing numbers that are *like* (to use Professor Kempthorne's useful word) the numbers we would get from a distribution that has a mean close to 712 and which is sort of normal looking in a fairly extensive region about 712. I don't wish to suggest that there are no problems about coming to agreement on the weight of A; on the contrary, I would suggest that there are very serious problems; that they are theoretical problems of inference; that, as evidenced by the practical agreement concerning measurement, they are not insoluble; and that the problems of deciding when any sample is sufficiently "like" a sample drawn from a certain hypothetical population (or a certain class of populations) are no more baffling in principle that the problem of deciding when to agree that the object A weighs 712 grams.

Why try to formalize what underlies this agreement? For the very reasons that Professor Kempthorne himself gives: "In a formal theory the rules of language manipulation and of adjoining symbols to make new sentences are very tight whereas in a purely verbal theory they may be so loose as to give rise to deep disagreements". I add: as they have in discussions of the foundations of statistical inference. And of course we may do this without making the absurd assumption that the language of science is eternally fixed!

J. W. Pratt:

The case of r successes in N trials is not typical. In particular it involves no nuisance parameters. Even this case is not as easy as Professor Kempthorne at times suggests. We need to know both r and N. A frequency statistician will want to know also the sampling scheme used. We want some kind of inferential statement: we wouldn't simply assume $p = r/N$.

If you are really only summarizing data why go to the trouble to do it by inverting a family of tests? Incidentally, I would not think it satisfactory or even possible to limit the interpretation of confidence intervals to their role as inverted tests, even if I considered tests a satisfactory instrument themselves.

Now a summary of Professor Kempthorne's main argument as it comes across to me: Fiducial and structural methods are nonsense. Jeffrey's Bayesian and subjective Bayesian methods are nonsense. Likelihood methods are nonsense. He doesn't say directly that orthodox methods are nonsense but he says it implicitly by his remarks on pages 481-482 and by Section 6 of his Addendum. In short, he says all methods are nonsense, therefore use orthodox methods.

At least Professor Kempthorne is fairer than those who argue that Bayesian methods are unsatisfactory because they contain a subjective element, but never notice that their own methods contain serious defects as well (equally serious, in my opinion) misleadingly cloaked in an aura of objectivity which is really tangential.

When it comes to priesthood, I cannot help remarking that orthodox statisticians are a priesthood too, as is evidenced by the number of practicing scientists who regard statistics as merely a kind of holy oil to pour over their data. (Many others disregard it altogether). As for brainwashing, there is plenty of orthodox brainwashing too. Furthermore, let me point out that there are Bayesians who do applied work too. I won't hang you because there are some orthodox statisticians who don't practice, if you won't hang me because there are some Bayesians who don't practice.

Professor Kempthorne may regard all schools as equally guilty of priesthood and brainwashing. And he appears to recognize that no methods are anywhere near justified by his lights. What I can't understand is why, under the circumstances, he comes out so purely orthodox.

It's time we all went beyond obvious criticism and even travesty of others' positions and tried to discuss the real issues and to do something about defects of our own methods. If you are going to convince the Bayesians, or anyone else, that they are wrong, you've got to deal with the best construction you can put on what they're doing, not the worst. Bayesian practice is no more limited by Bayesian theory than Professor Kempthorne's practice is limited by the kind of orthodox theory to which he also objects.

Finally, if you are going to reconvert heretics or convert intelligent heathens to orthodoxy, you've got to give it a convincing basis.

REPLY

It appears that I have little basic disagreement with Dr. Kyburg. I regard the quotation of Russell as too strong, because the discovery of the logical implications of a set of axioms is not a useless activity. But modern mathematics seems to validate part of what Russell is saying, in that mathematics consists partly of the development of sentences from primitives by inexorable rules. A super-mind would be able to see the primitives and the resulting complex statements all as one entity. But even this is misleading, because the vocabulary is growing all the time. The quotation from Gauss is also perhaps a bit unfair, because of what Dr. Kyburg says.

A statement that the weight of an object is 712 grams, clearly uses a conventional abbreviation and probability ideas as Dr. Kyburg says. It is based on the idea of a process of weighing that is under statistical control, giving a result of 712 grams with a standard deviation that is negligible with regard to the intended uses. It does not use, I believe, any idea such as "the probability that the true weight is in 712 ±1 grams is 0.99."

I applaud very strongly the activity of Dr. Kyburg as a philosopher in interacting with statistics. My reading of philosophers in general suggests to me that as a body, with almost no exception, they are not in touch with the real problems of accumulating knowledge.

In reply to Dr. Good, the quotation from L. J. Savage is from *the* book, *The Foundations of Statistics*. I hope, with Dr. Good, that the remark is misleading, and for the same reasons as Dr. Good. But we all must depend on the written word, and I find that I must always read the fine print. If one wants a final or posterior probability of a hypothesis one has to input a prior. The part of the Bayesian process I do not like is the use of a prior which makes the calculation easy. If the neo-Bayesian introspects and finds that a prior is approximately a particular beta distribution with certain parameters, it is reasonable to use this beta because the ensuing calculations are easy. But the bulk of neo-

Bayesians seem all to use the same prior for a particular model, such as that of independent Bernoulli trials. They all seem to have the same history and they all seem to take over a task on which they are not competent but on which the investigator, if anyone, is competent, because he has, presumably, the background. Also we see papers on a comparison of Bayesian conclusions with confidence conclusions, as though the former is a well-defined entity.

I agree with Dr. Good that Fisher's writings are most obscure and do not seem to be internally consistent. I find this to be true of *all* the writers, so we should not point the finger especially at Fisher.

As regards condemning a man without a probability that he is guilty, I do not think I am irresponsible. If my opinion is that the model that he is innocent is quite untenable and the model that he is guilty is totally tenable in the light of the evidence I shall vote for the guilty verdict.

It appears that some people need probabilities of hypotheses. I think I do not. I am at present highly convinced that my probability that the theory of relativity is correct is zero. A theory is just a model and I have no doubt that every model we have is just a stepping stone. So my probability for any scientific theory being true is zero.

With regard to a theory of probability being related to the structure of language, I am open to the possibility that there may be some fertility in the idea. But to extrapolate this to the feasibility of logical probabilities does not seem permissible to me, because it is not clear that language is totally in accord with any sort of logic.

As regards significance tests, the literature is really most obscure. I have tried to understand Fisher's overall idea of these without total success. I read Jeffreys who has a very different idea and am quite unconvinced, and I find Dr. Good to be close in some senses to Jeffreys. On the other hand, I find, perhaps incorrectly, that the idea of significance tests which appeals to me and I believe to scientists is not used in the foundational literature. This is the idea that a significance test is a standardized measure of *a* distance of the actual data from the set of data sets which the probability model implies. If this distance is large, I feel that the probability model is untenable. I may, however, "accept" an untenable model because I can envisage no other model that is consonant with the data.

A constant recurring difficulty in interpreting the present conflict is that the neo-Bayesians are not a homogeneous class. It is clear that Dr. Good is not a typical member if one defines the class by the writings of L. J. Savage, Lindley, and R. Savage for instance. I have great difficulty in envisaging the processes which any of the broad class of neo-Bayesians would follow with a real data interpretation problem. I have the impression, perhaps fallacious, that they do not interact with people in the "knowledge business". I would like to see case studies of any such interaction.

With regard to Dr. Pratt, I surmise that he regards Neyman-Pearson theory (with emphasis on Neyman) as "orthodox". I am in the position that I regard some of Neyman-Pearson theory as relevant but not totally forcing. The notion of behaviour of a procedure under some class of repetitions is essential in my opinion. But the problem of defining an appropriate class of repetitions is very difficult, as the work of Basu, for example, on the theory of ancillaries exhibits. I see some merit in an honest Bayesian approach, but not in the use of "intellectual" priors like those of Jeffreys. It would be wonderful, perhaps, if a reasonable set of postulates led to a unique prior. But even in this case, the posterior probabilities would not have demonstrable predictive value.

I have to agree with Dr. Godambe on the failure of the symposium to address itself to a problem which I regard as completely fundamental.

With regard to the Bureau of the Census, I was using this organization as an example of a critically important public body which uses random sampling or probability sampling. I am not expert on the full spectrum of its activities, but, I surmise, as knowledgeable as Dr. Godambe. There may be defects in some of their processes and I certainly do not wish to make any sort of "blanket" approval. However, I regard the comment of Dr. Godambe as irresponsible *of the highest order.* As I said in my own essay, we are not, or should not be, playing games. If Dr. Godambe would work harder to be in touch with real problems and would make a much greater effort to present his ideas intelligibly, he might aid the field. Dr. Godambe has acted as a "spur", but this is not enough. He would probably make the same sort of statement about me. My reply would be to stand by the record as it unfolds.

*In order that the readers can make sense of this exchange between Professors Kempthorne and Godambe it is necessary to point out that during the previous discussions Professor Godambe had said that "unfortunately most sampling bureaus, Census Bureau included indulge many times under the name of the applications of statistics in 'unthinking rituals'. And it is ridiculous on the part of Professor Kempthorne to refer to these rituals as instances of genuine applications of statistics."—Editors.

CRITICAL LEVELS, STATISTICAL LANGUAGE, AND SCIENTIFIC INFERENCE

Irwin D.J. Bross
Roswell Park Memorial Institute

1
Critical Levels and Complacency

Most of us have become so accustomed to the traditional *critical levels* of statistical methods that we tend to take them for granted. When we read an article in the scientific literature which uses statistical techniques we expect to see the familiar *5% probability level* or the *1% probability level* and usually this is what we find. Of course, there may be variant terminology — *95% level* or *99% level* — or there may be calculated values of probabilities given, but in nearly all current scientific studies which use anything more than the most elementary statistics, the procedures employ the concepts and the language associated with critical levels.

From time to time the notion of critical levels has come under heavy attack — generally on rather abstract theoretic grounds. However, most statistical practitioners have tended to be quite complacent about these attacks and it is only rarely that anyone bothers to answer the objections that have been made to critical levels. Of late the attacks on critical levels (and the statistical methods based on these levels) have become more frequent and more vehement. For example, in a recent paper in the *American Statistician* (1), Jerome Cornfield asserts that the "biostatistician's responsibility for providing biomedical scientists with a satisfactory explication of inference" cannot be met if statistical methods are based on critical levels. The only good answer to such an objection is to provide a *satisfactory explication* for the use of critical levels in scientific inference — and this is the main purpose of this article.

In one sense Cornfield has a legitimate complaint: it is not easy to find a realistic and convincing explanation of the role of critical levels in scientific inference. Elementary statistics texts are not equipped to go into the matter; advanced texts are too preoccupied with the latest and fanciest statistical techniques to have space for anything so elementary. Thus the justifications

500

for critical levels that are commonly offered are flimsy, superficial, and badly outdated. When the Neyman-Pearson and Fisherian theories were first advanced in the 1920's and 1930's, their original conceptions were based on limited experience in the use of the new statistical methods (with a focus on agricultural applications). The justifications offered reflected the *philosophy* or way of speaking about inference which was popular in those days. But time has not stood still. In the past generation there has been a further diffusion of statistical methods through the sciences — the biomedical sciences in particular — accompanied by a vast accumulation of experience in the utilization of these methods in actual scientific inference. The statisticians who have worked with critical level procedures have learned from their experience. These lessons of experience have been incorporated in that body of knowledge called *good statistical practice*. A satisfactory explication of critical levels is therefore closely tied to current statistical practice.

2

Changes in Statistical Practice

The precepts of good statistical practice tend to be passed along by word of mouth from one practitioner to another; they rarely appear in textbooks and they are hard to teach in a classroom because a certain amount of practical experience is needed in order for the precepts to be meaningful. In the first decades of this century the statisticians who were developing new theories — including Sir R.A. Fisher — were in close contact with good statistical practice (and were not ashamed to admit it). Unfortunately, many of the theorists nowadays have lost touch with statistical practice and as a consequence, their work is mathematically sophisticated but scientifically very naive. As Cornfield acknowledges "Biostatisticians have tended to regard the(se) theoretical developments...as unduly abstract and perhaps of no great relevance to statistical practice." This divorce from practice leads to sterile algebraic exercises. Most of the fruitful concepts in statistics (including the concept of a critical level) represents ideas which were implicit in statistical practice and which theorists then formulated into a more explicit and precise language.

During the past 20 years there have been important changes in good statistical practice. If you will forgive me an autobiographical example, I can illustrate this point. During the late 1940's when I was a graduate student at the Institute of Statistics at North Carolina State College, almost all of the staff were ardent Fisherians. In those days it would have been little short of heresy to say that a distribution-free test — such as the Wilcoxon test — was preferable to Fisher's two-sample *t*-test. The justifications of Fisherian methods that were given at the time tended to stress that these were the best possible procedures and any other methods were, at best, *quick and dirty* substitutes. On grounds of theoretic *efficiency* the distribution-free tests were rejected out of hand. The tone was sharply prescriptive: *One must reject inefficient methods.* At that time the *normal distribution* reigned supreme and any skeptic was annihilated by asymptotic proofs and the central limit theorem. This is how things were 20 years ago. Since then, good statistical practice has greatly changed — so have the justifications for the statistical methods that are used. It is true that many of the outmoded notions still persist in theoretical discussions of statistical methods. But a satisfactory explication of statistical inference does not have to rely on these old arguments.

Today a conservative (but thoughtful) Fisherian such as Donald Mainland can recommend rank tests for biomedical studies with the remark that "We are losing little in efficiency, and perhaps gaining some, when we employ these tests instead of *t*-tests" (2). To a theorist this comment might seem contradictory: How can we both lose and gain in efficiency? The answer, of course, is that when we use a statistical method in an actual scientific study we never know what the *true* distribution is. We never know what is the unique *best test*. As working statisticians we have stopped pretending that we know these things. We are willing to admit that in an actual application we do not know whether we are dealing with a nearly normal distribution (where the rank test is a little less efficient than the *t*-test) or a non-normal distribution (where the rank test may be slightly more efficient). If the issue seemed important, then many practitioners nowadays would simply run both tests and see if they gave different results. We have become a little more mature; we no longer have to rely on exaggerated claims of *bestness* or *uniqueness* to provide a satisfactory explication of scientific inference.

3
Communication Networks in the Sciences

Now let us consider what kind of justification can be given for the use of critical levels and for the standard statistical methods that employ the customary 5% and 1% probability levels. We would like this justification to convince two different groups of people — the biomedical scientists (and others) who use statistical methods in their research and the biostatisticians who are familiar with good statistical practice in the contemporary sense of this term. Therefore a satisfactory explication would have to be meaningful both in the scientific context of research studies and in the context of statistical practice. To anchor this discussion to experimental realities, it will be useful to take a specific actual study as a continuing example. Since the issue was raised in connection with sequential analysis, let us therefore consider one of the first sequential trials that was ever run — an analgesic study by Newton and Tanner (3) which is discussed at some length by Armitage in his *Sequential Medical Trials* (4).

The study was a test of a new analgesic (N.A.P.A.P.) against a standard one, codeine. A randomized crossover design was used with each patient serving as his own control. Each drug was given for a week and at the end of the two week trial the patient was asked to state his preference. A second replication was carried out in the next two weeks and patients with consistent preferences were plotted on the sequential chart for Bross's Plan B (5). After 42 patients were entered in the study, 23 consistent preferences were found with 18 favoring codeine and 5 favoring the new drug. In accordance with the stopping rule of Plan B the test was terminated and it was concluded that the preference was for codeine. The critical level here was 1% for the Type I error and the subsequent discussion will be confined to Type I errors (though much of it applies equally well to the Type II error specification of the sequential plan).

What was the scientific justification for specifying the critical level here? As a first step toward an answer let us consider the scientific objectives of a clinical pharmacologist or other scientist who is concerned with the testing of analgesic agents. Clearly his purpose is to reach scientific conclusions concerning the relative efficacy (or safety) of the drugs under test. From the investigator's standpoint, he would like to make statements which are reliable and informative. From a somewhat broader standpoint, we can consider the larger communication network that exists in a given research area — the network which would connect a clinical pharmacologist with his colleagues and with the practicing physicians who might make use of his findings. Any realistic picture of scientific inference must take some account of the communication networks that exist in the sciences.

This is not the place for any comprehensive discussion of communication networks in the sciences. However there are 3 key points which should be noted. First, messages generate messages. For example, when an investigator reports on an analgesic there are likely to be repercussions. A colleague may repeat the study. Physicians may use the new drug. The original message may thus lead to further messages which may or may not agree with the original message. A second key point is: discordant messages produce *noise* in the network. Several different kinds of noise can be distinguished. Some noise comes from random sources and would be often called *sampling variation* or experimental error or *biological variation*. Other noise comes from non-random sources. These sources are sometimes called *artifacts* or *biases* or *extraneous variables*. A third key point is: statistical methods are useful in controlling the noise in the network. Different methods would be used for different sources of noise. Critical levels are used to control random noise. Other procedures (e.g., the randomization and crossover in the sequential trial) are used to control non-random noise. The multiple sources of noise constitute one of the *facts of life* of scientific research and no satisfactory explication of inference can ignore this problem.

4

Control of the Noise Level

The noise in a communication network obscures and obliterates the informative messages. When the *noise level* becomes high enough, communication breaks down. This is not a hypothetical matter. Prior to the advent of controlled clinical trials, the noise level in the analgesic testing area was high enough to seriously impede progress. The statistical control of noise in controlled trials has brought a marked improvement — an improvement which has been recognized by most clinical pharmacologists. To be sure, there are still incorrigible optimists who issue enthusiastic reports about drugs irrespective of what data — if any — they may have. But reputable medical journals are increasingly insistent on adequate statistical analyses and even when the editorial processing slips up, a careful reader can tune out the noise. Thus the specification of the critical levels (e.g., the 1% Type 1 error in the sequential trial) has proved *in practice* to be an effective method for controlling the noise in communication networks. It should be understood, however, that the critical level does not do the job all alone — it is only effective in conjunction with the other protocol features of a modern controlled clinical trial.

How does the critical level control the noise in a communication network? The basic principle is relatively simple. It might be called the principle of extended local control. The strategy here is a recurrent theme in modern statistics: by exercising error-control in each segment we can achieve error-control in a broad area. A narrow technical example of this strategy is to be found in the construction of many conditional significance tests. Thus if x is the conditioning parameter and for each value of x the error-rate is kept under 5%, then the rate will be controlled for all values of x. In the communication network, we can get a clearer picture of the underlying mechanism by considering a hypothetical example with specific numerical values.

Let us first suppose that we have an uncontrolled drug testing network where all of the investigators issue a favorable report on every new drug they test (some years ago this would not have been an entirely unrealistic model for certain networks). Suppose that 4 out of 5 drugs tested are relatively ineffective (i.e. compared to a standard agent). Then the noise level in the system would be so high that communication would be chaotic — 80% of the reports circulating in the system would be false and it would be impossible to place any reliance on the messages. Now consider what happens if each member of the network exercises self-discipline and employs standard statistical methods to keep his proportion of false positive claims at a low level — say 5%. Even with this self-discipline there would still be false reports. In 100 studies there would be on the average 2 false positive claims coming from the 80 relatively ineffective drugs. There would also be 2 claims of advantage for the standard agent which might be false (but which would not be likely to produce much confusion). There would be a notable improvement in communication since the vast majority of the false reports would *not* be circulating. The proportion of false reports in the total of circulating reports would still be a little high — 4 out of 24 or 17%. But the self-discipline would reduce the noise level to a point where a member of the communication network could place some reliance in the claims for new drugs.

This simple example serves to bring out several features of the usual critical level techniques which are often overlooked although they are important in practice. Clearly, if each investigator holds the proportion of false positive reports in his studies at under 5%, then the proportion of false positive reports from all of the studies carried out by the participating members of the network will be held at less than 5%. This property does not sound very impressive — it sounds like the sort of property one would expect any sensible statistical method to have. But it might be noted that most of the methods advocated by the theoreticians who object to critical levels lack this and other important properties which facilitate control of the noise level in the network. It is also worth noting that confidence intervals have an important network property which significance tests lack. If the basic unit is taken as the study then the 5% level for significance tests is a *network* specification. But as this example shows, the 5% specification does *not* apply if the unit is the *circulating report*. However for confidence intervals, the specification applies whether the unit is taken as the study or as the circulating report.

5

The Role of Statistics in Communication

Once we become accustomed to considering statistical techniques in the context of communication networks in the sciences, we immediately gain new insight into the role of these methods. The concept of *statistical inference,* which tends to be a pale and esoteric abstraction in much of the statistical literature, takes on a lively and practical aspect when considered in the context of scientific communication. Instead of hypothetical samples from idealized probability distributions (which can never be observed), the focus shifts to the conflicting claims and other noise of an actual communication network — entities which are not only observable but also of immediate importance to the statistical practitioner. It has been my own experience that an explication of inference in these terms is meaningful both to biomedical scientists and to students of biostatistics.

We quickly become aware of a very important point when we take the communication network viewpoint. In these networks, statistics is used as a means of communication between the members of the network. In modern linguistics the term *living language* is broadly construed to refer to any signal system that is regularly used for communication between human beings. Hence we can say that statistics has become one of the major living languages of modern science. Even a casual survey of the scientific literature will show the widespread usage of statistical jargon such as *significant difference* or *5% level.* From the standpoint of statistics as a language it is natural to ask: Why have scientists adopted these specialized statistical sub-languages as a means of communicating with each other? Do they just happen to like the sound of statistical jargons?

From my own experience with biomedical scientists, I would say that they definitely do not like statistical jargons — they are often outspoken in their dislike of this language. I do not blame them for this; these jargons have many idioms which are very baffling and irritating. Nevertheless, despite their unattractive features the statistical languages are widely used in the communication networks and the question is: Why?

I believe there is one main reason: scientists use these languages because they have proved effective in communicating the kinds of factual information that are important in the sciences. Perhaps the greatest appeal is this: the statistical jargons used in scientific networks are *strongly fact-limited.* In other words, there is a direct and explicit relationship between the original reports recorded in the question schedules or laboratory notebooks of the scientist and the subsequent statistical statements that this scientist makes in his paper.

The sequential trial of the new analgesic (N.A.P.A.P.) provides a good illustration of the linguistic habits of a strongly fact-limited language. For example, a strong effort was made to insure that the conclusions of the study would be determined by facts — and *only* by facts. The authors may have had preconceived opinions about the chances that the test drug would be effective or they might have been able to guess at hypothetical costs but they did not use such subjective items in arriving at the conclusion that codeine was preferable. The authors had some choice in selecting their statistical procedure — they might, for instance, have used a fixed sample plan instead of Bross Plan B. But once this choice was made, they were led step-by-step by their clinical reports to a decision to stop the study and to an assertion of a particular conclusion. Furthermore, any reader who wishes to do so can follow step-by-step in the footsteps of the authors by marking the clinical results onto the sequential chart until he comes to exactly the same decisions and conclusions. Thus in a very direct way, the statement of the conclusions is *limited* by the facts. This is a linguistic performance characteristic which has great appeal for a working scientist (6).

By and large the linguistic habits of a conscientious user of statistical methods based on critical levels are such that he makes assertions only when his data permits him to and hence this jargon is strongly fact-limited. Since fact-limited methodologies are a tactical application of the prime strategic directive of modern science — a theory must fit the facts — it is hardly surprising that this kind of language is attractive to a good scientist.

6

Justification of Traditional Critical Levels

I now want to turn to the basic question: how can we justify the use of the traditional 5% and 1% probability levels? If we think of statistics as a major living language of the sciences, we can give an answer to this question which is more penetrating and more realistic than the usual answers. We are also rewarded with fresh insights on the performance of familiar statistical methods.

From a linguistic standpoint, the 5% level is a *convention*. We are accustomed to thinking of linguistic conventions as things which are very arbitrary. However, the studies of living languages of modern linguistics have revealed the role of evolutionary processes which shape a living language in somewhat the same way that corresponding processes shape a living organism. At first sight, a convention such as the 5% level may appear entirely arbitrary but a closer look leads to a different impression.

The general acceptance of the 5% level as a convention in scientific languages reflects the influence of Karl Pearson and Ronald A. Fisher. But while these statisticians were responsible for introducing the 5% level and other *critical values* jargon (and hence it might seem an arbitrary decision on their part), incorporation in a living language requires their tacit acceptance by most of the speakers of this language (and hence is an evolutionary process). In this latter process, practical aspects of communication systems play an important role in acceptance (or rejection) of a linguistic innovation. For instance, a new convention must ordinarily be compatible with previous conventions, it must also fit in with the deeper linguistic habits of the language (e.g., the fact-limited characteristics), and ordinarily the innovation must serve the needs of the speakers of a language and be useful to them. Hence, however arbitrary the convention may have been originally, its incorporation in a living language depends on the practical value of the convention in a communication context.

Consider, for instance, the point about compatibility with other conventions. Once investigators were willing to acknowledge that experimental error or sampling variation or biological variability precluded *absolutely certain* conclusions, they were willing to settle for the *reasonably certain* conclusions derived by statistical methods. It was then consistent with existing conventions to require that false positive claims be made only *rarely* or *infrequently*. The *5% level* was proposed as a probability level which would serve to *define* the term rarely (as this term was used by experimenters). The general acceptance of the 5% level indicates general agreement that this is an appropriate specification for rarely and that this nominal level of protection seemed satisfactory to a majority of scientists. This suggests that the original proposal was *not* arbitrary since it reflected prevailing usage in the scientific languages. It might be noted that acceptance of innovations is quite selective. Only a few of the definitions or specifications that are proposed ever become part of a living language. To see this point for yourself, consider what would have probably happened had the *20% level* been proposed.

The continuing usage of the 5% level is indicative of another important practical point: it is a feasible level at which to do research work. In other words, if the 5% level is used, then in most experimental situations it is feasible (though not necessarily easy) to set up a study which will have a fair chance of picking up those effects which are large enough to be of scientific interest. If past experience in actual applications had not shown this feasibility, the convention would not have been useful to scientists and it would not have stayed in their languages. For suppose that the 0.1% level had been proposed. This level is rarely attainable in biomedical experimentation. If it were made a prerequisite for reporting positive results, there would be very little to report. Hence from the standpoint of communication the level would have been of little value and the evolutionary process would have eliminated it.

From a linguistic standpoint, the fact that the 5% level has become firmly established as a convention in scientific languages tells us a good deal about the practical value of the convention. Moreover we can see how the choice of possible critical values is restricted to a limited numerical range by practical aspects of communication. But is there any reason to single out one particular value, 5%, instead of some other value such as 3.98% or 7.13%?

7
Setting of Numerical Conventions

There is no logical reason to fix on 5% but there are a number of practical reasons to choose one particular number as a convention. It would produce a lot of unnecessary confusion in communication if some investigators used 3.98%, others 5%, and still others 7.13%. Again, in conformity with the linguistic patterns in setting conventions it is natural to use a *round number* like 5%. Such round numbers serve to avoid any suggestion that the critical value has been gerrymandered or otherwise picked to prove a point in a particular study. From an abstract standpoint, it might seem more logical to allow a range of critical values rather than to choose one number but to do so would be to go contrary to the linguistic habits of fact-limited languages. Such languages tend to minimize the freedom of choice of the speaker in order to insure that a statement results from factual evidence and not from a little language-game played by the speaker.

Anyone familiar with certain areas of the scientific literature will be well aware of the need for curtailing language-games. Thus if there were no 5% level firmly established, then some persons would stretch the level to 6% or 7% to prove their point. Soon others would be stretching to 10% and 15% and the jargon would become meaningless. Whereas nowadays a phrase such as *statistically significant difference* provides some assurance that the results are not merely a manifestation of sampling variation, the phrase would mean very little if everyone played language-games. To be sure, there are always a few folks who fiddle with significance levels — who will switch from two-tailed to one-tailed tests or from one significance test to another in an effort to get positive results. However such gamesmanship is severely frowned upon and is rarely practiced by persons who are *native speakers* of fact-limited scientific languages — it is the mark of an amateur.

8
Justification of Statistical Methods

In the previous section I have given the beginnings of a justification for the traditional 5% level. To go into the justification more deeply it would be necessary to consider specific performance characteristics (e.g., the remarkable *robustness* of two-tailed techniques at the 5% level) as well as data on actual performance in scientific communication networks. I will not try to give a more comprehensive justification here. However the aim of the justification would be to show how standard critical value statistical methods are useful to working scientists who want to communicate their findings to their colleagues.

While this would be regarded as a *justification* by speakers of scientific fact-limited languages, some readers will not consider that the foregoing remarks represented any justification for statistical methods. These individuals will be native speakers of a different family of statistical jargons — jargons which for contrast I will call *fact-free* (7).

Although the same words and phrases are often used in both families of statistical languages, these words have a very different meaning in a fact-free jargon than in a fact-limited jargon. For instance, justification in a fact-limited jargon requires factual evidence whereas no such evidence is needed — or even pertinent — in a fact-free jargon. This conflicting usage has led to a great deal of unnecessary confusion and controversy in the statistical literature. In particular, most of the criticisms of standard statistical methods and critical levels has been phrased in the fact-free statistical jargons that are commonly used by theorists. Therefore it will be worthwhile to consider briefly the broader question: how can statistical methods be justified?

Since the answer to the above question depends on *which* language we are using, let us look into the meaning of justification in a fact-free statistical jargon. In such jargons the linguistic habits are patterned after those in purely mathematical languages (e.g., set theoretic usage). The process of justification in a fact-free jargon consists of the application of a series of formal manipulations to a set of abstract statements (which is often called an *axiom set* or *underlying assumptions*). The abstract statements, *theorems* or *conclu-*

sions which are generated by this process are then considered to be *proved* or *justified*.

No factual information comes into this process (though occasionally it may be argued that an axiom is *reasonable* or *self-evident*). The permissible manipulations are noninformative (i.e. cannot add any information). So it is natural for the speaker of a fact-limited language to wonder: how can these abstract algebraic manipulations tell us anything about the use of statistical methods in actual studies? The answer is simple: they cannot. This procedure cannot justify (in a fact-limited sense) *any* statement concerning events in the world about us or *any* recommendation for action in dealing with such events.

I am not saying anything new and different here — I am merely making a point which has been clearly established for at least two generations. There is an ancient myth — dating back to Pythagoras and Plato — that mathematical languages have mysterious powers to reveal the truth. But it has now been clearly shown — by studies in both fact-free languages (e.g., symbolic logic) and fact-limited languages (e.g., modern linguistics) — that a fact-free language cannot, in and of itself, generate informative statements about affairs in the world about us. At this late date I shouldn't have to stress this well known point. Unfortunately, however, there are some theorists (in statistics and other areas) who insist on making pronouncements about statistical practices or scientific methods on the basis of their manipulations of a fact-free language. Indeed, much of the criticism of standard statistical methods has come from these hopelessly confused critics.

9
Unscientific Criticisms of Standard Methods

These hapless critics are the victims of an insidious linguistic trap which arises because a statement which is *true* in a fact-free language can be *false* in a fact-limited language and vice-versa. For example, suppose that a critic says (in fact-free jargon): "There is absolutely no justification for the use of *critical value* statistical methods." This sounds like a devastating criticism. But the statement is innocuous enough when translated into fact-limited language: "I (the critic) have not come across a sequence of algebraic manipulations which I would be willing to call a *justification* of *critical values.*" As has been noted, such manipulations cannot justify (in a fact-limited sense) the use of statistical methods. Hence their absence is not a matter of much consequence. What is important, however, is this: the critic's assertion is flatly contrary to fact if it is read as a statement in a fact-limited jargon (and this is how most statisticians would read it). Consequently, while the critic has the best of intentions and (in his own jargon) is speaking the *truth,* he is merely misinforming and confusing most of his audience. The critic cannot escape from this trap very easily because he is a prisoner of his own jargon — he will reject arguments to factual evidence since these have no force in a fact-free jargon (8).

When, on occasion, I have pointed out this linguistic trap to my more theoretical friends, I have tended to get this kind of reply: "You do us an injustice;

you have overlooked the fact that the statements of an axiom set provide a direct connection to the real world. If an appropriate axiom set is chosen, the underlying assumptions will be valid for actual experimentation. Indeed, if the assumptions weren't valid, then standard statistical methods — which are based on very similar assumptions — would fail completely. But you yourself have claimed that they work in practice." This bit of fact-free jargon sounds very plausible — and familiar. It is the kind of argument that is often heard in classrooms. In a fact-free jargon it sounds all right but in a fact-limited jargon it is all wrong.

The adequacy of underlying assumptions is a question which can often be put to a factual test in a given study. Some investigators do this routinely (when possible). Since this work is often a preliminary to the eventual analysis, the results are rarely reported in the literature. However I do not think that anyone with much first-hand experience with this discouraging chore would care to make any sweeping claims for the validity of the usual underlying assumptions. The commonly encountered axiom sets have been chosen to facilitate manipulations in the fact-free languages; at best they can be regarded as crude prototypes for actual experimental situations. For example, in the past 20 years I have usually seen more than 100 biomedical studies per year, many of which have involved a multiplicity of variables. I do not believe I have come across a single study where the familiar multivariate normal distribution provided an adequate description of the complicated relationships between strong variables that are typical of actual data.

If the underlying assumptions fail, how can standard statistical methods continue to work? The answer is simple: these methods are robust. The satisfactory performance of these methods does *not* require that the particular assumptions in the usual axiom sets be valid. However the robustness does depend on the choice of the traditional 5% level rather than some very different probability level. Looking back at the critics' reply, it is easy to see that it is based on the unwarranted presumption (typical of fact-free languages) that abstract manipulations are somehow essential to the performance of statistical methods. This is an attractive notion to a theorist since it magnifies the importance of his manipulations in a fact-free language, but from a scientific standpoint this notion is merely a lingering remnant of the ancient mythology of mathematics.

10

The Foundations of Statistical Inference are a Myth

My message is simple: the foundations of statistical inference are a myth. The best thing that we could do at this conference is to make an official statement to this effect and go home.

I do not expect this to happen. Myths are created and perpetuated because they serve some useful purpose for the proponents of the myths. If you recall the famous tale of "The Emperor's New Clothes", you will have a close analogy to what goes on in the foundations of statistical inference. If enough people agree that something exists and is important, then this entity, though entirely mythical, can survive. Like the emperor's new clothes, the foundations of statistical inference do not cover the subject. As in the story, however, no one is very anxious to point this out. However, I am going to play the role of the street urchin in the story who blurted out the truth: The emperor wasn't wearing any clothes at all.

It is probably hard for you to believe that the foundations of statistical inference are a myth — especially when we are assembled here at a conference to discuss this topic. This may seem like a hard pill to swallow. However the pill I want you to take is a much bigger one. What I have to say is not limited to statistical inference — it applies to all of the so-called *foundations* in the sciences. There is nothing unique about the foundations of statistical inference — all the foundations are mythical.

In ordinary language the foundations of a building are something that the building rests upon. If the foundations are undermined or destroyed, the building falls down. But this is not the case with the foundations that theorists talk about. For example, early in this century Russell and Whitehead set up an ambitious program for building the foundations of mathematics on a strictly logical basis. Some years later Gödel proved that it was impossible to carry out this program. The foundations of mathematics were supposedly in ruins. But neither the creation of the foundations nor their subsequent collapse had any noticeable effect on the work in the area of mathematics. It flourished throughout this period.

From my experience with half a dozen or more different areas of the sciences, I can report the curious fact that although the foundations of the subjects often appear to be in a shambles, the subjects themselves are flourishing. This is the case in linguistics, biology, and other areas. It is also true in statistics. In each of the disciplines, the scientists are convinced that the troubles in the foundations arise from the unique and especially difficult problems of the subject matter area. However, when we look at the overall picture, we can see a different explanation for what is happening. And this is simply that the language, approach, and what is talked about in the theoretical foundations of a subject, have all kinds of inherent troubles but, fortunately, have very little to do with the subject itself.

Once we appreciate that the troubles in the theoretic foundations of statistical inference have very little to do with the practical applications of statistical methods, we can stop worrying about them. We can concentrate instead on gaining a better understanding of the role of statistical methods in scientific inference. We can gain this understanding by carefully observing and *then* analyzing how statistical methods are actually used in scientific papers that report informative and useful findings. In other words if we want to lay *foundations of statistical inference* that actually deal with the subject, we will have to go about our task in much the same way that we would go about the development of any other theory in the sciences. That is, we have to take a look at what is actually going on in the real world and then try to describe this in some concise and communicable way — perhaps mathematically, perhaps not.

This means that we have to abandon the traditional *prescriptive* attitude and adopt the *descriptive* approach which is characteristic of the empirical sciences. If we do so, then we get a very different picture of what statistical and scientific inference is all about. It is very difficult, I believe, to get such a picture unless you have had some first hand experience as an independent investigator in a scientific study. You then learn that drawing conclusions from statistical data can be a traumatic experience. A statistical consultant who takes a detached view of things just does not feel this pain.

The discovery of new and important findings concerning, say, the origin or nature of human disease or health hazards is, at first, an exhilarating feeling. But then the reaction sets in — there are so many things that can go wrong. You recall exciting findings that turned out to be an artifact of the measuring system or which resulted from a clerical error, one of the many steps in the processing of the data. You suddenly remember major findings which vanished when the data was grouped a little differently or analysed in a more sophisticated way. When you announce a new finding, you put your scientific reputation on the line. You colleagues probably cannot remember all your achievements, but they will never forget any of your mistakes! Second thoughts like these produce an acute sense of insecurity.

What you want — and want desperately — is all the protection you can get against the "slings and arrows of outrageous fortune". You want to say something informative and useful about the origins and nature of a disease or health hazard. But you do not want your statement to come back and haunt you for the rest of your life. You do not really care whether you fail to meet the criteria of formal logic or theoretical optimality — this is not the danger. What you want is as much protection as possible against vicissitudes of experimentation which are ignored in the theoretic foundations of statistical inference. You look to the standard statistical methods to provide this protection.

The reason why the statistical methods found in *Snedecor and Cochran* have been so quickly accepted — and so widely accepted — in the sciences is that they give the investigator a much-wanted sense of security. There is good evidence that this confidence is warranted; fads may come and fads may go, but when statistical methods come into a science, they stay.

The standard techniques for statistical inference have become an accepted part of the living languages of the sciences. They are widely used in the communication networks of the scientific specialties for reporting the results of experiments and surveys. This is the real *statistical inference*. What is called statistical inference in discussions of the foundations is nothing but a myth.

As I said at the start, I do not expect many of you to abandon the mythology that we are gathered here to discuss. At best I would hope to reach those of you who are concerned with the formidable public problems that are threatening the very existence of our technological society. Perhaps you will ask yourselves: what is the best use of my time and talents? Is it to work on easy myths or on hard realities?

References

1. Cornfield, J., "Sequential Trials, Sequential Analysis and the Likelihood Principle," *The American Statistician 20* (2), 18-23, April, 1966.
2. Mainland, D., *Elementary Medical Statistics,* Philadelphia and London, W.B. Sauders Co., 1963.

3. Newton, D.R.L. and Tanner, J.M., "N-acetyl-para-aminophenol as an analgesic. A Controlled Clinical Trial Using the Method of Sequential Analysis," *British Medical Journal ii,* 1096-1099, 1956.
4. Armitage, P., *Sequential Medical Trials,* Oxford, Blackwell Scientific Publications, 1960.
5. Bross, I., "Sequential Medical Plans," *Biometrics 8* (3), 188-205, September, 1952.
6. Bross, I., "Linguistic Habits of Scientists," *Perspectives in Biology and Medicine,* VI (3), 322-345, Spring, 1963.
7. Bross, I., "Linguistic Analysis of a Statistical Controversy," *The American Statistician, 17* (1), 18-21, February, 1963.
8. Bross, I., "Prisoners of Jargon," *American Journal of Public Health, 54* (6), 918-927, June, 1964.

COMMENTS

I.J. Good:

Some of my remarks apply more to the mimeographed version of Dr. Bross's paper than to his spoken version.

In a conversation in about 1950, M.G. Kendall questioned whether tail-area probabilities should be reported as they are, for example, 3.9%, and not just *less than 5%*. My reply was that the extra information should be preserved, and that we might want to combine the information from several experiments. Moreover, if the tail-area probability is 4.9%, it is dishonest to say merely that the 5% significance level has been obtained. The reason significance levels are used is that tables were published by Fisher, from which we can easily tell whether the levels are attained. We are then often too lazy to work out the exact tail-area probability, especially as a factor of 2 is usually unimportant. When computers become more readily available the precise values will be used more often. Kendall mentioned that Fisher produced the tables of significance levels to save space and to avoid copyright problems with Karl Pearson, whom he disliked. Fisher's first two books on statistics dominated the market for over ten years so it became fashionable to use significance levels. Why the preference for 5%?; perhaps because this level is often obtained when the null hypothesis is approximately *true*. At least 5% of experiments lead automatically to a publication! A better justification is that a tail-area probability of 5% often corresponds to a Bayes factor of about 3 or 4. This often justifies a repetition of the experiment if it was worth performing originally. But this argument breaks down when it takes a long time to reach the 5% level. It also depends on the kind of experiment: sometimes the initial probability that a drug is effective is very small.

Dr. Bross's argument in favour of the use of the 5% significance level is subjective. I think he should argue his case by the only method he supports, namely, by an objectivistic argument.

Dr. Bross has emphasized that our arguments are often of the form: "ordinary language → jargon → ordinary language". I agree with this except that I usually express it: "judgments → theory → discernments" (see, for example, my 1950 book).

Dr. Bross is not a "street urchin". He is by no means the first person to suggest that philosophy does not help practice. I have heard it several times from experimental scientists and sometimes even from philosophers.

We all have a philosophy which we think is of some importance: it is the *other* philosophers who are mostly wrong. I cannot see why Dr. Bross wrote his other paper, in which he tried to defend the use of the 5% significance level, if he believes that the foundations of statistical inference are a myth.

Another proof that Dr. Bross is not a street urchin is that not everyone will at once agree that the "emperor is naked". I don't for one. For example, I believe it is significant that Poincaré and Einstein were both very much interested in philosophy. But the world also needs technicians and if anyone wants to be a pure technician good luck to him. In fact most people are better at technique than at philosophy although technique is complicated whereas the true philosophy of rationality can be condensed into about 27 principles. As in physics, the basic principles are few but somehow difficult to grasp, especially once a "false religion" has been embraced.

H. E. Kyburg:

Dr. Bross deplores taking what is true in some jargon and supposing that it is true in ordinary language; but it is difficult to know what is true in ordinary language. The line between jargon and ordinary language is not only vague but shifting in time. Ordinary language is as alive and changing as our jargons. There is no virtue in being faithful to ordinary language just for the sake of faithfulness.

This can be illustrated by a true story. Most symbolic logics have a principle of double negation. A distinguished British philosopher of ordinary language was giving a paper in New York once and remarked on the fact that the principle was true of most ordinary languages, but that there was no ordinary language which had the principle that two affirmitives yielded a negation. From the back of the auditorium came the comment "Yeah, Yeah!".

It is also illustrated by Hempel's Raven Paradox to which Dr. Bross has referred several times. This is not a paradox of formal languages, but a paradox in ordinary language. Given certain very natural propositions in ordinary language two contrary conclusions seem to follow by ordinary argument. It is to find out *exactly* what is going wrong that philosophers have often formalized the argument or rephrased it in formal languages. I find it a little hard to take such analyses very seriously; but the point is that the whole problem arises out of ordinary and not formalized languages.

The distinction between fact-limited and fact-free languages strikes me as a somewhat misleading one and one which leads to the confusion of several sorts of justification. Consider the justification of a statement in a fact free language, for example, that every polynomial of degree n has n roots. It is justified by producing a proof of it in that fact free language to which it is taken to belong. That language is characterized (up to a point, as Gödel showed) by a certain set of axioms and a certain set of rules of inference, which lead to an explicit notion of proof. Now consider the justification of a statement in geometry, for example, that the sum of the angles of a triangle is 180 degrees. If we consider Euclidean geometry to be a fact free language characterized by its axioms and rules of inference we are in the same situation as before. But now suppose we take such a language to have content: that is, we might take a straight line to be the path of a light ray *in vacuo* and so on. Now, with this interpretation, I take it that we have a fact limited language. To ask for the justification of the statement that the sum of the angles of a triangle is 180 degrees in this interpreted language is now ambiguous. It may still be a request for a proof from the axioms. Though the axioms now have empirical content, the concept of a proof is precisely the same as it was when the language was fact free. Indeed this is the whole importance and interest of fact free languages: the transformations of fact free languages do not add empirical content, and thus they are as acceptable in interpreted or fact limited languages having the same abstract structure as the fact free language as they were in the original language.

There is another way in which we may ask for the justification of the statement that the sum of the angles of a triangle is 180 degrees. One may say: "Granted, it follows from the axioms that the sum of the angles is 180 degrees, but is it really the case that the sum of the interior angles of a triangle composed of light rays in vacuo is 180 degrees?" If it is not the case, then of course under the assumed interpretation the axioms must be false; but since the axioms have empirical content, they may indeed be false. The justification that is now being asked for will not be provided by a proof of the statement from the axioms but must be anwered by evidence. This can have any number of forms. It may be direct: we can do a study of triangles. It may be indirect: perhaps we can provide evidence for the truth of the axioms which will then, by virtue of the fact that the axioms entail the statement in question, be transmitted to our statement about triangles. It may be even more indirect: perhaps (more realistically, in fact) the statement in question, together with a host of statements which we are not at the moment questioning, will entail yet another statement which is quite directly testable and which will provide evidence on the question.

Now it is perfectly clear that by our manipulations of the fact free version of Euclidean language we can find out whether or not the fact limited version of Euclidian language contains the factual statement that all triangles contain 180 degrees. Given certain statements about the world (the axioms) the fact free language does generate informative statements about the world.

The question is whether, in a perfectly clear and precise sense, the study of fact free languages can give us insight into the relation of evidence to hypothesis in the non-deductive case as in the deductive case. I admit that it is an open question; it is possible that every discipline must have its own standards of evidence. If so, it seems clear that every discipline must have its own statistical techniques: techniques that are all right in medicine need not be any good at all in biological research. I don't believe it. I believe that if some techniques are more important in one field than in another, it is because of a difference in the ways of acquiring evidence or because of a difference in background knowledge and not merely because of the difference in subject matter. Just as the deductive relation can be characterized in the abstract in fact free languages and can then be applied universally and correctly in the fact limited languages which are interpretations and specializations of these fact free languages, so I think that the inductive statistical relation between evidence and hypothesis can be characterized in fact free languages and applied universally and beneficially in fact limited languages. That it hasn't been done yet is not an argument that it can't be done.

J. W. Pratt:

There are situations where orthodox methods are problematical and rough Bayesian methods are certainly easier and more straightforward and perhaps even more convincing. For example, consider making inferences about the maximum second-order interaction in a multi-factor experiment or components of variance or treatment differences in unequal variance situations or regression of time series on lagged variables. I would lean on this more than theoretical coherence in making a case for Bayesian methods to practicing statisticians like Dr. Bross. (Incidentally, the Bayesian use of the word coherence seems to me as near its ordinary meaning as the orthodox use of such terms as significance and confidence).

Bayesians don't claim that what good practical statisticians are doing is terribly wrong by and large; furthermore, such practical devices as controls and double-blinding are just as important to Bayesians as to orthodox statisticians. I do think orthodox theory may be hurting practice some in certain kinds of situations. For example, if we must have a routine in the two-sample *t* situation, a better one might assume unequal variances. Preliminary tests I have serious misgivings about generally. Regarding goodness of fit tests, Berkson hit the nail on the head in 1938, and he is widely regarded, I believe (though perhaps not by Dr. Bross) as an excellent, practising, orthodox statistician.

Though it may not make a tremendous difference to practice, it still seems to me worthwhile to provide a better explanation and one nearer the practising scientist's understanding, of the terms we use, such as confidence level and P value. Also, orthodox methods hamstring us unnecessarily. We have confidence statements which have the form of probability distributions, but we can't do the obvious things with them. We can't use the obvious solutions to some problems; instead we must jump through hoops and even then must derive procedures based on the worst cases which are ridiculously pessimistic.

What really seems to be wanted for reporting experimental results is conventional procedures. For instance, no bizarre confidence procedures, shrinkage estimates, etc. Would Dr. Bross be satisfied with the use of strange tests and confidence intervals? They are empirically verifiable. So, for that matter, are the sampling properties of Bayesian methods.

If the foundations of statistics are in a shambles, why must we limit ourselves to a certain very confining class of methods? We might develop procedures in either a frequency or Bayesian framework and check that they make sense in the other.

Regarding the increased risk of leukemia, I really want to know: How sure are you? How sure should others be? How much risk? How sure are you of that? I want to know in this instance, not with reference to a communications network. This requires inference. Inference may not be everything or even the most important thing, but it is worth some effort. It is not a mythical problem, and the *status quo* should not satisfy us.

REPLY

Nothing that I have seen or heard at this meeting has provided any solid evidence against my original thesis: the *foundations* of statistical inference are a myth.

On the other hand there has been ample evidence at this conference that this myth arises from—and is sustained by—a misplaced faith in the mystic powers of mathematical languages. We are not likely to get rid of the myths about inference until we get rid of some widely prevalent superstitions about mathematical and logical languages.

Let me begin with Professor Kyburg's remarks on Hempel's Raven Paradox because the resolution of this paradox brings out a crucial limitation of logical languages. The paradox shows that there is a basic incompatibility between the jargons of symbolic logic and ordinary language: if we try to be coherent in the jargon, we will be incoherent in the mother tongue. There is a linguistic reason for this incompatibility—the linguistic manipulations which are habitual and acceptable in logical languages are markedly different from the corresponding manipulations in English.

Hempel's paradox is easily resolved if we focus on actual usage in the respective languages. As the standard for acceptability in the logical languages, let us take the precise formulas in Carnap's text (1). As the standard for acceptability in English, let us take the equally precise formulas in Zellig Harris' monumental work on "Mathematical Structures of Language" (2). If, following Carnap, certain linguistic manipulations are carried through, the results can be backtranslated into English as: "The observation of a white shoe supports the hypothesis that all crows are black". This seems an odd statement to a native speaker of English: What has a shoe got to do with crows? The answer to this question depends on whether we are talking about a logical jargon or about ordinary language.

In formal logic, we can speak of a "shoe" as a member of the class of "non-crows" and thus establish a connection. In ordinary English, this won't work. The statement "A non-crow is white" is like saying "A whatszit is white". Unless the previous discourse has already established an acceptable class, it violates the rules for connectivity in English discourse to speak of a "non-crow". If, for instance, we had previously been discussing birds which might be mistaken for crows, then we could say "The non-crow is white". Otherwise the negation merely produces confusion. There is a similar difficulty with "A shoe is a non-crow". In other words, the argument that leads to Hempel's paradox satisfies the rules for connected discourse in logical languages but violates the rules for English. The paradox is not *in* either language, it arises when we try to apply the rules of one language to the linguistic operations in a different language.

Once we can clearly see that there are differences between the linguistic structures of logical and ordinary languages, we can also clearly see that the claims commonly made for fact free jargons are largely mythical. Thus Professor Kyburg says that in ordinary or fact limited languages "the concept of a proof is precisely the same as it was when the language was fact free. Indeed this is the whole importance and interest of fact free languages:...(they have) ...the same abstract structure as...the original language." This claim rests on the mythical universality of linguistic structure and we've already seen that, for example, the rules for negation are drastically different. Since the work on the foundations of statistical inference relies on logical languages and their abstract mathematical relatives, the "whole importance and interest" of this work is open to serious question.

Professor Kyburg says "... I think that the inductive statistical relationship between evidence and hypothesis can be characterized in fact free languages, and applied universally and beneficially in fact limited languages. That it hasn't been done yet is not an argument that it can't be done." I would see little basis for these fond hopes unless the logical languages are restructured to eliminate the deviations from natural language which lead to paradoxical results (3). In the meantime there should be a moratorium on misleading claims such as the one made in conjunction with the example of Euclidean Geometry: "Given certain statements about the world (the axioms) the fact free language does generate informative statements about the world." This assertion is again assumed on a universality of linguistic structure which does not, in fact, exist.

Dr. Good has supplied some interesting historical background on the origins of the 5% probability level as a linguistic convention in the scientific reporting of statistical results. In the evolution of linguistic conventions, a fortuitous choice by someone who is an influential speaker of the language is often involved in innovations. Most innovations, however, are fairly short-lived. There is a natural selection of those linguistic changes which serve the needs of the speakers of the language. So while copyright laws or other happenstances may have led to the original choice of the 5% level, the incorporation of this level as an enduring convention of applied statistical languages is due to its usefulness and is not accidental.

My arguments in favor of the 5% level are based on the value of these procedures in actual scientific discourse—in controlling the noise level of the statements circulating in the communication networks. As one specific example I mentioned the network which connects the clinical pharmacologists who are concerned with analgesic testing with their colleagues and with practicing physicians. The improvement in the noise level in this network over the past 20 years—and after the controlled clinical trial became the method of choice in testing—is something which anyone who is interested can verify if he wants to take the trouble (4). This may not be absolutely objective —nothing in science is absolutely objective—but this is not the private and personal opinion of one individual. In ordinary scientific usage this argument would not be called "subjective".

Dr. Good wonders, "why Dr. Bross wrote his other paper, in which he tried to defend the use of the 5% significance level, if he believes that the foundations of statistical inference are a myth". First of all, I have stressed that statistical inference as an operational procedure for making statements is *not* a myth. In almost any scientific journal you pick up you can find standard statistical methods being used to produce useful and important scientific findings. I have challenged Dr. Good and his colleagues to produce a *single* important scientific finding which can be obtained by subjective techniques and not by standard statistical methods. I have waited a year for an answer and expect to wait for a lot longer. Second, I am in full agreement with Dr. Good and others that the justifications commonly offered in class or in textbooks for standard statistical methods are not at all convincing. They are as mythical as the justifications for the unorthodox techniques at this conference. To me, the fact that standard techniques do a useful and important job of quality control in actual scientific inference is a solid justification. Arguments based on the abstract languages used in foundations of statistical inference are, I think, tenuous and unreal.

My third reason for defending the 5% level and standard statistical methods is that they have come under very heavy and very unfair attacks in the name of *foundations of statistical inference*. We've had some examples at this conference. These attacks are based on incredibly naive notions of *ideal* inference—conceptions which have no resemblance to the difficult and complex process which scientists use to draw reliable conclusions from their data.

There is the silly notion, for example, that when an investigator gets a test of significance which reaches the 5% level he immediately rushes into print with some positive statement. No good scientist would dream of acting this way. A significance test is just one step in a long series of steps which any competent scientist uses to check and crosscheck his results before he goes out on a limb. Indeed, ruling out the hypothesis that his results are due to sampling variation alone—a common use of significance tests—may be the least of the investigator's worries. His big problem may be ruling out an extraneous factor which some colleague is likely to raise as a counterhypothesis (5). One of my main objections to the myths in "foundations" of statistical inference is that they are preventing us from getting on with the real problems of scientific inference.

References

1. R. Carnap, *Introduction to Symbolic Logic and Its Applications,* Dover Publications, Inc., New York, 1958.
2. Z. S. Harris, *Mathematical Structures of Language,* Interscience Publishers, New York, John Wiley & Sons, 1968.
3. J. Tucker, "Lectures on the Foundations of Mathematics," Waterloo, Ontario, 1970.
4. I. D. J. Bross, "What Is Pain? A Scientific Approach to Questions of Psychopharmacological Language," in *Psychopharmacology: Dimensions and Perspectives,* Edited by C. R. B. Joyce for Tavistock Publications, London, England, J. B. Lippincott Company, 1968.
5. I. D. J. Bross, "Pertinency of an Extraneous Variable," *Journal of Chronic Diseases,* 20, 487-495, 1967.